AF726610

Handbook of Petroleum Geoscience: Exploration, Characterization, and Exploitation of Hydrocarbon Reservoirs

Handbook of Petroleum Geoscience

Exploration, Characterization, and Exploitation of Hydrocarbon Reservoirs

Edited by

Soumyajit Mukherjee
Indian Institute of Technology Bombay
Powai, Mumbai, India

Swagato Dasgupta
Indian Institute of Technology (ISM) Dhanbad
Dhanbad, India

Chandan Majumdar
Schlumberger Ltd.
Shiravane, Nerui Navi Mumbai
Mumbai, India

Subhandip Mandal
Integrated Reservoir Solutions
Houston, TX, USA

Troyee Dasgupta
Indian Institute of Technology Bombay
Powai, Mumbai, India

WILEY

This edition first published 2023
© John Wiley & Sons Ltd

All rights reserved. No part of this publication may be reproduced, stored in a retrieval system, or transmitted, in any form or by any means, electronic, mechanical, photocopying, recording or otherwise, except as permitted by law. Advice on how to obtain permission to reuse material from this title is available at http://www.wiley.com/go/permissions.

The right of Soumyajit Mukherjee, Swagato Dasgupta, Chandan Majumdar, Subhandip Mandal and Troyee Dasgupta to be identified as the authors of the editorial material in this work has been asserted in accordance with law.

Registered Office(s)
John Wiley & Sons, Inc., 111 River Street, Hoboken, NJ 07030, USA
John Wiley & Sons Ltd, The Atrium, Southern Gate, Chichester, West Sussex, PO19 8SQ, UK

Editorial Office
Boschstr. 12, 69469 Weinheim, Germany

For details of our global editorial offices, customer services, and more information about Wiley products visit us at www.wiley.com.

Wiley also publishes its books in a variety of electronic formats and by print-on-demand. Some content that appears in standard print versions of this book may not be available in other formats.

Limit of Liability/Disclaimer of Warranty
In view of ongoing research, equipment modifications, changes in governmental regulations, and the constant flow of information relating to the use of experimental reagents, equipment, and devices, the reader is urged to review and evaluate the information provided in the package insert or instructions for each chemical, piece of equipment, reagent, or device for, among other things, any changes in the instructions or indication of usage and for added warnings and precautions. While the publisher and authors have used their best efforts in preparing this work, they make no representations or warranties with respect to the accuracy or completeness of the contents of this work and specifically disclaim all warranties, including without limitation any implied warranties of merchantability or fitness for a particular purpose. No warranty may be created or extended by sales representatives, written sales materials or promotional statements for this work. The fact that an organization, website, or product is referred to in this work as a citation and/or potential source of further information does not mean that the publisher and authors endorse the information or services the organization, website, or product may provide or recommendations it may make. This work is sold with the understanding that the publisher is not engaged in rendering professional services. The advice and strategies contained herein may not be suitable for your situation. You should consult with a specialist where appropriate. Further, readers should be aware that websites listed in this work may have changed or disappeared between when this work was written and when it is read. Neither the publisher nor authors shall be liable for any loss of profit or any other commercial damages, including but not limited to special, incidental, consequential, or other damages.

Library of Congress Cataloging-in-Publication Data
Names: Mukherjee, Soumyajit, editor. | Dasgupta, Swagato, editor. | Majumdar, Chandan, editor. | Mandal, Subhadip, editor. | Dasgupta, Troyee, editor.
Title: Handbook of petroleum geoscience : exploration, characterization, and exploitation of hydrocarbon reservoirs / edited by Soumyajit Mukherjee, Swagato Dasgupta, Chandan Majumdar, Subhadip Mandal, Troyee Dasgupta.
Description: Hoboken, NJ : Wiley, 2023. | Includes bibliographical references.
Identifiers: LCCN 2022026417 (print) | LCCN 2022026418 (ebook) | ISBN 9781119680031 (hardback) | ISBN 9781119680086 (adobe pdf) | ISBN 9781119680109 (epub)
Subjects: LCSH: Petroleum–Geology. | Hydrocarbon reservoirs.
Classification: LCC TN870.5 .H358 2023 (print) | LCC TN870.5 (ebook) | DDC 622/.1828–dc23/eng/20220713
LC record available at https://lccn.loc.gov/2022026417
LC ebook record available at https://lccn.loc.gov/2022026418

Cover Design: Wiley
Cover Image: © P.V.R.Murty/Shutterstock

Set in 9.5/12.5pt STIXTwoText by Straive, Pondicherry, India
Printed and bound by CPI Group (UK) Ltd, Croydon, CR0 4YY

C9781119680031_031022

S. Mukherjee dedicates this book to Dr. A. K. Dubey (Senior Scientist retired from Wadia Institute of Himalayan Geology, Dehradun), and to Prof. Dr. Wolf-Christian Dullo (retired from GEOMAR, Germany, former Chief Editor in the International Journal of Earth Sciences*).*

Contents

List of Contributors

Tuli Bakshi
Computational and Experimental Geomechanics Laboratory
Department of Earth Sciences
Indian Institute of Technology Bombay
Mumbai, Maharashtra, India
and
Clean Coal Technology Laboratory
Department of Mining Engineering
Indian Institute of Technology Kharagpur
Kharagpur, West Bengal, India

Devojit Bezbaruah
Department of Applied Geology
Dibrugarh University
Dibrugarh, Assam, India

Abhijeet Bhardwaj
Department of Industrial and Systems Engineering
University of Wisconsin-Madison
Madison, WI, USA

Soumi Chaki
Division of Signals and Systems
Department of Electrical Engineering
Uppsala University
Uppsala, Sweden

Chandreyi Chatterjee
Schlumberger
Navi Mumbai, Maharashtra, India

Juan Carlos Rabanal Chavez
Schlumberger
Quito, Pichincha, Ecuador

Guilherme Furlan Chinelatto
Geology Institute
Universidade Estadual de Campinas (UNICAMP)
São Paulo, Brazil

Nigel Clegg
Global Subject Matter Expert – Geosteering
Halliburton, Norway

Renzo D'souza
Department of Earth Sciences
Indian Institute of Technology Bombay (IIT Bombay)
Mumbai, Maharashtra, India

Mohuli Das
Department of Earth Sciences
Indian Institute of Technology Bombay (IIT Bombay)
Mumbai, Maharashtra, India

Sudipta Dasgupta
Department of Earth Sciences
Indian Institute of Technology Bombay (IIT Bombay)
Mumbai, Maharashtra, India

Swagato Dasgupta
Department of Applied Geophysics
Indian Institute of Technology (ISM)
Dhanbad, Jharkhand, India

Troyee Dasgupta
Department of Earth Sciences
Indian Institute of Technology Bombay
Mumbai, Maharashtra, India

Bronwyn Djefel
Epiroc
Kewdale, Western Australia, Australia

Deepali Gadkari
Department of Geography
University of Mumbai
Kalina, Santacruz (E)
Mumbai, Maharashtra, India

Tapos Kumar Goswami
Department of Applied Geology
Dibrugarh University
Dibrugarh, Assam, India

Eduardo Guareschi
Geology Graduate Program
Universidade do Vale do Rio Dos Sinos
(UNISINOS)
São Leopoldo, Brazil

Lenin Mora Guerrero
Schlumberger
Bogota, Colombia

S.L. Happy
HP PPS India Operations Pvt. Ltd.
Bengaluru, Karnataka, India

Bodhisatwa Hazra
CSIR-Central Institute of Mining and Fuel Research
Dhanbad, Jharkhand, India
and
Academy of Scientific and Innovative Research (AcSIR)
Ghaziabad, Uttar Pradesh, India

Rodrigo Scalise Horodyski
Geology Graduate Program
Universidade do Vale do Rio Dos Sinos
(UNISINOS)
São Leopoldo, Brazil

Delwar Hossain
Department of Geological Sciences
Jahangirnagar University
Dhaka, Bangladesh

Qinhong Hu
Department of Earth and Environmental Sciences
The University of Texas at Arlington
Arlington, TX, USA

M.S. Kalpana
Centre for Earth, Ocean and Atmospheric Sciences
University of Hyderabad
Hyderabad, Telangana, India

Kamruzzaman
Department of Geological Sciences
Jahangirnagar University
Dhaka, Bangladesh

Nihar Ranjan Kar
Centre for Earth, Ocean and Atmospheric Sciences
University of Hyderabad
Hyderabad, Telangana, India

Mahmuda Khatunand
Department of Geological Sciences
Jahangirnagar University
Dhaka, Bangladesh

Md Golam Kibria
Department of Earth and Environmental Sciences
The University of Texas at Arlington
Arlington, TX, USA

Marcos Antonio Klunk
Department of Mechanical Engineering
University of Vale do Rio dos Sinos
São Leopoldo, Brazil

Chandan Majumdar
Schlumberger
Navi Mumbai, Maharashtra, India

Subhadip Mandal
Freeport-McMoRan
Morenci Operation
Morenci, AZ, USA

Devleena Mani
Centre for Earth, Ocean and Atmospheric Sciences
University of Hyderabad
Hyderabad, Telangana, India

Subhobroto Mazumder
Remote Sensing and Geomatics
KDM Institute of Petroleum Exploration
Oil and Natural Gas Corporation
Dehradun, Uttarakhand, India

D.S. Mitra
Remote Sensing and Geomatics
Geology Group, KDM Institute of Petroleum Exploration
Oil and Natural Gas Corporation
Dehradun, Uttarakhand, India

William K. Mohanty
Department of Geology and Geophysics
Indian Institute of Technology Kharagpur
Kharagpur, West Bengal, India

Biplab Kumar Mukherjee
LTI
Mind Space SEZ (Serene Properties)
Navi Mumbai, Maharashtra, India

Soumyajit Mukherjee
Department of Earth Sciences
Indian Institute of Technology Bombay
Mumbai, Maharashtra, India

Rajesh Pandey
Oil and Natural Gas Corporation
Dehradun, Uttarakhand, India

K.K.S. Pangtey
Remote Sensing and Geomatics
Geology Group, KDM Institute of Petroleum Exploration
Oil and Natural Gas Corporation
Dehradun, Uttarakhand, India

Khanindra Pathak
Clean Coal Technology Laboratory
Department of Mining Engineering
Indian Institute of Technology Kharagpur
Kharagpur, West Bengal, India

Ravi Prakash
Remote Sensing and Geomatics
KDM Institute of Petroleum Exploration
Oil and Natural Gas Corporation
Dehradun, Uttarakhand, India

Deependra Pratap Singh
CSIR-Central Institute of Mining and Fuel Research
Dhanbad, Jharkhand, India
and
Academy of Scientific and Innovative Research (AcSIR)
Ghaziabad, Uttar Pradesh, India

B.K. Prusty
Clean Coal Technology Laboratory
Department of Mining Engineering
Indian Institute of Technology Kharagpur
Kharagpur, West Bengal, India

Karol Riofrio Rodriguez
Geosteering Focal Point
Halliburton
Norway

Aurobinda Routray
Department of Electrical Engineering
Indian Institute of Technology Kharagpur
Kharagpur, West Bengal, India

Somali Roy
Department of Earth Sciences
Indian Institute of Technology Bombay
Mumbai, Maharashtra, India

Sankhajit Saha
Computational and Experimental Geomechanics Laboratory
Department of Earth Sciences
Indian Institute of Technology Bombay
Mumbai, Maharashtra, India

Mizanur Rahman Sarker
Department of Geological Sciences
Jahangirnagar University
Dhaka, Bangladesh

Hugo Schmidt-Neto
Geology Graduate Program
Universidade do Vale do Rio Dos Sinos (UNISINOS)
São Leopoldo, Brazil

Md. Upal Shahriar
Department of Geological Sciences
Jahangirnagar University
Dhaka, Bangladesh

Ranjan Kumar Sarmah
Department of Applied Geology
Dibrugarh University
Dibrugarh, Assam, India

Pradeep K. Singh
CSIR-Central Institute of Mining and Fuel Research
Dhanbad, Jharkhand, India

Shubham Singh
McCombs School of Business
The University of Texas at Austin
Austin, TX, USA

Vivek Singh
CSIR-Central Institute of Mining and Fuel Research
Dhanbad, Jharkhand, India

Soyane Juceli Siqueira Xavier
Department of Geology
University of Vale do Rio dos Sinos
São Leopoldo, Brazil

Kuppili Meenakshi Sundaram
Petrophysics Consultant
Mumbai, Maharastra, India

Blecy Tep
Remote Sensing and Geomatics
Geology Group, KDM Institute of Petroleum Exploration
Oil and Natural Gas Corporation
Dehradun, Uttarakhand, India

Venkata Yasaswy Turlapati
Clean Coal Technology Laboratory
Department of Mining Engineering
Indian Institute of Technology Kharagpur
Kharagpur, West Bengal, India

Vikram Vishal
Computational and Experimental Geomechanics Laboratory
Department of Earth Sciences
Indian Institute of Technology Bombay
Mumbai, Maharashtra, India

Paulo Roberto Wander
Department of Mechanical Engineering
University of Vale do Rio dos Sinos
São Leopoldo, Brazil

Preface

This edited book provides recent updates in hydrocarbon geology and economic issues. Several chapters present scientific issues and provide reviews of a few specific disciplines of petroleum geology.

Refer to this book as:
Mukherjee S, Dasgupta S, Majumdar C, Mandal S, Dasgupta T. 2022. Handbook of Petroleum Geoscience: Exploration, Characterization, and Exploitation of Hydrocarbon Reservoirs. John Wiley & Sons, Hoboken, ISBN 978-1-119-68003-1.

Refer to an individual chapter as:
Singh S, Bhardwaj A. 2022. Petrophysical Predictions Using Regression and Advanced Machine Learning Algorithm. In: Mukherjee S, Dasgupta S, Majumdar C, Mandal S, Dasgupta T (Eds) Handbook of Petroleum Geoscience: Exploration, Characterization, and Exploitation of Hydrocarbon Reservoirs. Wiley-Blackwell. John Wiley & Sons, Hoboken, ISBN 978-1-119-68003-1, pp. 21–49.

Soumyajit Mukherjee, Swagato Dasgupta, Chandan Majumdar, Subhadip Mandal, Troyee Dasgupta **(Editors)**.

Acknowledgments

We thank the **book proposal reviewers** for making us cautious in the phase of planning this book. We acknowledge the **authors** for contributing the chapters, and the **reviewers** for improving the articles. **Merryl Le Roux**, **Frank Otmar Weinreich**, **Andrew Harrison**, and the proofreading team (**Wiley Blackwell**) are thanked for their assistance. We are additionally thankful to the authors and the publisher for patiently waiting during the COVID pandemic time, in giving us significant extra time, when all of us were struggling to fit with the *new normal* way of life and faced uncertainty/drastic changes in the job (responsibilities). SM acknowledges **Subhobroto Mazumder** (Oil and Natural Gas Corporation Limited, Dehradun), **Sukanta Goswami** (Atomic Mineral Directorate for Exploration and Research, Hyderabad), **Mohit Kumar Puniya** (Survey of India, Dehradun), **Bikramaditya Mondal**, and **Md. Haroon Shaikh** (IIT Bombay) for assistance.

Soumyajit Mukherjee, Swagato Dasgupta, Chandan Majumdar, Subhadip Mandal, and Troyee Dasgupta **(Editors)**

Introduction to "Handbook of Petroleum Geoscience: Exploration, Characterization, and Exploitation of Hydrocarbon Reservoirs"

Soumyajit Mukherjee[1,*], *Swagato Dasgupta*[2], *Chandan Majumdar*[3], *Subhadip Mandal*[4], *and Troyee Dasgupta*[5]

[1] *Department of Earth Sciences, Indian Institute of Technology Bombay, Mumbai, Maharashtra, India*
[2] *Department of Applied Geophysics, Indian Institute of Technology (ISM), Dhanbad, Jharkhand, India*
[3] *Geological Services, Schlumberger Asia Services Limited, P21, Navi Mumbai, Maharashtra, India*
[4] *Freeport-McMoRan, Morenci Operation, Morenci, AZ, USA*
[5] *Department of Earth Sciences, Indian Institute of Technology Bombay, Mumbai, Maharashtra, India*

[*] *Author for correspondence: soumyajitm@gmail.com, smukherjee@iitb.ac.in*

This edited book consists of 22 main chapters. Chaki et al. (2022a) present how signal processing and machine learning algorithms are applied to characterize hydrocarbon reservoirs. This review article includes algorithms for preprocessing, prediction, and postprocessing for such characterization. Singh and Bharadwaj (2022) present how regression and advanced machine learning algorithms can make petrophysical predictions. The predicted results are then verified using test wells. Chaki et al. (2022b) present an altered guided filter capable of denoising seismic data. Based on real data sets from a hydrocarbon field, performance evaluators using a real hydrocarbon field dataset have been assessed. Chatterjee et al. (2022) present few case studies of challenges faced in geomechanics. They cover drilling and hydraulic fracturing optimization from the Assam-Arakan Basin, production optimization from the basement reservoir in the Kutch Saurashtra Basin (India), and sanding prediction from a reservoir in Ecuador. Sundaram (2022) summarizes theoretical aspects of in situ stress estimation from log measurements. The article covers several inversion processes of geophysical parameters and also discusses the utility of such studies. Clegg and Rodriguez (2022) emphasize the need of 3D inversion and modeling of reservoirs. They elaborate this issue with the help of Azimuthal resistivity images. Djefel (2022) presents the key skill requirements for a geosteering engineer in oil industry. The article will be important for the geoscientist about to join a for geosteering research. Djefel and Clegg (2022) deal with a more specific issue – how the company Halliburton has been handling geosteering-related activities. These authors state that although several sophisticated geosteering techniques are available presently, gamma geosteering continues to be a the simple yet useful method. Saha and Vishal (2022) present how to undertake local calibration for the unconfined compressive strength on a carbonate rock sample. The work elaborates how to test and calibrate rock cores. Roy et al. (2021) present rapid method of documenting lithologic boundaries

Handbook of Petroleum Geoscience: Exploration, Characterization, and Exploitation of Hydrocarbon Reservoirs, First Edition. Edited by Soumyajit Mukherjee, Swagato Dasgupta, Chandan Majumdar, Subhandip Mandal and Troyee Dasgupta.
© 2023 John Wiley & Sons Ltd. Published 2023 by John Wiley & Sons Ltd.

using the gamma ray log. Amplitude spectrum of the seismic data was tied with the well-derived output.

Shale has been a very important topic of study in petroleum geosciences in last few decades (e.g. Mukherjee and Kumar (2018)). Singh et al. (2022) review organic porosity in shale reservoirs. Thermal maturity is considered the most crucial factor for the development of nanoporosity in shales. Kibria and Hu (2022) study experimentally reservoir nanopore structures and wettability characters of the USA shale. Hydrocarbon recovery seems to significantly control the connectivity of the matrix pores and the network of fractures. Bakshi et al. (2022) analyze pore characters of the Indian shales. Different kinds of pores are distinguished using scanning electron microscopy. Greater nacropore volume percentages are manifested in terms of lower tightness and better connectivity. Mesoporous samples are tighter.

Klunk et al. (2022) study diagenetic changes in a turbidite through a software-based model. The key point of study has been how fluid flows through the reservoir and non-reservoir sediments. Mazumder et al. (2022a) present an interesting work of using the plate fit model to predict hydrocarbon-bearing rock in the Indian mainland. Their key finding has been that the premised rift sediment can plausibly be targeted for exploration in the Cauvery and the Krishna-Godavari basins. The work is in line with the first author's previous work: Mazumder et al. (2017). Tectonics of the eastern Himalayas have been of great international attention (e.g. Goswami et al. 2018, 2020, 2022). In this context, Bezbaruah et al. (2022) discuss the origin of the Upper Assam shelf area in terms of tectonics amongst Indian, Eurasian, and Burmese plates. Being a petroliferous basin, the study has important implications in hydrocarbon geosciences. Mazumder et al. (2022b) present the basement tectonics of the Assam shelf through compiled geological and geophysical studies. In the authors' language, the work is important since *"Structural highs associated with such fractured areas form promising zones of basement exploration"*. Horodyski et al. (2022) discuss how taphofacies can be important in petroleum geoscience. They provide examples from a few Brazilian terrains. Pandey (2022) presents a literature review-cum-original work on the interpretation of seismic images from the petroliferous Jaisalmer Basin from western Rajasthan, India (also see Biswas et al. 2022). He comes up with a fine phase evolution of the basin. Field structural data from this basin has been sparse. Therefore, this work will be invaluable for the next geologist to work on this terrain. Kamruzzaman et al. (2022) present gravity and magnetic studies from the Habiganj gas field (Bangladesh) and evaluated the petrophysical parameters for a specific well from that field. The authors finally postulated two hydrocarbon-bearing zones. Gadkari and Mukherjee (2022) study the b-value in relation to the seismicity from Nepal, Sumatra, Japan, and Chile. In these studied cases the b-value dropped before major earthquakes. Continuous monitoring of b-values is recommended in seismic regions.

Acknowledgments

We thank the authors for contributing the chapters, and the reviewers for improving the articles. Merryl Le Roux, Frank Otmar Weinreich, Joss Everett, Andrew Harrison, and the proofreading team (Wiley Blackwell) are thanked for their assistance. CPDA grant (IIT Bombay) supported SM.

References

Bakshi, T., Turlapati, V.Y., Vishal, V. et al. (2022). Analysis of pore characteristics of select Indian shale samples and assessment of pore connectivity by conformance correction of mercury intrusion porosimetry results. In: *Handbook of Petroleum Geoscience: Exploration, Characterization, and Exploitation of Hydrocarbon Reservoirs* (ed. S. Mukherjee, S. Dasgupta, C. Majumdar, et al.). John Wiley & Sons, Hoboken, ISBN 978-1-119-68003-1.

Bezbaruah, D., Goswami, T.K., and Sarmah, R.K. (2022). The Upper Assam Basin, its evolution and modification: A Review. In: *Handbook of Petroleum Geoscience: Exploration, Characterization, and Exploitation of Hydrocarbon Reservoirs* (ed. S. Mukherjee, S. Dasgupta, C. Majumdar, et al.). John Wiley & Sons, Hoboken, ISBN 978-1-119-68003-1.

Biswas M., Gogoi M.P., Mondal B. et al. (2022). Geomorphic assessment of active tectonics in Jaisalmer basin (western Rajasthan, India). *Geocarto International.* DOI: 10.1080/10106049.2022.2066726.

Chaki, S., Routray, A., and Mohanty, W.K. (2022a). Application of machine learning algorithms for petroleum reservoir characterization. In: *Handbook of Petroleum Geoscience: Exploration, Characterization, and Exploitation of Hydrocarbon Reservoirs* (ed. S. Mukherjee, S. Dasgupta, C. Majumdar, et al.). John Wiley & Sons, Hoboken, ISBN 978-1-119-68003-1.

Chaki, S., Happy, S.L., Routray, A., and Mohanty, W.K. (2022b). A modified guided filter to denoise seismic attributes. In: *Handbook of Petroleum Geoscience: Exploration, Characterization, and Exploitation of Hydrocarbon Reservoirs* (ed. S. Mukherjee, S. Dasgupta, C. Majumdar, et al.). John Wiley & Sons, Hoboken, ISBN 978-1-119-68003-1.

Chatterjee, C., Majumdar, C., Guerrero, L.M., and Chavez, J.C.R. (2022). Geomechanics: A basic requirement for wells at every operational stage. In: *Handbook of Petroleum Geoscience: Exploration, Characterization, and Exploitation of Hydrocarbon Reservoirs* (ed. S. Mukherjee, S. Dasgupta, C. Majumdar, et al.). John Wiley & Sons, Hoboken, ISBN 978-1-119-68003-1.

Clegg, N. and Rodriguez, K.R. (2022). 3D inversion of ultra-deep azimuthal electromagnetic logging-while-drilling data. In: *Handbook of Petroleum Geoscience: Exploration, Characterization, and Exploitation of Hydrocarbon Reservoirs* (ed. S. Mukherjee, S. Dasgupta, C. Majumdar, et al.). John Wiley & Sons, Hoboken, ISBN 978-1-119-68003-1.

Djefel, B. (2022). Solving the puzzle: seven effective habits of geosteering team members. In: *Handbook of Petroleum Geoscience: Exploration, Characterization, and Exploitation of Hydrocarbon Reservoirs* (ed. S. Mukherjee, S. Dasgupta, C. Majumdar, et al.). John Wiley & Sons, Hoboken, ISBN 978-1-119-68003-1.

Djefel, B. and Clegg, N. (2022). Driving technology for geosteering decisions: Halliburton geosteering. In: *Handbook of Petroleum Geoscience: Exploration, Characterization, and Exploitation of Hydrocarbon Reservoirs* (ed. S. Mukherjee, S. Dasgupta, C. Majumdar, et al.). John Wiley & Sons, Hoboken, ISBN 978-1-119-68003-1.

Gadkari D, Mukherjee S. 2022. Assessment of Efficacy of 'b' Value as a Seismic Precursor for Select Major Seismic Events. In: *Handbook of Petroleum Geoscience: Exploration, Characterization, and Exploitation of Hydrocarbon Reservoirs* (ed. S. Mukherjee, S. Dasgupta, C. Majumdar, et al.). John Wiley & Sons, Hoboken, ISBN 978-1-119-68003-1.

Goswami, T., Gogoi, M., Mahanta, B.N. et al. (2022). Brittle tectonics in the western Arunachal frontal fold belt, India: change in stress regime from pre-collisional extension to collisional compression. *Geological Journal* https://doi.org/10.1002/gj.4393.

Goswami, T.K., Bezbaruah, D., Mukherjee, S. et al. (2018). Structures and morphotectonic evolution of the frontal fold-thrust belt, Kameng river section, Arunachal Himalaya. *Indian Journal of Earth System Science* 127: 88.

Goswami, T.K., Kalita, P., Mukherjee, S. et al. (2022). Basement cross-strike Bomdila fault beneath Arunachal Himalaya: deformation along curved thrust traces, seismicity and implications in hydrocarbon prospect of the Gondwana sediments. *Geological Journal*. https://doi.org/10.1002/gj.4514.

Goswami, T.K., Mahanta, B.N., Mukherjee, S. et al. (2020). Orogen-transverse structures in the eastern Himalaya: Dextral Riedel shear along the Main Boundary Thrust in the Garu-Gensi area (Arunachal Pradesh, India), implication in hydrocarbon geoscience. *Marine and Petroleum Geology* 114: 104242.

Horodyski, R.S., Chinelatto, G.F., Schmidt-Neto, H. et al. (2022). Taphonomy, petrophysics, and the relationship of dense shell-accumulation with reservoir quality. In: *Handbook of Petroleum Geoscience: Exploration, Characterization, and Exploitation of Hydrocarbon Reservoirs* (ed. S. Mukherjee, S. Dasgupta, C. Majumdar, et al.). John Wiley & Sons, Hoboken, ISBN 978-1-119-68003-1.

Kamruzzaman, M., Hossain, D., Sarker, M.R., and Mahmuda Khatunand, S.U. (2022). Improving insights into petrophysics using Geophysical Data for the Habiganj Structure, Surma Basin, Bangladesh. In: *Handbook of Petroleum Geoscience: Exploration, Characterization, and Exploitation of Hydrocarbon Reservoirs* (ed. S. Mukherjee, S. Dasgupta, C. Majumdar, et al.). John Wiley & Sons, Hoboken, ISBN 978-1-119-68003-1.

Kibria, M.G. and Hu, Q. (2022). Experimental understanding of pore structure and wettability of the unconventional reservoir. In: *Handbook of Petroleum Geoscience: Exploration, Characterization, and Exploitation of Hydrocarbon Reservoirs* (ed. S. Mukherjee, S. Dasgupta, C. Majumdar, et al.). John Wiley & Sons, Hoboken, ISBN 978-1-119-68003-1.

Klunk, M.A., Dasgupta, S., Das, M. et al. (2022). Geochemical modeling of diagenetic reactions in the Eocene sediment-gravity-flow deposit reservoirs influenced by salt tectonics: the Espírito Santo Basin, Brazil. In: *Handbook of Petroleum Geoscience: Exploration, Characterization, and Exploitation of Hydrocarbon Reservoirs* (ed. S. Mukherjee, S. Dasgupta, C. Majumdar, et al.). John Wiley & Sons, Hoboken, ISBN 978-1-119-68003-1.

Mazumder, S., Prakash, R., and Mitra, D.S. (2022a). A GIS-based approach to explore the possibility of a N-S Gondwana Rift in the south-eastern part of India. In: *Handbook of Petroleum Geoscience: Exploration, Characterization, and Exploitation of Hydrocarbon Reservoirs* (ed. S. Mukherjee, S. Dasgupta, C. Majumdar, et al.). John Wiley & Sons, Hoboken, ISBN 978-1-119-68003-1.

Mazumder, S., Tep, B., Pangtey, K.K.S. et al. (2017). Probable existence of a Gondwana transcontinental rift system in western India: Implications in hydrocarbon exploration in Kutch and Saurashtra offshore: a GIS-based approach. *Journal of Earth System, Science* 126: 81.

Mazumder, S., Tep, B., Mitra, D.S., and Pangtey, K.K.S. (2022b). Basement tectonics in the assam shelf and its implications in hydrocarbon exploration- a remote sensing and GIS based perspective. In: *Handbook of Petroleum Geoscience: Exploration, Characterization, and Exploitation of Hydrocarbon Reservoirs* (ed. S. Mukherjee, S. Dasgupta, C. Majumdar, et al.). John Wiley & Sons, Hoboken, ISBN 978-1-119-68003-1.

Mukherjee, S. and Kumar (2018). A first-order model for temperature rise for uniform and differential compression of sediments in basins. *International Journal of Earth Sciences* 107: 2999–3004.

Pandey, R. (2022). Tectonic evolution of Jaisalmer basin (Rajasthan, India). In: *Handbook of Petroleum Geoscience: Exploration, Characterization, and Exploitation of Hydrocarbon Reservoirs* (ed. S. Mukherjee, S. Dasgupta, C. Majumdar, et al.). John Wiley & Sons, Hoboken, ISBN 978-1-119-68003-1.

Singh, D.P., Hazra, B., Singh, V., and Singh, P.K. (2022). Review on organic porosity in shale reservoirs. In: *Handbook of Petroleum Geoscience: Exploration, Characterization, and Exploitation of Hydrocarbon Reservoirs* (ed. S. Mukherjee, S. Dasgupta, C. Majumdar, et al.). John Wiley & Sons, Hoboken, ISBN 978-1-119-68003-1.

Saha, S. and Vishal, V. (2022). Rock strength estimation from petrophysical logs through core data calibration in low porosity and low permeability carbonate rocks. In: *Handbook of Petroleum Geoscience: Exploration, Characterization, and Exploitation of Hydrocarbon Reservoirs* (ed. S. Mukherjee, S. Dasgupta, C. Majumdar, et al.). John Wiley & Sons, Hoboken, ISBN 978-1-119-68003-1.

Singh, S. and Bharadwaj, A. (2022). Petrophysical predictions using regression and advanced machine learning algorithm. In: *Handbook of Petroleum Geoscience: Exploration, Characterization, and Exploitation of Hydrocarbon Reservoirs* (ed. S. Mukherjee, S. Dasgupta, C. Majumdar, et al.). John Wiley & Sons, Hoboken, ISBN 978-1-119-68003-1.

Sundaram, K.M. (2022). In-situ stresses from log measurements. In: *Handbook of Petroleum Geoscience: Exploration, Characterization, and Exploitation of Hydrocarbon Reservoirs* (ed. S. Mukherjee, S. Dasgupta, C. Majumdar, et al.). John Wiley & Sons, Hoboken, ISBN 978-1-119-68003-1.

1

Application of Machine Learning Algorithms for Petroleum Reservoir Characterization

Soumi Chaki[1,*], *Aurobinda Routray*[2], *and William K. Mohanty*[3]

[1] *Division of Signals and Systems, Department of Electrical Engineering, Uppsala University, Uppsala, Sweden*
[2] *Department of Electrical Engineering, Indian Institute of Technology Kharagpur, West Bengal, India*
[3] *Department of Geology and Geophysics, Indian Institute of Technology Kharagpur, West Bengal, India*

[*] *Author for Correspondence: soumilab24@gmail.com*

Abstract

The aim of reservoir characterization (RC) is to find different petrophysical properties and facies information from integrated dataset of different domains, especially seismic attributes and well logs. The nonlinear and diverse subsurface properties pose several challenges to accurately estimating petrophysical properties. In order to rise above those challenges, RC has evolved as an interdisciplinary research domain relating to computational science, geostatistics, signal processing, and geophysics over the years.

This article provides an overall summary of the applications of signal processing and machine learning (ML) algorithms for the petroleum RC. We provide an account of various state-of-the-art algorithms essential for preprocessing, prediction, and, finally, "post-processing" stages of an RC framework. It is a highly data-driven research area. Experts decipher about the subsurface properties from integrated dataset of a large amount of seismic and well logs to identify the existence of a potential reservoir. The signal-processing tools are primarily applied for information matching, removing noise and artifacts, and post-processing for eliminating outliers in the predicted data; whereas, the ML tools are required to map the seismic data to well logs. This chapter offers a comprehensive account of the recent advances in RC involving seismic volumes and well logs.

Keywords *Artificial neural network (ANN); deep learning; machine learning (ML); prediction; pre-processing; post-processing; regularization; reservoir characterization (RC); seismic attributes; signal processing; well-logs*

1.1 Introduction

3D seismic survey is carried out at a prospective field before taking decisions regarding where to start drilling initial wells for oil exploration. Acoustic energy is sent to the subsurface by explosions or seismic vibrators and the reflected or refracted energy from different

Handbook of Petroleum Geoscience: Exploration, Characterization, and Exploitation of Hydrocarbon Reservoirs, First Edition. Edited by Soumyajit Mukherjee, Swagato Dasgupta, Chandan Majumdar, Subhandip Mandal and Troyee Dasgupta.
© 2023 John Wiley & Sons Ltd. Published 2023 by John Wiley & Sons Ltd.

layers is recorded by an array of geophones strategically placed at various locations over the study area. The seismic volume resulting from the 3D survey characterizes the subsurface more efficiently compared to its 2D counterpart (Vyas and Sharma 2016). Different attributes like impedance, instantaneous frequency, instantaneous phase, sweetness, etc., are estimated from the seismic amplitude cube (Schlumberger Oilfield Glossary n.d.). Most of the processed seismic variables are smooth lacking detailed information along the depth. Instead, the well logs, carrying high vertical resolution, are acquired at few numbers of well locations over a study area. Experienced scientists derive multiple lithological properties such as sand fraction, shale fraction, porosity, permeability, or facies from well logs. Otherwise, these logs are determined from core samples under a laboratory environment. These logs can indicate prospective hydrocarbon-enriched zones in a field (Chaki et al. 2015a). The sections are categorized in dry, water-bearing, and hydrocarbon-containing layers from the distributions of lithological properties to identify the prospective zones. As these properties are only available at the borehole locations, seismic attributes and existing well logs can be used to approximate them away from the boreholes (Bosch et al. 2010; Chaki et al. 2015a). However, there is no definitive approach to carrying out this task. Therefore, the calibration of the relationship between these data is an open challenge to the researchers. Literature survey shows that reservoir characterization (RC) has evolved as an interdisciplinary research domain relating to computational science, geostatistics, signal processing, and geophysics. In summary, the aim of RC is to find different petrophysical properties and facies information from an integrated dataset of different domains, especially seismic attributes and well logs (Chaki et al. 2018a, 2018b, 2019, 2022; Vallabhaneni et al. 2022). The nonlinear and heterogeneous subsurface characteristics are the major challenge in accurately estimating the reservoir properties. An overview of the integration of well logs and seismic data for RC from an ML and signal-processing perspective was presented in Chaki et al. (2018b). ML has also been used for proxy flow modelling (Chaki et al. 2021) and domain conversion (Roy et al. 2021).

In the present work, we have added additional details on the application of ML models, especially deep learning techniques for petroleum RC that were carried out mostly in the last two years. The application of different deep learning techniques is coming out rapidly in this field due to the availability of a huge amount of data as well as powerful computation systems.

1.2 The Importance of Preprocessing Stage

In petroleum RC, any framework starts with preprocessing of the dataset involving several stages, viz. integration, normalization, filtering, feature selection, etc. The seismic data is available over the reservoir area, whereas well logs are sparsely available at a limited number of well locations only. Due to the necessity of having an estimation of different reservoir properties away from the boreholes, seismic datasets are used as guidance to predict those properties over the study area. Figure 1.1 presents a petroleum reservoir where the seismic amplitude is collected over the complete field along inlines, xlines, and time-wise. A few number of wells marked with the red line are placed over the field.

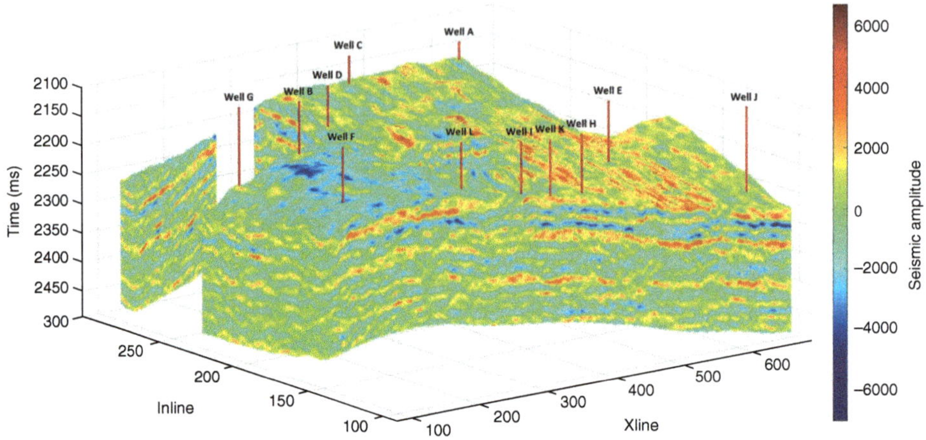

Figure 1.1 The distributions of seismic amplitude and a limited number of wells (12) in a petroleum reservoir.

1.2.1 Integration

The contrasts between the characteristic of seismic attributes and well logs are listed inside Figure 1.2. Due to the difference between the information content between these two sources, a regularization step is introduced to match high-information-carrying well logs to the low-information-carrying seismic attributes (Chaki et al. 2015a, 2018a, 2019). First, seismic attributes are extracted along well trajectories and well logs are converted from the depth domain to the time domain using velocity profiles provided at well locations (Chaki et al. 2015a). Different research groups have identified different solutions to address the disparity in resolution, frequency content, and sampling interval of these datasets during integration (Stright et al. 2009, Chaki et al. 2015a). For example, wavelet transforms (WTs) are used for interpolation (Balz and Schlumberger 2000). In Chaki et al. (2015a), a sinc interpolator is used to interpolate seismic at well log instants. However, no additional information is introduced in the interpolation step (Stright et al. 2009). So, seismic attributes would not be able to completely delineate the thin layers of a petroleum field.

1.2.2 Regularization Step

The deep contrast between the information content of the high-resolution well logs and low-resolution seismic data as indicated in Figure 1.2 necessitates a regularization step. Fig. 1.3 presents the regularization approach of the target signal by four alternative methods. An information-filtering scheme using three alternative approaches based on empirical mode decomposition (EMD), WT, and Fourier transform (FT) has been evaluated and is proposed (Chaki et al. 2015a). Later, a variational mode decomposition (VMD)-based option was also introduced in the scheme (Chaki et al. 2018a, 2019). The threshold of the regularization is decided based on the values of entropy and normalized mutual information. The detailed mathematical backgrounds for these approaches have been provided in Chaki et al. (2015a, 2019).

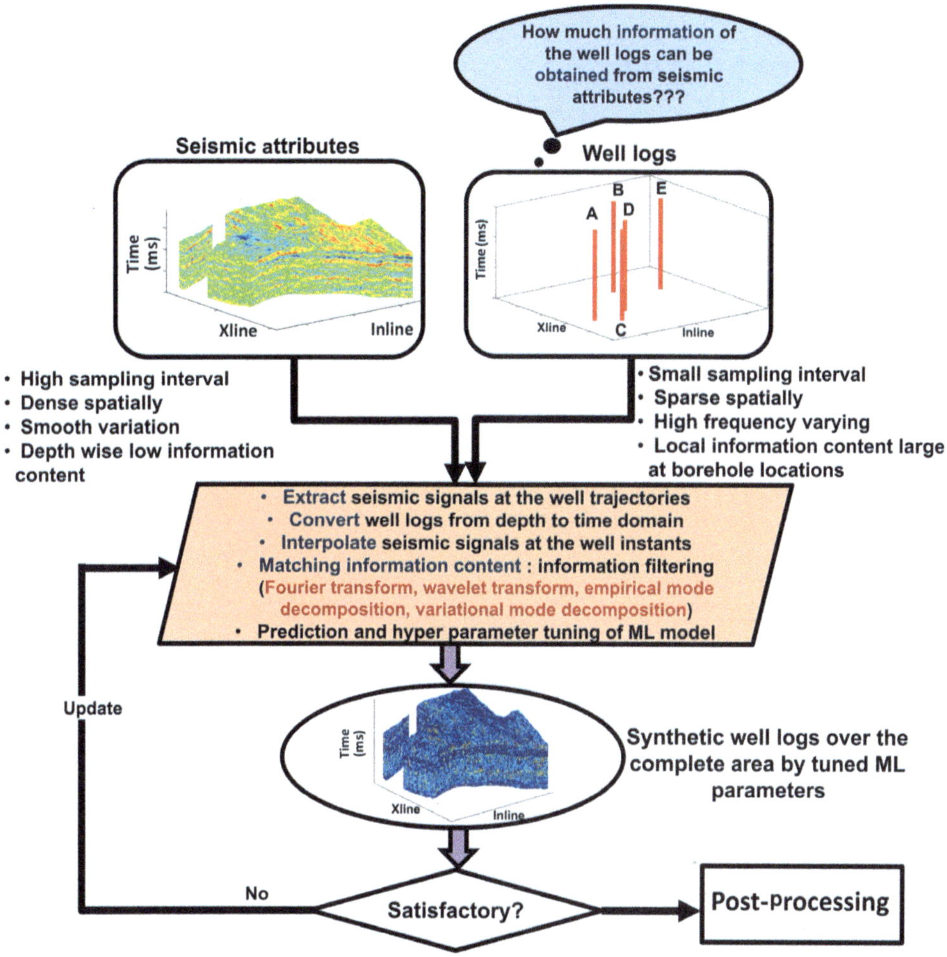

Figure 1.2 The integration of seismic attributes and well logs to generate pseudo logs over the reservoir. *Source:* Modifications are carried out in figure 2 of Chaki et al., (2018b).

As different predictors and target variables have different ranges, normalization of the dataset by z-score or min-max methods is a common preprocessing step (Baziar et al. 2014; Chaki et al. 2014a, 2015a). A user would select an appropriate normalization approach best suited for the working dataset and prediction algorithms.

1.2.3 Feature Extraction

In many cases, the values of seismic or well log attributes are directly used in modeling (Baziar et al. 2014; Chaki et al. 2015b). In some of the cases, feature selection is carried out in the preprocessing stage to extract maximum information from the dataset. For example, an alternative conditional expectation (ACE)-based nonparametric method is

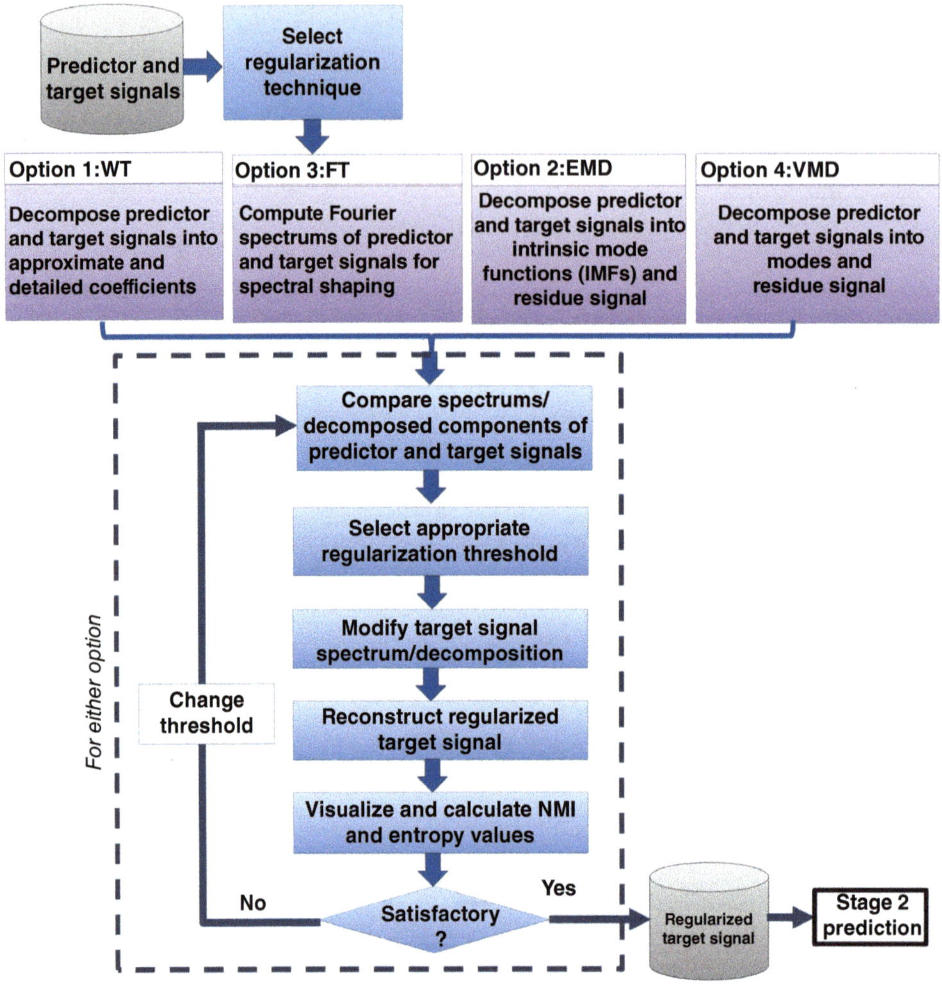

Figure 1.3 Regularization approach of the target signal by four alternative methods. *Source:* Modifications are carried out in figure 3 of Chaki et al. (2018b).

implemented to extract features from seismic attributes for porosity estimation by a committee model (Gholami and Ansari 2017). Rabbani et al. (2017) show the extraction of features from thin section images of rock to estimate permeability. Different feature selection schemes are implemented to assist classification-based workflows for lithology class label prediction from seismic and petrophysical attributes (Chaki et al. 2014b, 2014c, 2015b; Sebtosheikh et al. 2015; Sebtosheikh and Salehi 2015). Another such example is provided in Rabbani et al. (2017), where features are extracted from thin section images and used for predicting carbonate permeability. The purpose of preprocessing stage in a petroleum RC framework is to extract and comprehend all available information in the best possible manner before stepping into the model building and post-processing stages.

1.3 Relevance of Prediction Stage

The prediction stage can be summarized into two steps: selection of appropriate ML technique, and hyperparameter tuning via repeated training, validation, and testing to fine-tune the parameters of the selected ML model.

Depending on the target logs of an RC problem, two types of ML algorithms can be selected for the prediction stage of the framework. These are: (i) for continuous values such as porosity, permeability, sand fraction, etc., a regression model is built ("prediction"), and (ii) for discrete values like facies logs, a classifier model is built ("classification"). The error between the actual and the predicted targets or the misclassification error between them is used as a cost function and the aim is to reduce the cost function value so that the estimated target closely follows the actual one.

The hyperparameter tuning is carried out to obtain the best possible performance of an ML algorithm. The detailed steps of training-validation-testing and k-fold validation are presented in Chaki et al. (2018b). For regression-based work, different performance metrics such as root mean square error (RMSE), absolute error mean (AEM), correlation coefficient (CC), and scatter index (SI) are calculated between actual and predicted targets to quantify the performances of an ML technique (Chaki et al. 2015a). For classification-based frameworks, a confusion matrix is derived by comparing actual and predicted class labels and classification accuracy, sensitivity, specificity, and F-score are calculated for quantifying the classifier performance (Chaki et al. 2014b, 2014c).

We have briefly discussed artificial neural network (ANN), adaptive neuro-fuzzy inference system (ANFIS), and support vector regression (SVR) algorithms, which are well established for prediction tasks in RC frameworks over the last two decades.

1.3.1 Artificial Neural Network (ANN)

ANN and its variants (Hampson et al. 2001; Wong et al. 2002; Chaki et al. 2015a), neuro-fuzzy systems (Nikravesh and Aminzadeh 2001; Nikravesh 2004; Verma et al. 2014) are predominantly used ML techniques in RC. The learning of an ANN is carried out in two phases, such as: (i) forward phase: the input signal propagates from the input layer to the output layer through the hidden layers, and (ii) backward phase: the error signal between the target and predicted output is computed at the output layer and propagates back to the input layer through the hidden layers.

Figure 1.4 presents a workflow for addressing prediction and classification-based problems using an ANN, ANFIS, or SVM alternatively. The network structure and parameters, e.g. number of hidden layers, number of neurons, activation functions, etc., are calibrated to optimize its performance. Different activation functions such as hyperbolic tangent sigmoid, log sigmoid transfer functions, etc., play an important role in the performance of an ANN (Chaki et al. 2015a). The scaled conjugate gradient (SCG) method (Chaki et al. 2015a) is one of the optimization algorithms to update the network parameters. ANN can also be used for classification when the target logs are discrete class labels instead of continuous values.

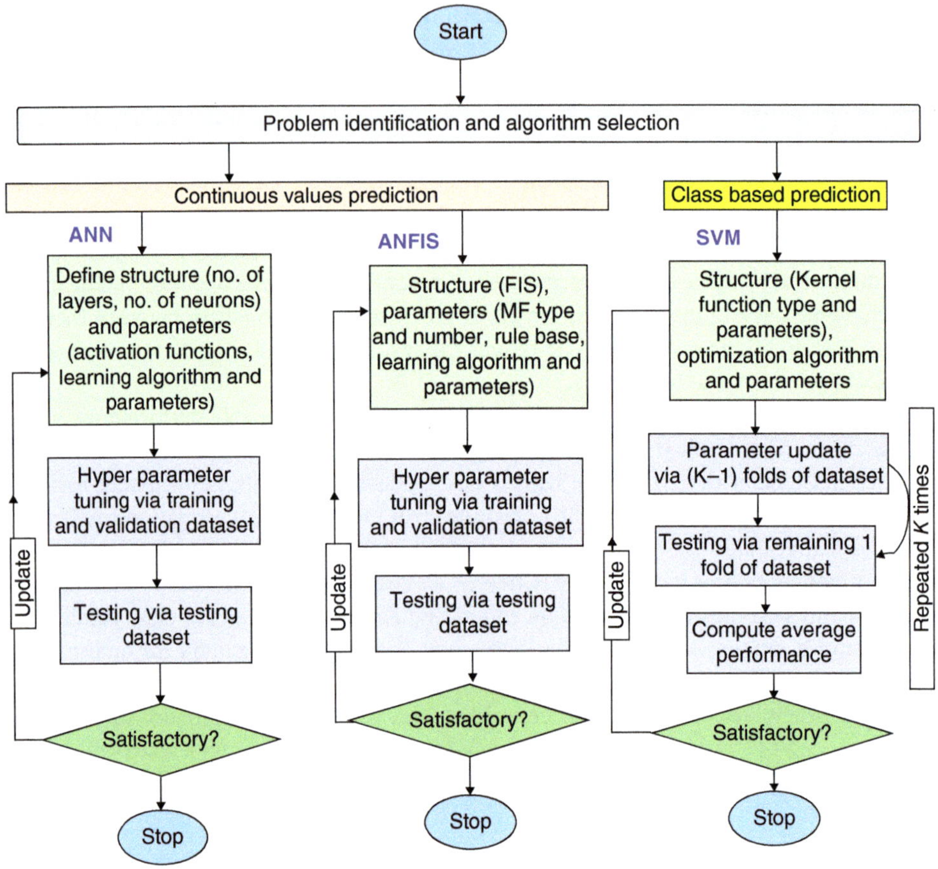

Figure 1.4 A prediction workflow for different types of target logs (e.g. continuous values and class labels) using ANN/ANFIS/SVM algorithms alternatively. *Source:* Modifications are carried out in figure 4 of Chaki et al. (2018b).

1.3.2 Adaptive Neuro-Fuzzy Inference System Approach (ANFIS)

The fuzzy approach has been used increasingly in geosciences (e.g. Anand et al. 2021). ANFIS is another ML algorithm largely used for RC (Nikravesh and Aminzadeh 2001; Nikravesh 2004; Verma et al. 2014). In Verma et al. (2014), the theory and application of ANFIS on a real petroleum reservoir are described. The inputs and target values are fuzzified by different membership functions (MFs) and those values are fed to an ANFIS; after the operations, the final output is defuzzified to obtain the predicted output. Figure 1.4 shows an ANFIS workflow that starts with the initialization of MFs, rule base, and fuzzy inference system (FIS). The selection of MFs, the FIS inference method, and the rule base is crucial to the performance of an ANFIS model.

1.3.3 Support Vector Regression (SVR)

SVR belongs to the family of support vector machines (SVMs) and is used to estimate the values of predicted petrophysical properties, e.g. porosity, permeability, etc. (Al-Anazi and

Gates 2010b, 2012; Naimi et al. 2014) from well logs or seismic attributes. SVR learns the functional relationship between the predictor and target attributes by means of a function in a higher-dimensional space. Literature reveals the effect of the number of available training samples and parameter values on the performance of SVR (Al-Anazi and Gates 2010b, 2012).

We have highlighted the workflows of the three of the most popular ML algorithms used for RC in a reservoir. Besides, recent publications account for the application of the latest ML tools such as Bayesian inversion (Fattahi and Karimpouli 2016), committee model (Gholami and Ansari 2017), and ensemble ML (Anifowose et al. 2017) to predict porosity, water saturation, etc.

1.3.4 Support Vector Machines (SVMs)

SVM is the most common algorithm used for classification in the RC domain (Al-Anazi and Gates 2010a; Sebtosheikh and Salehi 2015; Chaki et al. 2015c). Support Vector Data Description (SVDD)- and SVM-based algorithms are applied to categorize water saturation into different levels from well logs (Chaki et al. 2014b) and seismic attributes (Chaki et al. 2014c). As shown in Figure 1.4, K-fold cross-validation is carried out to evaluate the testing performance of an SVM. The most common values for K are 3, 5, and 10 (Sebtosheikh and Salehi 2015).

Apart from these three types of ML models, different ML algorithms such as genetic algorithms (Ahmadi et al. 2013) and hybrid systems (Ahmadi et al. 2013; Fattahi and Karimpouli 2016; Anifowose et al. 2017; Gholami and Ansari 2017) are also applied to carrying out RC. In order to delineate different microstructural constituents, scanning electron microscope (SEM) images play a crucial role as these images capture the distribution, topology, and morphology of the microstructural constituents effectively. In Wu and Misra (2020), the random forest has been used for this work with a wavelet and Hessian matrix calculation. A data-driven framework has been designed for subsurface geomechanical characterization using four alternative supervised algorithms such as ordinary least squares (OLS), partial least squares (PLS), elastic net (EN), least absolute shrinkage and selection operator (LASSO), multivariate adaptive regression splines (MARS), and ANN (He et al. 2019). In another study, shallow neural networks and classification methods are implemented to model subsurface fluid-filled pore size distribution (Misra and He 2019).

1.3.5 Application of Deep Learning for Reservoir Characterization

Research on the application of deep neural networks for RC is gaining attention and momentum in the last five years apart from ANN and other supervised algorithms. The convolutional layer, pooling layer, rectified linear unit (ReLU) layer, and fully connected layer are building blocks of convolutional neural network (CNN)-type deep neural works. The number of filters, the shape of filters in different layers along with dropout value, max-pooling shape, and the number of nodes in fully connected layers are some of the important hyperparameters. A user needs to design a network after analyzing the input and output variables for the specific problem, and then tune the hyperparameters to obtain the best outcomes. Deep neural network-based frameworks are built to model petrophysical characteristics for building an earth model of a big reservoir using all available information in Korjani et al. (2016a, 2016b). Multiple well logs (e.g. spontaneous potential (SP), gamma ray (GR), resistivity, neutron porosity, caliper logs) along with distance, dip, angle of nearest neighbor wells, and fuzzy kriging weights are integrated to create the feature set (Korjani et al. 2016a, 2016b). Here, multiple deep networks are built with 3–6 hidden layers with 100–250 numbers of hidden neurons in Korjani et al. (2016a). Different deep neural networks and recurrent neural network-based frameworks are used for proxy flow modeling to predict well rates from static reservoir properties (Chaki et al. 2020). The acquisition of numerical magnetic resonance (NMR) logs is more expensive and requires complex infrastructure compared to conventional well logs (Li and Misra 2017). However, NMR logs are valuable to characterize bound fluid volume, mobility of the pore-filling fluid phases, and "pore structure" of hydrocarbon-bearing sections in a reservoir (Li and Misra 2017). Different deep learning architectures, e.g. long-short term memory, convolutional neural network, and variational autoencoder have been used to model NMR logs (Li and Misra 2017, 2018; Li et al. 2020). Recurrent neural networks have also been used for the estimation of petrophysical properties from seismic data (Alfarraj and AlRegib 2018). Seismic interpretation also plays an important role to characterize the subsurface properly. It is a difficult and time-consuming task requiring years of experience. In recent years, several supervised and semi-supervised frameworks based on deep-learning architectures are implemented for this purpose (e.g. Alfarhan et al. 2020; Aribido et al. 2020; Di and AlRegib 2020; Soliman et al. 2020). These deep learning tools also eradicate the need for complex preprocessing including the feature extraction step. With the availability of a large amount of data and high computation power, the research on deep learning techniques is evolving at an unprecedented speed. At the same time, these deep learning architectures along with advanced image processing and signal-processing tools are being implemented for better performance of petroleum RC. It can be envisaged that in a few years, deep learning would enable users to automate many processes to a great extent, which now requires continuous human intervention and better RC will ultimately lead to reduced cost.

1.4 Post-Processing Stage

It is the final block of an RC framework. After calibration of ML models, the tuned parameters are used to estimate reservoir properties over the whole field from seismic attributes. Then the predicted volumes of lithological properties are visualized. As the distribution of these properties should not change abruptly in a neighborhood, 2D or 3D image processing applications are used to filter some of the irregularities from the generated outputs (Chaki et al. 2017, 2019).

Figure 1.5 shows one such example of filtering the predicted porosity by means of a 3×3 spatial filter. The predicted porosity is convolved with the filter coefficients where $*$ represents the convolution operator to generate the filtered porosity. The distribution of the filtered porosity is smoother compared to its predicted counterpart. A complete account of multiple 2D and 3D image processing algorithms and their application in the post-processing stage of an RC framework is discussed in (Chaki et al. 2017).

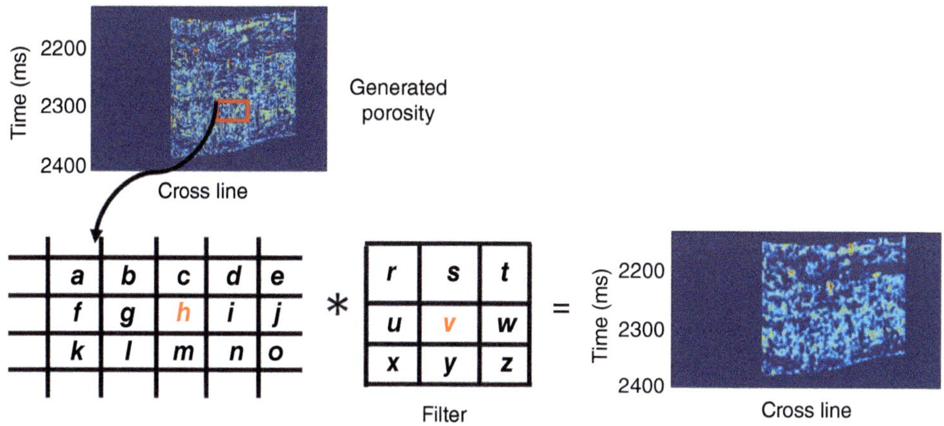

$h_{\text{filtered}} = vh + rn + sm + tl + wg + zb + yc + xd + ui$

Figure 1.5 A generic filtering example of the predicted lithological property (e.g. porosity). *Source:* Modifications are carried out in figure 5 (b) of Chaki et al. (2018b).

The final objective of the post-processing stage is the visualization of the predicted and filtered properties over the study area across an inline/xline or by generating time slices. The visualization across the study area helps understand the variation of the predicted properties for RC. Petroleum RC is interdisciplinary in nature requiring expertise from several domains (Figure 1.6).

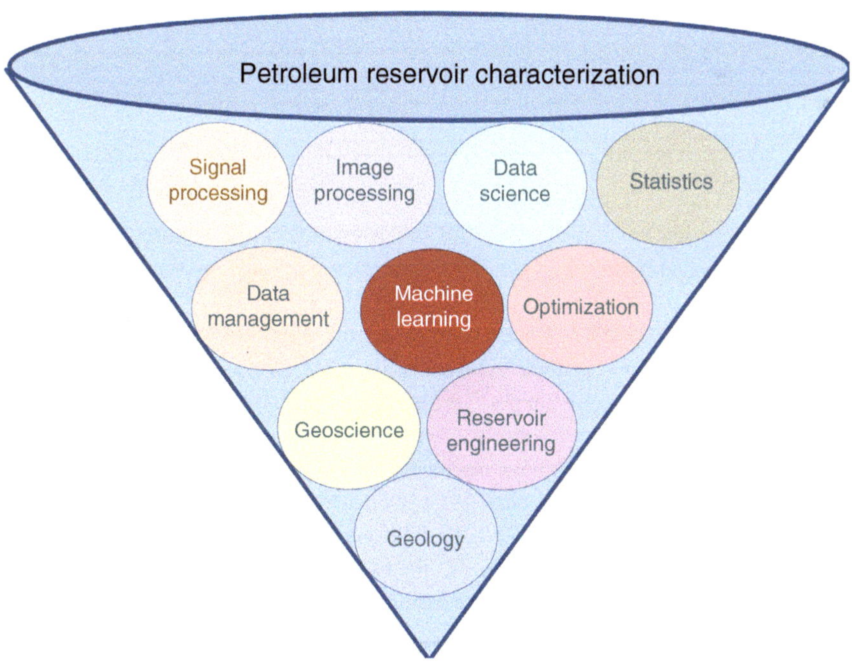

Figure 1.6 Different components of petroleum reservoir characterization.

1.5 Conclusions

An overview of the application of ML and signal-processing algorithms in petroleum RC is provided in this chapter. The motivation and significance of the work along with the associated challenges are discussed here. An account of different preprocessing steps and their importance is presented. We provide an overview of different ML techniques and especially deep learning models implemented to solve different tasks for RC. The chapter ended with a brief overview of the post-processing stage in an RC framework.

To efficiently design and implement an RC framework requires collaboration among researchers from domains of academia, petroleum engineers as well as subject matter experts working in the leading oil industries. New proof-of-concepts involving the latest ML and image-processing tools can be built in an academic environment and upon successful implementation, it can be transferred to industry experts for more testing and deployment on a larger scale. The seamless collaboration between industry and academia will enrich our knowledge in RC and assist to design frameworks that are more realistic. We sincerely hope that this chapter would provide an overall picture of petroleum RC and would encourage such interdisciplinary collaborations.

Acknowledgments

S. Mukherjee invited, handled, and reviewed this article. **Merryl Le Roux**, **Frank Otmar Weinreich**, **Joss Everett**, **Andrew Harrison**, and the proofreading team (**Wiley Blackwell**) are thanked for their assistance. The article has been summarized in Mukherjee et al. (2023).

References

Ahmadi, M.A., Zendehboudi, S., Lohi, A. et al. (2013). Reservoir permeability prediction by neural networks combined with hybrid genetic algorithm and particle swarm optimization. *Geophys. Prospect.* 61: 582–598.

Al-Anazi, A. and Gates, I.D. (2010a). A support vector machine algorithm to classify lithofacies and model permeability in heterogeneous reservoirs. *Eng. Geol.* 114: 267–277.

Al-Anazi, A.F. and Gates, I.D. (2010b). Support vector regression for porosity prediction in a heterogeneous reservoir: a comparative study. *Comput. Geosci.* 36: 1494–1503.

Al-Anazi, A.F. and Gates, I.D. (2012). Support vector regression to predict porosity and permeability: effect of sample size. *Comput. Geosci.* 39: 64–76.

Alfarhan, M., Deriche, M., Maalej, A. et al. (2020). Multiple events detection in seismic structures using a novel U-net variant. *IEEE International Conference on Image Processing (ICIP)* (25–28 October 2020). New York, NY, USA: Institute of Electrical and Electronics Engineers (IEEE), pp. 2900–2904. https://2020.ieeeicip.org/

Alfarraj, M. and AlRegib, G. (2018). Petrophysical property estimation from seismic data using recurrent neural networks. *SEG International Exposition and 88th Annual Meeting*, Anaheim, CA(14–19 October 2018). Tulsa, OK, USA: SEG, pp. 2141–2146.

Anand, P., Chakraverty, S., and Mukherjee, S. (2021). Fuzzy set concept in structural geology: example of ductile simple shear. *J. Earth Syst. Sci.* 130 (4): 1–9.

Anifowose, F.A., Labadin, J., and Abdulraheem, A. (2017). Ensemble machine learning: an untapped modeling paradigm for petroleum reservoir characterization. *J. Pet. Sci. Eng.* 151: 480–487.

Aribido, O.J., AlRegib, G. and Deriche, M. (2020). Self-supervised annotation of seismic images using latent space factorization. *IEEE International Conference on Image Processing (ICIP)* (25–28 October 2020). New York, NY, USA: Institute of Electrical and Electronics Engineers (IEEE), pp. 2421–2425. https://2020.ieeeicip.org/

Balz, O., and Schlumberger, D. (2000). "Upscaling well logs to seismic scale–comparison of a wavelet-transform based method with a geostatistical technique." In *EAGE Conference on Exploring the Synergies between Surface and Borehole Geoscience-Petrophysics meets Geophysics*, pp. cp-19. European Association of Geoscientists & Engineers, Paris, France (6–8 November 2000), pp. 1–9.

Baziar, S., Tadayoni, M., Nabi-Bidhendi, M., and Khalili, M. (2014). Prediction of permeability in a tight gas reservoir by using three soft computing approaches: a comparative study. *J. Nat. Gas Sci. Eng.* 21: 718–724.

Bosch, M., Mukerji, T., and Gonzalez, E.F. (2010). Seismic inversion for reservoir properties combining statistical rock physics and geostatistics: a review. *Geophysics* 75: 75A165–75A176.

Chaki, S., Verma, A.K., Routray, A. et al. (2014a). Well tops guided prediction of reservoir properties using modular neural network concept: a case study from western onshore. *India. J. Pet. Sci. Eng.* 123: 155–163.

Chaki, S., Verma, A.K., Routray, A. et al. (2014b). A one-class classification framework using SVDD: application to an imbalanced geological dataset. *IEEE Students' Technology Symposium (Techsym)*, Kharagpur, India (28 February–2 March 2014). New York, NY, USA: IEEE, pp. 76–81.

Chaki, S., Verma, A.K., Routray, A. et al. (2014c). A novel framework based on SVDD to classify water saturation from seismic attributes. *Proceedings of the Fourth International Conference on Emerging Applications of Information Technology (EAIT)*, Kolkata, India (19–21 December 2014). New York, NY, USA: IEEE. pp. 64–69.

Chaki, S., Routray, A., and Mohanty, W.K. (2015a). A novel preprocessing scheme to improve the prediction of sand fraction from seismic attributes using neural networks. *IEEE J. Sel. Top. Appl. Earth Obs. Remote Sens.* 8: 1808–1820.

Chaki, S., Routray, A., Mohanty, W.K., and Jenamani, M. (2015b). A novel multiclassSVM based framework to classify lithology from well logs : a real-world application. *12th IEEE India International Conference (INDICON)* (17–20 December 2015). New York, NY, USA: IEEE, pp. 1–5.

Chaki, S., Routray, A., Mohanty, W.K., and Jenamani, M. (2015c). Development of a hybrid learning system based on SVM, ANFIS and domain knowledge: DKFIS. *12th IEEE India International Conference (INDICON)*, New Delhi, India, pp. 1–6.

Chaki, S., Routray, A., and Mohanty, W.K. (2017). A diffusion filter based scheme to denoise seismic attributes and improve predicted porosity volume. *IEEE J. Sel. Top. Appl. Earth Obs. Remote Sens.* 10: 5265–5274.

Chaki, S., Routray, A. and Mohanty, W.K. (2018a). A variational mode decomposition based novel preprocessing method for reservoir characterization using support vector regression. *2018 IEEE International Geoscience and Remote Sensing Symposium*, Valencia, Spain (23–27 July 2018). Valencia, Spain: IEEE, pp. 1–4.

Chaki, S., Routray, A., and Mohanty, W.K. (2018b). Well log and seismic data integration for reservoir characterization from a signal processing and machine learning perspective: an overview. *IEEE Signal Process. Mag.* 35: 72–81.

Chaki, S., Routray, A., and Mohanty, W.K. (2019). A novel preprocessing method based on variational mode decomposition for reservoir characterization using support vector regression. *IEEE J. Sel. Top. Appl. Earth Obs. Remote Sens.* 12: 1–10.

Chaki, S., Zagayevskiy, Y., Shi, X. et al. (2020). Machine learning for proxy modeling of dynamic reservoir systems: Deep neural network DNN and recurrent neural network RNN applications. *International Petroleum Technology Conference 2020, IPTC 2020*, Dhahran, Kingdom of Saudi Arabia (13–15 January 2020).

Chaki, S., Zagayevskiy, Y., & Wong, T. (2021). A proxy flow modelling workflow to estimate gridded dynamic properties and well production rates by deep learning algorithms. In *SPE/IATMI Asia Pacific Oil & Gas Conference and Exhibition*. OnePetro (12–14 October 2021). Dubai, UAE: Society of Petroleum Engineers.

Chaki, S., Routray, A., and Mohanty, W.K. (2022). A probabilistic neural network (PNN) based framework for lithology classification using seismic attributes. *Journal of Applied Geophysics* 199 (104578): 1–11.

Di, H. and AlRegib, G. (2020). A comparison of seismic saltbody interpretation via neural networks at sample and pattern levels. *Geophys. Prospect.* 68: 521–535.

Fattahi, H. and Karimpouli, S. (2016). Prediction of porosity and water saturation using pre-stack seismic attributes: a comparison of Bayesian inversion and computational intelligence methods. *Comput. Geosci.* 20: 1075–1094.

Gholami, A. and Ansari, H.R. (2017). Estimation of porosity from seismic attributes using a committee model with bat-inspired optimization algorithm. *J. Pet. Sci. Eng.* 152: 238–249.

Hampson, D.P., Schuelke, J.S., and Quirein, J.A. (2001). Use of multiattribute transforms to predict log properties from seismic data. *Geophysics* 66: 220–236.

He, J., Li, H., and Misra, S. (2019). Data-driven in-situ sonic-log synthesis in shale reservoirs for geomechanical characterization. *SPE Reserv. Eval. Eng.* 22 (4): 1225–1239.

Korjani, M.M., Popa, A., Grijalva, E., et al. (2016a). Reservoir characterization using fuzzy kriging and deep learning neural networks. *SPE Annual Technical Conference and Exhibition,* Dubai, UAE (26–28 September 2016). Dubai, UAE: Society of Petroleum Engineers, pp. 1–15.

Korjani, M., Popa, A., Grijalva, E. et al. (2016b). A new approach to reservoir characterization using deep learning neural networks. *SPE Western Regional Meeting,* Anchorage, Alaska, USA (23–26 May 2016). Dubai, UAE: Society of Petroleum Engineers, pp. 1–16.

Li, H. and Misra, S. (2017). Prediction of subsurface NMR T2 distributions in a shale petroleum system using variationalautoencoder-based neural networks. *IEEE Geosci. Remote Sens. Lett.* 14: 2395–2397.

Li, H. and Misra, S. (2018). Long short-term memory and variational autoencoder with convolutional neural networks for generating NMR T2 distributions. *IEEE Geosci. Remote Sens. Lett.* 16: 192–195.

Li, H., Misra, S., and He, J. (2020). Neural network modeling of in situ fluid-filled pore size distributions in subsurface shale reservoirs under data constraints. *Neural Comput. Appl.* 32: 3873–3885.

Misra, S. and He, J. (2019). Shallow neural networks and classification methods for approximating the subsurface in situ fluid-filled pore size distribution. In: *Machine Learning for Subsurface Characterization* (ed. S. Misra, H. Li and J. He), 65–101. Oxford, United Kingdom: Gulf Professional Publishing.

Mukherjee, S., Dasgupta, S., Majumdar, C. et al. (2023). Introduction to: Handbook of Petroleum Geoscience: Exploration, Characterization, and Exploitation of Hydrocarbon Reservoirs. In: *Handbook of Petroleum Geoscience: Exploration, Characterization, and Exploitation of Hydrocarbon Reservoirs*, Petroleum Industry Updates: Geoscientific and Economic Issues (ed. S. Mukherjee, S. Dasgupta, C. Majumdar, et al.). John Wiley & Sons, Hoboken, ISBN 978-1-119-68003-1.

Naimi, S.R., Shadizadeh, S.R., Riahi, M.A., and Mirzakhanian, M. (2014). Estimation of reservoir porosity and water saturation based on seismic attributes using support vector regression approach. *J. Appl. Geophys.* 107: 93–101.

Nikravesh, M. (2004). Soft computing-based computational intelligent for reservoir characterization. *Expert Syst. Appl.* 26: 19–38.

Nikravesh, M. and Aminzadeh, F. (2001). Past, present and future intelligent reservoir characterization trends. *J. Pet. Sci. Eng.* 31: 67–79.

Rabbani, A., Assadi, A., Kharrat, R. et al. (2017). Estimation of carbonates permeability using pore network parameters extracted from thin section images and comparison with experimental data. *J. Nat. Gas Sci. Eng.* 42: 85–98.

Roy, S., Chaki, S., and Vallabhaneni, S. (2021). "Data-driven domain conversion using machine learning techniques." U.S. Patent Application 16/841,890, filed 7 October 2021.

Schlumberger Oilfield Glossary [WWW Document] (n.d.) http://www.glossary.oilfield.slb.com/ (accessed 1 June 2020).

Sebtosheikh, M.A. and Salehi, A. (2015). Lithology prediction by support vector classifiers using inverted seismic attributes data and petrophysical logs as a new approach and investigation of training data set size effect on its performance in a heterogeneous carbonate reservoir. *J. Pet. Sci. Eng.* 134: 143–149.

Sebtosheikh, M.A., Motafakkerfard, R., Riahi, M.A. et al. (2015). Support vector machine method, a new technique for lithology prediction in an Iranian heterogeneous carbonate reservoir using petrophysical well logs. *Carbonates Evaporites* 30: 59–68.

Soliman, M., Lehman, C., and AlRegib, G. (2020). S 6 : semi-supervised self-supervised semantic segmentation. *IEEE International Conference on Image Processing (ICIP)* (25–28 October 2020). New York, NY, USA: Institute of Electrical and Electronics Engineers (IEEE), pp. 1861–1865. https://2020.ieeeicip.org/

Stright, L., Boucher, A., Mukherji, T., and Derksen, R. (2009). Revisiting the use of seismic attributes as soft data for subseismicfacies prediction: proportions versus probabilities. *Lead. Edge* 28: 1460–1679.

Vallabhaneni, S., Roy, S., and Chaki, S. et al. (2022). "Automated reservoir model prediction using ml/ai intergrating seismic, well log and production data." U.S. Patent Application 17/016,075, filed 10 March 2022.

Verma, A.K., Chaki, S., Routray, A. et al. (2014). Quantification of sand fraction from seismic attributes using Neuro-Fuzzy approach. *J. Appl. Geophys.* 111: 141–155.

Vyas, M. and Sharma, A. (2016). Systems and methods for optimal stacking of seismic data. US9429667 B2.

Wong, P.M., Bruce, A.G., and Gedeon, T.T.D. (2002). Confidence bounds of petrophysical predictions from conventional neural networks. *IEEE Trans. Geosci. Remote Sens.* 40: 1440–1444.

Wu, Y. and Misra, S. (2020). Intelligent image segmentation for organic-rich shales using random forest, wavelet transform, and hessian matrix. *IEEE Geosci. Remote Sens. Lett.* 17: 1144–1147.

2

Petrophysical Predictions Using Regression and Advanced Machine Learning Algorithm

Shubham Singh[1] *and Abhijeet Bhardwaj*[2,*]

[1] *McCombs School of Business, The University of Texas at Austin, Austin, TX, USA*
[2] *Department of Industrial and Systems Engineering, University of Wisconsin-Madison, Madison, WI, USA*

[*] *Author for correspondence: abhijeet.bhardwajiitr@gmail.com*

Abstract

"Civilization is a progress from an indefinite, incoherent homogeneity toward a definite, coherent heterogeneity." —Herbert Spencer.

The similar philosophy of progress can be applied to methodologies employed in interpretation/processing of petrophysical data. The composition of the rock types drastically changes from one location to another, different physical attributes respond to these changes in variety of ways turning the entire process of interpretation complex.

One cannot use a "*one size fits all*" approach as a generic solution to all the problem sets. Instead, (s)he needs to classify her/his objectives, find best fit solution for each of the objectives, and subsequently build a coherent pipelined solution model. This approach has been extended in the instances of regression analysis of variable. It has been used to predict output variables (for example, porosities derived from compressional slowness) in cases of poor data quality acquisition, instrument failure, etc. The proposed methodology helps organize, as well as formulate, classification-based regression technique to deal with interpolation as well as prediction-based difficulties. In cases of univariate regression (where one only has got single input variable), the approach aids to introduce learned nonlinearity attributes to predict optimally. The efficiency of the predicted results is verified using test wells. In addition to this, time complexities as well as applicability of various subapproaches are compared and presented.

What you have solved it already?

Well that would be too much to say. I have discovered a suggestive fact that is all.

–Dr. Watson and Sherlock Holmes in The Sign of Four

Keywords *Generalised linear model; polynomial regression; spline interpolation; slowness vs density; bias-variance trade-off*

Handbook of Petroleum Geoscience: Exploration, Characterization, and Exploitation of Hydrocarbon Reservoirs, First Edition. Edited by Soumyajit Mukherjee, Swagato Dasgupta, Chandan Majumdar, Subhandip Mandal and Troyee Dasgupta.
© 2023 John Wiley & Sons Ltd. Published 2023 by John Wiley & Sons Ltd.

2.1 Introduction

Making inferences about essential parameters from observed data (e.g. Mukherjee et al. 2019) has been a common approach for geoscientists. In oil and gas industry, the life cycle of an exploration and production (E&P) activity begins by drawing inferences from the observed seismic data (e.g. Misra and Mukherjee 2018; Asl et al. 2019; Faghih et al. 2019). These inferences provide deep insights into the geoscientists to take important decisions in planning, exploration, and production. It is also highly desirable that the inferences made also quantify for the uncertainty in estimation of a parameter. Geoscientists use theoretical statistics in order to do the same. These methods still prove to be highly efficient in doing their jobs. Oil and gas industry has always generated and will continue to generate tremendous amount of subsurface data. Thus, it becomes inquisitive for geoscientists to tackle a broader level of problems to enhance decision making. Apart from inferring parameter tasks like classification of data into different buckets and quantifying similarity in data for generating insights have also emerged to be essential objectives for geoscientists. Electrofacies generation (Serra and Abbott 1982) was one such attempt to better interpret the subsurface structure from well-logging data. Machine Learning, thus, has gathered a high interest amongst geoscientists. The term "Machine Learning" (ML) was coined by Arthur Samuel in 1959 provided a more formal definition for the same. He states ML as: "A computer program is said to learn from experience E with respect to some class of tasks T and performance measure P if its performance at tasks in T, as measured by P, improves with experience E."

ML algorithms, as per this definition, can perform several tasks, which can be classified based on the type of objective, which the user is interested in. The broad categories of the task performed by ML algorithms classify into supervised and unsupervised learning (Sathya and Abraham 2013; Pan et al. 2013). As the name suggests, in supervised learning, the experiments have some kind of supervision. By supervision, we mean to say that every data point in some vector space χ is related to a single data point of a vector $\acute{\upsilon}$ of interest by some mathematical function f. The major purpose here is to be able to predict the values for vector $\acute{\upsilon}$ from observed data points in χ. Supervised learning problems further classify into Regression and Classification problem depending on the nature of vector $\acute{\upsilon}$. If $\acute{\upsilon}$ is a single dimensional vector, then the task of predicting values for $\acute{\upsilon}$ is termed as regression. But if the vector $\acute{\upsilon}$ is multidimensional having its own vector space χ', then the problem at hand is the classification problem. The major purpose of unsupervised learning is to quantify the similarity in the data. As the name suggests, unsupervised learning does not comprise any supervising vector $\acute{\upsilon}$ and the outcome here are the clusters of data set on vector space χ demonstrating the closeness or similarity between data (generation of electrofacies can be considered as an example of these types of algorithms, Lim et al. 1997).

Having concise knowledge of various applications of ML algorithm can prove very handy for geoscientists in order to perform their activities more efficiently. Numerous activities performed in E&P industries can be divided into broad sections. The insights from each section are highly beneficial in decision making for the overall industry and for the sections as well. It is difficult to draw boundaries, but the major sections can broadly consist of activities related to exploration phase, reservoir characterization (RC) phase during appraisal and development, and, lastly, the production phase. Geoscientists have started using ML in most of the phases and, thus, it is essential to be aware of the same.

This article reviews application of ML in oil and gas industry along with various data sources. A case study is also presented on the Volve dataset (free dataset provided by Equinor) of the North Sea offshore basin for petrophysical evaluation. Here crucial insights about various steps to be performed for a supervised learning algorithm are demonstrated along with their advantages. The chapter finally closes with the discussion on adaption of ML algorithms for various other geoscientific activities.

2.2 State of the Art

We provide a brief overview of the adoption of machine learning algorithms for various activities in oil and gas domains. Major activities related to oil and gas industry can be broadly classified into three categories. Drawing hard boundaries between these activities is difficult as often the study of one supports the other. The three main categories are activities related to: (i) subsurface exploration, (ii) development, i.e. characterization of reservoir and petrophysical evaluations, and (iii) production and drilling.

2.2.1 Machine Learning Application for Exploration Activities

Exploration activities are conducted in order to study and evaluate the subsurface structure of the earth. Applied physical methods involving seismic prospecting, gravitational prospecting, magnetic prospecting, and electrical and electromagnetic (EM) prospecting are applied to measure various response characteristics that aids in providing deep understanding about subsurface structure of the earth (Telford et al. 1990). Although all the physical methods mentioned earlier have high potential in providing understanding of the earth's subsurface structure, the seismic prospecting methods have been studied extensively in oil and gas industry. The reflection of seismic methods (Yilmaz 2001) has been extensively used to delineate the subsurface geology. Reflection seismology involves the study of convoluted seismic reflection signals in order to derive inferences about the true reflectivity series of the earth's subsurface. However, the data collected for reflection seismic studies have to be processed before making it useful for deriving inferences about subsurface geology.

Seismic signal processing is a research area of key importance for oil and gas industry with well-developed theoretical methods. The theoretical methods based on physical and mathematical models have been extensively studied in literature (Indira and Gupta 1998; Yilmaz 2001; Sambridge and Mosegaard 2002). The primary steps involved in seismic signal processing are deconvolution, stacking, and migration. The major purpose of the seismic processing is to identify the reflectivity series of the earth's subsurface, which gets convoluted with the seismic waves to produce the measured seismic response traces. However, the measured traces also contain high amount of noise on account of reverberations (in marine seismic) and weathered layers (in land seismic). Instruments (geophones) measuring the seismic traces can also be affected by environmental disturbances, which may increase the signal to noise ratio (a criterion to identify the amount of noise in the measured seismic traces). Apart from that, the local geology of the area may make it difficult to carry out seismic exploration activities, thus resulting in the problem of missing seismic traces in the acquired data.

Active applications of ML algorithms in geology and geophysics (G&G) can be traced back to 1991 (McCormack 1991) in which the author described the use of neural computing for various G&G applications. Since then, researchers have used ML algorithms for almost all the activities related to G&G. The initial applications of ML algorithms include the work done by Hansen (1993), who used neural networks to identify primary reflections from seismic data. Machine learning algorithms have found many applications in processing reflection seismic data. For instance, Kononov et al. (2007) demonstrated the use of neural networks for predicting travel time in a 3-D model. Jiang et al. (2019) used deep convolutional neural networks in order to improve the seismic geometry quality control. This work would help make the process of estimating correct positions of source and receiver more robust by avoiding the use of the first-break travel times. Denoising seismic signal is another area where ML algorithms have been used regularly. Some recent examples could include the work by Deng et al. (2010) that used the least squares support vector regression (SVR) with implicit bias term (ILS-SVR) to reduce random noise and remove drifts from the seismic data. Another example can be taken from Li et al. (2019). They applied hybrid learning-based framework for denoising the seismic data by attenuating both incoherent noise (environmental noise) and coherent noise (ground roll and scattered noise). The problem of missing seismic traces also offers a great potential for the application of the ML algorithms. Jia and Ma (2017) proposes the use of SVR for reconstructing seismic data from under-sampled or missing traces. Here the SVR method constructs a regression hyperplane from training data. Then it interpolates missing seismic traces for other input data by using the learned equation. Jia et al. (2018) used Monte Carlo-based intelligent reduction of training set for this task and thus provided an improvement by avoiding redundancy and computational cost.

Application of reflection of seismic data involves the study of Amplitude Variation with Offset (AVO) (Aki and Richards 1980; Avesth et al. 2005). AVO analysis is used by geophysicists to evaluate various subsurface properties like rock fluid content, its elastic properties, etc. AVO is the study of the relationship between reflection coefficient and angle of incidence and is derived by Zoeppritz equations (Aki and Richards 1980; Shuey 1985). The process of using the observed variation of reflection amplitude with angle to estimate elastic parameters is referred to as AVO inversion (Zong et al. 2012). AVO inversion has been studied extensively to estimate various key parameters and thus involve the estimation techniques of the classical frequentist as well as Bayesian approach (Riedel et al. 2003; Ball et al. 2016). Traditional approach includes inverting various parameters from AVO using least-square inversion techniques. Buland and Omre (2003) provided a Bayesian framework for performing the AVO inversion. In this work, the inversion algorithm is based on the convolutional model and a linearized weak contrast approximation of the Zoeppritz equation. The objective is to then obtain posterior distributions for P-wave velocity, S-wave velocity, and density. ML model like SVR and neural networks have also been applied in estimating parameters from AVO inversion. Kuzma and Rector (2004) identified that model/data pairs computed using already existing geophysical relations can be used to train support vector machines (SVMs) to approximate inverse relationships. They demonstrated that if an SVM is trained using data having the required appropriate assumptions, it can be used to produce the required estimates from AVO inversion. A more recent work (She et al. 2018) provides a data-driven approach to solve the ill-posed AVO inversion problem, which reduces the dependence of inversion problem on assumed prior information about the actual geology.

Earth is a low-pass filter and thus with increasing depths, the resolution of seismic traces reduces greatly. Hence, for greater depths, seismic waves for large frequencies are only available, which cause it difficult to delineate thin reflectors. Various techniques are applied to increase the spectral resolution of the seismic traces. The traditional methods use spectral decomposition and wavelet transformation for enhancing the resolution of seismic data (Miao et al. 2007). Various time-frequency spectral models involving the deconvolution method are also proposed for enhancing the spectral resolution of the seismic data (Zhou et al. 2014). Neural networks have also been used for enhancing the spectral resolution (Canning et al. 2017). Advanced convolutional networks are also proposed by Choi et al. (2019) to increase the spectral resolution of seismic data for better detection of thin beds. Multiple energy attenuation model (Karrenbach et al. 2000) also utilizes a neural network to learn about multiple energy from the training data and remove it from the test data. Various other applications of predictive ML methods by using seismic data also exist like automation of fault prediction and microseismic event prediction (Huang et al. 2018; Li et al. 2019).

2.2.2 Machine Learning Application for Reservoir Characterization and Petrophysical Evaluation

RC is the act of developing a reservoir model such that it incorporates all the characteristics of the reservoir that provide us insights into its ability to store hydrocarbons and to produce them. RC has been defined by Kelkar (2000) as the process of integrating various qualities and quantities of data in a consistent manner to describe reservoir properties of interest at inter-well locations. The major objective of RC is to understand and predict the fluid flow behavior in the reservoir over age. Brownfields which have been producing over time undergo a substantial change in their substructural properties with time. Thus, time lapse seismic data provides a substantial insight into the changes in the reservoir properties over time. In order to understand the fluid flow of the reservoir, it is essential to estimate various properties of the reservoir. These properties are of two classes (Fanchi 2002): (i) geologic – which includes the texture, mineralogy, sedimentary structure, bedding contact, and permeability barriers, and (ii) petrophysical – which includes porosity, permeability, compressibility of underlying rock structures, and fluid saturation. As seen in Section 2.1, in order to estimate rock properties from data, it is necessary to solve the inverse problem where we estimate model parameters from the given observation. As already identified, an inverse problem can be treated as the maximum likelihood estimation problem where we would draw inference about the model parameter from the data provided to us. Classical textbooks (e.g. Menke (1989)) detail various methods for geophysical inversion. There are other rigorous statistical treatment books on statistical inference as well (Casella and Berger 1990). Tarantola (1984) is also a pioneer for studying nonlinear seismic inversion problem.

The solution of inversion problem can be found by geophysical as well as ML inversion. For geophysical methods, it is essential to know the physics behind the data collection process to invert it successfully (for example, using Zoeppritz equations for AVO inversion). However, machine learning methods rely on the data and learn the functional form from the data. As given in Yui and Nora (2018), "To implement geophysical inversion, a forward operator should be given to investigators based on parametric equations describing a specific physical relation of data and model space." ML uses statistical techniques and makes decision boundaries for classification based on data distribution and density for unsupervised learning or human intervention and/or a priori information for supervised learning. Geophysical inversions involve techniques such as impedance inversion, amplitude-variation-with-offset inversion, and travel time tomography to yield the final estimates (Kim and Nakata 2018). The convergence for the estimate is achieved by utilizing the least-squares method, which minimizes the sum of the squared error difference between the desired and the actual signal. Bosch et al. (2010) provides a good reference for geophysical inversion of seismic data using Bayesian estimator. They define the posterior distribution for the estimate (elastic property) to have the Bayesian estimator relation of the form Posterior $=$ Constant $*$ Prior $*$ Likelihood. The problem is solved sequentially. First, the seismic data is inverted into elastic properties, and then rock-physics models transform those elastic properties into the reservoir properties.

Data-driven methods for RC are not new to the G&G research group. An and Moon (1993) provided the use of feed forward neural network to learn the functional mapping between seismic and well log data, thus providing a method to identify very thin reflector layers. Using 4-D seismic or time-lapse seismic data for understanding the reservoir changes has also been carried out effectively by the ML algorithms. Tiwari et al. (2018) demonstrated the feasibility of using ML methods for modeling reservoir property changes using 4D pre- and post-stack seismic data. Osman et al. (2001) used artificial neural networks (ANNs) for predicting the formation volume factor at the bubble point pressure. Daniele Colombo et al. explored the use of EM inversion method for reservoir monitoring. They explore the use of ML techniques in the form of deep learning neural networks for implementing EM-based reservoir monitoring coupled with a dynamic fluid flow simulator. ML has been extensively used to estimate the rock properties from the available data. Gong et al. (2019) utilized an assembling of data mining techniques to predict elastic properties of the rocks using various ML algorithms, e.g. multivariate linear regression, multivariate adaptive regression, SVMs, and neural networks. Methods making use of seismic attributes for estimating reservoir parameters have also been demonstrated (Yuan et al. 2018). Fuzzy logic algorithm is used for feature selection and, finally, uses SVR to predict the required rock parameter (Yuan et al. 2018). The task of RC could also be extended to study subsurface facies. Qi et al. (2020) used the Gaussian mixture models to best estimate seismic attribute to be used to differentiate one class of chaotic seismic facies from another. Jin (2018) used deep learning networks to predict seismic facies and invert reservoir properties to provide better results than the existing methods.

There is a regular use of machine learning methods for making petrophysical evaluations (Ahmadi and Chen 2019; Xu et al. 2019; Silva et al. 2020). Depending on the type of data available, petrophysical evaluations can be made at a point on the field, for example, missing well log curve or it can be made for the entire reservoir. Evaluating petrophysical properties for entire reservoir is attained by combing well log data with the seismic data. However, the frequency of measurement for both these types of data is different and thus combining them together involves either loss of information from high frequency well log data due to downsampling or introduction of more noise in seismic data due to upsampling. Chaki et al. (2018) discussed the use of seismic attributes to predict sand fraction by using conjugate gradient network. The author also showed the method for information filtering to be performed in order to have data from well logs and seismic on the same scale. Another very regular application for ML is the use of ML algorithm to predict information about missing well logs from other suits of well logs available. Researchers have used Bayesian approach to predict missing sonic log from other routinely acquired well logs. A work for predicting missing sonic log using short long-term convolutional neural network has been done by Pham and Wu (2019) using well log data from 177 wells.

2.2.3 ML Applications in Production

Production activities in oil and gas industry produce a huge amount of data. The real-time sensor data produced at each field is used for monitoring the health of the equipment involved in drilling and production activities. Sensors are even located in the borehole to provide continuous monitoring of the drilling well (e.g. Dasgupta et al. (2019a, 2019b)). Such a wide variety of data provides great opportunities for ML methods to be applied for various drilling-related activities. The major activity of interest for the G&G people can be to monitor and predict real-time sensor data for incidents like blowouts and gas kicks and to provide key indicators for the same (Noshi and Schubert 2018). Various physical models are already deployed for achieving the same. However, data reduction techniques like principal component analysis (PCA) can help significantly to create a customized indicator that can be used for monitoring the state of the system. An example can be given from the work done by Natarajana and Srinivasan (2010) where they develop multi-model-based process condition monitoring for offshore oil and gas production systems. Another example can be cited from Yang et al. (2019) where they make use of PCA for dimensionality reduction along with neural networks to make predictions for gas kicks.

Another important task can be to monitor lost circulation, which can cause serious leaks in the well bores, thus making the borehole more susceptible to failure. To predict lost circulation, Abbas et al. (2019) applied ANNs and SVM models along with constrained parameters. Al-Hameedi et al. (2018) also demonstrated the use of machine learning algorithms in predicting lost circulation. Another important application of predictive machine learning models is to predict drilling equipments' health. Health of a system is extensively studied in survival analysis. Classical parametric and non-parametric models have already been proposed for the same (Sikorska et al. (2011)). Cox (1984) proposed a classical model for studying survival data based on biostatistical applications, which have been translated for equipment health diagnosis as well. Non-parametric models from Kaplan and Meier (1958) have also been used extensively to model time to failure for various equipments. Researchers have variously attempted to design ML models to provide predictive capabilities for estimating system failures. However, most of the approaches so far have made use of data reduction techniques for monitoring system failure, for example, electrical submersible pumps (ESPs) rather than providing predictive capabilities to the models.

Bhardwaj et al. (2019) showed an attempt to model ESP failure from statistical process control perspective to generate real-time alarms for a failure incident using sensor data for ESP. Various other applications of ML algorithms for drilling process also include the prediction of rate of penetration (ROP) for drill bit in order to optimize the drilling process and reduce the nonproductive time of the drilling operation. Hegde et al. (2015) make use of bagged trees and random forest algorithms to predict the ROP of a drill bit. Hedge et al. (2017) propose optimization methods for input parameters to predict maximum possible ROP for a formation.

2.2.4 Type of Data in the Oil and Gas Industry

The oil and gas industry produces voluminous data for each of the activities indicated earlier. Seismic data amounts to be the one with huge storage spaces. A simple 3D volume with 400 in-lines and 300 traces per in-lines would need ~1.4 gigabytes of storage space. However, for most ML applications, raw seismic traces are seldom used. The processed seismic data is used to generate seismic attributes by traditional geophysical software. These attributes are already processed and thus are regular as compared to other sources of data like well logs or sensor data and they also include the bias induced by the processing operations. Seismic traces after processing (time-depth conversion) become variable with depth for the entire reservoir space. Other types of depth-dependent data (e.g. Mukherjee (2018)) are acquired to make petrophysical inferences about the geology of the area as the well log data. Well log data are not preprocessed; they occur as a response to a physical phenomenon used for their activation. Well log data is thus random in nature and provides good statistical basis for its treatment. The third and the most modern type of data collected by the oil and gas industry is the real-time monitoring sensor data. This type of data can be assumed to be completely random in nature as it records the response of an equipment to a measurement over time and hence statistical time-series methods are applied to model long- and short-term trends of the sensor signal. This helps infer key information on system age and its health (Nguyen et al. 2017). Other unconventional data sources for oil and gas industry include various image data and free text data that can be mined to gain relevant information like nonproductive time analysis using rig maintenance records or fault prediction using seismic images.

2.3 Case Study

2.3.1 Background

The prime requirement of the oil and gas service industry is to acquire process and interpret the petrophysical data. The logs associated with the petrophysics (e.g. Dasgupta et al. (2019); Dasgupta and Mukherjee (2020)) are considered as the most important elements for reservoir assessment and they are frequently employed to evaluate reservoir parameters. The degree of certainty of the interpreted information much depends on the quantity of the data available. For most of the time, the entire log packages of gamma ray (GR), density, resistivity, sonic, and many more are available for characterization and parametric evaluation processes. The sonic log, amongst all, is utilized to evaluate the sonic porosity (to get a preliminary idea of the formation type), to evaluate the geomechanical parameters (to comment on the well bore stability and much more), to delineate (validated using formation microimager logs) the fracture zones, etc. But there can be an unavailability of data, primarily the sonic data, from the comprehensive log package. This unavailability can be of two major kinds: (i) Either the entire sonic log is eliminated from the log package of the well section because of some financial or contractual concern (sonic tool is not run for the section). This leads to estimation of a predefined empirical relationship for computing the sonic velocity in the given section. (ii) Or there is an unavailability of the sonic data in some depth ranges within a section. This issue can surface because of multiple reasons associated with incomplete logging, instrument fault/failure and/or poor-quality recordings.

In either cases, the interpreter strives to build a continuous slowness, and to do so, one employs several regression-based techniques to precisely compute the missing output variable (for example, compressional slowness) utilizing the available independent input variables (density, porosity, resistivity, etc.). The algorithms of regression may differ in types subjected to (i) the number of input variables present; (ii) the correlation between input-output curves; and (iii) the degree of consistency within individual input variables (variance).

Favorably, at times, one can have several independent input variables to predict the output. But if that is not the case always (as discussed earlier), then, at such times, reliance of our equation depends entirely on one or, at max, two input variables and authenticity of the model is subjected to a kind of algorithm employed (Whitmore 1985). The output then depends on the logic of the algorithm when one has a few independent input variables. One needs to start with analyzing the reliability of the input curve and question – how much the input variable explains about the formation type, how much is the extent of its variability, and, most importantly, to what degree is the input curve related with the output curve, etc.

We will discuss a common, though a notable, case of a cased hole logging where one needs to process the compressional slowness using only the provided (logged) GR curve. In fast formation types, the slowness values of steel casing fall in the same range as that of fast compressional waves. Subsequently, it gets difficult to delineate the opposite values of compressional slowness because of its high level of interference with the casing waves. Therefore, a regression-based empirical formula had to be devised to predict the slowness values in places of contentions. Other times because of mud contamination and/or tool flexural interruptions, it turns tough to demarcate the correct values of slowness. Rate of logging, data acquisition techniques, and tool types also play a major role in the determination of the quality of the logged data that further influence the processing.

GR log deals with the assessment of the natural GR emissions. The signatures of GR (GR) emission for the sand and shale bodies differ. So, the log is used as a classifier of a reservoir and an estimator of the shale volume. Different formations though can demonstrate divergent values of GR reading subjected to the degree of GR counts that themselves are dependent on the K, U, and TH abundance in the formation type. The sonic log, on the other hand, is the direct measurement of the travel time of a multitude of sonic waves, especially P-wave from source to formation and back to the receiver. Sonic log values depend on the moduli and the density of the material it is travelling through. The moduli depend on the material type of the formation rock and the density of the material is, to some extent, dependent on the porosity of the rocks. Consequently, an associative relationship can be formulated between the two log readings and a model can be accordingly trained on the provided sets of two values and that model can be further used to predict any of the missing variables out of the two contributors.

The methodology, as well as the logic of the algorithms discussed here, differs for different kinds of losses in data. We will present two case studies, devised for two different scenarios. The first one deals with gaps in the output slowness data (DTCO) within the same well (interpolation using the predictive method). The second one deals with the prediction of compressional slowness values (DTCO) for a completely new well from the same field. The major motivation of the case study is the scarcity of the available explanatory variables in statistical studies. We try to overcome this by generating dummy variables from the ones available to us to explain the model variance more. Briefly, in the first case study, we try to resolve the problem of interpolating missing DTCO values for certain regions of well bore where DTCO has not been measured during logging while drilling (LWD). We try to tackle the issues of scarcity of explanatory variables for these tasks by employing a polynomial regression over a single variable (GR) by observing its behavior for different ranges of DTCO values. To briefly explain the second case study, we try to predict the value of DTCO in a well where it has not been measured at all by only the training model for the wells where DTCO has been measured. In this case however the Depth and the GR values of the training wells are used to develop, a random forest regressor (RFR). Along with this, a composite hyperparameter tuning workflow is also introduced where the effect of each hyperparameter over the model accuracy is quantified (Probst et al. 2019).

2.3.2 Preprocessing of DATA

The loaded data need to be analyzed besides being manipulated in a variety of ways because of various reasons ranging from computational efficiency and data applicability to the efficacy of the generated results. On similar lines, a number of preprocessing techniques, of which a few apply to all the model-sets and the others were model-specific, were employed. The general preprocessing steps included normalization of the data and application of log-normal transformation (for the skewed datasets). The box plots (Figure 2.1a and b) of GR variable from all three wells were analyzed. The purpose of the plots was to first analyze the spread/ variation of the GR curve and then remove any outlier points if present in the dataset. The values of the GR curve were themselves scaled, on a min-max type scalar, in the range of 0–150 (0 represents the minimum GR value and 150 the maximum GR value). This unit is based on the calibration standard used for an artificially radioactive concrete block at the University of Houston, Texas, USA, that is defined to have a radioactivity of 200 (double of typical shale values), American Petroleum Institute (API, formed 1919) units. However, many standard petrophysical scales use 0–150 range for representation and further interpretation purposes (Glover (2011)).

The depth values of the well (in actual measured depth – MD) were normalized using the maximum-minimum scalar function. The maximum and the minimum depth values of the combined dataset (all three wells) were determined. The values were then used to normalize the MD variable of all three datasets. In order to analyze the Normality of the variable quantile-quantile (Q-Q) plots were also plotted. The Q-Q plot (Figure 2.2a and b) is a powerful

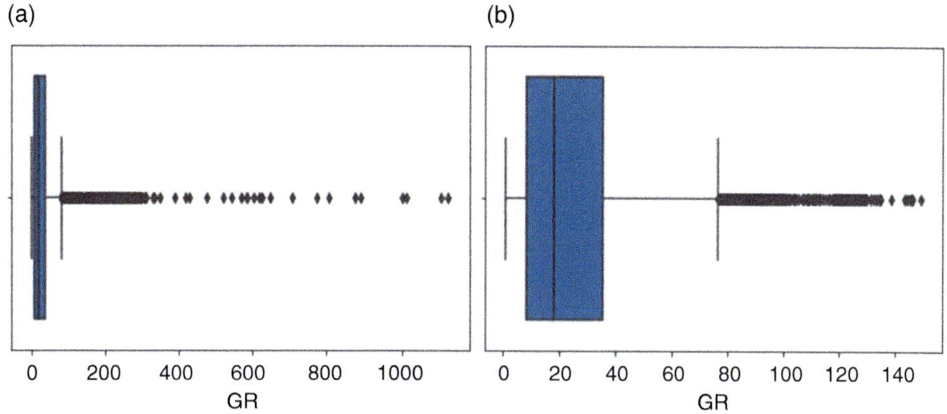

Figure 2.1 (a) GR Boxplot for 15_9_F_11 before preprocessing; and (b) GR Boxplot for 15_9_F_11 after preprocessing.

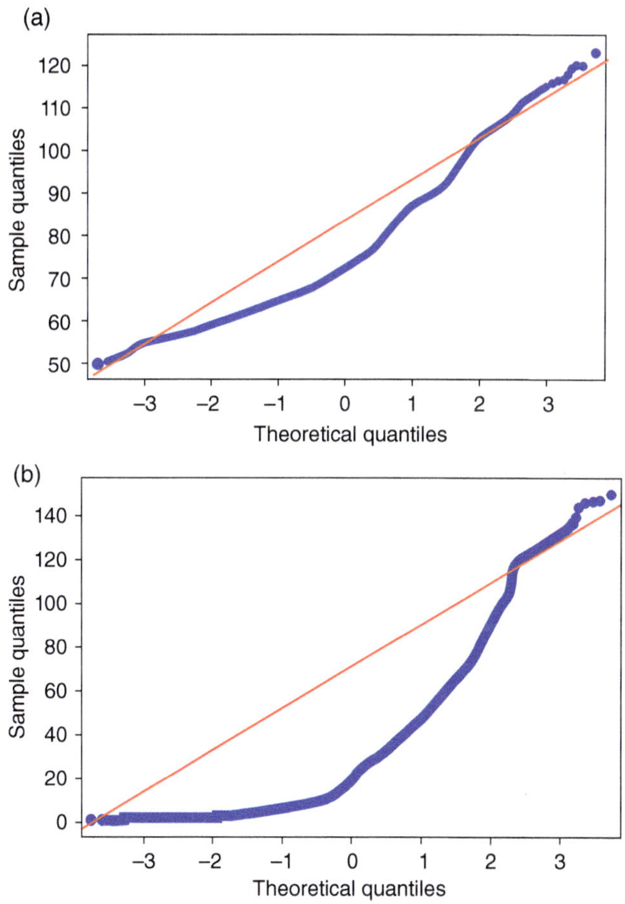

Figure 2.2 (a) QQ plot for DTCO in F_11 well; and (b) QQ plot for GR in F_11 well.

tool that plots two sets of quantile against each other and if the quantiles come from similar distribution then a straight line is achieved. In the case of testing the normality of the distribution, the quantile of the reference curve comes from the standard normal distribution. Based on the Q-Q plot, the DTCO values are interpreted to follow the Normal distribution while the GR value had a heavy left tail distribution (reasonably so as most values belong to (Ball et al. 2016; Telford et al. 1990).

2.3.3 Data Description (Training, Test, and Validation Sets)

Three distinct wells (15_9-F-5, 15_9-F-11, 15_9-F-15) were used from the Volve oil field, located ~200 km west of the Stavanger at the southern end of the Norwegian coast. The data has been made available by Equinor on their open data source website (https://www.equinor.com/en/how-and-why/digitalisation-in-our-dna/volve-field-data-village-download.html). The petrophysical data set majorly comprised the GR curve along with compressional slowness (DTCO). For the first case, the data gaps present in the DTCO curves of F-5 and F-11 well had to be interpolated. Hence, for the first case study, the GR values of the two wells were used independently to develop two different models (each for one well) to predict (interpolate) the DTCO values, in both the wells respectively. The datasets (GR and DTCO) from the two well were distributed into training and test sets, and the training set was further divided into training and validation set to tune the hyperparameters of the model. The accuracy of the model and the correctness of the predicted/interpolated DTCO were then measured on the test set (detail in Section 3.6.1).

In the second case, we also included depth measurement(MD/TDEP – Measured Depth/Tool Depth) curve to develop the models (where both DTCO and GR values were present) from the F-5 and F-11 wells. Then the entire data set was divided into training and test sets (as discussed in the description of the case studies). Further the training set was again divided into training and validation set. The training set was used for estimation of the model, whereas the validation set was employed to tune the hyperparameter and avoid overfitting. The test set was finally used to evaluate the final model (Section 3.6.2).

2.4 Methodology

With the aim to predict/interpolate missing values, a three-step approach was implemented. First, a model was trained using elementary logic(s), following that the results (accuracy of the model) were assessed on validation datasets in order to highlight the most desirable features and the most dependable parameters. The model with the best parameters was employed to predict the results.

2.4.1 Case 1

As described, this case study provides a workflow for interpolating missing sonic data in a given well (to fill the data gaps in 15_9_F-11 and in F-5 wells, respectively). The provided log data was primarily segmented into the training and the testing part. The ratio of training to test sets was kept as 3:1. The correlation value between the input GR and the output DTCO

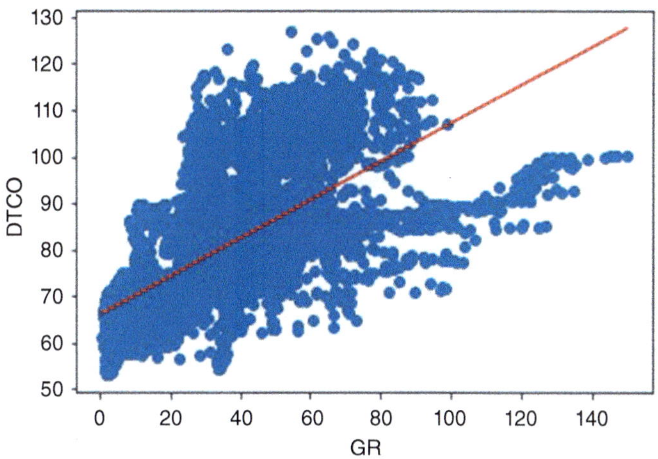

Figure 2.3 Linearly predicted DTCO curve overlaid on the scatter plot of GR vs. DTCO variables for Well F_11.

(compressional slowness) was measured for chunks of datasets from a series of wells mentioned earlier. Then a cross plot between two values from the training datasets was generated. Subsequently, a best fit linear line that can empirically relate the two sets of variables was estimated (on the same training dataset): for that, we simply employed a basic simple regression technique and evaluated DTCO using GR variable. Subsequently, the predicted DTCO was overlaid (red line in Figure 2.3) on the cross plot of logged (true) GR and DTCO (blue points in Figure 2.3) curves.

As can be observed (Figure 2.3), the one-degree equation curve was unable to take care of the scope of the variation of DTCO as well as the GR variable. The variability of the two datasets cannot be explained (as well as assessed) employing a simple linear empirical equation. The simple linear model, to a great extent, underfits the dataset. It was evident that the trend in variation of the two curves may not get explained well using a one-degree equation but may get better described using a higher degree polynomial equation. Thus, instead of applying regression only on the GR variable, higher-order (degree) terms of GR were also regressed to explain DTCO. Such a choice of model was motivated by observing the cross plot of DTCO vs. GR in Figure 2.4, where one must employ a higher-degree equation to relate two variables. The linear regression equation ($\text{DTCO} = 63.21 + 0.23{*}\text{GR}$), as seen in the right panel of Figure 2.4, is not able to explain the variation of the dataset properly. Whereas, a higher degree polynomial regression equation, order 5 ($\text{DTCO} = 42.4 + 2.52 * \text{GR} - 6.3e^{-2} * \text{GR}^2 + 6.6e^{-4} * \text{GR}^3 - 2.6e^{-6} * \text{GR}^4 + 3.0e^{-9} * \text{GR}^5$) as demonstrated in the left panel of Figure 2.4, was able to explain the DTCO values better for different values of GR.

The usage of polynomial regression method engenders sudden shoot up in number of features. For that reason, P-values were utilized to sort out and choose the degree of explanatory variable that are related to the output (DTCO). It was also observed that the number of datapoints in discrete sections of the crossplot were significantly different. A data density

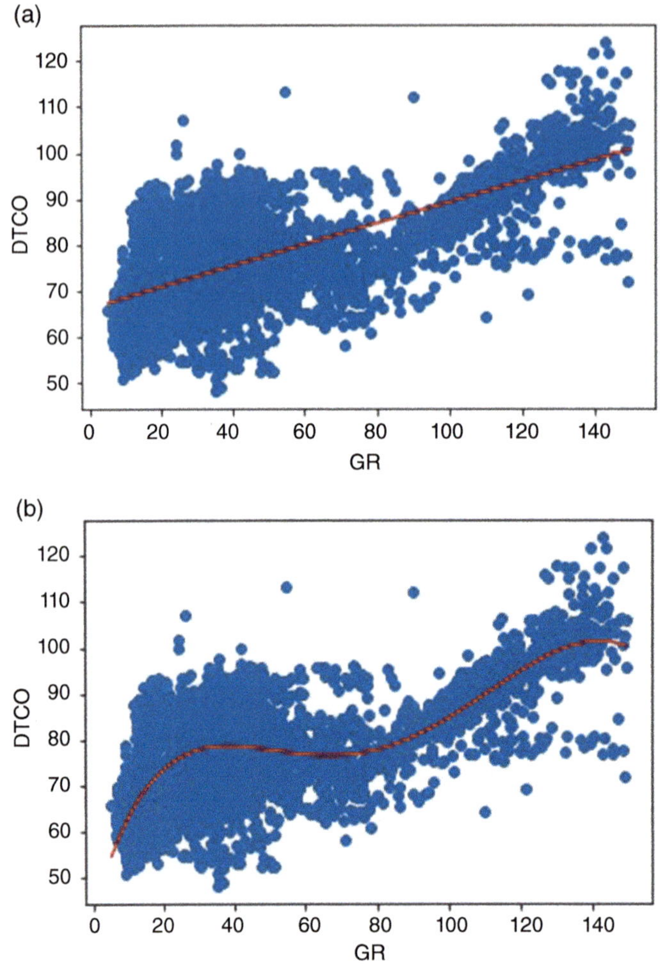

Figure 2.4 (a) Different degrees of polynomial regression curves (1 degree) overlay GR vs. DTCO scatter plot for F_11 well; and (b) Different degrees of polynomial regression curves (5 degrees) overlay GR vs. DTCO scatter plot for F_11 well.

plot (Figure 2.5, demonstrating the number of sample points present for different values of DTCO and GR) was prepared. It was noted that the density of data points was quite high (1270 data points present in the zone where Gr values lie between 10 and 20 and DTCO values between 60 and 70 us/ft for well F_11) in a few square spaces/grids as compared to other grid/zones (48 data points present in the zone where Gr values lie between 80 and 90 and DTCO values lie between 110 and 120 us/ft for well F_11). Subsequently, the dataset (each training set) was compartmentalized into batches and analyzed separately. The GR variable was initially divided into four groups or quantiles (where each quantile signified sandy, shaly-sand, sandy-shale, shaly formation in increasing order). As shown in Figure 2.6b, the interdependability of GR and DTCO variables is different in different quantiles. Each quantile (in this case quartile) was studied separately and the relationship

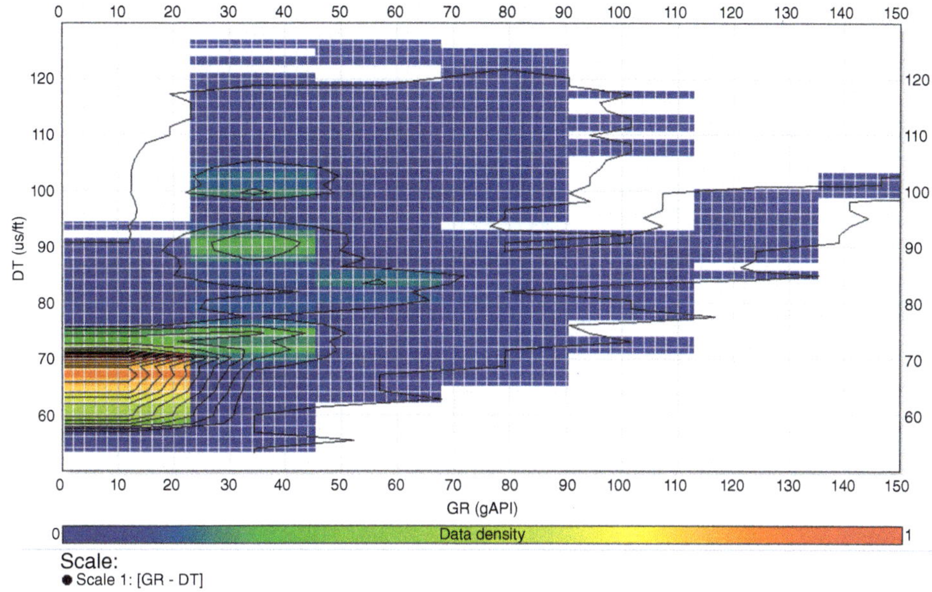

Figure 2.5 Data density crossplot between GR and DTCO, demonstrating number of readings/datapoints present for different values of GR and DTCO.

between DTCO-GR variables was derived one by one. It was noticed that the correlation of DTCO and GR in different quantiles was not the same. Besides, the variation in the values of DTCO was distinctly related to the values of GR in separate compartments. For example, in one of them, the DTCO follows a near-linear relationship with the GR and in other the variability of DTCO can only be comprehended using the higher order function of GR. It can be inferred that separate regression techniques need to be applied for distinct quantiles of GR values as within the same training set DTCO shows different degrees of dependency for distinct groups of GR values. It was thus decided not to restrict computation to four quartiles as there was a possibility of getting a better fit (results) with more numbers of segments. Consequently, it was observed that the more the compartments (of input GR variable), the better was the learning of the employed model.

Additionally, the mean square error of predicted and real DTCO values in the test set was used as selective criterion of the function (the criterion or the basis on which the algorithm selects the number of compartments/quantiles as well as the order of polynomial regression for each compartment). The snippet (Figure 2.7) of a table shows how the order and the respective error are tabulated in the final output. Besides, in order to avoid a one-to-one function situation (where one can only get single output for a single input variable), the measured depth variable, that is readily available with all logs, was included as the auxiliary input variable. The values of depth ranges were normalized using a min-max scalar and were then incorporated as an addendum input variable for increasing the precision of our predicted DTCO results. Conclusively, an automated function (based on *itertools* module of python) was iteratively used. Firstly, it determined the number of segments/zones the

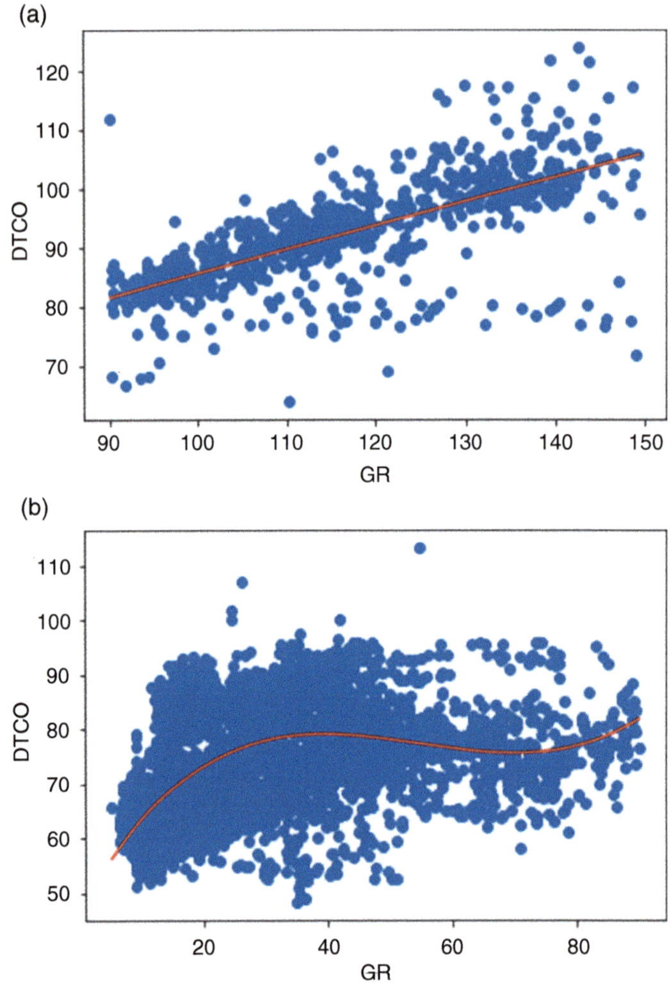

Figure 2.6 (a) DTCO vs. GR crossplot for subsection values of F_5 well; and (b) DTCO vs. GR crossplot for other subsection values of F_5 well.

GR function should be divided into. The function then decides the optimum order of the polynomial equation that fits each segment. Figure 2.7b. highlights the major sub-functions that are employed first to divide the data set into required compartments and second to assign an order to each compartment. Simultaneously, the function documents the mean squared error (MSE) for both training and validation sets (Figure 2.7c). This error works as the selecting criterion for the function. The function provides the number of compartments/quantiles and the respective order used in each segment. The function provides with the final predicted DTCO values in the end. The trained algorithm then can be used to predict DTCO where there are data gaps, the algorithm will first assess in which segment the value of the input variable (GR) falls and then apply the respective polynomial regression equation to provide the opposite DTCO result.

2.4.2 Case 2

Here, the DTCO has to be predicted for a well where it has not been measured completely. For this case in the training part, the data from the 15_9-F-5 and 15_9-F-11 were concatenated together. The data for the 15_9-F-15 well was not used for training and was kept as a blind (left out) well to compute the final model error. A total of 16 000 sample points of GR

(a)

Validation error of the model

Degree of Polynomial used for each quantile

Index	Error	Order
0	88.521 31	1111
1	86.968 77	1112
2	86.341 93	1113
3	86.253 10	1114
4	86.033 86	1115
5	85.711 97	1116
6	81.788 24	1121
7	80.235 70	1122
8	79.608 86	1123
9	79.520 03	1124
10	79.300 79	1125

(b)

W is the list quantile/compartments we wish to divide our data into

Y is the list degree of polynomial that one wish to use

This defined function helps to divide the data into number of compartments

The function assigns polynomial degree to the respective compartment

```
I = list(itertools.product((w,x,y)for w in [2,3,4,5,8] for x in range(w) for y in [1,3,5,8,10] ))
for i in range(len(I)):
    sub_quantile.append(I[i][0][1])

    Z = compartment(d_2,I[i][0][0])



    Z[I[i][0][1]] = assign_degree(Z[I[i][0][1]],I[i][0][2])
    Linreg_1 = LinearRegression()
    Linreg_1.fit(Z[I[i][0][1]].drop(['DTCO','L'],axis=1), Z[I[i][0][1]]['DTCO'])
    Z[I[i][0][1]]['Poly']= Linreg_1.predict(Z[I[i][0][1]][Z[I[i][0][1]].drop(['DTCO','L'],axis=1).columns])
    e = mean_squared_error(Z[I[i][0][1]].DTCO,Z[I[i][0][1]].Poly)
    qua.append(I[i][0][0])
    Err.append(e)
        #sub_quantile.append(I[i][0][1])

    deg.append(I[i][0][2])

 H=pd.DataFrame()
 H['quantile']= qua
 H['sub_quantile']=sub_quantile
 H['Degree']= deg
 H['Error']= Err
```

Figure 2.7 (a) The table demonstrating the validation error for different order of polynomial used for different compartments in a four-quantile model. The dataset was divided into four quantiles; (b) Snippet of the code that has been employed for prediction case; and (c) Snippet from the code explaining the computation of validation and training errors.

(c)

The training and validation sets are used independently

```
error(D_train,D_validation,6,6,6,6)
    for w in range(1,a+1):
        for x in range(1,b+1):
            for y in range(1,c+1):
                for z in range(1,d+1):
                    T_1,T_2,T_3,T_4 = quantile(T)
                    t_1,t_2,t_3,t_4 = quantile(t)

                    T_1=hel(T_1,w)
                    T_2=hel(T_2,x)
                    T_3=hel(T_3,y)
                    T_4=hel(T_4,z)
                    t_1=hel(t_1,w)
                    t_2=hel(t_2,x)
                    t_3=hel(t_3,y)
                    t_4=hel(t_4,z)
                    Linreg_1 = LinearRegression()
                    Linreg_1.fit(T_1.drop(['DTCO'],axis=1), T_1['DTCO'])
                    t_1['Poly']= Linreg_1.predict(t_1[t_1.drop(['DTCO'],axis=1).columns])
                    Linreg_2 = LinearRegression()
                    Linreg_2.fit(T_2.drop(['DTCO'],axis=1), T_2['DTCO'])
                    t_2['Poly']= Linreg_2.predict(t_2[t_2.drop(['DTCO'],axis=1).columns])
                    Linreg_3 = LinearRegression()
                    Linreg_3.fit(T_3.drop(['DTCO'],axis=1), T_3['DTCO'])
                    t_3['Poly']= Linreg_3.predict(t_3[t_3.drop(['DTCO'],axis=1).columns])
                    Linreg_4 = LinearRegression()
                    Linreg_4.fit(T_4.drop(['DTCO'],axis=1), T_4['DTCO'])
                    t_4['Poly']= Linreg_4.predict(t_4[t_4.drop(['DTCO'],axis=1).columns])

                    Te_f= pd.concat([t_1[['DTCO','Poly']],t_2[['DTCO','Poly']],t_3[['DTCO','Poly']],t_4[['DTCO','Poly']]], ignore_index=False)
                    Tot = pd.merge(left=t,left_index=True,right=Te_f[['Poly']],right_index=True,how='inner')
                    e= mean_squared_error(Tot.DTCO,Tot.Poly)
                    P.append(e)
                    L.append(str(w)+str(x)+str(y)+str(z))

    E=pd.DataFrame({'Error':P, 'Order':L})
    print(E.sort_values(by='Error').head(5))
```

The model is made to learn on the training set and error is calculated on validation set

Output errors are sorted in ascending order and the best fit model is thus selected

Figure 2.7 (Continued)

and DTCO variables from the two wells were assembled. Subsequently, the data was split randomly into training (12 000) and test (4000) sets taking a train to test ratio of 0.25. Successively we devised our models using the GR (at times MD) values as the input and DTCO values as the output from the training set. The training set was further divided into validation sets where one-third of that training data was isolated for validation. Thus, we had a training set of 8000 values, a validation set of 4000 values and a test set of 4000 values.

While generating an algorithm for predicting DTCO in different wells, a robust polynomial based RFR technique was employed. Prediction of porosity and permeability alteration is based on ML algorithms (Erofeev et al. 2019). The depth values of the new dataset (of the blind well) differ. Therefore, depth (MD) was also incorporated alongside the GR values, as the input variable. We included the polynomial-featuring technique to make our model better understand the correlation between the input GR and the output DTCO variable. In addition to this, bootstrap-backed multi-estimator RFR algorithm (Ho 1998 provided full references in full references list and not here) was employed to train our model. In simple terms, the model adopts the bootstrapping technique to sample the subset of datapoints and instead of generating a single decision tree, it develops aggregated estimator based on multiple trees. Therefore, the bagging-based RFR technique was used to aggregate different combinations of decision trees to predict the data as per Grömping (2012).

Primarily, a step-by-step methodical approach was employed to select the best hyperparameters of the RFR model. In this, the concatenated data was used to build the model, subsequently, the training error as well as the validation error for range of values of one hyperparameter were evaluated and plotted (Figure 2.8). The values of hyperparameters

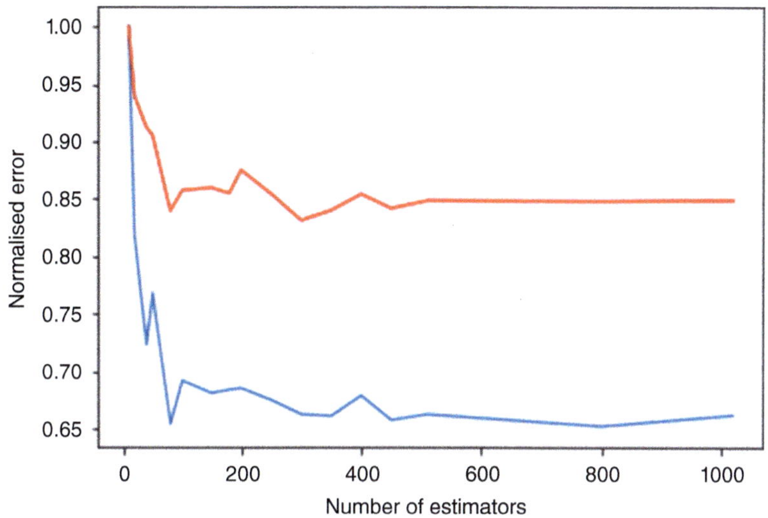

Figure 2.8 Red line represents the normalized validation error and the blue one represents the normalized training error of the model for different values of 'number of estimators'.

were tuned using a step by step procedure rather than a grid search approach as it showed that this method aided in the reduction of errors quite well as compared to grid search for practical purposes. The training and validation errors, for a set of values, of a single hyperparameter (for example, "number of estimators") were computed and, subsequently, plotted (Figure 2.8). The value of the hyperparameter ("number of estimators") for which the validation set error was the minimum was selected for the final model. The training error was analyzed to verify whether the model is functioning properly. Simultaneously, the minima of the validation set error were determined in order to avoid the overfitting of the model. These steps were repeated on an iterative basis for determining the most appropriate value of each hyperparameter used in the model. Thus, an optimal best-fit model was compiled that can be employed to predict the DTCO values for the wells from the same field.

As per Figure 2.8, when more than 300 numbers of estimators were used the training error does come down, but the validation error increases. Therefore, it can be concluded that the "number of estimators" had to be 300 in number. Step by step, the opposite values of each hyperparameter were evaluated using a similar technique. Finally, a terminal RFR model was computed with the most reliable set of hyperparameter values.

2.5 Results

2.5.1 Case 1

The quantization polynomial regression method employed for interpolation provided a root mean squared error (RMSE) of 4.58 (55.62% improvement over simple regression results, which gave an RMSE of 10.32) and 6.75 (39% improvement), respectively for training sets and test sets for the F_11 well. For the F_5 well, the training set RMSE error was 4 (55.28% improvement) and the test set RMSE error was 6.54 (30% improvement). The operation/running time of the method depends on the number of quantiles we are dividing the dataset into as well as on the range of degrees we are providing as input for the polynomial regression to work on. For computational purposes, the number of quantiles was restricted to 4. In addition to this, the upper and lower limits of the degree of Gr variable were assigned as 1 and 15, respectively for the model.

The predicted and the real DTCO values of the test set were plotted side by side and compared (Figure 2.9). The compressional slowness obtained from the model was compared with the real DTCO values of the F_11 (Figure 2.9a) and F_5 (Figure 2.9b) well,

(a)

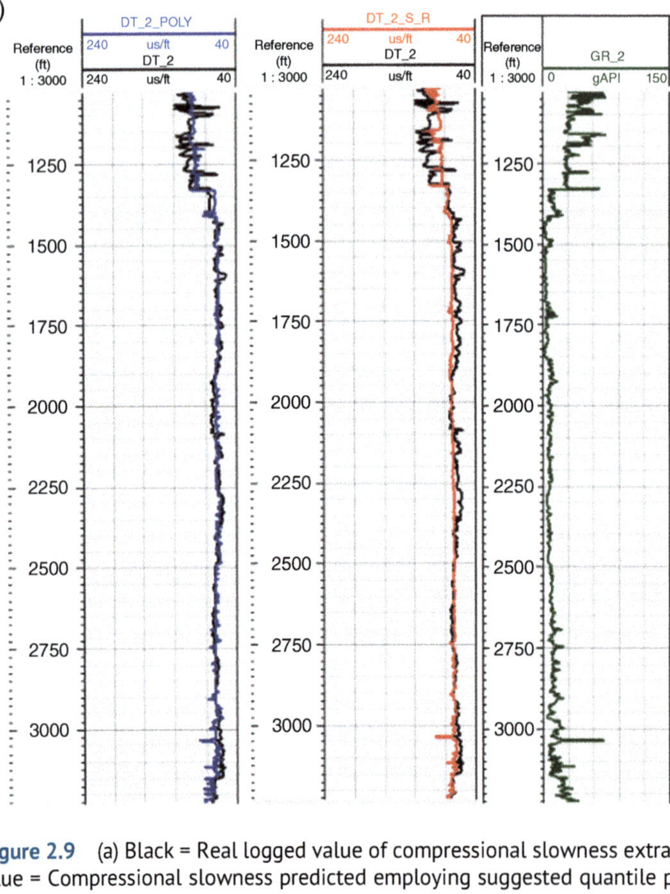

Figure 2.9 (a) Black = Real logged value of compressional slowness extracted from **15-9-F-11 well**. Blue = Compressional slowness predicted employing suggested quantile polynomial regression method. Red = Compressional slowness generated using traditional simple regression method. Green = Gamma Ray Log. (b) Black = Real logged value of compressional slowness extracted from **15-9-F-5 well**. Blue = Compressional slowness predicted employing suggested quantile polynomial regression method. Red = Compressional slowness generated using traditional simple regression method. Green = Gamma Ray Log. (c) Compression generated from the proposed quantile polynomial regression technique demonstrates more consistency with GR than the consistency displayed by the compressional slowness generated using simple regression equation in the F_5 well. (d) The model used to predict the DTCO values in the regions where the real DTCO values were missing for the **15_9_F-11 well**.

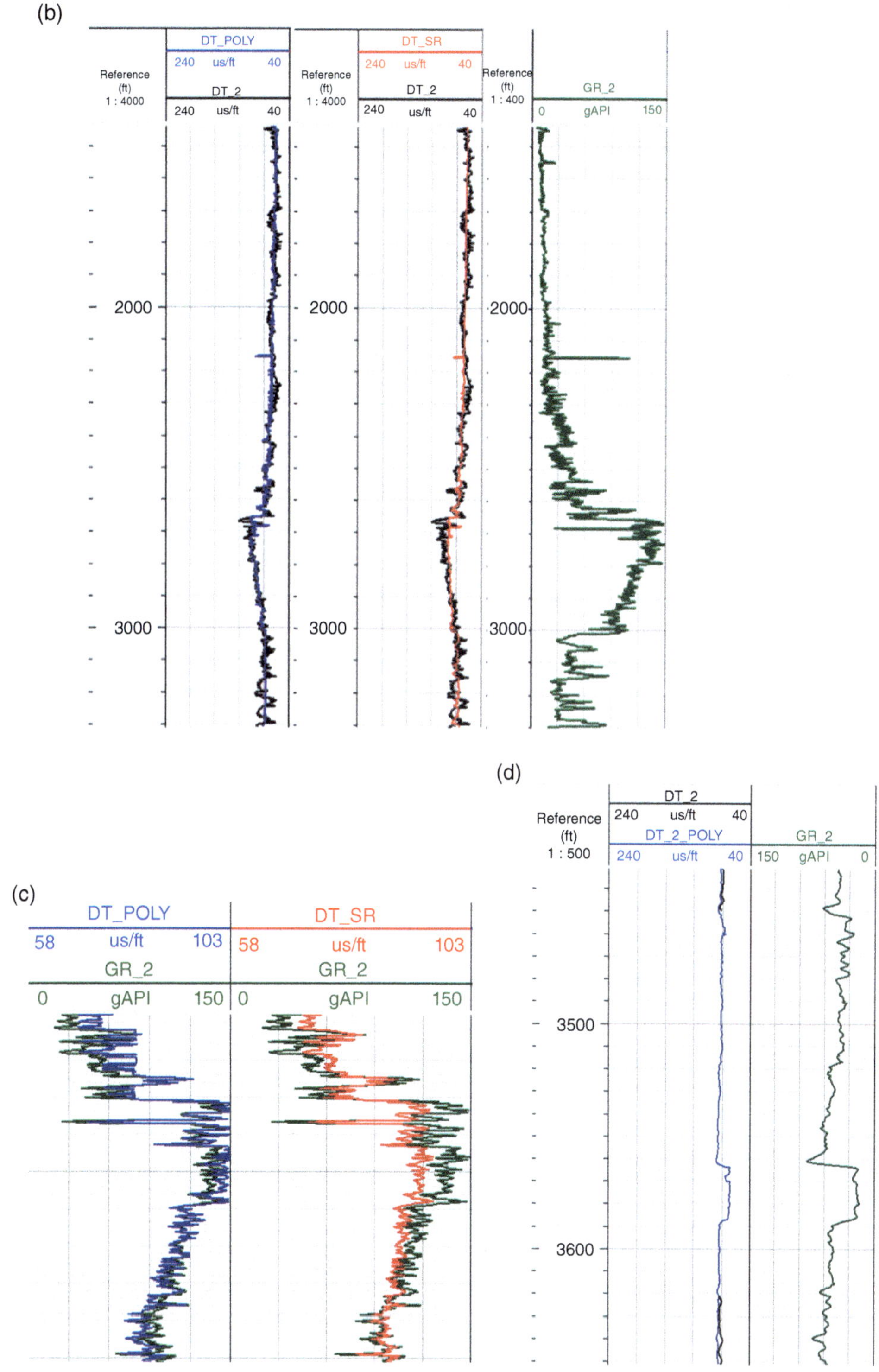

Figure 2.9 (Continued)

Table 2.1 MSE errors obtained from the simple and the proposed regression technique for training and test sets of contributing wells.

Well name	Training set MSE error from quantile regression	Training set MSE error from simple regression	Test set MSE error from quantile regression	Test set MSE error from simple regression
Volve_15_9_F_11	21.23	106.45	45.58	116.64
Volve_15_9_F_5	15.92	80.28	42.73	85.95

respectively. In addition, the DTCO values computed using a simple regression model was also plotted alongside for comparison purpose and the "train" and "test" error for simple regression was also computed (Table 2.1).It was observed that the values of the DTCO predicted using the proposed polynomial quantile regression were closer to the real DTCO values in the test sets than the one predicted with the simple linear regression. This can also be verified from the decrease in MSE error of the real and predicted DTCO values.

The predicted DTCO values demonstrated corresponding trends with real DTCO values. Furthermore, the predicted DTCO values were in conformance with the GR values (DTCO was slowing down in shaly regions and becoming faster in sandy zones as indicated by the low GR zones) as in Figure 2.9c. Successively, the trained model was utilized to predict the DTCO values in the depth regions where there was absence of logged DTCO (depth 3449–3621 m in F_11 well) values (Figure 2.9d). The predicted values not only displayed harmonious consistency with the GR values but also were in tune with the real DTCO values at the boundaries of data gap regions. Thus, the method proved viable for in house prediction/ interpolation of data gaps.

2.5.2 Case 2

The polynomial backed RFR technique used for computation of DTCO in the blind well produced a final optimized RMSE of 9.59 (when applied on the blind (left out) well F_15) that was an improved score over the ones obtained for simple RFR regressor 11.66 or a linear regressor RMSE of 14.07 (Figure 2.10). The composite running time for the single search method was ~600 seconds. The most suitable values of each hyperparameter were computed employing a step-by-step validation vs. training error analysis. A final RFR model was instantiated using the validated hyperparameter values (Table 2.2). This model was employed to compute the blind well (F_15) DTCO. The DTCO results predicted from the model were consistent with the GR values; the DTCO has different sets of values for different types of formation/zones. The error part (Table 2.3) of the predicted results was also reduced considerably employing the proposed model.

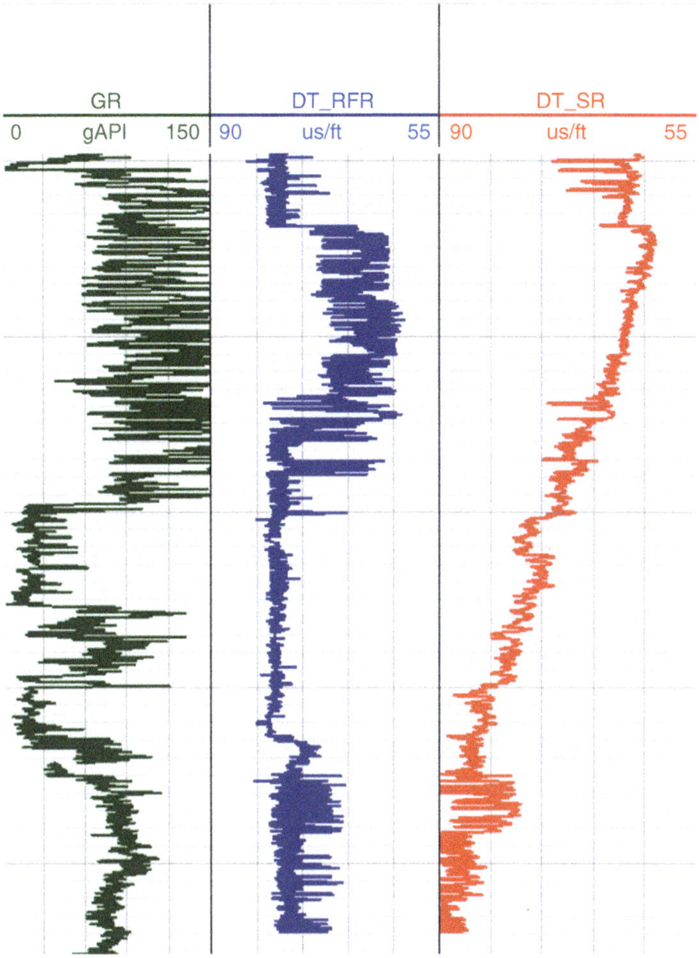

Figure 2.10 The compression evaluated using the recommended RFR technique is more consistent with the input GR feature than the compression generated using simple regression technique. Thus, we suggest to employ our proposed polynomial featuring backed RFR-centered empirical relation to predict values of DTCO for wells that are at different depth levels than the training wells.

Table 2.2 Optimum values of hyperparameters computed from the proposed approach.

Hyperparameter/parameter	Value of parameter
Number of estimators in RFR	300
Maximum depth in RFR	8
Minimum samples at leaf in RFR	2
Bootstrapping of samples in RFR	True
Degree of polynomial of GR used	5

Table 2.3 MSE, RMSE, and percentage errors for different models.

Well name	MSE error	RMSE error	Percentage improvement over linear regression model (MSE Error)
Linear regression equation	198	14.07	—
Simple RFR equation	136	11.66	31.31%
Tuned RFR model	92	9.59	53.53%

2.6 Conclusion and Way Forward

A composite model based on principle of segmentation and local-regression was formulated. The robustness of the model helps provide a solution for demanding cases when there is a scarcity of the number of input features (like in the one presented where DTCO had to be predicted using only GR values). The model helps to create and select dummy variables for each segment (quantile) and thus construct a comprehensive model for predicting the data.

For the first case, in which the simple regression model could not explain the variability of the DTCO well, a quantile-/zonation-based polyregression technique not only realized the variation of the DTCO in a better way but also predicted more accurate results.

Whereas, for the second case, it was observed that a single GR variable with a simple RFR model was not able to provide apposite results. Additionally, owing to the limitation of the dataset available, feature engineering methods had to be employed in order to explain the DTCO values better than explained by simple GR values. Therefore, a degree of nonlinearity was introduced by incorporating polynomial GR values in the RFR model.

The method of dividing a dataset into segments can be extended by further dividing the segments into the subsegments and carrying out this iteration till an optimal/balanced solution is obtained. The bias variance tradeoff has to be taken into consideration while formulating such an iterative algorithm. The proposed methodology of segmentation can even be applied when we are using more than one input variable, in which we can create sub-matrices of the large dataset. The submatrix will comprise data points with values in subrange of the input variables. For example, if we have two input variables, porosity and GR, we can create a submatrix for a formation type that is highly porous and shows low GR values. Then we can train our model independently for many similar cases. Further, more efficient ways of performing feature engineering along with methods to have better hyperparameter are to be explored.

Acknowledgments

Soumyajit Mukherjee (IIT Bombay) and Swagato Dasgupta annotated. Merryl Le Roux, Frank Otmar Weinreich, Joss Everett, Andrew Harrison, and the proofreading team (Wiley Blackwell) are thanked for their assistance. The article has been summarized in Mukherjee et al. (2023).

References

Abbas, A.K., Al-haideri, N.A., and Bashikh, A.A. (2019). Implementing artificial neural networks and support vector machines to predict lost circulation. *Egyptian Journal of Petroleum* 28: 339–347.

AliAhmadi, M. and Chen, Z. (2019). Comparison of machine learning methods for estimating permeability and porosity of oil reservoirs via petro-physical logs. *Petroleum* 5 (3): 217–284.

Aki, K. and Richards, P.G. (1980). *Quantitative seismology: Theory and methods*, vol. 1. W.H. Freeman and Co.

Al-Hameedi, A.T.T., Alkinani, H.H., Dunn-Norman, S. et al. (2018). Using machine learning to predict lost circulation in the rumaila field, Iraq. *Society of Petroleum Engineers.* https://doi.org/10.2118/191933-MS.

An, P. and Moon, W.M. (1993). Reservoir characterization using feedforward neural networks. SEG Technical Program Expanded Abstracts, 258–262. Society of Exploration Geophysicists.

Asl, M.E., Faghih, A., Mukherjee, S., and Soleimany, B. (2019). Style and timing of salt movement in the Persian Gulf Basin, offshore Iran: insights from halokinetic sequences adjacent to the Tonb-e-Bozorg salt diapir. *Journal of Structural Geology* 122: 116–132.

Avesth, P., Mukerji, T., and Mavko, G. (2005). *Quantitative Seismic Interpretation.* Cambridge, UK: Cambridge University Press.

Ball, V., Nasser, M., and Kolbjørnsen, O. (2016). Introduction to this special section: AVO inversion. *The Leading Edge* 35: 399–404.

Bhardwaj, A.S., Saraf, R., Nair, G.G., and Vallabhaneni, S. (2019). Real-time monitoring and predictive failure identification for electrical submersible pumps. *Society of Petroleum Engineers* https://doi.org/10.2118/197911-MS.

Bosch, M., Mukerji, T., and Gonzalez, E.F. (2010). Seismic inversion for reservoir properties combining statistical rock physics and geostatistics: a review. *Geophysics* 75: 75A165–75A176.

Buland, A. and Omre, H. (2003). Bayesian linearized AVO inversion. *Geophysics* 68: 185–198.

Canning, A., Moulière-Reiser, D., Weiss, Y. et al. (2017). Neural networks approach to spectral enhancement. SEG Technical Program. Society of Exploration Geophysicists.

Casella, G. and Breger, G. (1990). *Statistical Inference Duxbury Advanced Series.* Brooks/Cole Cengage Learning.

Chaki, S., Routray, A., and Mohanty, W.K. (2018). A novel preprocessing scheme to improve the prediction of sand fraction from seismic attributes using neural networks. *IEEE Journal of Selected Topics in Applied Earth Observations and Remote Sensing* 8: 1808–1820.

Choi, Y., Jee Seol, S., Byun, J., and Kim, Y. (2019). Vertical resolution enhancement of seismic data with convolutional U-net. SEG Technical Program. Society of Exploration Geophysicists.

Cox, D.R. (1984). *Analysis of Survival Data.* New York: Chapman and Hall/CRC https://doi.org/10.1201/9781315137438.

Dasgupta, T. and Mukherjee, S. (2020). *Sediment Compaction and Applications in Petroleum Geoscience*, Series: Advances in Oil and Gas Exploration & Production. Springer ISSN: 2509-372X.

Dasgupta, S., Dasgupta, T., Singh, V.N., and Mukherjee, S. (2019a). A numerical study of epithermal neutron log response and application of image log for porosity determination. In: *Problems and Solutions in Structural Geology and Tectonics*, Developments in Structural

Geology and Tectonics Book Series. Series Editor: Mukherjee S (ed. A. Billi and A. Fagereng), 237–258. Elsevier ISBN: 9780128140482.

Dasgupta, T., Dasgupta, S., and Mukherjee, S. (2019b). Image log interpretation of geomechanical issues. In: *Teaching Methodologies in Structural Geology and Tectonics* (ed. S. Mukherjee), 237–251. Springer. ISBN: 978-981-13-2781-0.

Deng, X., Yang, D., Peng, J. et al. (2010). Noise reduction and drift removal using least-squares support vector regression with the implicit bias term. *Geophysics* 75: V119–V127.

Dr. Paul Glover. (2011). Petrophysics course notes. http://homepages.see.leeds.ac.uk/~earpwjg/PG_EN/CD%20Contents/GGL-66565%20Petrophysics%20English/New%20Folder/Chapter%2011.PDF

Erofeev, A., Orlov, D., Ryzhov, A., and Koroteev, D. (2019). Prediction of porosity and permeability alteration based on machine learning algorithms. *Transport in Porous Media* 128: 677–700.

Faghih, A., Asl, E.M., Mukherjee, S., and Soleimany, B. (2019). Characterizing halokinesis and timing of salt movement in the Abu Musa salt diapir, Persian Gulf, offshore Iran. *Marine and Petroleum Geology* 105: 338–352.

Fanchi, J.R. (2002). Fundamentals of Reservoir Characterization. *Shared Earth Modeling* 170–181.

Gong, Y., Mehana, M., El-Monier, I et al. (2019). Machine learning for estimating rock mechanical properties beyond traditional considerations. SEG Global Meeting Abstracts. 466–480, Unconventional Resources Technology Conference Society of Exploration Geophysicists.

Grömping, U. (2012). Variable importance assessment in regression: linear regression versus random forest. *The American Statistician* 63: 308–319.

Hansen, K.V. (1993). *Neural Networks For Primary Reflection Identification*. Society of Exploration Geophysicists.

Hegde, C. and Gray, K.E. (2017). Use of machine learning and data analytics to increase drilling efficiency for nearby wells. *Journal of Natural Gas Science and Engineering* 40: 327–335.

Hedge, C., Daigle, H., Millwater, H., and Gray, K. (2017). Analysis of rate of penetration (ROP) prediction in drilling using physics-based and data-driven models. *Journal of Petroleum Science and Engineering* 159: 295–306.

Ho, T. (1998). The random subspace method for constructing decision forests. *Pattern Analysis and Machine Intelligence* 20 (8): 832–844.

Huang, L., Li, J., Hao, H., and Li, X. (2018). *Micro-seismic event detection and location in underground mines by using Convolutional Neural Networks (CNN) and deep learning, Tunnelling and Underground Space Technology*, vol. 81, 265–276. ISSN 0886-7798.

Indira, N.K. and Gupta, P.K. (1998). *Inverse Methods: General Principles and Applications to Earth System Sciences*. Narosa Publishing House.

Jia, Y. and Ma, J. (2017). What can machine learning do for seismic data processing? An interpolation application. *Geophysics* 82: V163–V177.

Jia, Y., Siwei, Y., and Ma, J. (2018). Intelligent interpolation by Monte Carlo machine learning. *Geophysics* 83: V83–V97.

Jiang, W., Zhang, J., and Lee, B. (2019). 3D seismic geometry quality control and corrections by applying machine learning. *Geophysics* 84: P87–P96.

Long Jin. (2018). Machine learning approaches for seismic-facies prediction and reservoir-property inversion. SEG Technical Program Expanded Abstracts. 2147–2151. Society of Exploration Geophysicists.

Kaplan, E.L. and Meier, P. (1958). Nonparametric estimation from incomplete observations. *Journal of the American Statistical Association* 53 (282): 457–481. https://doi.org/10.2307/2281868. JSTOR 2281868.

Karrenbach, M., Essenreiter, R., and Treitel, S. (2000). Multiple attenuation with attribute-based neural networks. *Society of Exploration Geophysicists.*

Kelkar, M. (2000). "Application of geostatistics for reservoir characterization – accomplishments and challenges", distinguished author series. *Journal of Canadian Petroleum Technology* 39 (7).

Kim, Y. and Nakata, N. (2018). Geophysical inversion versus machine learning in inverse problems https://doi.org/10.1190/tle37120894.1. *The Leading Edge* 37: 894–901.

Kononov, A., Gisolf, D., and Verschuur, E. (2007). Application of neural networks to travel-times computation. *Society of Exploration Geophysicists.*

Kuzma, H.A. and Rector, J.W. (2004). Non-linear avo inversion using support vector machines. SEG Technical Program. Society of Exploration Geophysicists

Li, S., Yang, C., Sun, H., and Zhang, H. (2019). Seismic fault detection using an encoder–decoder convolutional neural network with a small training set. *Journal of Geophysics and Engineering* 16 (1): 175–189. https://doi.org/10.1093/jge/gxy015.

Lim, J.-S., Kang, J.M., and Kim, J. (1997). Multivariate statistical analysis for automatic electrofacies determination from well log measurements. *Society of Petroleum Engineers* https://doi.org/10.2118/38028-MS.

McCormack, M.D. (1991). Neural computing in geophysics. *The Leading Edge* 10: 11–15.

Menke, W. (1989). *Geophysical Data Analysis: Discrete Inverse Theory*. Academic Press, Inc.

Miao, X., Todorovic-Marinic, D., and Klatt, T. (2007). Enhancing seismic insight by spectral decomposition. SEG Technical Program. Society of Exploration Geophysicists.

Misra, A.A. and Mukherjee, S. (2018). *Atlas of Structural Geological Interpretation from Seismic Images*. Wiley Blackwell ISBN: 978-1-119-15832-5.

Mohammad, A., Ahmad, R., Hamed, A., and Seyed, A.H. (2019). Machine learning technique for the prediction of shear wave velocity using petrophysical logs. *Journal of Petroleum Science and Engineering* 174: 306–327.

Mukherjee, S. (2018). Locating center of gravity in geological contexts. *International Journal of Earth Sciences* 107: 1935–1939.

Mukherjee, S., Bose, N., Ghosh, R. et al. (2019). *Structural Geological Atlas*. Springer ISBN: 978-981-13-9825-4.

Mukherjee, S., Dasgupta, S., Majumdar, C. et al. (2023). Introduction to: Handbook of Petroleum Geoscience: Exploration, Characterization, and Exploitation of Hydrocarbon Reservoirs. In: *Handbook of Petroleum Geoscience: Exploration, Characterization, and Exploitation of Hydrocarbon Reservoirs*, Petroleum Industry Updates: Geoscientific and Economic Issues (ed. S. Mukherjee, S. Dasgupta, C. Majumdar, et al.). John Wiley & Sons, Hoboken, ISBN 978-1-119-68003-1.

Natarajan, S. and Srinivasan, R. (2010). Multi-model-based process condition monitoring of offshore oil and gas production process. *Chemical Engineering Research and Design* 88: 572–591.

Nguyen, H.H., Mirza, F., Naeem, M.A., and Nguyen, M. (2017). A review on IoT healthcare monitoring applications and a vision for transforming sensor data into real-time clinical feedback. 2017 IEEE 21st International Conference on Computer Supported Cooperative Work in Design (CSCWD), Wellington, 257–262, https://doi.org/10.1109/CSCWD.2017.8066704. https://www.equinor.com/en/how-and-why/digitalisation-in-our-dna/volve-field-data-village-download.html

Noshi, C.I. and Schubert, J.J. (2018). *The Role of Machine Learning in Drilling Operations: A Review*. Society of Petroleum Engineers doi:10.2118/191823-18ERM-MS.

Osman, E., Abdel-Wahhab, O., and Al-Marhoun, Muhammad. (2001). Prediction of oil PVT properties using neural networks. Paper presented at the SPE Middle East Oil Show, Manama, Bahrain, March 2001. Paper Number: SPE-68233-MS. https://doi.org/10.2118/68233-MS.

Pham, N. and Wu, X. (2019). Missing sonic log prediction using convolutional long short-term memory. SEG Technical Program Expanded Abstracts. 2403–2407. Society of Exploration Geophysicists

Probst, P., Wright, M.N., and Boulesteix, A.-L. (2019). Hyperparameters and tuning strategies for random forest. *WIREs Data Mining and Knowledge Discovery* 9: 201905.

Qi, J., Zhang, B., Lyu, B., and Marfurt, K. (2020). Seismic attribute selection for machine-learning-based facies analysis. *Geophysics* 85: O17–O35.

Riedel, M., Dosso, S.E., and Beran, L. (2003). Uncertainty estimation for amplitude variation with offset (AVO) inversion. *Geophysics* 68: 1485–1496.

Sambridge, M. and Mosegaard, K. (2002). Monte Carlo methods in geophysical inverse problems. *Reviews of Geophysics* 40: 1009. AGU.

Sathya, R. and Abraham, A. (2013). Comparison of supervised and unsupervised learning algorithms for pattern classification. *(IJARAI) International Journal of Advanced Research in Artificial Intelligence* 2 (2).

Serra, O. and Abbott, H.T. (1982). The contribution of logging data to sedimentology and stratigraphy. *Society of Petroleum Engineers Journal* 22 (1): 117–130.

She, B., Wang, Y., Liang, J. et al. (2018). A data-driven amplitude variation with offset inversion method via learned dictionaries and sparse representation. *Geophysics* 83: R725–R748.

Shuey, R.T. (1985). A simplification of the Zoeppritz equations. *Geophysics* 50 (9): 609–614. https://doi.org/10.1190/1.1441936.

Sikorska, J.Z., Hodkiewicz, M., and Ma, L. (2011). Prognostic modelling options for remaining useful life estimation by industry. *Mechanical Systems and Signal Processing* 25 (5): 1803–1836. ISSN 0888-3270.

Silva, A.A., Tavares, M.W., and Carrasquilla, A. (2020). Roseane Misságia; Petrofacies classification using machine learning algorithms. Marco Ceia. *Geophysics* 85 (4): WA101–WA113.

Tarantola, A. (1984). Inversion of seismic reflection data in the acoustic approximation. *Geophysics* 49 (8): 1259–1266.

Telford, W., Geldart, L., and Sheriff, R. (1990). *Applied Geophysics*. Cambridge: Cambridge University Press https://doi.org/10.1017/CBO9781139167932.

Tiwari, U., Roy, B., and Cardozo, L.E. (2018). SAGD dynamic reservoir property characterization using machine learning. SEG Technical Program Expanded Abstracts. 2031–2035. Society of Exploration Geophysicists.

Whitmore, G.A. (1985). Prediction limits for a univariate normal observation. *The American Statistician* 141–143.

Xu, C., Misra, S., Srinivasan, P., and Ma, S. (2019). *When Petrophysics Meets Big Data: What can Machine Do?* Society of Exploration Geophysicists.

Yang, J., Sun, T., Zhao, Y. et al. (2019). *Advanced Real-Time Gas Kick Detection Using Machine Learning Technology*. International Society of Offshore and Polar Engineers.

Yilmaz, Öz (2001). *Seimsic Data Analysis (Processing, Inversion and Interpretation of Seismic Data) - Investigations in Geophysics*. Society of Exploration Geophysicists ISBN-1-56080-098-4.

Yuan, Z., Huang, H., Jiang, Y., and Tang, J., (2018). Multiattribute reservoir parameter estimation based on a machine learning technique. SEG Technical Program Expanded Abstracts: 2266–2270. Society of Exploration Geophysicists.

Zhou, H., Wang, C., Marfurt, K.J. et al. (2014). Enhancing the resolution of seismic data using improved time-frequency spectral modeling. SEG Technical Program. Society of Exploration Geophysicists

Zong et al. (2012). Reflection coefficient equation and pre-stack seismic inversion with Young's modulus and Poisson ratio. *Chinese Journal of Geophysics – Chinese Edition* 55 (11): 3786–3794. https://doi.org/10.6038/j.issn.0001-5733.2012.11.025.

3

A Modified Guided Filter to Denoise Seismic Attributes

Soumi Chaki[1,*], *S.L. Happy*[2], *Aurobinda Routray*[3], *and William K. Mohanty*[4]

[1] *Division of Signals and Systems, Department of Electrical Engineering, Uppsala University, Uppsala, Sweden*
[2] *HP PPS India Operations Pvt. Ltd., Bengaluru, Karnataka, India*
[3] *Department of Electrical Engineering, Indian Institute of Technology Kharagpur, Kharagpur, West Bengal, India*
[4] *Department of Geology and Geophysics, Indian Institute of Technology Kharagpur, Kharagpur, West Bengal, India*

Author for correspondence: soumilab24@gmail.com

Abstract

In this article, we propose a modified guided filter to denoise seismic data corrupted with random noise. The traditional guided filter uses the input image or another image as a "guidance" to filter the noise and transfer the structure of the guidance image to the filtered image. Seismic data are inherently three-dimensional (3D). The 3D seismic attributes are helpful to depict the structure of the subsurface accurately. The main aim of seismic denoising is to retain the structure of the data intact while removing the noise from it. These facts lead us to extend the traditional two-dimensional (2D) guided filter to 3D and apply it to the seismic dataset. The nature of the traces cannot change abruptly from one trace to its subsequent neighbor. This fact is also considered in the proposed filter while designing the guidance volume. The performance of the proposed filter is compared with several existing filters such as mean filter, median filter, bilateral filter, and traditional guided filter based on the four evaluators, such as root mean square error (RMSE), normalized RMSE (NRMSE), signal-to-noise ratio (SNR), and peak-signal-to-noise ratio (PSNR). The designed filter has yielded superior performance compared to its counterparts for various noise levels. Experimental results based on the qualitative assessment as well as values of the performance evaluators using a real hydrocarbon field dataset have established the modified guided filter as a suitable method for efficient seismic data denoising.

Keywords *Denoising; dynamic time warping (DTW); filters; guided filter; root mean square error (RMSE); seismic attributes; signal-to-noise-ratio (SNR); window size.*

Handbook of Petroleum Geoscience: Exploration, Characterization, and Exploitation of Hydrocarbon Reservoirs, First Edition. Edited by Soumyajit Mukherjee, Swagato Dasgupta, Chandan Majumdar, Subhandip Mandal and Troyee Dasgupta.
© 2023 John Wiley & Sons Ltd. Published 2023 by John Wiley & Sons Ltd.

3.1 Introduction

The seismic amplitude and other derived seismic attributes, such as seismic impedance, instantaneous amplitude, and instantaneous sweetness are 3D volumes characterized by their locations (inline and crossline) and time (or depth) information (Chaki et al. 2018). To understand the structure of the subsurface and delineate the thin layers, 3D seismic attributes play a critical role. The random noise corrupts the seismic data misleading the reservoir engineers with misplaced layer information. Therefore, the denoising of seismic attributes is of enormous importance. On the other hand, the use of filtering algorithms attenuates the signal strength, especially higher frequency components, along with the noise, thus affecting its quality. Therefore, it is essential to develop denoising algorithms to improve the signal-to-noise ratio (SNR) of seismic records while preserving their trends.

Literature study reveals that several 2D filtering algorithms, such as median filter (Liu et al. 2006), bandelet transform (Wang et al. 2015), bilinear filter (Ferahtia et al. 2013), and diffusion filter (Baddari et al. 2011; Chaki et al. 2017; Ferahtia et al. 2013). These have been applied to denoise corrupted seismic data. Different seismic denoising methods have been developed to improve the SNR of corrupted seismic signals by singular value decomposition (Gan et al. 2015), f-k filtering, stacking, principal component analysis, Karhunen–Loeve transform (Baddari et al. 2011), time-frequency peak filtering and its variants (Lin et al. 2014; Zhuang et al. 2015), low-rank approximation (Chen et al. 2017), wavelet sharpening (Chen and Jin 2016), radon transform (Latif and Mousa 2017), etc. These algorithms are implemented on different types of datasets. To the best of my knowledge, none of the publications has provided a comparative empirical review of all these algorithms on a single dataset so far. Therefore, it is impossible to identify a single algorithm which will work well for all types of datasets. The removal of spike noises (Ferahtia et al. 2013) along with difficulty in implementation and parameter selection poses a challenge for any filter to perform well. Though few nonlinear filters such as median filters (Gonzalez and Woods 2002), bilateral filters (Hale 2011), and varieties of ANNs- (Djarfour et al. 2008) based (Hashemi et al. 2008) algorithms surpassed these limitations, they sometimes modified the actual data (Ferahtia et al. 2013). These factors encourage researchers to apply advanced image-processing algorithms to denoise seismic attributes by exploiting the analogy between a seismic section and an image (Ferahtia et al. 2013). In a recent paper (Chaki et al. 2017), a set of diffusion and non-diffusion image-based algorithms have been compared and an improved complex adaptive diffusion filter (ICADF) has outperformed other filters in terms of several performance metrics as well as visual inspection. However, for 2D filtering, the 3D seismic cubes are first converted to the 2D matrix by "matricization"; then, 2D filters are applied to the 2D matrix, and finally, the filtered 2D data is converted back to 3D volume. The inherent relation between different dimensions of a multidimensional dataset is lost in the case of 2D filtering due to the matricization step. Intuitively, the 3D structure of the subsurface may be captured and retained in a better way if the algorithm is 3D itself, which leads us to extend the guided filter from 2D to 3D for better outcomes.

The guided filter is a recently published novel explicit filter (He et al. 2013), derived from a local linear model. The guided filter output is computed by taking the content of a guidance image into consideration. The input image itself or a different image can be used as a guidance image. The guided filter has shown the capacity of preserving the edges of an image much like the bilateral filter. It has been gaining its popularity in computer vision and graphics applications for image filtering, dehazing, smoothing, compression, etc., due to its effectiveness as well as computational efficiency for different kernel sizes and intensity ranges (He et al. 2013). Guided filter can transfer the structure of the guidance image to the filtered image along with smoothing. Therefore, we have envisaged that the guided filter can be effective for seismic data denoising as well as it can be extended to (applied) 3D seismic cubes directly for improved filtering performance.

The traditional guided filter, a 2D filtering algorithm, is used on images for noise reduction while retaining the textural structure of the image in spatial domain. In this study, we propose a modified guided filter that is extended to 3D to accommodate all the three dimensions of seismic cube, i.e. location (inline and crossline information) and depth (time information) wise. A regularization parameter was also introduced in the algorithm to accommodate the fact that the nature of seismic traces would not change abruptly in spatial domain. Rather, there exists a local continuity among traces. We evaluated the performance of the designed filter using a real dataset of an onshore hydrocarbon field from Western India and compared with the existing filtering algorithms (e.g. mean filter, median filter, bilateral filter, ICADF, and traditional guided filter) used in seismic signal processing. The seismic attributes are available from 0 to 3000 ms at and interval of 2 ms. The well logs are around 3000 m in depth and zone of interest of each well varies between 2700 and 2950 m. If we added noise of different levels (low – 5%, medium – 20%, high – 50% Gaussian noise) to the seismic amplitude volume, then the filter performances are evaluated in terms of four metrics such as root mean square error (RMSE), normalized RMSE (NRMSE), SNR, and peak signal-to-noise ratio (PSNR) along with visual inspection following the guidelines in (Chaki et al. 2017). Therefore, the contributions of the research can be summarized as:

- Propose a modified guidance filter to improve noise cancellation while retaining the local trend of seismic data along depth and location
- Design of the guidance image byexploiting the local continuity in the seismic data
- Comparison of the proposed filter with the existing algorithms by a systematic framework
- Appropriate parameter selection for the selected filters based on performance metrics as well as visual inspection

3.2 Theory

3.2.1 Guided Filter

Let us define the guided filter involving an input image p, a guidance image I, and an output filtered image q as per the equations provided in the main paper on 2D guided filter (He et al. 2013). The guided filter is a local linear model with the key assumption that q is a linear transform of guided image I in a local window ω_k centered at the pixel k:

$$q_i = a_k I_i + b_k, \quad \forall i \in \omega_k, \tag{3.1}$$

where linear coefficients (a_k, b_k) are considered to be constant in ω_k. Here, ω_k can be considered as a square window of size $r \times r$. This model confirms that the output image q has an edge only if the guidance image I has one since because $\nabla q = a \nabla I$.

The coefficients (a_k, b_k) are determined using constraints from the filtering input p. The output is modeled by subtracting some undesired components n like noise from the input p:

$$q_i = p_i - n_i \tag{3.2}$$

The difference between q and p is to be minimized while maintaining the model (3.1). Thus, the guided filter seeks the optimum value of coefficients (a_k, b_k) by minimizing the cost function in the window ω_k, given by:

$$\mathcal{L}(a_k, b_k) = \sum_{i \in \omega_k} \left((a_k I_i + b_k - p_i)^2 + \varepsilon a_k^2 \right) \tag{3.3}$$

where ε is a regularization parameter penalizing large a_k. The solution of the *linear ridge regression model* in (3.3) is given by:

$$a_k = \frac{(1/|\omega|)\sum_{i \in \omega_k} I_i p_i - \mu_k \overline{p_k}}{\sigma_k^2 + \varepsilon}, \tag{3.4}$$

$$b_k = \overline{p_k} - a_k \mu_k \tag{3.5}$$

where, μ_k and σ_k^2 are the mean and variance of I in ω_k, respectively. $|\omega| = r^2$ is the number of pixels in ω_k, r is the local window radius, and $\overline{p_k} = (1/|\omega|)\sum_{i \in \omega_k} p_i$ is the mean of p in ω_k.

The filter output q_i can be computed using equation (1) after obtaining coefficients (a_k, b_k). However, the computation of q_i is not unique, rather each pixel q_i is computed multiple times because of the overlapping sliding window ω_k. Therefore, the filtered output $q_i$is computed by taking into account all q_i values and averaging it

$$q_i = \frac{1}{|\omega|} \sum_{k | i \in \omega_k} (a_k I_i + b_k). \tag{3.6}$$

As $\sum_{k | i \in \omega_k} a_k = \sum_{k \in \omega_i} a_k$, Eq. (3.6) can be rewritten as

$$q_i = \overline{a_i} I_i + \overline{b_i}, \tag{3.7}$$

Here, the averaged coefficients of all the overlapping windows can be computed as $\overline{a_i} = \frac{1}{|\omega|} \sum_{k \in \omega_i} a_k$ and $\overline{b_i} = \frac{1}{|\omega|} \sum_{k \in \omega_i} b_k$. Equations (3.5)–(3.7) define the guided filter. Coefficients (a_i, b_i) vary spatially and ∇q is not a scaling of ∇I. However, their gradients are much smaller near strong edges compared to that of I. Therefore, we can still consider $\nabla q \approx \overline{a} \nabla I$ reflecting abrupt changes in I in q.

3.2.2 Modified Guided Filter and Guidance Image

We extend the traditional 2D guided filter to 3D to accommodate the 3D of the seismic data. The modified guided filter is designed based on the assumptions of the original guided filter for the volumetric data, i.e. the volumetric seismic data is a linear transformation of the guidance image in local neighborhood. We derived the 3D guided filter and observed that extending the window ω_k to 3D with size $r \times r \times r$ while operating on 3D seismic data does not significantly alter the derivation of the traditional guided filter.

Considering the continuity of the traces in the local neighborhood, we considered the following assumption to improve the seismic denoising:

$$q_i = a_k * \left(\frac{1}{|w|} \sum_{j \in w_k} I_j * \text{sim}(I_j, p_i) \right) + b_k \tag{3.8}$$

where, the $\text{sim}(I_j, p_i)$ is a function which computes the similarity between I_j and p_i. To avoid the nonlinearity introduced by the similarity computation, we assume the perfect linear correlation in local neighborhood, thereby resulting in linear combination of local information, given by $q_i = a_k * \left(\frac{1}{|w|} \sum_{j \in w_k} I_j \right) + b_k$. This is equivalent to using the 3D mean filtered output of the seismic data as the guidance image. Thus, the cost function to be minimized in this case is equivalent to the loss function defined in (3.3). However, the terms associated with the equation carry different meanings due to the 3D nature of the input and guidance image. Here, the minimization of $\mathcal{L}(a_k, b_k) = \sum_{i \in w_k} (a_k I_i + b_k - p_i)^2 + \epsilon a_k^2$ is performed with respect to the 3D window w_k, and Eqs. (3.4) and (3.5) hold if we consider μ_k and σ_k^2 to be the 3D mean and variance of I in the local window. And the solution to the coefficients would lead to:

$$a_k = \frac{\frac{1}{|w|} \sum_{i \in w_k} \overline{p_i} p_i - \mathbb{E}_{w_k}\left[\overline{p_k}\right] \overline{p_k}}{\mathbb{E}_{w_k}\left[(\overline{p_k})^2\right] - \left(\mathbb{E}_{w_k}\left[\overline{p_k}\right]\right)^2 + \epsilon} \tag{3.9}$$

$$b_k = \overline{p_k} - a_k \mathbb{E}_{w_k}\left[\overline{p_k}\right] \tag{3.10}$$

where, $\overline{p_k} = \frac{1}{|w|} \sum_{i \in w_k} p_i$ and $\mathbb{E}_{w_k}[.]$ represents the expected value in the local window w_k. Thus, the modified filter, described in Algorithm 3.1, is applicable on a volume instead of an image and the mean filters associated with algorithm are 3D.

Algorithm 3.1 Modified Guided Filter

```
Input: Input volume p, guidance volume I, radius r,
regularization parameter ε
Output: Filtered volume v_out
1. Calculate mean and correlation of the input and guidance
volumes
// Assume that f_mean is a 3D mean filter of radius r
      mean_I  = f_mean(I)
      mean_p  = f_mean(p)
      corr_I  = f_mean(I. * I)
      corr_Ip = f_mean(I. * p)
```

(Continued)

Algorithm 3.1 (Continued)

```
2. Calculate variance and covariance of the input and guidance
volumes
      var_I  = corr_I - mean_I. * mean_I
      cov_Ip = corr_Ip - mean_I. * mean_p
3. Calculate a and b
      a = cov_Ip. /(var_I + ε)
      b = mean_p - a. * mean_I
4. Calculate mean of a and b
      mean_a = f_mean(a)
      mean_b = f_mean(b)
5. Calculate output volume v_out
      v_out = mean_a * I + mean_b
```

3.2.3 Design of the Guidance Image

Design of the guidance volume considering local neighborhood of a voxel inside the seismic volume is a major contribution of this study. Lack of suitable guidance image is the primary reason for reduced denoising performance. In case of unavailability of proper guidance image, the input image can act as its own guidance. As we have discussed earlier, the mean filtered input can also act as a suitable guidance in our application. However, the similarity of the traces can be considered during design of guidance image to further improve the performance. Introducing nonlinearity in the modified guided filter makes its derivation difficult. However, nonlinearity can be introduced in the guidance image itself to enhance the denoising performance, which has been studied here. The guidance volume is created from a smoother version of the original noisy volume by modifying the trace of a particular location with respect to its immediate neighbors in the inline, crossline, and depth directions.

In case of seismic amplitude volume, the trace segments (i.e. the trace signal under certain depth interval) maintain continuity along nearby inline and crossline traces. With this assumption, we created the guidance image as a weighted linear transformation of neighborhood traces. At each location and depth, the guidance image intensity is computed by considering the similarity of the trace segment centered at that point with its immediate neighbor segments. Figure 3.1 demonstrates the computation of $I_{k,\ c}$ (the location and depth-wise central pixel of kth window). Here p_1, p_2, ..., $p_8 \in R^d$ represents the trace segments in a locality. The similarity between the central trace segment $p_k \in R^d$ and the surrounding traces are computed and used as the contribution of nearby pixel intensities for the computation of $I_{k,\ c}$. However, this local linearity assumption is void if the trace segment is too large because of the noise interference. Therefore, we used small depth interval, i.e. the trace segments of length 9 in our experiments.

We used dynamic time warping (DTW) as the similarity measure between two local trace segments. DTW is a nonlinear algorithm to measure similarity between two time series from different domains such as geophysical datasets (Herrera and van der Baan 2014;

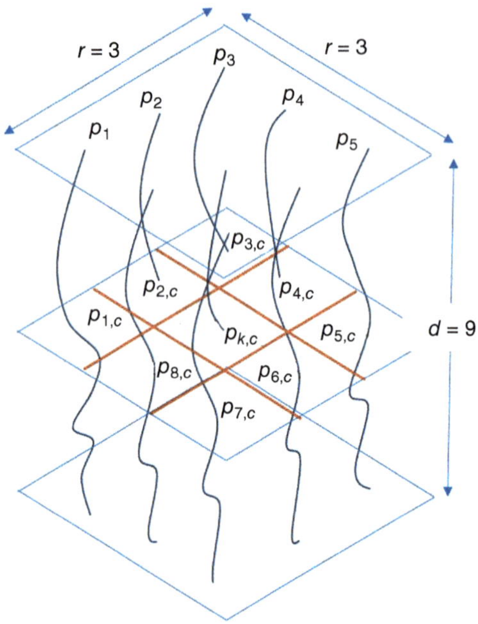

Figure 3.1 Computation of guidance image.

Muller 2007), voice biometrics, etc. The similarity between local segments of adjacent traces is quantified by DTW as described in Algorithm 3.2, and then used to generate the guidance volume.

Algorithm 3.3 presents the steps associated with the creation of the proposed guidance volume. As can be seen in step 2 of Algorithm 3.3, the trace segments of noisy input (I_N) are used to compute the similarity score with DTW followed by the weight assignment and computation of guidance image. In our implementation, we used a simple distance-based linear weight decay function for weight assignment. However, other variants of weight assignment are also possible in computation of guidance image.

Algorithm 3.2 Dynamic Time Warping

```
Input: Two time-series signals s and t with n and m time points
Output: Distance d
1. Initialize a data matrix S of size (n, m) storing similarity
measures.
S(0, 0) = 0
FOR i = 1 to m DO LOOP
      S(0, i) = ∞
END
FOR i = 1 to n DO LOOP
      S(i, 0) = ∞
END
```

(*Continued*)

Algorithm 3.2 (Continued)

2. Fill in S with the differences between s and t using pairwise method

```
FOR i = 1 to n DO LOOP
     FOR j = 1 to m DO LOOP

S(i, j) = distance(s(i), t(j)) + MIN(S(i-1, j),      //increment
                                 S(i, j-1),          //decrement
                                 S(i-1, j-1))        //match
     END
END
d = S(n, m)
```

Algorithm 3.3 Creation of the Guidance Image

Input: Noisy image I_N, window size = (r, r, d)

Output: Guidance image I_G, same size as I_N

1. Create a smoother version of I_N

// Use 3D median or mean filter

$I_N \leftarrow f_{mean}(I_N)$

2. Compute the similarity of central trace segment (p_k) with neighboring segments (p_i, for i = 1, 2, …, 8)

FOR $i \in w_k$ DO LOOP

$dist_i \leftarrow DTW(p_i, p_k)$

END

3. Normalize the distance measures and compute the weights. Here G can take multiple forms, such as radial distance weights, exponential distance weights, power distance weights, etc.

$$\rho_i \leftarrow G\left(\frac{dist_i}{\sum_i dist_i}\right)$$

4. Compute the guidance image

$$I_{k,c} \leftarrow \frac{1}{r^2}\left(p_{k,c} + \sum_{i \in w_k} \rho_i{}^* p_{i,c}\right)$$

3.3 Experimental Results

The seismic dataset used in this study has been acquired over a hydrocarbon field located at western India. The ".sgy" files are converted to MATLAB platform compatible ".mat" format to easily access the data.

In the case of the real dataset, we assume that the seismic amplitude used in this study does not carry any inherent noise. Three different sets of noisy datasets have been created by adding 5, 20, and 50% Gaussian noise to test the proposed as well as existing filters.

3.3.1 Parameter Settings

The parameters of the filters are adjusted to obtain the best possible results. For example, the window size defines the neighborhood around a particular pixel and controls the filtered output for the window-based algorithms such as smoothing, 2D median filtering, and 3D median filtering. Let us assume $N_{i,j}^{w}$ is the neighborhood of size $w \times w$ centered at pixel $x_{i,\,j}$ for an image I (here, seismic amplitude). The filtered value of each pixel can be represented as

$$x_{i,j}^{filt,w} = \text{median}\left(\text{pixel values in } N_{i,j}^{w}\right) \tag{3.11}$$

or

$$x_{i,j}^{filt,w} = \text{mean}\left(\text{pixel values in } N_{i,j}^{w}\right) \tag{3.12}$$

for the 2D median filter and smoothing using a mean filter, respectively. Similarly, $w \times w \times w$ represents the neighborhood $N_{i,j,k}^{w}$ centered at $x_{i,\,j,\,k}$ location for seismic amplitude volume in the case of the 3D median filter. The filtered seismic amplitude can be obtained by carrying out 3D median filtering for individual pixels throughout the volume using

$$x_{i,j,k}^{filt,w} = \text{median}\left(\text{pixel values in } N_{i,j,k}^{w}\right) \tag{3.13}$$

Here, the window sizes for the three algorithms – smoothing, 2D, and 3D median filtering – are calibrated to yield the best results in terms of performance evaluators and distribution of denoised seismic amplitude at a particular inline/crossline.

A bilateral filter can be defined as

$$\begin{aligned} (I_p)_{\text{BF}} &= \frac{1}{W_p}\left(\sum_{q\in S} G_{\sigma_s}(\|p-q\|)G_{\sigma_r}(|I_p - I_q|)I_q\right) \\ \text{where, } W_p &= \sum_{q\in S} G_{\sigma_s}(\|p-q\|)G_{\sigma_r}(|I_p - I_q|) \end{aligned} \tag{3.14}$$

where, $\|\cdot\|$, $|\cdot|$, I_p, and $(I_p)_{\text{BF}}$ represent L_2 norm, absolute value, noisy, and filtered pixels at location p, respectively. The set S and set R represent space and range domain, respectively, whereas, G_{σ_r} and G_{σ_s} are Gaussian kernels associated in those domains. The window size w and standard deviations of the Gaussian kernels are tuned for the best result.

In the case of a multidimensional function I (here, 2D seismic amplitude), the general nonlinear diffusion process solves for

$$\frac{\partial I}{\partial t} = \nabla\cdot(d(s)\nabla I) \tag{3.15}$$

where t denotes time, ∇, $\nabla\cdot$, $d(s)$ are the gradient, divergence operators, and a general diffusivity function, respectively. In (Gilboa et al. 2004), the diffusivity is presented as

$$d = \frac{\exp(i\theta)}{1 + ((\mathrm{Im}(I))/\kappa\theta)^2} \tag{3.16}$$

where, $\theta \in (-\pi/2; \pi/2)$ is a phase angle, κ regulates the spread of the diffusion coefficient, and Im(I) represents the imaginary part of a complex function. Equation (3.12) can be discretized by a *forward in time and centered in space* (FTCS) method as

$$I_{n+1} = I_n + \Delta t_n\left(d_n\nabla^2 I_n + \nabla d_n\cdot\nabla I_n\right) \tag{3.17}$$

where, ∇^2 and Δt are the Laplacian operator and a discrete time step, respectively (Bernardes et al. 2010; Rodrigues and Bernardes 2012). In the case of the ICADF filter, the time step is adapted and can be expressed as

$$\Delta t_n = \frac{1}{\alpha}\left[a + b\exp\{-\max(|\mathrm{Re}(\partial I_n/\partial t)|/|\mathrm{Re}(I_n)|)\}\right] \tag{3.18}$$

where, α is 4 for 2D images, a and b govern the time step and are generally set as $a + b \leq 1$. The other adaptive parameter κ can be updated as

$$\kappa = \kappa_{\max} + (\kappa_{\min} - \kappa_{\max})\frac{g - \min(g)}{\max(g) - \min(g)} \tag{3.19}$$

where min(g) and max(g) refer to the minimum and maximum values of g, (Bernardes et al. 2010) with g defined as

$$g = G_{N,\sigma} * \mathrm{Re}(I) \tag{3.20}$$

Here, $G_{N,\sigma}$ is an $N \times N$ local Gaussian kernel with variance σ and $*$ is the convolution operator. Literature study reveals that ICADF performs better than other diffusion filters for denoising optical coherence tomography as well as filtering corrupted seismic data and predicted lithological properties (Bernardes et al. 2010; Chaki et al. 2017; Gilboa et al. 2004).

Several algorithms such as median filter, bilateral filter as well as diffusion algorithms, e.g. Perona Malik diffusion filter, complex diffusion filter, ICADF, etc., have been used to denoise seismic signals (Baddari et al. 2011; Bednar 1983; Chaki et al. 2017; Hale 2011; Liu et al. 2006, 2008). In Chaki et al. (2017), ICADF has outperformed different diffusion as well as nondiffusion filters in terms of metric values and visual inspection of the distribution of seismic attributes. For the present work, the performances of the selected filters have been evaluated using a real hydrocarbon field dataset acquired from a field in western India.

The performance metrics – RMSE, NRMSE, SNR, and PSNR – are calculated along with a visual comparison of the noisy and denoised seismic amplitude at a particular

Table 3.1 The values of the tuned parameters associated with the selected algorithms for seismic amplitude filtering.

Filters	Parameters
2D median filtering	Window size = 5×5
3D median filtering	Window size = $5 \times 5 \times 5$
Smoothing	Window size = 5×5
Bilateral filter	Window size = 5×5, spatial standard deviation = 3, range standard deviation = 0.1
ICADF	The constants in Eq. (3.18), $a = 0.25$, $b = 0.75$ In Eq. (3.19), $\kappa_{\min} = 2$ and $\kappa_{\max} =$ mean(attribute) In Eq. (3.20), local Gaussian kernel, window size = 3×3, variance = 10, maximum diffusion time = 0.7 seconds, number of iterations = 6
2D guided filter	Local window radius: $r = 1$, Regularization parameter: $\varepsilon = 0.01$
Proposed filter	Local window radius: $r = 1$, Regularization parameter: $\varepsilon = 0.0001$
Proposed filter + guidance image	Local window radius: $r = 1$, Regularization parameter: $\varepsilon = 0.0001$, Window size = $3 \times 3 \times 9$(9: depth size)

inline/crossline to evaluate the performance of individual algorithms. Table 3.1 represents the values of the tuned filter parameters at their best configurations. These parameters are optimally set in terms of performance metrics and visual inspection of the filtered results as well. Table 3.2 reports the comparative results for seismic data denoising using the selected filters at their optimal parameter configurations.

Figure 3.2 shows the variation of performance metrics with varying regularization parameters associated with the proposed guided filter with the creation of the guided image. It can be seen from Figure 3.2 that the proposed filter obtains its best performance at $\varepsilon = 0.0001$. The superior performance of the proposed guided filter compared to the other filters is evident in Table 3.2. The seismic signal denoised by the proposed filter shows a higher correlation coefficient (0.99) with the clean (actual) seismic signal. Figure 3.3 shows the outputs of the top four filters – 3D median filter, ICADF, 2D guided filter, and the proposed filter, along with actual and noisy seismic attributes at a single trace. We have marked the peaks and troughs of the signals with small rectangles to highlight the similarity between the output of the proposed filter and the original amplitude. Though the difference between the filter outputs is insignificant at a particular location, it contributes to a large performance variation when computed over the study area in terms of the metrics – RMSE, NRMSE, SNR, and PSNR (Table 3.2). In Figure 3.3, the performance of the proposed filter with the proposed guided image overlapped very closely with the performance of the proposed filter alone. Therefore, it has been excluded from the figure.

As per Table 3.2, the performance of the proposed filter without the proposed guidance image is suitable when the data is less noisy. However, the use of the guidance image improves the performance in the presence of heavy noise as it improves the RMSE value

Table 3.2 The performance of different filtering algorithms used on noisy seismic amplitude.

Method\noise level	5%				20%				50%			
	RMSE ($e^{\wedge}10^5$)	NRMSE	SNR (dB)	PSNR (dB)	RMSE (e^5)	NRMSE	SNR (dB)	PSNR (dB)	RMSE (e^5)	NRMSE	SNR (dB)	PSNR (dB)
Noisy signal	1.011	7.348	13.01	26.53	4.0460	29.391	6.99	20.51	10.113	73.461	3.01	16.53
2D median filtering	0.926	6.725	13.34	26.91	1.245	9.045	12.11	25.63	2.144	15.574	9.75	23.27
3D median filtering	0.591	4.290	15.35	28.86	0.8604	6.25	13.72	27.23	1.619	11.760	10.97	24.49
Smoothing	1.002	7.275	13.06	26.57	1.122	8.148	12.56	26.08	4.352	31.615	6.68	20.19
Bilateral filter	0.553	4.023	15.63	29.14	0.921	6.694	13.42	26.93	3.059	22.227	8.21	21.72
ICADF	0.57	4.15	15.5	29.01	0.74	5.4	14.35	27.87	1.095	7.953	12.67	26.19
2D guided filter	0.334	2.424	17.83	31.35	0.856	6.219	13.37	27.25	1.064	7.731	12.79	26.31
Proposed filter	**0.243**	**1.766**	**19.20**	**32.72**	**0.424**	**3.081**	**16.79**	**30.3**	**0.776**	**5.64**	**14.16**	**27.68**
Proposed filter + guidance image	**0.263**	**1.914**	**18.85**	**32.37**	**0.42**	**3.08**	**16.79**	**30.3**	**0.734**	**5.34**	**14.4**	**27.92**

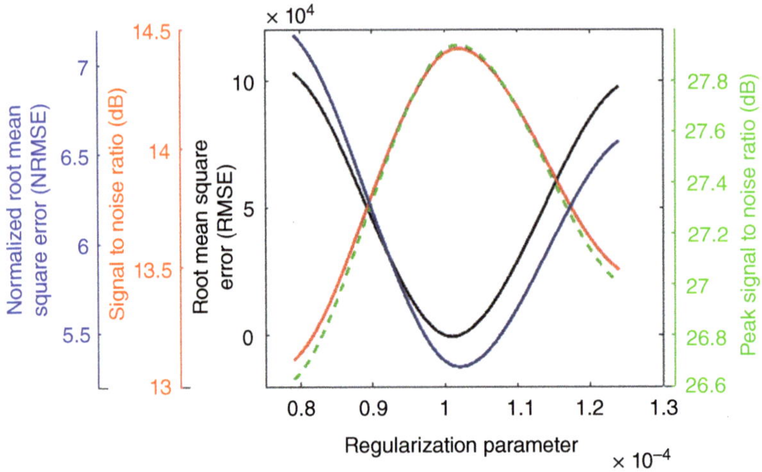

Figure 3.2 The variation of the performance evaluators with varying parameter values for noisy seismic amplitude (corrupted with 50% Gaussian noise) filtering using the proposed filter with guidance image.

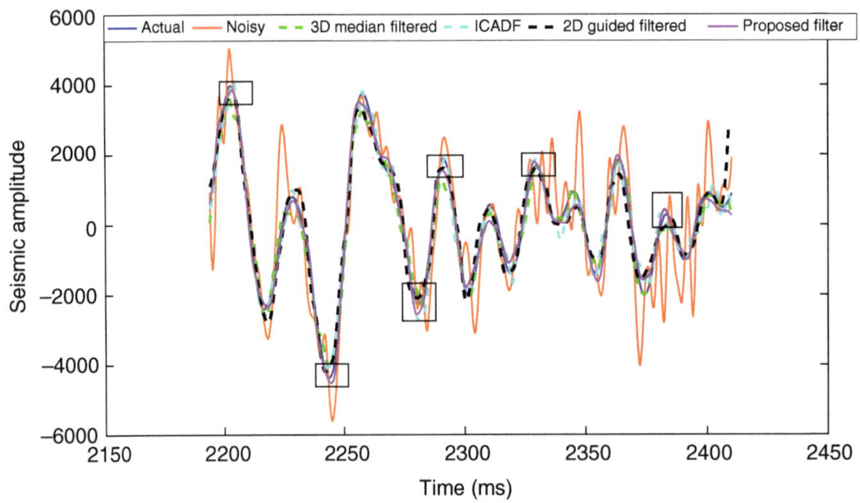

Figure 3.3 The superimposed plot of the outputs of different filtering algorithms along with actual and noisy seismic amplitude at a particular location.

from 0.776 to 0.734 at 50% noise level. Similarly, the performance boost in SNR (14.14–14.4) and PSNR (27.68–27.92) can also be observed. Moreover, the performance with and without the proposed guided image is very close in less noisy data and almost similar when 20% noise is added to the data. This implicates the robustness of the guidance image to noise. The performances of the proposed filter at a particular inline are presented in Figure 3.4. It can be seen that the distribution of the filtered seismic amplitude closely resembles

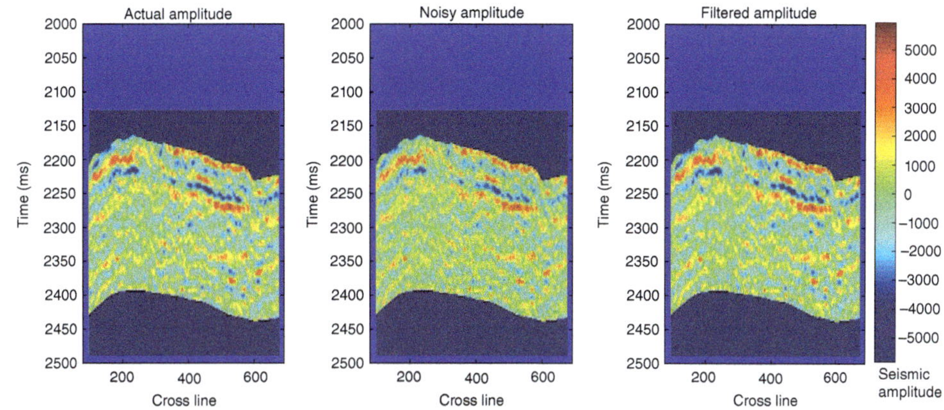

Figure 3.4 The performance of the proposed filter with a guidance image on noisy seismic amplitude corrupted with 50% Gaussian noise at a particular inline.

the actual amplitude distribution. Similar results have been obtained for other seismic attributes like seismic impedance, instantaneous frequency, etc.

3.4 Conclusions

This study proposes a novel filtering algorithm to denoise 3D seismic attributes. The local continuity among seismic traces has been taken into account by designing the guided image in the process. The comparison of the proposed filter with the existing algorithms in terms of performance metrics and visual inspection establishes it as a suitable candidate for preprocessing noisy seismic datasets. For each case, the parameters related to the particular technique have been tuned to obtain the best possible outcomes. The results achieved using a real dataset encourage us to apply this algorithm to filter the predicted lithological property volumes. The predicted lithological volume or the predictor seismic attributes can be used for guiding the filtering process. The guidance filter can transfer the structural information from the guidance image to the input image and, hence, can produce the output image by integrating this information along with removing artefacts and nonlinearity generated due to nonlinear mapping between seismic data and lithological properties. Thus, the proposed filter would be effective for the postprocessing of the predicted lithological property volumes. Another important research direction can be to extend this filter for probable application to denoising of other multidimensional data such as video and hyperspectral images in future.

Acknowledgments

Soumyajit Mukherjee (IIT Bombay) invited and reviewed this article. Merryl Le Roux, Frank Otmar Weinreich, Joss Everett, Andrew Harrison, and the proofreading team (Wiley Blackwell) are thanked for their assistance. The article has been summarized in Mukherjee et al. (2023).

References

Baddari, K., Ferahtia, J., Aïfa, T., and Djarfour, N. (2011). Seismic noise attenuation by means of an anisotropic non-linear diffusion filter. *Comput. Geosci.* 37: 456–463.

Bednar, J.B. (1983). Applications of median filtering to deconvolution, pulse estimation, and statistical editing of seismic data. *Geophysics* 48: 1598–1610.

Bernardes, R., Maduro, C., Serranho, P. et al. (2010). Improved adaptive complex diffusion despeckling filter. *Opt. Express* 18: 24048–24059.

Chaki, S., Routray, A., and Mohanty, W.K. (2017). A diffusion filter based scheme to denoise seismic attributes and improve predicted porosity volume. *IEEE J. Sel. Top. Appl. Earth Obs. Remote Sens.* 10: 5265–5274.

Chaki, S., Routray, A., and Mohanty, W.K. (2018). Well-log and seismic data integration for reservoir characterization from a signal processing and machine learning perspective: an overview. *IEEE Signal Process. Mag.* 35: 72–81.

Chen, Y. and Jin, Z. (2016). Simultaneously removing noise and increasing resolution of seismic data using waveform shaping. *IEEE Geosci. Remote Sens. Lett.* 13: 102–104.

Chen, Y., Zhou, Y., Chen, W. et al. (2017). Empirical low-rank approximation for seismic noise attenuation. *IEEE Trans. Geosci. Remote Sens.* 55: 4696–4711.

Djarfour, N., Aïfa, T., Baddari, K. et al. (2008). Application of feedback connection artificial neural network to seismic data filtering. *Comptes Rendus Geosci.* 340: 335–344.

Ferahtia, J., Aïfa, T., Baddari, K. et al. (2013). Image-based processing techniques applied to seismic data filtering. *J. Pet. Sci. Eng.* 104: 17–26.

Gan, S., Chen, Y., Zu, S. et al. (2015). Structure-oriented singular value decomposition for random noise attenuation of seismic data. *J. Geophys. Eng.* 12: 262–272.

Gilboa, G., Sochen, N., and Zeevi, Y.Y. (2004). Image enhancement and denoising by complex diffusion processes. *IEEE Trans. Pattern Anal. Mach. Intell.* 26: 1020–1036.

Gonzalez, R. and Woods, R. (2002). *Digital Image Processing*, 2e. Upper Saddle River, NJ: Prentice Hall.

Hale, D. (2011). Structure-oriented bilateral filtering of seismic images. SEG Annual Meeting, San Antonio, TX (18–23 September 2011).

Hashemi, H., Javaherian, A., and Babuska, R. (2008). A semi-supervised method to detect seismic random noise with fuzzy GK clustering. *J. Geophys. Eng.* 5: 457–468.

He, K., Sun, J., and Tang, X. (2013). Guided image filtering. *IEEE Trans. Pattern Anal. Mach. Intell.* 35: 1–14.

Herrera, R.H. and van der Baan, M. (2014). A semiautomatic method to tie well logs to seismic data. *Geophysics* 79: V47–V54.

Latif, A. and Mousa, W.A. (2017). An efficient undersampled high-resolution radon transform for exploration seismic data processing. *IEEE Trans. Geosci. Remote Sens.* 55: 1010–1024.

Lin, H., Li, Y., Yang, B. et al. (2014). Seismic random noise elimination by adaptive time-frequency peak filtering. *IEEE Geosci. Remote Sens. Lett.* 11: 337–341.

Liu, C., Liu, Y., Yang, B. et al. (2006). A 2D multistage median filter to reduce random seismic noise. *Geophysics* 71: V105–V110.

Liu, Y., Liu, C., and Wang, D. (2008). A 1D time-varying median filter for seismic random, spike-like noise elimination. *Geophysics* 74: V17–V24.

Mukherjee, S., Dasgupta, S., Majumdar, C. et al. (2023). Introduction to: Handbook of Petroleum Geoscience: Exploration, Characterization, and Exploitation of Hydrocarbon Reservoirs. In: *Handbook of Petroleum Geoscience: Exploration, Characterization, and Exploitation of Hydrocarbon Reservoirs*, Petroleum Industry Updates: Geoscientific and Economic Issues (ed. S. Mukherjee, S. Dasgupta, C. Majumdar, et al.). John Wiley & Sons, Hoboken, ISBN 978-1-119-68003-1.

Muller, M. (2007). Dynamic time warping. In: *Information Retrieval for Music and Motion*, 69–84. Berlin, Heidelberg: Springer Berlin Heidelberg.

Rodrigues, P. and Bernardes, R. (2012). 3-D adaptive nonlinear complex-diffusion despeckling filter. *IEEE Trans. Med. Imaging* 31: 2205–2212.

Wang, X., Gao, J., Chen, W., and Yang, C. (2015). The seismic random noise attenuation method based on enhanced bandelet transform. *J. Appl. Geophys.* 116: 146–155.

Zhuang, G., Li, Y., Wu, N., and Tian, Y. (2015). Curvature-varying hyperbolic trace TFPF for seismic random noise attenuation. *IEEE Geosci. Remote Sens. Lett.* 12: 2252–2256.

4

Geomechanics: A Basic Requirement for Wells at Every Operational Stage

Chandreyi Chatterjee[1], *Chandan Majumdar*[1,*], *Lenin Mora Guerrero*[2], *and Juan Carlos Rabanal Chavez*[3]

[1] *Schlumberger, Navi Mumbai, Maharashtra, India*
[2] *Schlumberger, Bogota, Colombia*
[3] *Schlumberger, Quito, Pichincha, Ecuador*

*Author for correspondence: cmajumdar@slb.com

Abstract

As the oil and gas industry progresses, exploration and development are becoming more challenging every day as the easily recoverable reservoirs are already exhausted. Presently the industry is focusing on tectonically active terrains, unconventionals, deep-water and extended reach drilling is becoming a common practice. Operational activities in these areas are associated with high risks, thus incur high costs for mitigation measures. To reduce the drilling and completion risks, industries worldwide are relying on geomechanical solutions specific to the field. This paper will provide an insight into the geomechanical applications specific to a few challenges through four case studies.

Keywords *Exploration challenges; Risk reduction; geomechanical solutions*

4.1 Introduction

Earth in the subsurface undergoes stress variations due to a variety of factors, viz. gravitational loading, tectonics uplift, pressure inflation or deflation, thermal effects etc. Thus, stresses in the Earth are not static and change gradually over time. Similarly, the strength of formations alters throughout the geological time, from their time of deposition up to the present day.

Rock mechanical properties and strength control the behavior of rock under stress. A geological formation or rock fails when these stresses exceed the strength of the rock. Geomechanical studies predict when and where the rock will fail, and the risks associated with it. Geomechanical analysis determines the stresses, strengths, and pressure profiles of formations or reservoirs. Thus, understanding the geological history of the formation is crucial for its reliable geomechanical characterization.

Handbook of Petroleum Geoscience: Exploration, Characterization, and Exploitation of Hydrocarbon Reservoirs, First Edition. Edited by Soumyajit Mukherjee, Swagato Dasgupta, Chandan Majumdar, Subhandip Mandal and Troyee Dasgupta.
© 2023 John Wiley & Sons Ltd. Published 2023 by John Wiley & Sons Ltd.

As the oil industry is moving toward more complex environments e.g. tectonically active region, deep water (e.g. Misra and Mukherjee 2018), overpressure regime (e.g. Dasgupta and Mukherjee 2020), unconventional resources, etc., the need to optimize exploration and production operations becomes critical (Dasgupta et al. 2020). Geomechanics applies right across the E&P industry, from exploration to abandonment, and from the single wellbore to the modeling of reservoirs and field-level study.

Geomechanics to oil and gas exploration and development leads to significant improvements in the economic performance of the asset. It is used to optimize the mud weight, casing, and trajectory planning in the predrill exploration stage. This further identifies the risks associated with drilling and ensures proper mitigation plans. In the post-drilling development stage, geomechanics helps in completion optimization by defining the hydraulic fracturing model – height, length, operational planning, and execution. Predicting the critical drawdown pressure in a reservoir for preventing sand production stands high among other applications. With every passing day, new applications of geomechanics are surfacing including novel techniques of identifying producible zones in the fractured reservoir, caprock integrity for enhanced oil recovery (EOR) techniques, etc.

4.2 Procedures/Workflow

To conduct any aforementioned geomechanics studies, the first step is to construct a one-dimensional (1D) mechanical earth model (MEM). The fundamental approach for this is to integrate all available data or indicators of rock strength, deformability, in situ stress, pore pressure, overburden etc. It is a 10-step workflow (Figure 4.1) which uses drilling history, basic open hole logs like gamma ray, density, neutron, resistivity, sonic, images, caliper, reservoir pressure and production data. The MEM is calibrated using available measurements and observations (well events) while drilling the well, in order to create a representative subsurface model. These key geomechanical events and parameters are analyzed for engineering designs (Plumb et al. 2000). With more data being available, the MEM is revised and updated (i.e. during drilling in real-time, or after new logs or field measurements are acquired). It may include some time-dependent components such as pore pressure changes with production and injection operations (Dasgupta and Mukherjee 2020), which might be the reason for geomechanical changes occurring over the life of a field. A series of steps must be followed in order to extract the value out of the dataset and to draw conclusions with a reasonable estimation of uncertainty associated with it. Skipping or ignoring any one of these steps can lead to a nonrepresentative model of the subsurface and inconsistent results.

4.2.1 Data Audit and Drilling Event Review

The process starts with a data audit. It involves steps 1 and 2 of Figure 4.1. In this phase, all available input data are collected, reviewed and conditioned to be used as input data in the geomechanics model. The main part of this step is the review of drilling records to capture all geomechanically relevant drilling events, viz. stuck pipe, losses, kicks, and tight hole. A time vs. depth plot is prepared capturing all the relevant information like drilling events, mud weight, casing shoe depths, etc., to analyze the hazards encountered during drilling

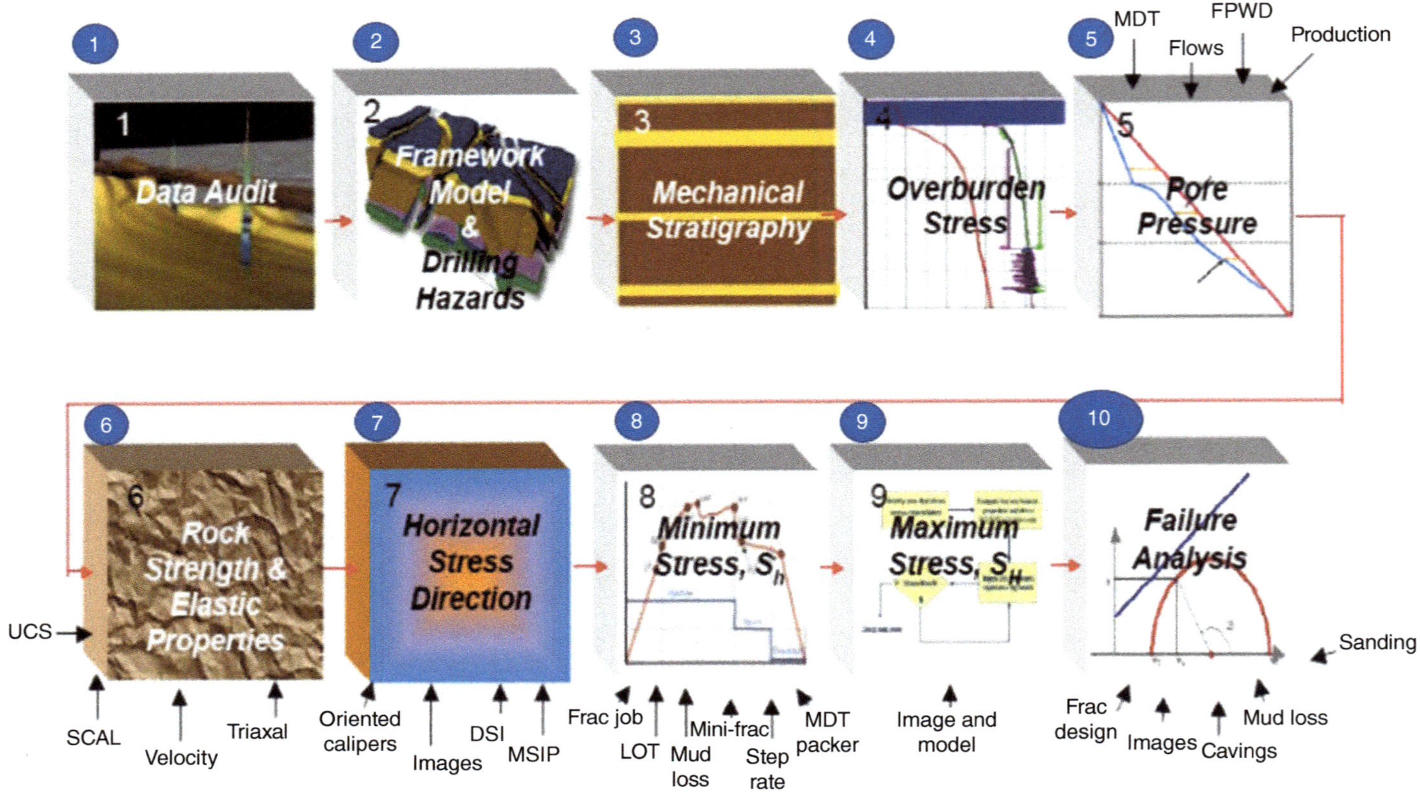

Figure 4.1 MEM building workflow: This is a 10-step workflow used to construct a geomechanical model. It starts with a data audit and drilling hazard review of a well (1 and 2). The next step is the construction of mechanical stratigraphy using open hole logs to differentiate the different rock types (3). Then comes overburden and pore pressure calculation (4 and 5). After that rock strength and elastic properties are computed using density, compressional and shear sonic logs (6). Several correlations are in place, which suit each rock type in a field and core data helps calibrate the computations and validate the results. After that horizontal stress magnitudes and direction are derived (7, 8 and 9). All the outputs from each workflow are taken as input to failure analysis or wellbore stability (WBS) model which is the last step (step 10). This will generate a stable mud weight window with breakout and loss limit which is referred for mud weight and casing optimization.

with time loss information. Good-quality data, when used with the right insight, replicates the drilling events in the same interval and, in turn, ensures a representative subsurface model.

4.2.2 Mechanical Stratigraphy

The actual MEM reconstruction process starts with mechanical stratigraphy construction marked as the third step in Figure 4.1. The concept of mechanical stratigraphy is based on the observation that elastic rock, when under stress, follows different trends depending on the type of load-bearing material present in the formation. A material where clay minerals are the predominant phase will show considerably different rock mechanical behavior than clean sandstone. Thus, it is often necessary to use different empirical correlations depending on the load-bearing mineral phase. As a result, it is crucial to define the mechanical stratigraphy by delineating and differentiating clay-supported facies from grain-supported ones using open hole log responses, e.g. gamma ray, shale volume, neutron, or density, etc.

4.2.3 Overburden Stress

After mechanical stratigraphy, the next step is to compute overburden stress marked as step 4 in Figure 4.1. It is a generic industry practice to compute overburden stress before rock mechanical properties as, in pore pressure and stress calculation, overburden is taken as an input. So, it should be calculated before that. The overburden stress at a given depth below the surface is the weight of overlying Earth material and can be calculated by integrating the density log for the above formations given by the following equation (e.g. Elebute et al. 2018):

$$\sigma_z = \int_0^z \rho(z) g dz \tag{4.1}$$

Where σ_z is overburden stress, ρ is formation density, and g is gravitational constant. Mukherjee (2017a, 2018a, 2018b, 2018c) presented how to tackle the situation where density varies in three directions following some equation.

4.2.4 Pore Pressure

Pore pressure is the fifth step in Figure 4.1, which comes after the overburden stress calculation. Pore pressure is defined as the pressure exerted by fluids present in the pore spaces of a rock. It increases with an increase in pore spaces which is directly proportional to an increase in porosity and a decrease in compressional wave velocity. Pore pressure is generally hydrostatic if the fluids escape from the sediment while it is getting compacted. But rapid burial (undercompaction), uplift due to tectonic event, diagenesis of sediments or rock, fluid migration, and other factors can lead to a situation where pore pressure can be abnormally high above the hydrostatic gradient. Velocity versus effective stress plot helps determine the type of increase in pore pressure, which is caused by undercompaction or other reasons. See Mukherjee and Kumar (2018) for compaction-related modeling.

Several methods are there to determine pore pressure, e.g. those by Eaton, Bower, and Traugott using compressional wave velocity and resistivity. Pore pressure is calibrated with formation pressure measurements such as modular dynamic testing (MDT), repeated formation test (RFT), and drill stem test (DST).

4.2.5 Elastic Properties

Computation of elastic properties is a vital step in the 1D-MEM workflow. It generally comes after pore pressure analysis in step 6 of Figure 4.1. In this step, dynamic elastic moduli are calculated from compressional, shear, and density logs (e.g. Elebute et al. 2018) with the following equations:

$$G_{\text{dyn}} = (13\ 474.45)\frac{\rho_{\text{b}}}{(\Delta t_{\text{shear}})^2} \tag{4.2}$$

$$K_{\text{dyn}} = (13\ 474.45)\rho_{\text{b}}\left[\frac{1}{(\Delta t_{\text{comp}})^2}\right] - \frac{4}{3}G_{\text{dyn}} \tag{4.3}$$

$$E_{\text{dyn}} = \frac{9G_{\text{dyn}} \times K_{\text{dyn}}}{G_{\text{dyn}} + 3K_{\text{dyn}}} \tag{4.4}$$

$$\nu_{\text{dyn}} = \frac{3K_{\text{dyn}} - 2G_{\text{dyn}}}{6K_{\text{dyn}} + 2G_{\text{dyn}}} \tag{4.5}$$

Here G_{dyn} is dynamic shear modulus computed from ρ_{b}, which is bulk density and Δt_{shear}, which is shear slowness (inverse of velocity) as shown in Eq. 4.2. K_{dyn} is the dynamic bulk modulus computed from the shear modulus, density, and compressional slowness (Δt_{comp}) (Eq. 4.3). E_{dyn} is the dynamic Young's modulus, which is computed from shear and bulk modulus (Eq. 4.4). V_{dyn} is the dynamic Poisson's ratio computed from the shear and the bulk moduli (Eq. 4.5).

The elastic moduli determined from Eqs. (4.2–4.5) are termed dynamic since sonic measurements are acquired with very high frequencies (~10 kHz), involving very small strains and the formations exhibit what is essentially an unstrained response. Under these dynamic conditions, the rock exhibits a stiffer response than it would under static loading such as in a rock mechanical laboratory test, or under static loading in the ground, or near a wellbore or completion. In fact, the weaker and more plastic the rock, the larger is the difference between the dynamic and the static properties.

Wellbore deformation and failure are a relatively slower process in comparison to high-frequency wave propagation. Thus, static data is needed for any geomechanical analysis. The best correlation to be used for transform will be those established from the laboratory test data on the study of well cores, if available. When no core test data is available, a large number of published and proprietary correlations are available to estimate the static behavior of rock from the dynamic properties. However, it is generally assumed that the static Poisson ratio equals the dynamic Poisson ratio.

The general form of the static Young's Modulus correlation is shown in the following equation (Bradford et al. 1998):

$$E_{\text{sta}} = aE_{\text{dyn}}^{b} \tag{4.6}$$

Here E_{sta} is the static Young's modulus, E_{dyn} is the dynamic Young's modulus, $a = 0.032$ and $b = 1.632$.

4.2.6 Rock Strength

It is a part of step 6 in Figure 4.1. The standard measure of rock strength is unconfined compressive strength (UCS), which can be routinely measured in a lab by loading cylindrical rock samples axially until failure. A large number of empirical correlations to determine UCS from log measurements are available in the literature, as well as proprietary ones used by service and oil companies. Friction angle is the ability of a rock to withstand shear stress (Mukherjee 2017b; Mukherjee and Khonsari 2017, 2018). It is an important parameter in defining the stability of the formation. The angle is computed using correlations based on gamma ray, shale volume, or porosity. Tensile strength is the maximum stress that a material can withstand before breaking. It is usually considered 10% of UCS. Ideally, laboratory measurements should be available to calibrate log-derived rock strength properties.

4.2.7 Horizontal Stress (σ_h and σ_H)

The parameters discussed in the previous sections are used as input in calculating horizontal stresses. The poro-elastic horizontal strain model (Fjaer et al. 1992) is a popular one (specific to Schlumberger) to model the magnitudes of the minimum (σ_h) and the maximum horizontal stresses (σ_H; steps 8, 9 in Figure 4.1).

$$\sigma_h = \frac{\nu}{1-\nu}\sigma_V + \frac{1-2\nu}{1-\nu}\alpha P_{\text{p}} + \frac{E}{1-\nu^2}\varepsilon_x + \frac{\nu E}{1-\nu^2}\varepsilon_y \tag{4.7}$$

$$\sigma_H = \frac{\nu}{1-\nu}\sigma_V + \frac{1-2\nu}{1-\nu}\alpha P_{\text{p}} + \frac{E}{1-\nu^2}\varepsilon_y + \frac{\nu E}{1-\nu^2}\varepsilon_x \tag{4.8}$$

Here ν is the Poisson's Ratio, E is the static Young's Modulus, σ_V is the overburden stress, P_{p} is the pore pressure, and α is the Biot's coefficient. Horizontal strains 'ε_x' and 'ε_y' (orthogonal to each other) may be compressional (i.e. positive, e.g. Mukherjee 2019a) or extensional (i.e. negative, to represent lateral spreading). The strains can be treated simply as calibration factors that can be adjusted to best match the resulting stress estimates to any leak-off test data or specific modes of rock failure seen in image logs (e.g. Dasgupta et al. 2020), etc.

4.2.8 Horizontal Stress Direction

Stress direction estimation is an independent step and can be calculated before stress magnitude estimation (step 7, Figure 4.1). Where the stress regime has remained the same over protracted time till the present, paleostress studies (e.g. Vanik et al. 2018; Dutta et al. 2019; Shaikh et al. 2020) can also give information about the present-day stress directions. Horizontal stress direction can be determined from either the drilling-induced fractures or breakouts identified on image log data. Shear wave anisotropy processing of dipole sonic measurements can provide similar information. Breakout direction in a vertical well indicates the minimum horizontal stress direction, whereas drilling-induced fractures and fast shear azimuth indicate direction of maximum horizontal stress. Breakouts from oriented calipers can identify stress direction as well.

Many field examples have proved that geomechanical analysis can open opportunities for drilling and completion in challenging environments, which looked intangible otherwise. In this context, we present four case studies.

Case 1 Drilling Optimization by Addressing Wellbore Stability Issues: Assam-Arakan Basin (N–E India)

Exploration in the Assam shelf, also known as the Belt of Schuppen (Gogoi et al. submitted), has been very challenging due to the complexity of structures, lithology, and built-up of stresses. This led to various drilling-related issues, e.g. stuck pipe, lost in hole, drill breaks, kicks, etc. This resulted in bad hole condition, poor log quality, and incomplete log acquisition. At instances, well total depth (TD) has to be declared before reaching target depth and premature cancellation of the drilling program. Multiple sidetracking of wells is a common phenomenon.

Well A in the Geleki field of Assam was not an exception and the drilling operation was suffering from tight hole and stuck pipe issues in 12.25 in section, which sidetracked three times but with no luck (Figure 4.2). Finally, an iterative geomechanics-based approach was adapted to tackle the WBS issues. The root causes for the tight holes and pipe stuck were identified by a post-drill WBS (Figure 4.3) and fault stability analysis. It was found that the well has crossed critically stressed fractures and a thrust fault trending NE–SW. The apparent displacement of the thrust fault was toward NW (Figure 4.4). A predrill model was built for the remaining section. Mud weight was recommended based on the analysis (Figure 4.5). By following the mud weight of 1.28–1.32 g/cc, a 542 m section could be drilled in ~21 days. Reduction in wellbore instability issues considerably reduced the time spent in reaming, mud circulation, cleaning tight spots, etc. The wellbore is in a much better shape as can be seen from the caliper data in the 6" section from various offset wells (Figure 4.6).

Case 1 (Continued)

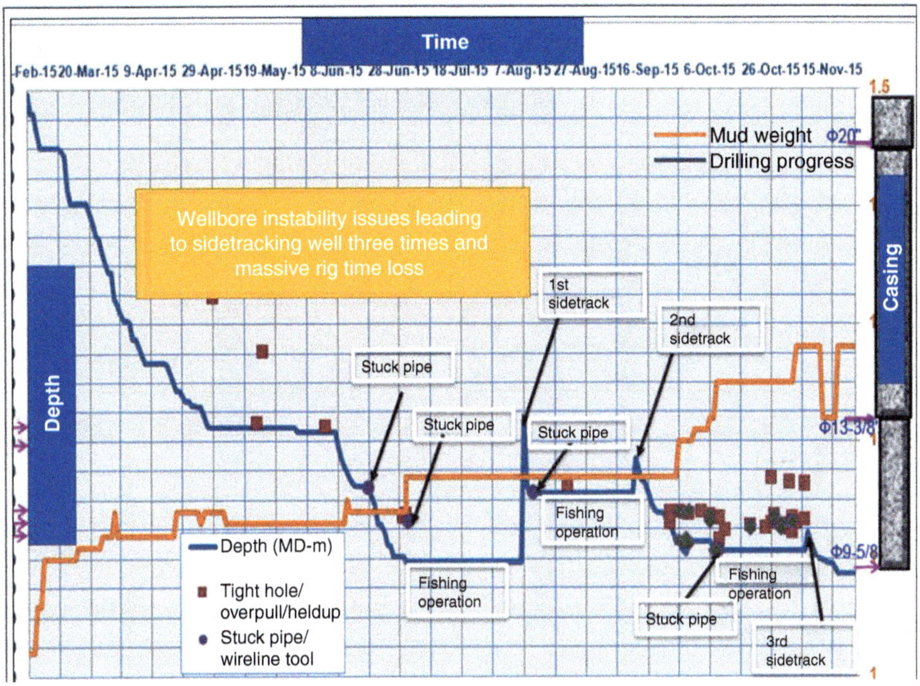

Figure 4.2 Drilling history plot showing time in *X*-axis and depth in *Y*-axis. It plots the progress of the drilling with time (blue curve). Orange curve: increasing trend from left to right is the mud weight used to drill. The drilling events or hazards encountered are plotted with corresponding depth and time. The casing is plotted on the right-hand side showing the depths of the casing. From the plot, 17.5 and 12.25 in sections showed drilling-related issues like tight holes, stuck pipe, sidetrack, and others. The fishing operation led to severe loss in time and inflated the rig cost. *Source:* Modified after Chatterjee et al. (2017).

(*Continued*)

Case 1 (Continued)

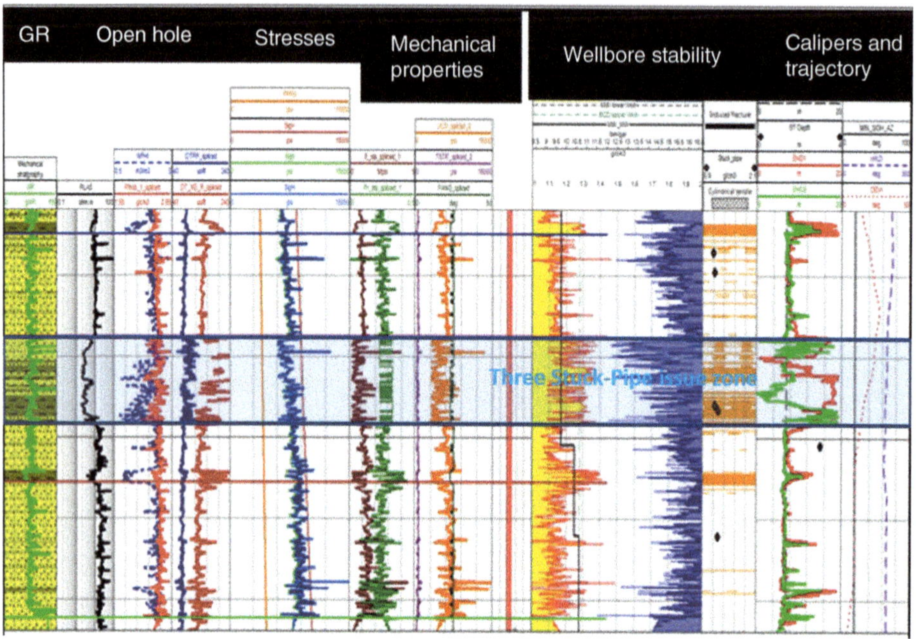

Figure 4.3 Post-drill wellbore stability (WBS) analysis for 12.25″ section. 1D MEM was constructed using open hole logs like GR, density, sonic compressional, and shear slowness. The first track shows gamma ray, which is used to construct mechanical stratigraphy as plotted in the same track showing sand and shale beds. The second track shows resistivity and the third one contains density and neutron curves. The triple combo logs helped identify the lithology and reservoir zones. The fourth track shows the compressional in blue and shear wave slowness in red. The fifth track shows the pore pressure and three principal stresses (minimum in green, maximum in blue, and vertical in brown). The sixth track shows rock elastic properties (Young's Modulus in brown and Poisson's ratio in green). The seventh track shows rock strength properties with UCS in orange, friction angle in green and tensile strength in purple. The eighth track shows casing. The ninth track shows the WBS model with breakout (in red) and loss (in blue). The curve in black is mud weight used to drill the well. The tenth track shows the failure image from the WBS model. The eleventh track shows a multi-arm caliper in red and green showing overguaged or bad hole zones. The twelfth track shows well trajectory with deviation in red and azimuth in pink.

Case 1 (Continued)

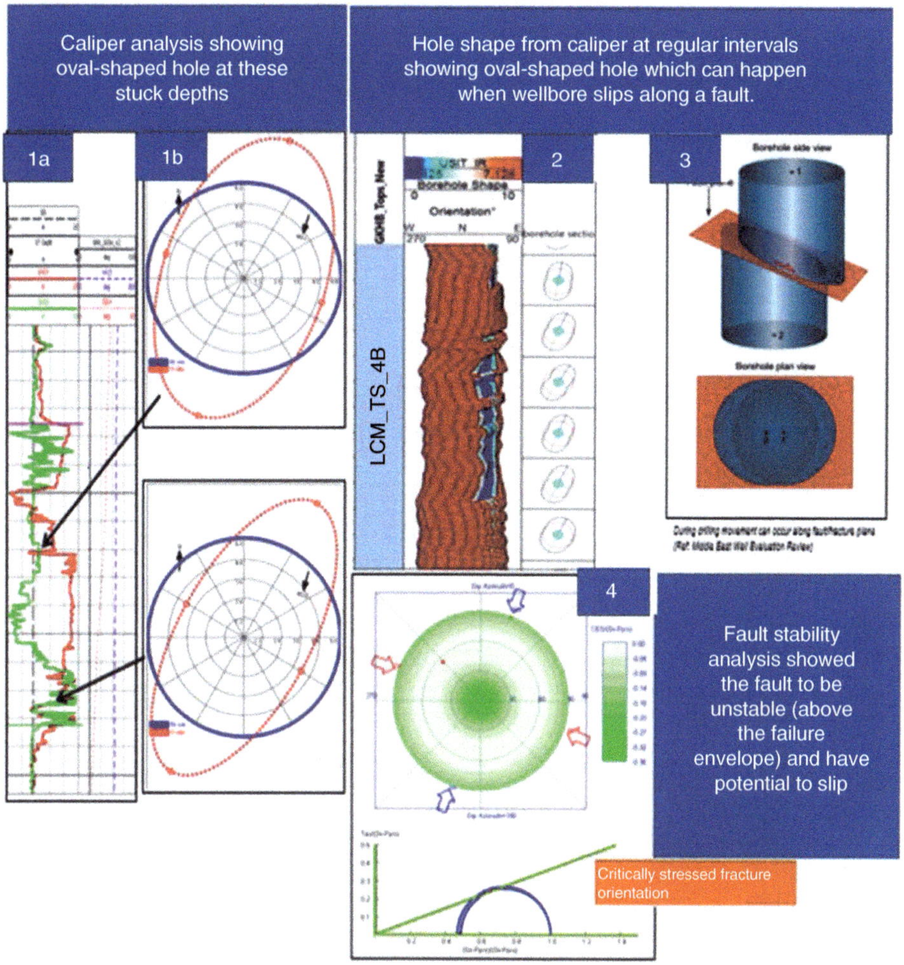

Figure 4.4 Root cause analysis of stuck events in 12.25″ section. The first figure (1a) shows the caliper in red and green with borehole section plot adjacent to it (1b). The calipers show one arm overgauge and another underguage, which is also evident from the bore hole section plot where an ovalized hole is seen. The second figure (2) shows the typical shape of the bore hole at these intervals, which is quite correlatable with fault slippage diagram in the third figure (3). The fourth figure (4) shows that the fault encountered at this depth is unstable when fault stability analysis is performed using 1D geomechanics model and fault orientation. In the Mohr–Coulomb diagram, the fault is just above the failure envelope which shows that it is unstable and has the potential to slip. Thus, the stuck events are concluded to be happened due to fault slip at these depths. *Source:* Modified after Chatterjee et al. (2017).

(*Continued*)

Case 1 (Continued)

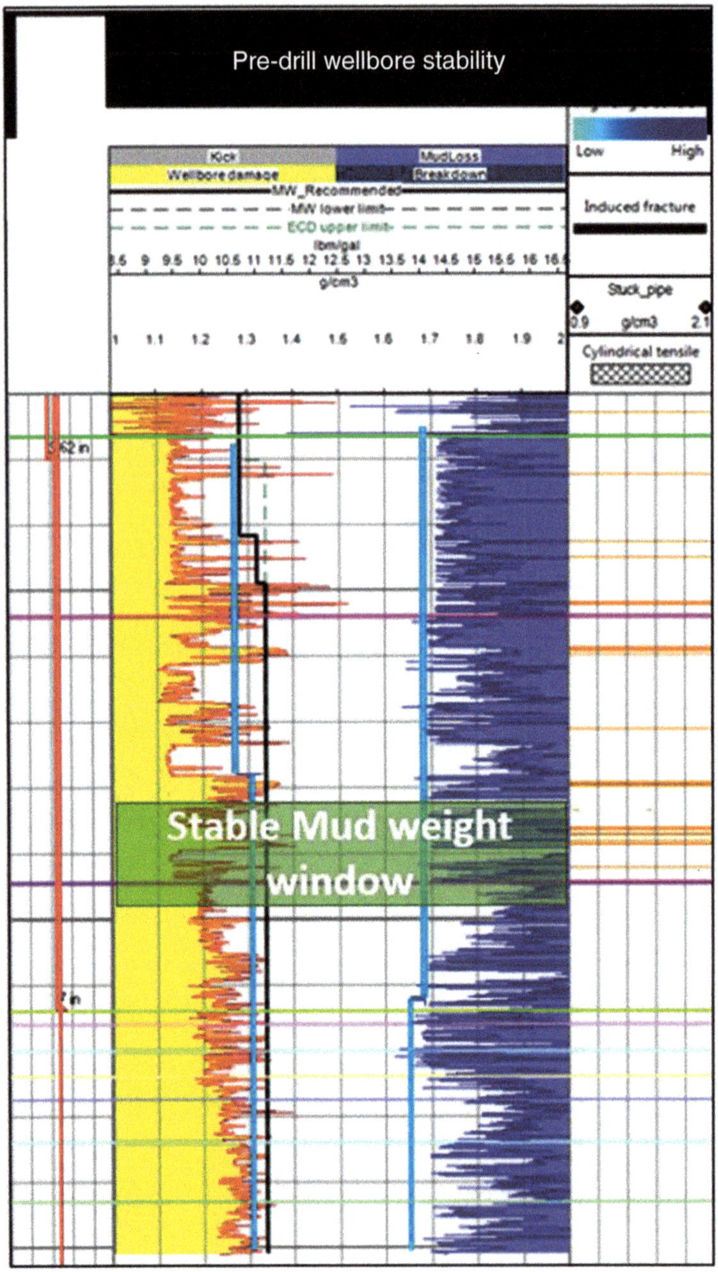

Figure 4.5 The predrill WBS analysis performed for 8.5" and 6" sections using data from nearby offset well. The first track shows casing depths of 8.5" and 6" sections. The second track shows the WBS model with breakout (in red) and loss (in blue). The curve in black is mud weight recommended to drill these sections. The third track shows the failure image from the WBS model.

(Continued)

Case 1 (Continued)

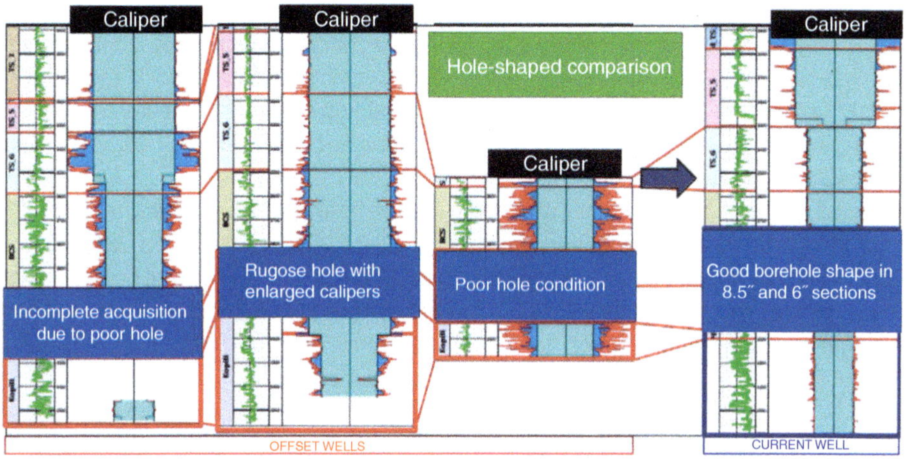

Figure 4.6 Borehole shape comparison from caliper between offset well and the current well where geomechanics study was conducted and mud weight recommended for drilling. Calipers for each well are plotted in red with blue shading in the overgauged portions where it is greater than bit size. The first three-hole shapes on the left are from offset wells where severe enlargement is seen from the caliper. The first hole shape from the left shows the incomplete acquisition of logs due to poor hole condition. The second hole shape shows a rugose hole and poor hole condition. The third hole shape is very poor with huge enlargements, which led to early termination of the well and could not reach till target depth. The fourth hole shape is of the current well where optimum mud weight was recommended, and the wellbore shows a very good hole shape with less caliper enlargement. *Source:* Modified after Chatterjee et al. (2017).

The post-drill WBS model constructed showed failure at certain intervals where the shear failure or breakout limit exceeds the mud weight. The zones showing failure from the WBS model match well with caliper and drilling events plotted on the tenth track as points (Figure 4.3). The well encountered major stuck issues at the interval where one caliper in red indicates completely overguaged hole whereas the other in green is undergauged. This has led to sidetracking the well three times to finally bypass the interval. Source: Modified after Chatterjee et al. (2017).

The predrill WBS plots the breakout and loss in red and blue, respectively in the second track showing the stable mud weight window (Figure 4.5). Based on the analysis, mud weight of 1.28–1.32 g/cc was recommended to drill these sections as shown in black. With this mud weight, the WBS model shows very less or no failure in the failure image track (third track). And this mud weight perceived not cause any losses since the loss gradient (in blue) is much higher and the stable mud weight window is wider. Source: Modified after Chatterjee et al. (2017).

Case 2 Hydraulic Fracturing Optimization: Assam-Arakan Basin (N–E India)

In a tight reservoir, hydraulic fracturing can accelerate production. However, not all wells/zones are suitable for hydraulic fracturing. Further the effectiveness of hydrofracturing determines the producibility from the interval of interest. Rock mechanical properties and stresses play crucial roles in determining the fracture geometry (height, width, and length), which eventually help in planning and executing the hydrofracturing operation.

Well B of the Bhogpara field of the Assam-Arakan basin was a candidate for hydraulic fracturing to boost production from this field. Geomechanical studies were performed to understand whether enough stress contrast is there to contain the fracture height within the target reservoir. It is observed that the overlying and the underlying shale layers have enough stress difference to act as good barriers for fracture containment (Figure 4.7). Wells with poor/no inflow prior to hydraulic fracturing operations exceeded operator expectations in post-fracturing production (Figure 4.8). This is an example of effective reservoir management through hydraulic fracturing that rejuvenated ceased wells with incremental oil production.

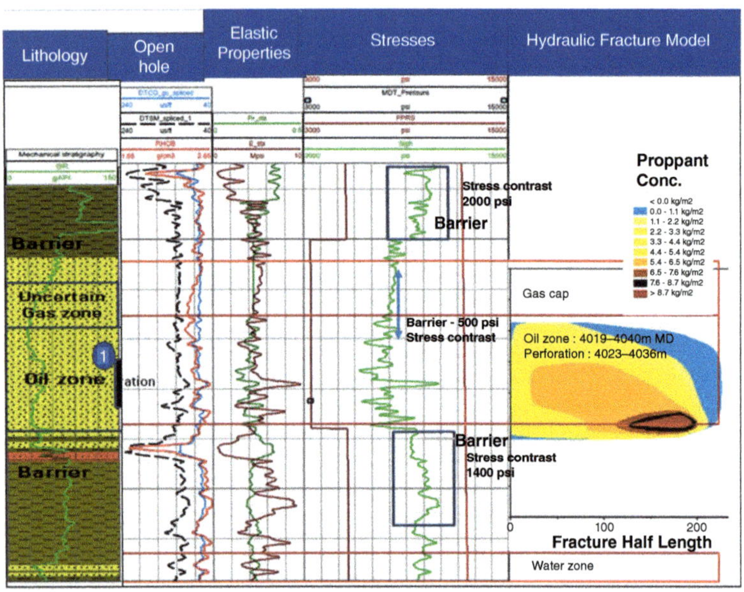

Figure 4.7 Geomechanical analysis for hydraulic fracturing design for well B. The first track shows lithology with gamma ray. Perforation 1 is also plotted depicting the zone of interest. The second track: density in red, compressional slowness in blue, and shear slowness in black which are used in geomechanical analysis. The third track: Young's modulus in brown and Poisson's ratio in green, used as an input to the hydraulic fracturing model design. The fourth track: minimum horizontal stress or fracture gradient computed using pore pressure (brown), overburden stress (red), and Young's modulus and Poisson's ratio in the third track. Hydraulic fracture model simulated using stress model and mechanical properties are displayed in the fifth track.

(Continued)

Case 2 (Continued)

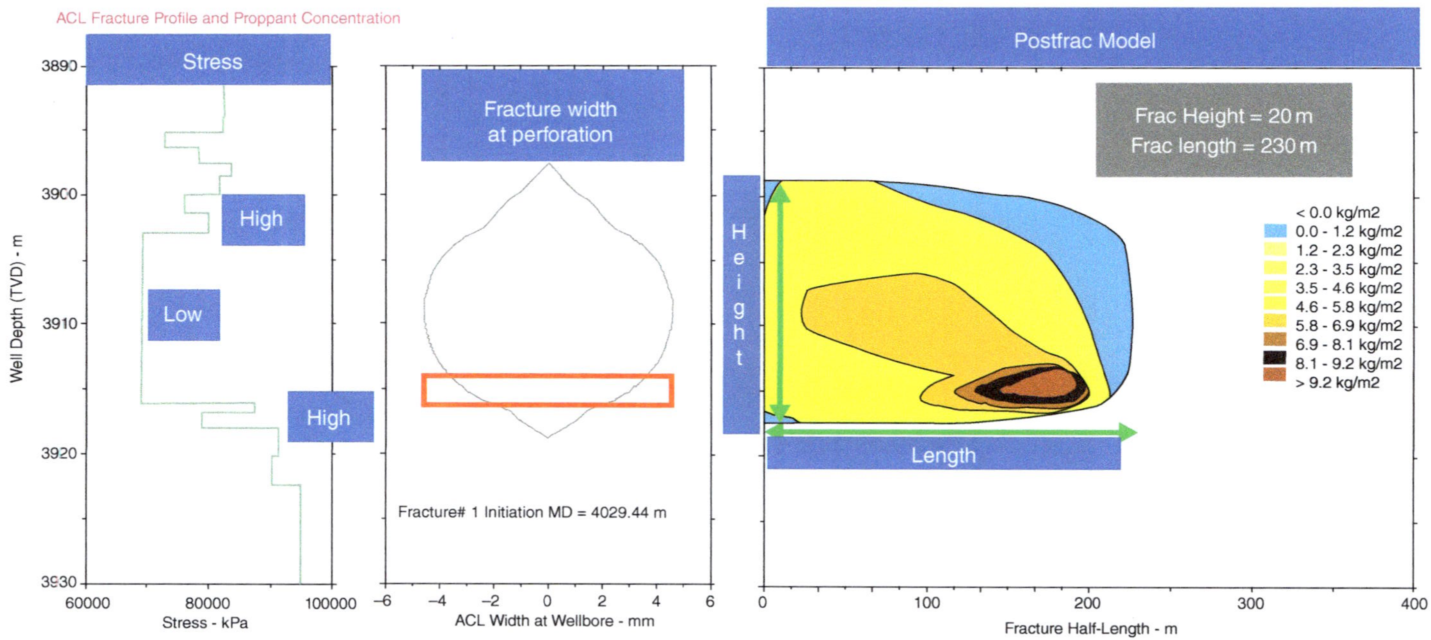

Figure 4.8 Post-frac analysis of well B. Treatment pressure history match showed, hydraulic fracturing was contained in the oil zone, generating a frac half-length of 230 and 20 m frac height. Well B responded with 14 m^3 day^{-1} post frac production rate on self-pressure from shut-in status (pre-frac). The geomechanical analysis helped in proper frac execution by restraining the frac from entering the gas cap zone. *Source:* Mukku et al. (2019)/with permission of Society of Petroleum Engineers.

From the geomechanical analysis, it is found that the fracture gradient is lower in the oil-bearing sand zone (in the perforation interval) than the overlying and underlying shale zones. The total stress contrast between sand and shale zone is ~2000 psi. There is a gas zone above the oil zone marked as gas cap, which is to be avoided for oil production. The stress contrast between the overlying gas layer vs. that of the oil-bearing sand is ~500–700 psi. Using this approximation, the hydraulic fracture model is designed such that it does not grow in height into the gas cap zone. Thus, permitted fracture height growth of ~21 m and length growth of 225 m made it a good candidate for fracturing. Source: Mukku et al. (2019)/with permission of Society of Petroleum Engineers.

Case 3 Production Optimization in Fractured Basement Reservoir: Kutch Saurashtra Basin, Western India

Over the last few years, unconventional reservoirs have gained significant importance in the exploration of oil and gas, owing to the depleting reserves in the conventional reservoirs. Basement rocks (tectonic review: Misra and Mukherjee (2015); Dasgupta and Mukherjee (2017)) depending on the fracture intensity can be considered as one of the promising unconventional reservoirs. This study shows an integrated methodology combining geology (images), petrophysics (acoustics), and geomechanics that can help characterize the permeability of the fracture network in the basement reservoirs (Nandi et al. 2019).

It has been noticed in multiple basement wells that merely the presence of non-sealing fracture is insufficient to aid hydrocarbon flow. In situ stress plays a critical role to guide fracture mobility. Knowing the fracture network (Mukherjee 2019b), which will dilate (further open) under the changing stress condition and are susceptible to flow, can optimize the production. Thus, a geomechanical study was carried out applying the Mohr–Coulomb method using stress and mechanical properties of fractures along with its orientation (Figure 4.9). The stress state of each of the fractures (network) was determined. This improved the understanding of the zones where fractures are susceptible to failure (i.e. it exceeds the failure envelope in Mohr–Coulomb circle) and have negative values of Coulomb excessive stress (CES). These fractures are termed as the *critically stressed fractures*, which finally aid in production.

The workflow starts with identifying natural conductive fractures from images and acoustics to identify open fractures and come out with a geomechanical output (Figure 4.10).

Case 3 (Continued)

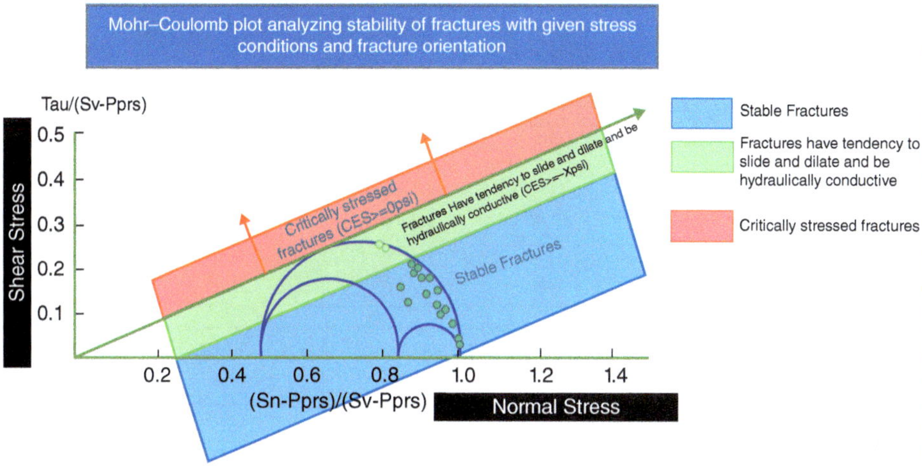

Figure 4.9 Classification of open natural fractures based on their proximity to Mohr–Coulomb failure envelop. Normal stress is plotted along the *X*-axis and shear stress along the *Y*-axis. A Mohr–Coulomb circle is derived with minimum and maximum principal effective stress and a failure envelope marked as a green line is obtained from the analysis, which is tangential to the circle with Cohesion 0 or reaching the (0, 0) coordinates. The region above the failure envelope (red) is unstable and is called critically stressed, whereas the region (green) is very close to the unstable zone and has tendency to slip and be hydraulically active under certain circumstances, e.g. depletion, change in stresses, etc. The area below green zone (blue) is the stable zone and less likely to fail. *Source:* Modified after Nandi et al. IPTC-12345-MS, (2019).

(*Continued*)

Case 3 (Continued)

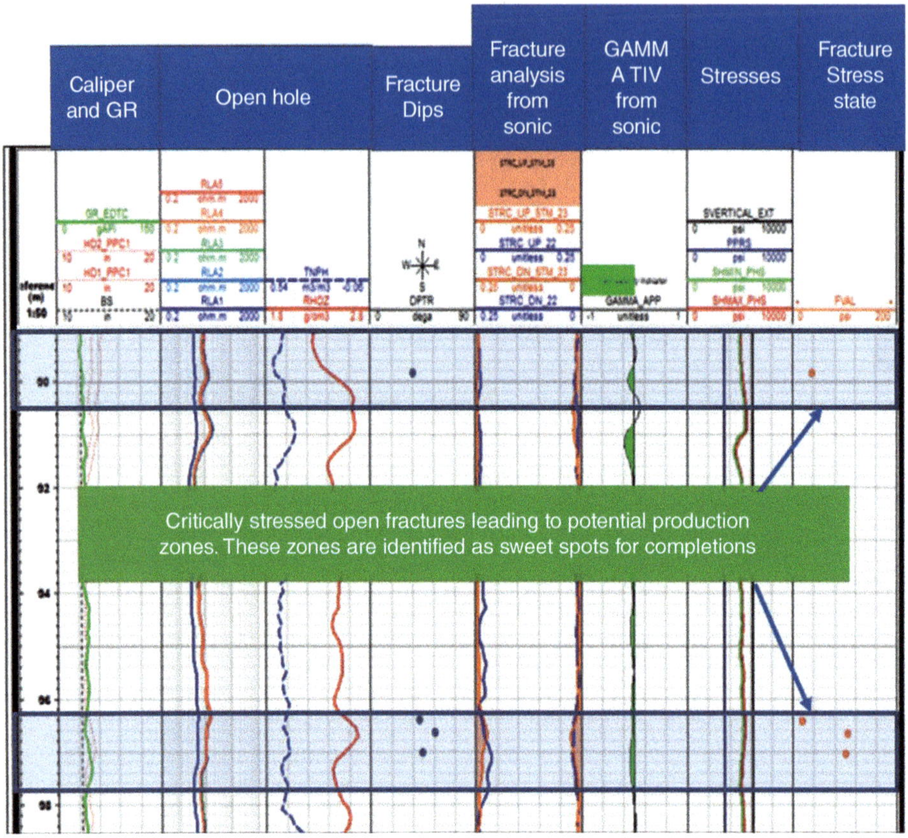

Figure 4.10 A composite plot integrating image, sonic, and geomechanics to understand the stress state of fractures for production. The first track shows depth. The second track shows caliper (HD1PPC1, HD2PPC1), bit size, and gamma ray (GR_EDTC). The third and fourth tracks show resistivity (RLA-1 to RLA-5) and density in red and neutron in blue. The fifth track shows conductive dips picked from an image. The sixth track shows fracture analysis from Stoneley shear. The seventh track shows GAMMA which determines type of anisotropy (due to the layers or fractures). If anisotropy is due to layering as in shales, GAMMA is positive, whereas if anisotropy is due to fractures, GAMMA is negative. The eighth track shows pore pressure in blue and three principal stresses, overburden stress in black, minimum horizontal stress in green and maximum horizontal stress in red. The ninth track shows FVAL or Coulomb excessive stress. It is computed to understand the stress state of each fracture. The positive FVAL value suggests fracture to be critically stressed and have potential to slip and produce whereas negative FVAL depicts stable fractures.

In this case, conductive dips were interpreted (from image log) along with Stoneley fractures to analyze the zones of openness due to presence of fractures. Gamma obtained from Stoneley shear showed negative value suggesting anisotropy due to fractures. After that, stresses were computed as a part of geomechanical analysis (1D-MEM). Combining the stresses computed with the fracture orientation, stress state were evaluated. Computation of FVAL showed positive value at some fractures, indicating these fractures were critically stressed and more likely to produce. This helped selecting the sweet spots for perforation among the other zones and considered will contribute to production. Source: Modified after Nandi et al. (2019).

Case 4 Sanding Prediction in M1 Reservoir of Apaika Field, Ecuador

Sand production is a common problem in loose/unconsolidated clastic reservoir beyond certain drawdown. It is the greatest hindrance in HC production in such reservoir as it often chokes the producing string. Sanding analysis is done to optimize HC productivity by managing sand production over the life cycle of a reservoir. It computes the critical drawdown pressure at which the first sand grain is produced, i.e. rock failures caused by drawdown in a depleted reservoir (Figure 4.11). Factors that control the critical drawdown are rock mechanical properties, stresses, pore pressure, grain diameters, perforation diameters, orientation of perforation, and completion design. Accurate estimation of the above parameters decides the future of a well in terms of production and recovery of HC.

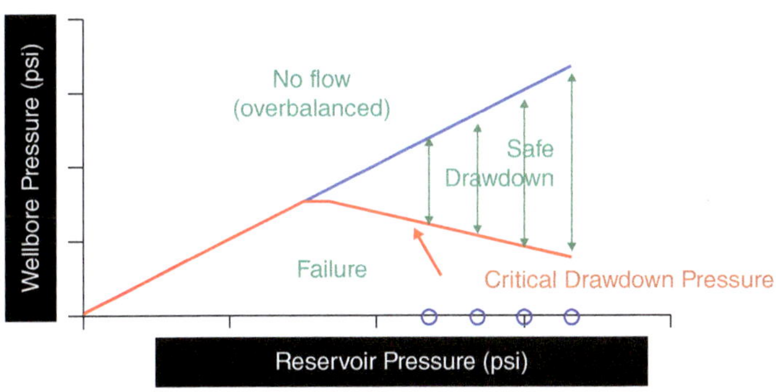

Figure 4.11 Sand management analysis showing reservoir pressure along the *X*-axis and wellbore pressure along the *Y*-axis. This graph shows how the critical drawdown pressure is analyzed for different input parameters discussed in this study. The pressure above the critical drawdown is termed as the safe drawdown, which does not have the potential to produce sand during production, and is safe for recovery. The pressure below the critical drawdown pressure is termed as failure where sand production will occur. Critical drawdown pressure is up to where production is viable. *Source:* Bravo et al. (2015)/with permission of Society of Petroleum Engineers.

(*Continued*)

Case 4 (Continued)

A case study from the Apaika field (Ecuador) is presented. Sanding analysis is performed geomechanically (1D-MEM). For a well, 1D-MEM is integrated with petrophysics logs to evaluate the reservoir quality and rock strength. The reservoir sand seemed to be soft with low rock strength, high porosity, and high permeability. Several granulometric tests were performed with cores to analyze the particle size distribution classifying the reservoir sand to be medium-sized grains. Sanding analysis was performed using all the available information to evaluate the critical drawdown pressure in the M1 reservoir (Figure 4.12). Several simulations were run taking into consideration the sensitivity of the parameters along with their associated uncertainty.

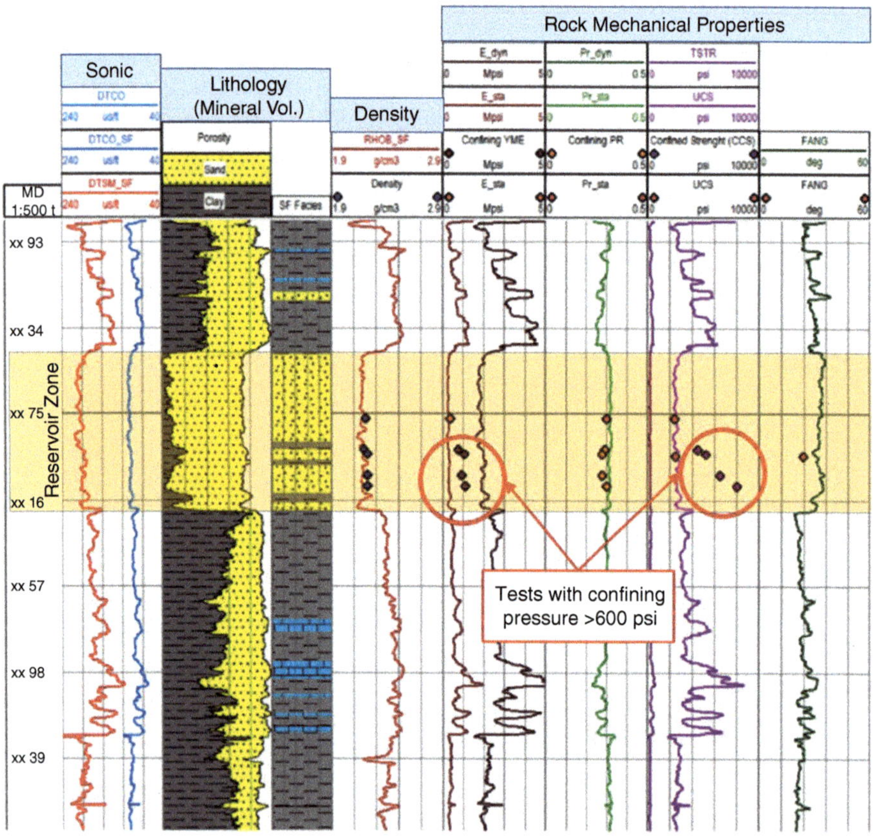

Figure 4.12 Geomechanical analysis showing the rock mechanical properties. First track: compressional slowness in blue and shear slowness in red. Second track: mineral volumes obtained from the petrophysical evaluation. Third track: mechanical stratigraphy derived from open hole log responses. Fourth track: density in the red curve which matches well with the core data (black points). Fifith track: Young's modulus dynamic in dark brown and Young's modulus (YME) static in light brown. Static YME is calibrated with the core test results. Sixth track: dynamic and static Poisson's ratio (PR), which matches well with the core data. UCS is shown in the seventh track (purple curve). Eighth track: friction angle, which matches reasonably with core data as well.

Case 4 (Continued)

Critical drawdown pressure was calculated for this medium-sized grain with grain diameter <200 um, bottom hole pressure of 1100 psi, reservoir pressure of 2288 psi, drawdown pressure is 1188 psi, which, in turn, concluded that the reservoir is at low risk of sanding. But if the bottom hole pressure, after renewed production over years, becomes ≤750 psi, sanding risk will prevail. So, the critical drawdown recommended to commence the well is 1190 psi (Figure 4.13).

By following the recommendation, the well was put to production with 1190 psi drawdown pressure for five months with 1065 BOPD of oil until it started sand production at the bottom-hole pressure <891 psi. Pressure measurements were not taken. Thus, the analysis helped in successful production and recovery from the M1 reservoir for a significant interval of time.

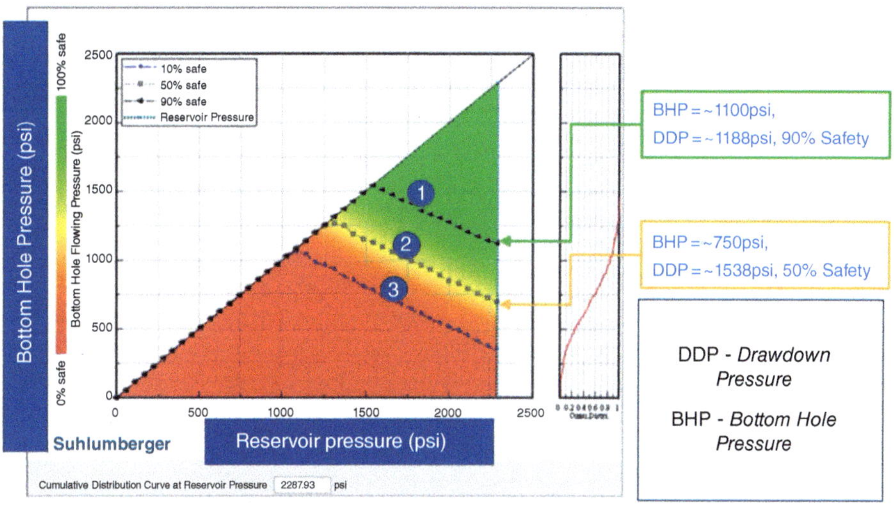

Figure 4.13 Sanding window illustrating reservoir pressure in X-axis and bottom hole pressure in Y-axis for M1 reservoir. The green zone shows a safe drawdown limit where sand will not produce. Line 1 gives a certainty limit of 90% safe production. Yellow zone: critical drawdown limit with decreasing bottom-hole pressure. Line 2: 50% safe production with decreased BHP. Red zone: unsafe limit at which sand will be produced. Line 3: 10% certainty of safe production.

Overall the rock mechanical properties match well with the core data. Some core was tested with >600 psi confining pressure and therefore shows a high value of YME and UCS. Therefore, YME and UCS are not calibrated to those values. M1 reservoir sand shows a low value of YME and UCS and high PR, which represents a weak formation. Thus, this zone is evaluated for sanding due to low rock strength. Source: Bravo et al. (2015)/with permission of Society of Petroleum Engineers.

In M1 sand, BHP of 1100 psi is evaluated with DDP of 1188 psi at which safe production is viable which can increase the risk of sanding at 750 psi BHP with a 50% chance of sand production. But it was recommended to commence the well at 1190 psi drawdown. (Source: Modified after Bravo et al. 2015).

4.3 Conclusions

Geomechanics plays a significant role in the oil and gas industry in reducing well construction costs, maximizing production, and increasing the life of the well. In recent times, the E&P industry is focusing on exploiting challenging reservoirs requiring proper estimation of the formation properties for efficient drilling and production campaigns. This can be achieved by constructing MEM using the basic open hole logs, including sonic velocities along with drilling data. The MEM, in turn, is an input to all the applications of geomechanics, e.g. WBS for drilling optimization, completion quality for hydraulic fracturing, sanding analysis, fracture or fault stability analysis, and others. These studies help in successful operational execution in challenging environments leading to a reduction in rig time and costs.

Acknowledgments

Thanks to Soumyajit Mukherjee for the invitation to submit, handle, and review this article in detail in two rounds. We thank Anindya Nandi, V. Mukku, Somessh Bahuguna, Joseph Zacharia, Sourav Das, Subhadeep Sarkar, Priyanuz Bhuyan, and Arvind Kumar for their support. Merryl Le Roux, Frank Otmar Weinreich, Joss Everett, Andrew Harrison and the proofreading team (Wiley Blackwell) are thanked for their assistance. The article has been summarized in Mukherjee et al. (2023).

References

I.D.R. Bradford;J. Fuller; P.J. Thompson; T.R. Walsgrove 1998, *Benefits of Assessing The Solids Production Elastoplastic Modeling*. Paper presented at the SPE/ISRM Rock Mechanics in Petroleum Engineering (8–10 July 1998). Paper Number: SPE-47360-MS.

Bravo, O. et al. (2015). Integrated reservoir analysis for optimizing production when there is high potential for sanding. SPE Latin American and Caribbean Petroleum Engineering Conference, (18 November 2015). SPE-177130-MS.

Chatterjee, C. et al. (2017). India, geomechanical solutions incorporating asdvanced acoustic measurements for achieving & improving wellbore stability in structurally complex and tectonically active Geleki field. SPWLA 4th Symposium (11 February 2017).

Dasgupta, S. and Mukherjee, S. (2017). Brittle shear tectonics in a narrow continental rift: asymmetric non-volcanic Barmer basin (Rajasthan, India). *The Journal of Geology* 125: 561–591.

Dasgupta T, Mukherjee S. (2020). *Sediment Compaction and Applications in Petroleum Geoscience*. Springer. Series: Advances in oil and gas exploration & production. ISSN: 2509-372X.

Dasgupta, S., Dasgupta, T., and Mukherjee, S. (2020). Image log interpretation of geomechanical issues. In: *Teaching Methodologies in Structural Geology and Tectonics* (ed. S. Mukherjee), 237–251. Springer.

Dutta, D., Biswas, T., and Mukherjee, S. (2019). Arc-parallel compression in the NW Himalaya: evidence from structural and palaeostress studies of brittle deformation from the clasts of the Upper Siwalik, Uttarakhand, India. *Journal of Earth System Science* 128: 125.

Elebute, O., Chatterjee, C., and Das, B. (2018). Successful drilling camwpaign using geomechanics aided solution in Opuama field, Niger Delta, Nigeria. The 2nd SPWLA Asia Pacific Technical Symposium, Indonesia (7–8 November 2018).

Fjaer, E., Holt, R.M., Horsrud, P. et al. (1992). Petroleum related rock mechanics. In: *Developments in Petroleum Science*, 33. Elsevier.

Gogoi, M.P., Gogoi, B., and Mukherjee, S. (2022). Tectonic instability of the petroliferous upper Assam valley (NE India): a geomorphic approach. *Journal of Earth System Science*. https://doi.org/10.1007/s12040-021-01752-6.

Misra, A.A. and Mukherjee, S. (2015). *Tectonic Inheritance in Continental Rifts and Passive Margins*. Springerbriefs in Earth Sciences ISBN 978-3-319-20576-2.

Misra, A.A. and Mukherjee, S. (2018). *Atlas of Structural Geological Interpretation from Seismic Images*. Wiley Blackwell ISBN: 978-1-119-15832-5.

Mukherjee, S. (2017a). Airy's isostatic model: a proposal for a realistic case. *Arabian Journal of Geosciences* 10: 268.

Mukherjee, S. (2017b). Shear heating by translational brittle reverse faulting along a single, sharp and straight fault plane. *Journal of Earth System Science* 126 (1).

Mukherjee, S. (2018a). Moment of inertia for rock blocks subject to bookshelf faulting with geologically plausible density distributions. *Journal of Earth System Science* 127: 80.

Mukherjee, S. (2018b). Locating center of gravity in geological contexts. *International Journal of Earth Sciences* 107: 1935–1939.

Mukherjee, S. (2018c). Locating center of pressure in 2D geological situations. *The Journal of Indian Geophysical Union* 22: 49–51.

Mukherjee S. (2019a) Kinematics of pure shear ductile deformation within rigid walls: new analyses. In: Billi A, Fagereng A (Eds) *Problems and Solutions in Structural Geology and Tectonics*. Series Editor: Mukherjee S. Developments in structural geology and tectonics book series. Elsevier. pp. 81–88. ISSN: 2542-9000.

Mukherjee S. (2019b) Using graph theory to represent brittle plane networks. In: Billi A, Fagereng A (Eds) *Problems and Solutions in Structural Geology and Tectonics*. Developments in structural geology and tectonics book series. Vol. 5. Series Editor: Mukherjee S. Elsevier. pp. 259–272. ISSN: 2542-9000. ISBN: 9780128140482.

Mukherjee, S. and Khonsari, M.M. (2017). Brittle rotational faults and the associated shear heating. *Marine and Petroleum Geology* 88: 551–554.

Mukherjee, S. and Khonsari, M.M. (2018). Inter-book normal fault-related shear heating in brittle bookshelf faults. *Marine and Petroleum Geology* 97: 45–48.

Mukherjee, S. and Kumar, N. (2018). A first-order model for temperature rise for uniform and differential compression of sediments in basins. *International Journal of Earth Sciences* 107: 2999–3004.

Mukherjee, S., Dasgupta, S., Majumdar, C. et al. (2023). Introduction to: Handbook of Petroleum Geoscience: Exploration, Characterization, and Exploitation of Hydrocarbon Reservoirs. In: *Handbook of Petroleum Geoscience: Exploration, Characterization, and Exploitation of Hydrocarbon Reservoirs*, Petroleum Industry Updates: Geoscientific and Economic Issues (ed. S. Mukherjee, S. Dasgupta, C. Majumdar, et al.). John Wiley & Sons, Hoboken, ISBN 978-1-119-68003-1.

Mukku, V. et al. (2019). Increase in oil production: methodology & best practices for hydraulic fracturing candidate selection and execution in Assam-Arakan Basin. Paper presented at the SPE Middle East Oil and Gas Show and Conference (18–21 March 2019). SPE-195026-MS.

Nandi, A., Das, S., and Chatterjee, C. (2019). Characterizing fracture basements using an integrated approach – 3D acoustic anisotropy analysis and formation micro-imager. *IPTC* https://doi.org/10.2523/IPTC-19355-MS.

Plumb, R. Edwards, S. Pidcock, G. and Lee, D. (2000). The Mechanical Earth Model Concept and its Application to High-risk Well Construction Projects. SPE Paper 59128 presented at the IADC/SPE Drilling Conference, New Orleans (23–25 February 2000).

Shaikh, M., Maurya, D.M., Mukherjee, S. et al. (2020). Tectonic evolution of the intra-uplift Vigodi-Gugriana-Khirasra-Netra fault system in the seismically active Kachchh Rift Basin, India: implications for the western continental margin of the Indian plate. *Journal of Structural Geology* 140: 104124.

Vanik, N., Shaikh, H., Mukherjee, S. et al. (2018). Post-Deccan trap stress reorientation under transpression: Evidence from fault slip analyses from SW Saurashtra, western India. *Journal of Geodynamics* 121: 9–19.

5

In Situ Stresses from Log Measurements

Kuppili Meenakshi Sundaram

Petrophysics Consultant, Mumbai, Maharashtra, India
Author for correspondence: sndrmkm.sir@gmail.com

Abstract

The technology of dipole excitation and flexural waveform recording by arrays of dipole receivers allows radial-profiling of C_{44}, C_{55} along the principal horizontal stress directions. The modulus C_{66} is reliably estimated by exciting a monopole acoustic source and, inverting the Stoneley waveforms data recorded by an array of monopole receivers. Using robust forward models, the radial profiles are inverted for recovering the principal horizontal stress magnitudes. The inversion is carried out subject to realistic constraints built around the well bore-stability criteria. The principal vertical stress is modeled using formation bulk density log data. Alternate approaches to the problem of principal stress-characterization are also discussed for the cases where radial profiling mentioned here is not feasible/unavailable. The techniques discussed apply to the case of data acquired in vertical boreholes penetrating formations within which the stress field has a single vertical principal component and two principal horizontal components. The concepts are easily extendible for wellbores inclined to the principal stress directions.

Keywords *Stress anisotropy; acoustoelastic theory; flexural wave; stoneley wave; fast shear wave; slow shear wave*

5.1 Introduction

Estimation of the initial state of stress is the key first step for numerous geoscientific applications, e.g. assessing borehole stability, planning a well trajectory, sanding analysis, and hydraulic fracture design. It is also a key step while planning long-term reservoir-flooding programs that ensure avoidance of reservoir subsidence due to depletion of reservoir pressure. Inversion of log data to in situ stress magnitudes is the only method available, which yields reliable results representative of entire rock masses.

This work aims at (i) introducing a few available techniques for computing the state of stress through inversion of log data, and (ii) technical details.

Handbook of Petroleum Geoscience: Exploration, Characterization, and Exploitation of Hydrocarbon Reservoirs, First Edition. Edited by Soumyajit Mukherjee, Swagato Dasgupta, Chandan Majumdar, Subhandip Mandal and Troyee Dasgupta.
© 2023 John Wiley & Sons Ltd. Published 2023 by John Wiley & Sons Ltd.

5.2 Stress Field Within a Formation Penetrated by a Borehole

Consider one of the principal stress directions to be vertical. The principal stresses along the vertical and the two horizontal directions are S_V, $S_{H\max}$, and $S_{h\min}$, respectively. The magnitude of the effective stress along the directions of the principal stresses are σ_V, σ_H, and σ_h. A rectangular coordinate system is set with X_1, X_2, and X_3, respectively along the directions of the principal stresses, the maximum horizontal stress, the minimum horizontal stress, and the overburden/lithostatic stress.

Consider a vertical borehole and a cylindrical coordinate system (r, θ, z) with the z axis along the axis of the borehole, and the reference direction for θ, the X_1 direction. The radial, azimuthal, and the vertical components of the effective stress field are notated here, as σ_{rr}, $\sigma_{\theta\theta}$, σ_{zz}.The stress state near-wellbore is denoted as $\sigma_{\text{nearfield}}$.

$$\sigma(r,\theta,z) = \begin{bmatrix} \sigma_{rr} & \sigma_{r\theta} & \sigma_{rz} \\ \sigma_{\theta r} & \sigma_{\theta\theta} & \sigma_{\theta z} \\ \sigma_{zr} & \sigma_{z\theta} & \sigma_{zz} \end{bmatrix} \tag{5.1}$$

As per Jaeger and Cook (1969):

$$\sigma_{rr} = \left(\frac{\sigma_H + \sigma_h}{2}\right)\left(1 - \frac{a^2}{r^2}\right) + \left(\frac{\sigma_H - \sigma_h}{2}\right)\left(1 - \frac{4a^2}{r^2} + \frac{3a^4}{r^4}\right)\cos 2\theta - P_w\left(\frac{a^2}{r^2}\right) \tag{5.2}$$

$$\sigma_{r\theta} = -\left(\frac{\sigma_H - \sigma_h}{2}\right)\left(1 + \frac{2a^2}{r^2} - \frac{3a^4}{r^4}\right)\sin 2\theta \tag{5.3}$$

$$\sigma_{rz} = 0 \tag{5.4}$$

$$\sigma_{\theta r} = \sigma_{r\theta} \tag{5.5}$$

$$\sigma_{\theta\theta} = \left(\frac{\sigma_H + \sigma_h}{2}\right)\left(1 + \frac{a^2}{r^2}\right) - \left(\frac{\sigma_H - \sigma_h}{2}\right)\left(1 + \frac{3a^4}{r^4}\right)\cos 2\theta + P_w\left(\frac{a^2}{r^2}\right) \tag{5.6}$$

$$\sigma_{\theta z} = 0 \tag{5.7}$$

$$\sigma_{zr} = \sigma_{rz} \tag{5.8}$$

$$\sigma_{z\theta} = \sigma_{\theta z} \tag{5.9}$$

$$\sigma_{zz} = \sigma_V - \nu(\sigma_H - \sigma_h)\left(\frac{2a^2}{r^2}\right)\cos 2\theta \tag{5.10}$$

The shear modulus of elasticity of the formation at any location in space is a function of the intrinsic elastic modulus of the formation and the stress field at that location. In general, the values of the elements C_{44}, C_{55}, C_{66} of the stiffness matrix are functions of position.

5.3 Birefringence of Flexural Waves in Presence of Horizontal Stress Anisotropy

In the presence of horizontal stress anisotropy, excitation by a dipole source within a liquid-filled borehole results in the twin-set of flexural wave packets. The phase components of the packets are polarized in the $X_1 - X_3$ plane and the $X_2 - X_3$ plane, respectively. For any given frequency, the phase velocity of the phase components polarized in the $X_1 - X_3$ plane is more than that of the phase components polarized in the $X_2 - X_3$ plane.

5.4 Radial Profiles of the Fast-Shear Velocity, Slow-Shear Velocity and C_{44}, C_{55}, and C_{66}

Shear waves polarized in the $X_1 - X_3$ plane (fast-shear) travel faster than shear waves polarized in the $X_2 - X_3$ plane (slow-shear).

The inversion of the dipole wave dispersion data to estimate the radial profile of shear wave velocity of propagation changes is inherently a nonunique inverse problem (Tang and Patterson 2010). This nonuniqueness is taken care of by applying the Backus–Gilbert inverse theory (Backus and Gilbert 1970; Burridge and Sinha 1996) or through the technique of constrained inversion (Tang and Patterson 2010). Inversion of the fast flexural and slow flexural dispersion using the above-mentioned inversion technique yields the required radial profiles along the X_1 and the X_2 directions, respectively.

The extraction of slow flexural and fast flexural wave packets from the recorded in-line and crossline receiver waveform input to the inversion involves the application of the four-component rotation technique, known as the Alford four-component rotation (Alford 1986, detail in Briggs et al. 2002). The output of the inversion of the slow flexural wave packet is the radial profile $V_{\mathrm{S}}^{\mathrm{Slow}}(r)$ of the slow-shear wave velocity. The output of the inversion of the fast flexural wave packet is the radial profile $V_{\mathrm{S}}^{\mathrm{Fast}}(r)$ of the fast-shear wave velocity. An example of the methodology of the inversion of flexural wave packet data to radial profiles of formation shear wave velocity can be found in Sinha (2007). The interested reader can also refer Sinha and Burridge (2001).

Formation of Stoneley wave slowness through nondispersive semblance processing of Stoneley data recorded by an array of receivers (Kimball and Marzetta 1984) is used to compute the value of C_{66}.The shear wave velocity for waves polarized in the $X_1 - X_2$ plane is independent of the radial distance. Hence, the radial profile of the shear wave velocity is a straight line.

It is implicitly assumed that there is no plastic yielding of the rock even close to the borehole surface. It is further assumed that the deformation falls within the region of linear elasticity.

Let ρ_b stand for the bulk density of the formation. Let $C_{44}(r)$, $C_{55}(r)$, and $C_{66}(r)$ stand for the radial profiles of C_{44}, C_{55}, and C_{66}, respectively. We have (Gholami et al. 2015):

$$C_{55}(r) = \rho_b\left[V_{\mathrm{S}}^{\mathrm{Fast}}(r)\right]^2 \tag{5.11}$$

$$C_{44}(r) = \rho_b\left[V_{\mathrm{S}}^{\mathrm{Slow}}(r)\right]^2 \tag{5.12}$$

$$C_{66}(r) = \rho_b\left[V_{\mathrm{Stoneley}}\right]^2 \tag{5.13}$$

The value of θ for the direction of maximum horizontal stress is 0°. Eqs. (5.2), (5.6), and (5.10) become, after rearranging,

$$\sigma_{rr} = \frac{\sigma_H}{2}\left(2 - \frac{5a^2}{r^2} + \frac{3a^4}{r^4}\right) + \frac{\sigma_h}{2}\left(\frac{3a^2}{r^2} - \frac{3a^4}{r^4}\right) - P_w\left(\frac{a^2}{r^2}\right) \tag{5.14}$$

$$\sigma_{\theta\theta} = \frac{\sigma_H}{2}\left(\frac{a^2}{r^2} - \frac{3a^4}{r^4}\right) + \frac{\sigma_h}{2}\left(2 + \frac{a^2}{r^2} + \frac{3a^4}{r^4}\right) + P_w\left(\frac{a^2}{r^2}\right) \tag{5.15}$$

$$\sigma_{zz} = \sigma_V - \nu(\sigma_H - \sigma_h)\left(\frac{2a^2}{r^2}\right) \tag{5.16}$$

σ_V is computed as (Sinha et al. 2005, Shannon et al. 2016):

$$\sigma_V(z) = S_V(z) - \alpha P_p(z) \tag{5.17}$$

$$S_V(z) = S_V(z_0) + \int_{z_0}^{z} g\rho_b(z)dz \tag{5.18}$$

Here $S_V(z)$ stands for the vertical principal stress at a subsurface location situated at a vertical depth z below the topmost level of the sediment column. $S_V(z_0)$ stands for the vertical principal stress at the topmost level of the sediment column and z_0 stands for the depth of the topmost level of the sediment column. All depths are usually referenced to the mean sea level (MSL). Mukherjee (2017) presented how to calculate the weight of the lithologic column when density varies in three orthogonal directions.

The anisotropic stressed state is considered the result of a perturbation to a reference state, which is a state of isotropic stress field within the formation.

5.5 Computation of C_{44}, C_{55}, and C_{66} as per Acoustoelastic Theory

The shear modulus of elasticity in the reference state is denoted by μ. In the reference state of isotropic stress, $C_{44} = C_{55} = C_{66} = \mu$ when the rock is isotropic. When the rock is intrinsically a TI medium, μ denotes the horizontal shear modulus. In the actual stressed state,

$$C_{44}(r) = \mu + \Delta C_{44} \tag{5.19}$$

$$C_{55}(r) = \mu + \Delta C_{55} \tag{5.20}$$

$$C_{66}(r) = \mu + \Delta C_{66} \tag{5.21}$$

The theoretical model of $\Delta C_{55}(r)$, $\Delta C_{44}(r)$, and $\Delta C_{66}(r)$ can be found in Sinha (2007). Using this theoretical model in Eqs. (5.19)–(5.21) and further analysis leads to the following equations (Sinha et al. 2007):

$$\begin{aligned} C_{44}(r) = \mu &+ \frac{1}{2\mu(1+\nu)}\left[\frac{\sigma_H}{2}(A_2C_2 + A_3C_1 + \nu(A_3 - A_4)C_3)\right.\\ &\left.+ \frac{\sigma_h}{2}(A_4C_2 + A_1C_1 - \nu(A_3 - A_4)C_3) + \sigma_V C_3\right]\\ &+ \frac{P_w}{2\mu(1+\nu)}\left[\left(\frac{A_3 - A_4}{2}\right)C_1 + \left(\frac{A_3 - A_4}{2}\right)C_2\right] - \frac{\sigma_{V0}}{2\mu(1+\nu)}(C_1 + C_2 + C_3) \end{aligned} \tag{5.22}$$

$$\begin{aligned} C_{55}(r) = \mu &+ \frac{1}{2\mu(1+\nu)}\left[\frac{\sigma_H}{2}(A_1C_1 + A_4C_2 - \nu(A_3 - A_4)C_3)\right]\\ &+ \frac{\sigma_h}{2}(A_3C_1 + A_2C_2 + \nu(A_3 - A_4)C_3) + \sigma_V C_3\Big]\\ &+ \frac{P_w}{2\mu(1+\nu)}\left[-\left(\frac{A_3 - A_4}{2}\right)C_1 + \left(\frac{A_3 - A_4}{2}\right)C_2\right] - \frac{\sigma_{V0}}{2\mu(1+\nu)}(C_1 + C_2 + C_3) \end{aligned} \tag{5.23}$$

$$C_{66} = \mu + \frac{(\sigma_H + \sigma_h)}{2\mu(1+\nu)}[C_{66R} - \nu C_{144} + (1-\nu)C_{155}] + \frac{\sigma_V}{2\mu(1+\nu)} [-(1+2\nu)C_{66} + C_{144} - 2\nu C_{155}] \quad (5.24)$$

Here,

$$A_1\left(\frac{r}{a}\right) = \left(2 - \frac{5a^2}{r^2} + \frac{3a^4}{r^4}\right) \quad (5.25)$$

$$A_2\left(\frac{r}{a}\right) = \left(2 + \frac{a^2}{r^2} + \frac{3a^4}{r^4}\right) \quad (5.26)$$

$$A_3\left(\frac{r}{a}\right) = \left(\frac{3a^2}{r^2} - \frac{3a^4}{r^4}\right) \quad (5.27)$$

$$A_4\left(\frac{r}{a}\right) = \left(\frac{a^2}{r^2} - \frac{3a^4}{r^4}\right) \quad (5.28)$$

$$C_1 = (C_{55R} - \nu C_{144} + (1-\nu)C_{155}) \quad (5.29)$$

$$C_2 = (-(1+2\nu)C_{55R} + C_{144} - 2\nu C_{155}) \quad (5.30)$$

$$C_3 = (C_{55R} - \nu C_{144} + (1-\nu)C_{155} + 2\mu(1+\nu)) \quad (5.31)$$

5.6 Relation Between Stress Field Components and Elastic Moduli

A porous rock can be considered solid-fluid composite with the fluid as a continuous phase within the composite. A generalization of the standard "small on large" acoustoelastic theory of solid-fluid composites, applied to the propagation of acoustic waves through rocks (Sinha 1982, Norris et al. 1994) leads to the following relationships (Sinha et al. 2007, 2008):

$$C_{44} - C_{66} = A_E(\sigma_V - \sigma_H) \quad (5.32)$$

$$C_{55} - C_{66} = A_E(\sigma_V - \sigma_h) \quad (5.33)$$

$$C_{55} - C_{44} = A_E(\sigma_H - \sigma_h) \quad (5.34)$$

A_E is called the acoustoelastic coefficient. It is given by (Pistre et al. 2009):

$$A_E = \left[\frac{E}{(1+\nu)} + 2C_{44R} + (C_{155} - C_{144})\right]\frac{(1+\nu)}{E} = \left(2 + \frac{C_{155} - C_{144}}{2\mu}\right) \quad (5.35)$$

The last equality in the RHS of Eq. (5.38) arises from the following reasoning. The quantity C_{44R} is the value of C_{44} for reference state of isotropic stress, and is nothing but the shear modulus of the rock when it is in the reference state of isotropic stress. Further, the elastic moduli E, ν also pertain to the reference state of isotropic stress, and since $\frac{E}{(1+\nu)} = \mu$, $\left[\frac{E}{(1+\nu)} + 2C_{44R} + (C_{155} - C_{144})\right]\frac{(1+\nu)}{\mathrm{E}}$ simplifies to $\left(2 + \frac{C_{155} - C_{144}}{2\mu}\right)$.

From Eqs. (5.17) and (5.18):

$$\sigma_V(z) = S_V(z_0) + \int_{z_0}^{z} g\rho_b(z)dz - \alpha P_p(z) \tag{5.36}$$

In Eq. (5.36), the values of $S_V(z_0)$, α and $P_p(z)$ are known for all depths z. Hence, $\sigma_V(z)$ is known for all z values. Hence, the value of σ_V which exists in Eqs. (5.36) and (5.37) is a known parameter at any depth.

5.6.1 Nondimensional Parameters α_1 and α_2

Two nondimensional parameters α_1 and α_2 used in the computation of the far-field stresses from the radial distribution of the shear modulus of the formation are now defined (Sinha 2011) with the quantities notated as A_1 and A_2 in the reference state, notated as α_1 and α_2 here:

$$\alpha_1 = -\nu C_{144} + (1-\nu)C_{155} \tag{5.37}$$

$$\alpha_2 = C_{144} - 2\nu C_{155} \tag{5.38}$$

The acoustoelastic coefficient is expressed in terms of α_1, α_2 (Sinha (2011) with quantities notated as A_1 and A_2 in the reference state, notated as α_1 and α_2 here) as:

$$A_{\mathrm{E}} = \left(2 + \frac{\alpha_1 - \alpha_2}{2\mu(1+\nu)}\right) \tag{5.39}$$

5.7 Computation of Principal Horizontal Stresses from the Radial Profiles $C_{44}(r)$ and $C_{55}(r)$

The key to the computation of the far-field horizontal stress magnitudes is the evaluation of the value of A_E. This is achieved as follows. The values of α_1 and α_2 are first computed. The value of A_E is then computed by substituting the values of α_1 and α_2 in Eq. (5.39).

The methodology of the computation of the values of α_1, α_2 and, therefrom, the value of A_E can be found at Sinha (1997a) (Sinha 2011).

With A_E evaluated as indicated here, σ_H and σ_h are now evaluated using Eqs. (5.32) and (5.33) (Sinha et al. 2007):

$$\sigma_H = \sigma_V - \frac{(C_{44} - C_{66})}{A_{\mathrm{E}}} \tag{5.40}$$

$$\sigma_h = \sigma_V - \frac{(C_{55} - C_{66})}{A_{\mathrm{E}}} \tag{5.41}$$

With S_V already evaluated, using Eq. (5.18), from the formation density log data, σ_V stands evaluated from Eq. (5.17).

We have the following relation between the components of the effective stress tensor and the total stress tensor (Sinha et al. 2007)

$$\sigma_{ij} = S_{ij} - \alpha P_p \tag{5.42}$$

In this notation, $i = j = 1, 2, 3$ denotes vertical stress, maximum horizontal stress and minimum horizontal stress, respectively. Using Eq. (5.42):

$$S_{\mathrm{Hmax}} = \sigma_H + \alpha P_p = \sigma_V - \frac{(C_{44} - C_{66})}{A_{\mathrm{E}}} \tag{5.43}$$

$$S_{\mathrm{Hmin}} = \sigma_h + \alpha P_p = \sigma_V - \frac{(C_{55} - C_{66})}{A_{\mathrm{E}}} \tag{5.44}$$

5.8 Stress Field Characterization Through Inversion of Flexural Wave Dispersion

The term "stress" is meant to convey effective stress from here onward. The velocity of propagation of the fast flexural wave and the slow flexural wave through the rock in the current state of stress are denoted by $V_{\mathrm{Flex}}^{\mathrm{F}}$ and $V_{\mathrm{Flex}}^{\mathrm{S}}$, respectively. The current stress field is a superposition of a stress perturbation over the isotropic stress field for the reference state.

Application of perturbation theory based on the perturbation model of Sinha (1997b) and Sinha et al. (2006) leads to (Sinha 2006):

$$\begin{aligned}
\left(\frac{\left[\left(V_{\mathrm{Flex}}^{\mathrm{F}}\right)_{k=k_i} - \left(V_{\mathrm{Flex}}^{\mathrm{S}}\right)_{k=k_i}\right]}{V_{\mathrm{Flex}}^{\mathrm{R}}}\right) = \frac{1}{V_{\mathrm{Flex}}^{\mathrm{R}}}\Bigg\{ & \left[\left(L_{V\mathrm{Fast}}^{k_i} + M_{V\mathrm{Fast}}^{k_i}\frac{C_{111}}{\mu} + N_{V\mathrm{Fast}}^{k_i}\frac{C_{155}}{\mu} + P_{V\mathrm{Fast}}^{k_i}\frac{C_{144}}{\mu}\right)\right. \\
& \left. - \left(L_{V\mathrm{Slow}}^{k_i} + M_{V\mathrm{Slow}}^{k_i}\frac{C_{111}}{\mu} + N_{V\mathrm{Slow}}^{k_i}\frac{C_{155}}{\mu} + P_{V\mathrm{Slow}}^{k_i}\frac{C_{144}}{\mu}\right)\right]\Delta\sigma_V \\
& + \left[\left(L_{H\mathrm{Fast}}^{k_i} + M_{H\mathrm{Fast}}^{k_i}\frac{C_{111}}{\mu} + N_{H\mathrm{Fast}}^{k_i}\frac{C_{155}}{\mu} + P_{H\mathrm{Fast}}^{k_i}\frac{C_{144}}{\mu}\right)\right. \\
& \left. - \left(L_{H\mathrm{Slow}}^{k_i} + M_{H\mathrm{Slow}}^{k_i}\frac{C_{111}}{\mu} + N_{H\mathrm{Slow}}^{k_i}\frac{C_{155}}{\mu} + P_{H\mathrm{Slow}}^{k_i}\frac{C_{144}}{\mu}\right)\right]\Delta\sigma_H \\
& + \left[\left(L_{h\mathrm{Fast}}^{k_i} + M_{h\mathrm{Fast}}^{k_i}\frac{C_{111}}{\mu} + N_{h\mathrm{Fast}}^{k_i}\frac{C_{155}}{\mu} + P_{h\mathrm{Fast}}^{k_i}\frac{C_{144}}{\mu}\right)\right. \\
& \left. - \left(L_{h\mathrm{Slow}}^{k_i} + M_{h\mathrm{Slow}}^{k_i}\frac{C_{111}}{\mu} + N_{h\mathrm{Slow}}^{k_i}\frac{C_{155}}{\mu} + P_{h\mathrm{Slow}}^{k_i}\frac{C_{144}}{\mu}\right)\right]\Delta\sigma_h\Bigg\}
\end{aligned} \tag{5.45}$$

Here, $\Delta\sigma_V$, $\Delta\sigma_H$, and $\Delta\sigma_h$ are the stress-perturbations relative to a chosen reference stress state.

The L, M, and N that appear in the RHS of Eq. (5.45), the first subscript indicates the principal stress direction. The second subscript connotes whether the stress coefficient pertains to the perturbation of the fast flexural velocity or the slow flexural velocity. The superscript indicates the wavenumber reference for the quantity concerned.

In Eq. (5.45), the value of LHS for any wavenumber k_i is known. The value of the contents of each pair of square braces in the RHS is known. The unknowns in Eq. (5.45) are the quantities C_{111}, C_{155}, C_{144}, $\Delta\sigma_H$, $\Delta\sigma_h$. C_{111}, C_{155}, and C_{144}.

Selecting more than five values of the wavenumber results in more than five equations. Equation (5.45) is one such equation. Such equations can be inverted for the values of the unknowns. The input data for the inversion would be the set of values of $\left(\left[\left(V^{\text{F}}_{\text{Flex}}\right)_{k=k_i} - \left(V^{\text{S}}_{\text{Flex}}\right)_{k=k_i}\right]/V^{\text{R}}_{\text{Flex}}\right)$ at the respective values of the wavenumber chosen. However, low values of the sensitivity coefficients M cause large variance in the inverted C_{111} magnitude when C_{111} is a part of the model to be recovered through the inversion. Hence, usually, C_{111} is first evaluated from the compressional velocity gradient with respect to depth in a chosen log interval. With the value of C_{111} known at this point, the model to be recovered through inversion would comprise C_{155}, C_{144}, $\Delta\sigma_H$, and $\Delta\sigma_h$. Four values of k_i are chosen and the four equations that result as mentioned here are inverted for the unknowns. The inversion is carried out subject to certain constraints (Sinha et al. 2007) built into the inversion. These constraints, parts of the inversion forward model, are built around the stipulations that σ_H and σ_h must conform to the bounds based on the following. (i) Observed tensile failure of borehole wall during mini-frac tests or XLOT; (ii) Leak off pressure during leak-off tests and fracture fracture-closure pressure values during mini-frac tests or XLOT; (iii) Borehole breakout episodes seen from high-resolution image logs. The constraints thereby set bounds on the $(\sigma_H - \sigma_h)$ values.

Since $\Delta\sigma_V$, $\Delta\sigma_H$, and $\Delta\sigma_h$ are the stress perturbations relative to a chosen reference stress state (Sinha et al. 2006),

$$(\sigma_H - \sigma_h) = (\Delta\sigma_H - \Delta\sigma_h) \tag{5.46}$$

The constraint here, therefore, effectively sets bounds on the value of $(\Delta\sigma_H - \Delta\sigma_h)$. The constrained inversion assures that the recovered model leads to results that always conform with the abovementioned data. This renders the methodology quite robust.

Note that, at any interim stage of the inversion, A_E is related to the inverted model of C_{144} and C_{155} through Eqs. (5.37)–(5.39). The values of the far-field shear moduli C_{44}, C_{55}, and C_{66} are known from the processing of the fast-flexural mode, slow-flexural mode, and Stoneley mode waveforms (Kimball 1998; Kimball and Marzetta 1984; Kimball et al. 1996). The values of σ_H and σ_h are related to the values of C_{44}, C_{55}, C_{66}, A_{E}, and σ_V (this value is evaluated outside the inversion and hence is known), through Eqs. (5.40) and (5.41), respectively. Thus, the values of σ_H and σ_h are related to the values of C_{44}, C_{55}, C_{66}, C_{144}, C_{155}, and σ_V.

5.9 Computation of the Principal Stress Magnitudes Under Plane Strain

Under this consideration, strain tensor due to the tectonic stress has no components along the vertical direction.

The symbols ε_H and ε_h denote the principal rock strains in the final stress state of the tectonic (residual) stress-field superimposed over the initial stress state of the rock. These directions are also the principal strain directions in the horizontal plane. No strain is present in the horizontal plane in the initial state. We then have (Plumb 1994, Amadei et al. 1988):

$$S_{\text{Hmax}} = (S_V - \alpha P_p)\left(\frac{\nu_{\text{Static}}}{1-\nu_{\text{Static}}}\right) + \alpha P_p + \frac{E_{\text{Static}}}{\left(1-\nu_{\text{Static}}^2\right)}\varepsilon_H + \frac{\nu_{\text{Static}}E_{\text{Static}}}{\left(1-\nu_{\text{Stat}}^2\right)}\varepsilon_h \tag{5.47}$$

$$S_{\text{hmin}} = (S_V - \alpha P_p)\left(\frac{\nu_{Static}}{1-\nu_{\text{Static}}}\right) + \alpha P_p + \left(\frac{E_{\text{Static}}\nu_{\text{Static}}}{1-\nu_{\text{Static}}^2}\right)\varepsilon_H + \left(\frac{E_{\text{Static}}}{1-\nu_{\text{Static}}^2}\right)\varepsilon_h \tag{5.48}$$

Here E_{Static} stands for Young's modulus measured along the x or y direction under static loading conditions. ν_{Static} denotes the magnitude of the ratio of strain along the $y-$ direction, to the strain along x-direction when an elongational/compressional stress is applied along the x-direction. This is also equal to the magnitude of the ratio of strain along the x-direction to that along the y-direction when an elongational/compressional stress is applied along the y-direction.

Using Eqs. (5.46) and (5.47), the following equations can be stated for the effective stresses:

$$\sigma_H = \sigma_V\left(\frac{\nu_{\text{Static}}}{1-\nu_{\text{Static}}}\right) + \frac{E_{\text{Static}}}{\left(1-\nu_{\text{Static}}^2\right)}\varepsilon_H + \frac{\nu_{\text{Static}}E_{\text{Static}}}{\left(1-\nu_{\text{Stat}}^2\right)}\varepsilon_h \tag{5.49}$$

$$\sigma_h = \sigma_V\left(\frac{\nu_{\text{Static}}}{1-\nu_{\text{Static}}}\right) + \left(\frac{E_{\text{Static}}\nu_{\text{Static}}}{1-\nu_{\text{Static}}^2}\right)\varepsilon_H + \left(\frac{E_{\text{Static}}}{1-\nu_{\text{Static}}^2}\right)\varepsilon_h \tag{5.50}$$

Dynamic elastic moduli are derived from the shear wave and compressional wave slowness data obtained from logs. These moduli are then converted to the static elastic moduli through correlations available in the literature (Lacy 1997, Wang 2000; Zhang and Bentley 2005).

At this point of analysis, the unknowns are σ_h, σ_H, ε_h, andε_H, respectively. For any pair of values of ε_H and ε_h, the values of S_H and S_h are evaluated. Through a proper choice of the pair of values of ε_H and ε_h mentioned earlier, the horizontal stresses are fixed in relation to the calibration points (e.g. Archer and Rasouli 2012).

5.10 Discussions

The methods of stress field characterization are valid for mud-supported and grain-supported sedimentary rocks. The methods can also be applied to log data acquired against non-sedimentary rock intervals. In general, it is important that it is ensured that the input log data do not include data acquired against intervals of excessive breakout or spallation of the borehole. This is so because these are evidences of plastic deformation and yielding of rock. Plastic deformation should not extend beyond one or preferably half the borehole radius beyond the borehole face. It is also desirable that the borehole is not too much oval-shaped.

Best results are obtained when the inversion is a constrained inversion honoring data such as LOT, XLOT, and hydro-fracturing data.

The elastic moduli computed from log data are dynamic moduli. Forward models are best served when adequate backing of laboratory data on the correlation between the dynamic and static elastic moduli are available and built into the forward models.

5.11 Conclusions

Inversion of flexural wave dispersion data to radial profiles of shear wave velocity using the Backus–Gilbert inverse theory is demonstrated. Inversion of radial profiles of shear moduli along the principal horizontal stress directions to principal horizontal stress magnitudes is demonstrated. Inversion of flexural wave data to the magnitude of the third-order elastic coefficients is demonstrated. The applicability of the generalization of the standard "small on large" acoustoelastic theory of solid-fluid composites, to the propagation of acoustic waves through rocks, is demonstrated through its usage in the different inversion processes.

The utility of perturbation models, in conjunction with the acoustoelastic theory of solid-fluid composites, is demonstrated. The utility of approaches based on the concepts of linear poro-elasticity applied under the plane strain assumption in stress field characterization is demonstrated.

Acknowledgments

Soumyajit Mukherjee invited, handled, and reviewed this article. An anonymous reviewer made minor comments and requested to shorten the article significantly. Merryl Le Roux, Frank Otmar Weinreich, Joss Everett, Andrew Harrison, and the proofreading team (Wiley Blackwell) are thanked for their assistance. The article has been summarized in Mukherjee et al. (2023).

References

Alford, R.M. (1986). Shear data in the presence of azimuthal anisotropy: Dilley,TX. *Expanded Abstracts, 56th SEG Annual International Meeting and Exposition*, Houston, TX, USA (2–6 November 1986). Houston, TX: Society of Exploration Geophysicists, pp. 476–479.

Amadei, B., Swolfs, H.S., and Savage, W.S. (1988). Gravity induced stresses in stratified rock masses. *Rock Mechanics and Rock Engineering* 21 (1): 1–20.

Archer, S. and Rasouli, V. (2012). A log-based analysis to estimate mechanical properties and in-situ stresses in a shale gas well in North Perth Basin. *WIT Transactions on Engineering Sciences* 81: 163–174.

Backus, G.E. and Gilbert, F. (1970). Uniqueness in the inversion of inaccurate gross Earth data. *Geological Journal International* 266 (1173): 123–192.

Briggs, V., Huang, X., Rama Rao, V.N., and Burns, D.R. (2002). Simultaneous inversion of fast azimuth and dispersion of flexural waves using cross-dipole data. Appendix B of Semi-Annual Technical Progress Report for the Period 22 May – 21 Nov 2002. (Authors: D. Burns and

M. Nafi Toksoz) Earth Resources Laboratory, Department of Earth, Atmospheric, and Planetary Sciences, Massachusetts Institute of Technology, Cambridge, Massachusetts, MA 02139.

Burridge, R. and Sinha, B.K. (1996). Inversion for formation shear modulus and radial depth of investigation using borehole flexural waves. *SEG Technical Program Expanded Abstracts* 1996: 158–161.

Gholami, R., Rasouli, V., Aadnoy, B., and Mohammadi, R. (2015). Application of in-situ stress estimation methods in well bore stability under isotropic and anisotropic conditions. *Journal of Geophysics and Engineering* 12: 657–673.

Jaeger, J.C. and Cook, N.G.W. (1969). *Fundamentals of Rock Mechanics*. New York, N.Y. and Methuen, London: Barnes and Noble.

Kimball, C.V. (1998). Shear slowness measurement by dispersive processing of the borehole flexural mode. *Geophysics* 63 (2): 337–344.

Kimball, C.V. and Marzetta, T.L. (1984). Semblance processing of borehole acoustic array data. *Geophysics* 49: 274–281.

Kimball, C.V., Scheibner, D.J. and Saiki, Y. (1996). Dispersive wave processing of the borehole flexural mode. SEG 66th Annual Meeting, Denver, CO (November 1996). Extended Abstract.

Lacy L.L. (1997). Dynamic rock mechanics testing for optimized fracture designs. SPE 38716, *The 1997 SPE Annual Technical Conference and Exhibition*, San Antonio, TX (October 1997). Houston, TX: Society of Petroleum Engineers.

Mukherjee, S. (2017). Airy's isostatic model: a proposal for a realistic case. *Arabian Journal of Geosciences* 10: 268.

Mukherjee, S., Dasgupta, S., Majumdar, C. et al. (2023). Introduction to: Handbook of Petroleum Geoscience: Exploration, Characterization, and Exploitation of Hydrocarbon Reservoirs. In: *Handbook of Petroleum Geoscience: Exploration, Characterization, and Exploitation of Hydrocarbon Reservoirs*, Petroleum Industry Updates: Geoscientific and Economic Issues (ed. S. Mukherjee, S. Dasgupta, C. Majumdar, et al.). John Wiley & Sons, Hoboken, ISBN 978-1-119-68003-1.

Norris, A.N., Sinha, B.K., and Kostek, S. (1994). Acoustoelasticity of solid/fluid composite systems. *Geophysical Journal International* 118: 439–446.

Pistre, V. Schlumberger, G., R. Yan, et al. (2009). Determining stress regime and Q factor from sonic data. SPWLA 50th Annual Symposium, The Woodlands, TX (21–24 June 2009).

Plumb, R.A. (1994). Influence of composition and texture on the failure properties of clastic rocks. *Eurocks 94, Rock Mechanics in Petroleum Engineering Conference*, Delft, The Netherlands (August 1994). Paper Number: SPE-28022-MS. Houston, TX: Society of Petroleum Engineers, pp. 13–20.

Shannon, H.B., Sitchler, J., and Bratton, T. (2016). Geomechanics for unconventional reservoirs, Chapter 7. In: *Unconventional Oil and Gas Resources Hand Book*, 1ee (ed. Y.Z. Ma and S. Holditch), 199–213. Amsterdam, The Netherlands: Elsevier.

Sinha, B.K. (1982). Elastic waves in crystals under a bias. *Ferroelectrics* 41: 61–73.

Sinha, B.K. (1997a). Inversion of borehole dispersions for formation stresses. In: *1997 IEEE Ultrasonics Symposium*, Toronto, Ontario, Canada (5–8 October 1997), IEEE Catalogue No. 97CH36118. Ny, USA: IEEE, pp 781–786.

Sinha, B.K. (1997b). Sensitivity and inversion of borehole flexural dispersions for formation parameters. *Geophysical Journal International* 128: 84–96.

Sinha, B.K. (2006). Determination of stress characteristics of earth formations. US Patent 7042802.

Sinha, B.K. (2011). Estimating formation stresses using the three shear moduli. WO Patent 2011/01/09 89.

Sinha, B.K. and Burridge, R. (2001). Radial profiling of formation shear velocity from borehole flexural dispersions. *2001 IEEE Ultrasonics Symposium*, Atlanta, Georgia, USA (7–10 October 2001).

Sinha, B.K., Bratton, T., Cryer, J. et al. (2005). Near-wellbore alteration and formation stress parameters using borehole sonic data. *SPE Annual Technical Conference*, Dallas, TX, (9–12 October). SPE 95841-PA.

Sinha, B.K., Vissapragada, B., Renlie, L. and Skomedal, E. (2006). Horizontal stress magnitude estimation using the three shear moduli – a Norwegian case study. SPE Annual Technical Conference and Exhibition, San Antonio, TX (24–27 September 2006). SPE 103079.

Sinha, B.K., Vissapragada, B., and Kongslien, M. et al. (2007). Estimation of formation stresses using radial variation of three shear moduli – a case study from a high pressure and high-temperature reservoir in Norwegian continental shelf. *SPE Annual Technical Conference and Exhibition*, Anaheim, California, USA, (11–14 November 2007), Paper SPE 109842.

Sinha, B.K., Wang, J., Kisra, S. et al. (2008). Estimation of formation stresses using borehole sonic data. SPWLA 49th Annual Logging Symposium.

Tang, X.M. and Patterson, D. (2010). Mapping radial shear-velocity variation by a constrained inversion of borehole flexural-wave dispersion data. *Geophysics* 75 (6).

Wang, H.F. (2000). *Theory of Linear Poroelasticity*. Princeton, NJ: Princeton University Press.

Zhang, J.J. and Bentley, L.R. (2005). Factors determining poisson ratio. *CREWES Research Report* 17: 1–5.

6

3D Inversion of Ultra-Deep Azimuthal Electromagnetic Logging-While-Drilling Data

Nigel Clegg[1,*] *and Karol Riofrio Rodriguez*[2]

[1] *Global Subject Matter Expert – Geosteering, Halliburton, Norway*
[2] *Geosteering Focal point, Halliburton, Norway*

Author for correspondence: Nigel.Clegg@Halliburton.com

Abstract

The increase of the depth of investigation provided by ultra-deep electromagnetic (EM) logging-while-drilling (LWD) tools to the 100+ ft radius range has significantly impacted well placement. Early identification of variations in formation position or fluid boundaries, resulting from this very large depth of investigation, allows for timely geosteering decisions, helps to decrease wellbore tortuosity, and allows for mapping of features far from the wellbore, enabling identification and mapping of targets for later wells.

Traditionally, data from ultra-deep resistivity tools have largely been presented as one-dimensional (1D) inversions of resistivity field data stitched together to form pseudo-two-dimensional (2D) models (curtain plots) plotted along the wellbore. These 1D inversions depict a resistivity profile representing the formation or fluid above and below the wellbore only. Because of the large depth of detection, these plots have provided more reservoir insight than possible with standard deep azimuthal resistivity tools, which have an approximately 20-ft depth of investigation. However, 1D inversions, by definition, do not identify or account for any features to the sides of the wellbore.

Subsurface geology is three-dimensional (3D), and variations in formation dip, lithology, and boundary position caused by structural dislocations or erosion can occur in all directions around the wellbore as well as variations in fluid saturations. These variations have implications for hydrocarbon distribution and therefore potential production. Additionally, 1D inversions assume that there are no lateral changes in resistivity in the volume being imaged. If the lateral resistivity is not constant, it can have a negative impact on the 1D inversion result. Any actual 2D or 3D variations in the resistivity field can manifest as ambiguous artifacts on the pseudo-2D inversion canvas, which can obscure the geology and limit reservoir understanding. Using synthetic and field examples, this work demonstrates the potential for 3D inversion of ultra-deep LWD data to help improve both reservoir understanding and well placement.

A true 3D inversion of EM LWD data allows operators to evaluate variations in resistivity in all directions around the wellbore. This provides the potential to incorporate changes in well

Handbook of Petroleum Geoscience: Exploration, Characterization, and Exploitation of Hydrocarbon Reservoirs, First Edition. Edited by Soumyajit Mukherjee, Swagato Dasgupta, Chandan Majumdar, Subhandip Mandal and Troyee Dasgupta.
© 2023 John Wiley & Sons Ltd. Published 2023 by John Wiley & Sons Ltd.

azimuth into well placement decisions and to assess the impact that lateral changes in resistivity have on well production.

Keywords *Geophysical techniques; well logging; petrophysics; sub-surface geology; 3D inversion*

6.1 Introduction

Deep azimuthal EM tools have a depth of investigation of ~20 ft (Bittar et al. 2007), while the depth of investigation of ultra-deep EM LWD tools is approximately 100+ ft (Constable et al. 2012; Tilsley-Baker et al. 2013; Wu et al. 2018). The increase in the depth of investigation is achieved using very long transmitter-receiver spacings (Figure 6.1) and lower transmission frequencies (Wu et al. 2018).

Measurements from these tools are generally interpreted using 1D inversions to produce models that represent the position of multiple formation and fluid boundaries (Constable et al. 2012; Tilsley-Baker et al. 2013; Wu et al. 2018). 1D inversion of ultra-deep azimuthal EM LWD data has proven to be a useful tool to aid optimal well placement. Additionally, the large depth of investigation allows geo-mapping of formation and fluid boundaries more than 100 ft from the wellbore. However, these inversions only represent a cross section along the well path and only identify variations in resistivity above and below the well path (Figure 6.2).

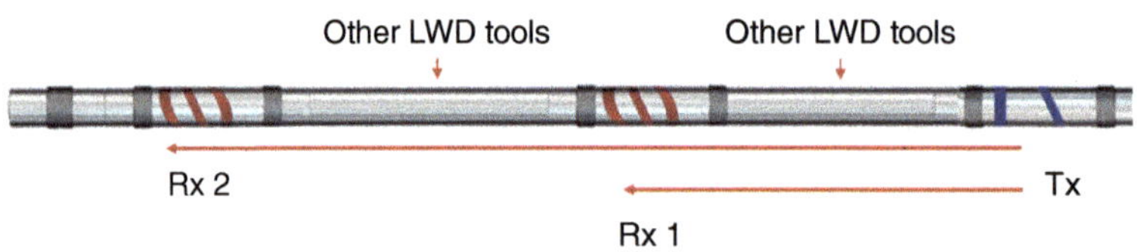

Figure 6.1 Ultra-deep resistivity tool configuration : one coaxial and one tilted antenna on the transmission sub and three tilted antennas on each receiver sub. Generally, Rx 1 is placed 25–50 ft from the transmitter and Rx 2 approximately 100 ft from the transmitter.

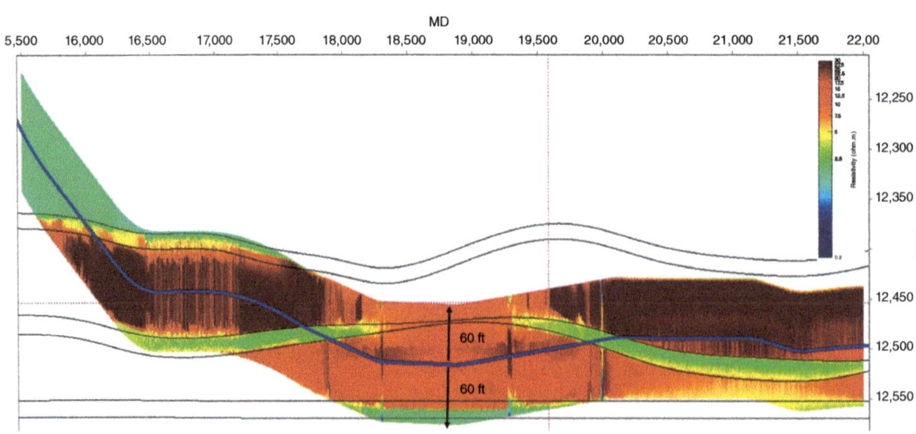

Figure 6.2 Synthetic 1D inversion; depth of inversion: 60 ft; only the resistivity profile above and below the wellbore is displayed, plotted along the well path.

Figure 6.3 Sandstone outcrop, Colorado, US. Courtesy of Halliburton.

Geology is inherently 3D (Figure 6.3). Significant lithological changes related to the depositional environment and the position of formations because of post-depositional structural changes could result in complex subsurface geology.

Where there has been production or fluid injection, the resultant resistivity distribution becomes even more complex because of the migration of low-resistivity water across the reservoir. This fluid migration is not only controlled by the proximity to injection wells but also by formation permeability and impermeable or semi-impermeable barriers, which can result in an extremely complex fluid distribution.

For reservoirs where lateral changes are minimal, 1D inversion of ultra-deep EM LWD data provides a valuable aid to well placement. However, where lateral variations in formations or fluid occur, 1D inversions might not only fail to show them but might be adversely affected by them, potentially resulting in inversion artifacts that can distort the 1D inversion canvas. Any actual 2D or 3D variations in the resistivity field can manifest as ambiguous artifacts on the pseudo-2D inversion canvas, which can obscure the geology, limiting reservoir understanding.

It is evident for 1D inversion canvasses (Figure 6.2) how a well inclination can be adjusted to achieve optimal well placement. However, there is no indication of the position of the formations on either side of the well path. For a depositional environment such as channel sand, there can be considerable lateral lithological variation, and a change in well azimuth could be more advantageous than simply changing the inclination of the well.

Ultra-deep azimuthal resistivity images using the coaxial antenna allow assessment of lateral changes in resistivity (Clegg et al. 2018). These images show the distribution of resistivity 360° around the wellbore. The depth of investigation of these images is determined by the transmitter/receiver spacing, frequency of transmission, and formation resistivity. Using modeling, it is possible to estimate the distance to boundaries where the geology is relatively straightforward; but, for more complex geology, although the distribution of resistivity changes is represented by these images, determining the actual structure becomes difficult. A 3D inversion of ultra-deep azimuthal EM LWD data is an optimal solution, allowing the visualization of resistivity changes in three dimensions. The azimuthal images from the coaxial antenna remain useful because, along with shallow LWD measurements, they allow verification of the 3D inversion results.

6.2 3D Inversion Process

A 3D inversion involves more variables than a 1D inversion, but results are required in a similar timeframe (a matter of minutes) so that well placement decisions can be made quickly.

To achieve a pseudo real-time update of the inversion requires the minimization of the computational complexity of the problem without compromising the accuracy of the computed solution. This is achieved through two different approaches: adaptive octree meshing and decoupling the forward and inverse problems (Clegg et al. 2019).

During the 3D inversion process, the reservoir is discretized into cells, within which the electrical conductivity is assumed constant, using an octree mesh (Haber et al. 2007). To represent the sensitivity of the EM LWD tool, the mesh uses extremely small cells in the volume directly around the well and larger cells farther away (Figure 6.4). Increasing the size of the cells farther away from the wellbore decreases the computational burden and reflects the decreasing sensitivity of the tool while the distance from the wellbore increases.

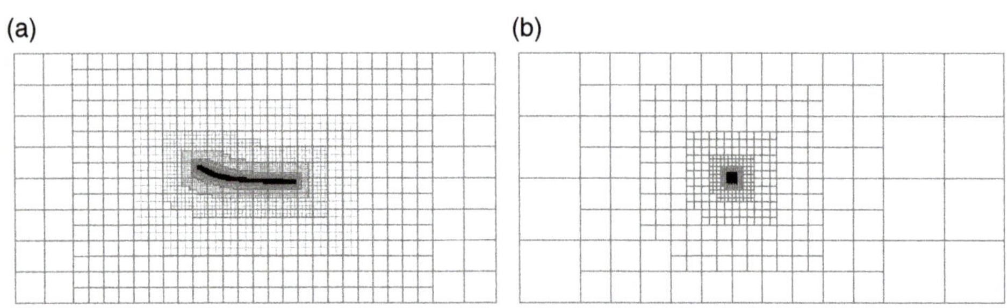

Figure 6.4 Cross sections of two octree meshes: (a) global mesh that discretizes the entire well path and is used to store the inverse model; (b) local mesh used to simulate the response and sensitivities of a single transmitter location at a single frequency.

Mesh decoupling involves addressing the forward and inverse problems on different meshes. The forward problems (and sensitivity calculations) at every frequency and location are discretized and addressed on their own specifically designed mesh (Figure 6.4); this breaks costly computations down into a series of smaller problems. Each mesh can then be addressed using direct methods in parallel; the overall time to process the data can therefore be decreased using multiple processors. With the availability of cloud resources, the number of processors used for an inversion can easily be scaled up or down, depending on the number of local meshes derived from the number of measurement points.

Once sensitivity calculations are performed on the local meshes, the results are interpolated back to the full reservoir global mesh on which the inversion is performed. This mesh-to-mesh interpolation is straightforward using octree meshes. The interpolated sensitivities are then used to calculate model updates on the global mesh.

The conductivity model is calculated through the minimization of an objective function that comprises a measure of the data misfit and a regularization function that penalizes rapid variations in the resulting model. The optimization problem is addressed using the Gauss–Newton method (Wilson et al. 2019). A single Gauss–Newton step requires a few seconds, and the 3D inversion is complete in a matter of minutes. The relative importance of fitting the component data and producing a model that is spatially smooth is balanced by the use of a trade-off parameter initialized with a value that favors a smooth solution over data fit, which is slowly changed until a model is produced that fits the data to the desired level.

6.3 Verification of the 3D Inversion

An inversion generates a geological model that produces a resistivity field matching the field measured by the ultra-deep EM tool; therefore, it is important to verify the result with independent data sources to confirm that it is a reasonable representation of the geology. This can be achieved using multiple methods (Clegg et al. 2019):

- Direct comparison with the azimuthal resistivity images from the coaxial antenna on the ultra-deep resistivity tool. These measurements are not used in the 3D inversion and therefore provide independent verification to a similar distance from the wellbore. If the 3D inversion exhibits a high resistivity zone to the right side of the wellbore, it should also be evident on the image logs. Because the depth of investigation of each image log is determined by the spacing between the transmitter and receivers and the frequency of transmission, a resistivity boundary farther away from the well path will not be evident on the shorter spacing/higher frequency azimuthal EM images; therefore, it is possible to verify the approximate distance of resistivity boundaries from the wellbore in the 3D inversion results (Clegg et al. 2018).
- Shallow LWD measurements (i.e. gamma ray, density, neutron porosity, and shallow EM logs) should match the 3D inversion close to the wellbore. The point that a well drills across a boundary should match on the 3D inversion and the shallow LWD data.

- Where significant boundaries are crossed and it is possible to pick dips from density, gamma, and resistivity images, the dip and dip direction of these boundaries can be compared to the dip of the structure as defined in the 3D results.
- The large depth of investigation of these tools bridges the gap between near-wellbore measurements and the seismic scale. Larger features, such as the major truncation of a reservoir by a fault, can be defined on both the 3D EM inversion and seismic images; therefore, seismic images can be used to verify the results.

6.4 3D Inversion Synthetic Example

3D inversion is not necessary where there is no lateral variation in resistivity. However, many geological environments exhibit considerable lateral variations in lithology and the distribution of related resistivity boundaries. Channel sands are a classic example of a depositional environment in which the formation can change laterally (Figure 6.5). This represents a good test case to demonstrate how 3D inversion of ultra-deep EM data can be a great benefit to well placement operations and increasing reservoir understanding.

Where a well approaches and then exits a channel sand through the side of the channel (Figure 6.5), identification of the boundary and changes in its position to the side of the wellbore would allow the potential to geosteer laterally to remain within the reservoir. Where the maximum dogleg permitted does not allow the well to remain within the sand because of severe changes in the orientation of the channel, mapping the lateral change in the channel increases reservoir understanding and allows for optimal planning of any potential sidetracks.

A 1D inversion of the EM data could potentially indicate thinning of the channel as the well approaches the boundary. However, it would not be possible to determine if the channel continued to the left or right of the wellbore or if the sand had just pinched out (Figure 6.6).

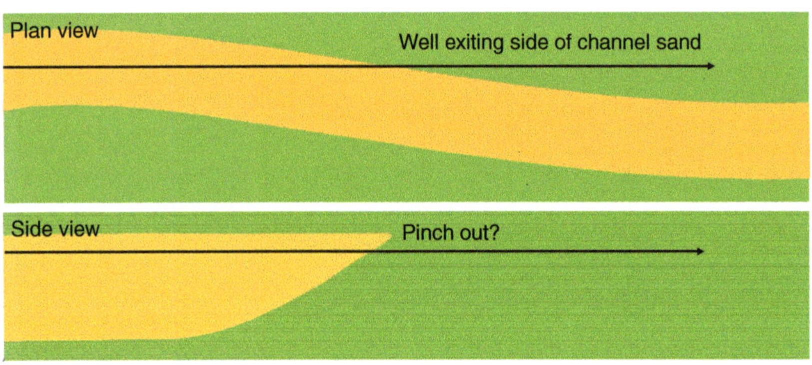

Figure 6.5 Simplified illustration of a channel sand deposit; plan view shows the well exiting the side of the channel, and the side view only shows that the well has exited the reservoir; this could be interpreted as a pinch out.

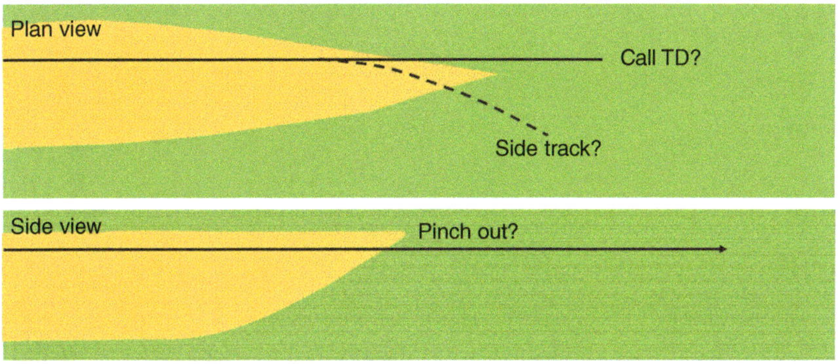

Figure 6.6 Simplified illustration of a pinch out; side view only shows that the well has exited the reservoir, and the plan view shows that the reservoir does not continue to the side of the well path.

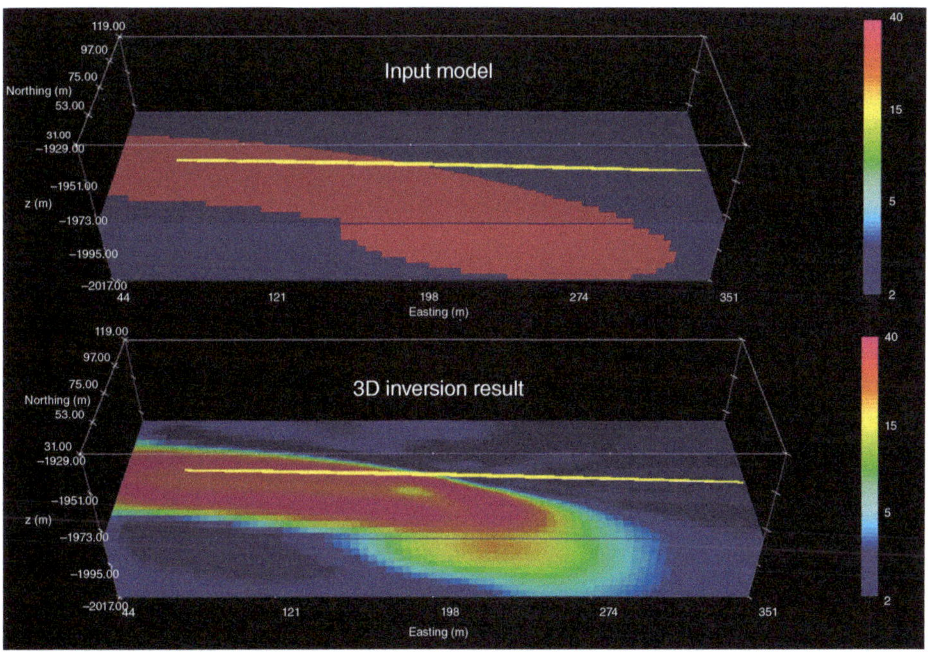

Figure 6.7 (Top) Input model for the synthetic inversion; (bottom) 3D inversion results for the synthetic data. Both volumes have been cut along the horizontal plane of the well, with the top half removed to aid visualization.

Where the reservoir pinches out completely, as in Figure 6.6, costly sidetracks looking for the continuation of the reservoir can be avoided if the lateral extent of the reservoir is mapped along with its vertical extent.

Using simple models, it is possible to recreate various scenarios to understand how this new information can be used. Figure 6.7 illustrates a scenario wherein the well exits the side of a channel and shows how the channel boundary can be tracked before and after the exit.

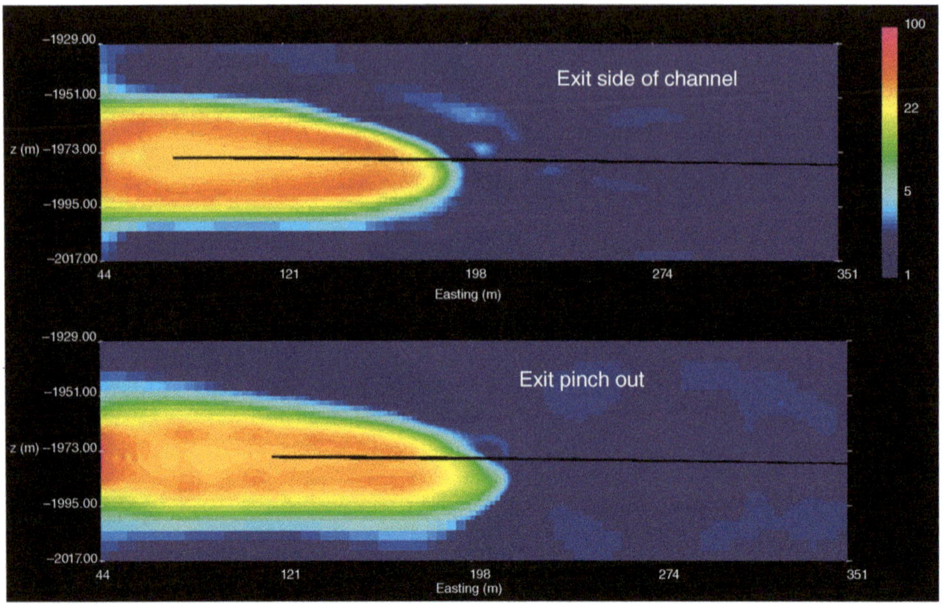

Figure 6.8 (Top) Slice along the wellbore for the channel exit scenario; (bottom) slice along the wellbore for a pinch-out scenario. Both images are the conventional 2D curtain slice, with MD on the *x*-axis and TVD on the *y*-axis.

From this model, it is possible to synthesize the expected EM field at measurement points along the well path. The 3D inversion result evidently shows that, in this situation, channel boundaries can be tracked both while the well is within the reservoir and after the exit point. Uncertainty increases as the well path moves farther from the channel because of the decreased sensitivity in the lower-resistivity medium. However, it is still possible to map the channel and optimize the well path to track it after an exit. Obviously, a proactive geosteering decision could be made earlier in measured depth, while the well approaches the side of the channel, to optimize well placement within the reservoir.

1D inversion results for a similar scenario would not exhibit adequate information to identify if the well had exited a channel or if it had pinched out (Figure 6.8).

A synthetic 3D inversion for a pinch-out scenario shows that the lateral resistivity distribution is obviously different from the channel exit scenario (Figure 6.9). For these scenarios, it would be difficult to achieve optimal well placement using 1D inversions alone.

6.5 3D Inversion Case Study

The case study is a trilateral well from the Norwegian continental shelf. The target reservoir is a good test case for 3D inversion because it consists of a heterolithic turbidite deposit, which has the significant potential for lateral changes in resistivity resulting from the depositional environment.

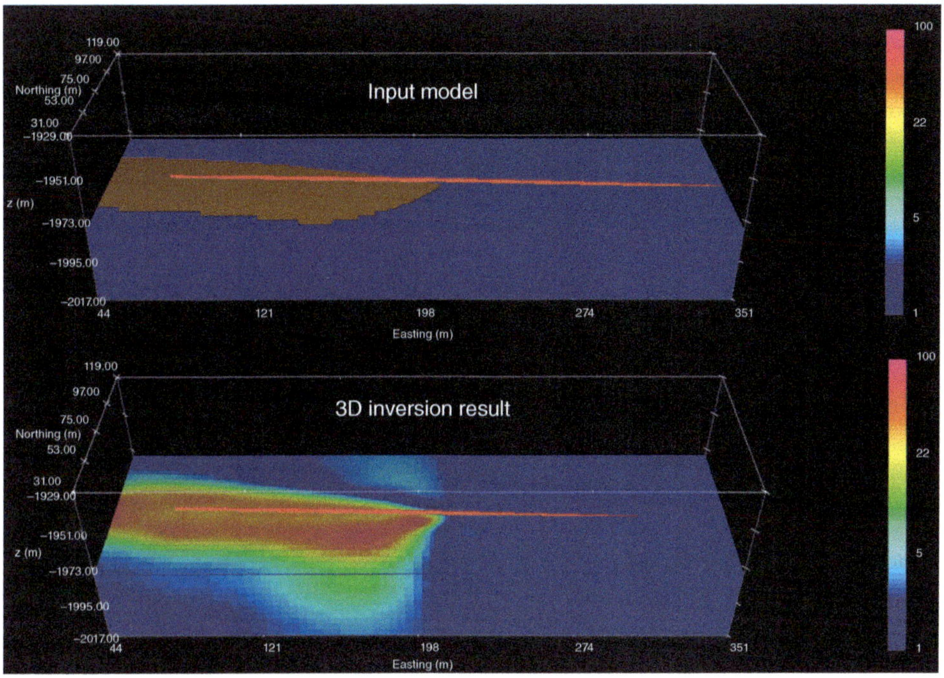

Figure 6.9 (Top) Input model for a pinch-out scenario; (bottom) 3D inversion result for the pinch-out model. Both volumes have been cut along the horizontal plane of the well, with the top half removed to aid visualization.

A full LWD suite, including gamma ray, density, neutron porosity, and both deep and ultra-deep resistivity tools, was deployed to assist in accurate well placement and reservoir mapping. A 1D inversion was run in real time, and the data was transmitted to surface using wired pipe, which provided excellent real-time data density (Figure 6.10). The expected lateral variation was to be investigated using azimuthal density and resistivity tools, allowing early assessment of lateral variation in formation positions and resistivity boundaries and therefore the suitability of this project as an early test case for 3D inversion of the ultra-deep EM LWD data (Figure 6.10).

Azimuthal images illustrate resistivity variation around the wellbore. Where the images are symmetrical around the center of the image, there is no lateral variation. Where the images are skewed, there is a lateral variation in resistivity, which is not directly evident on the 1D inversion image along the wellbore. The 1D inversion can be influenced by these lateral variations, creating artifacts often observed as flaring of the 1D inversion (Figure 6.10). These azimuthal images can be used to identify lateral variation in the resistivity and verify the 3D inversion results. Because the images are not an input in the inversion, they are an independent verification of the results.

3D inversion of the ultra-deep EM LWD data was run for all sections of the trilateral well to a radius of 25 m around the wellbore (Figure 6.11); 25 m represents the limit of the tool sensitivity based on the transmitter-to-receiver spacings of 7 m (Rx 1) and 14 m (Rx 2) used for drilling these wells.

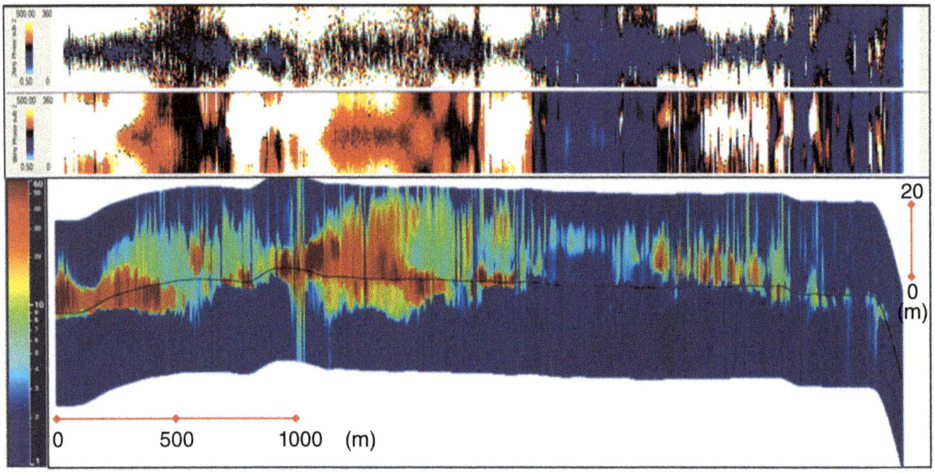

Figure 6.10 1D inversion and azimuthal resistivity images from the ultra-deep resistivity tool; (top) 2-kHz, phase-shift, azimuthal resistivity image from receiver two referenced to high side; (middle) 8-kHz, phase-shift azimuthal resistivity image from receiver two referenced to high side; (bottom) 1D inversion with a 23-m depth of inversion, 16- and 32-kHz data from receiver 1, 7.3 m behind the transmitter, 8- and 16-kHz data from receiver 2, 13.7 m behind the transmitter. *Source:* Picture from Wilson et al. (2019).

This discussion focuses on the main bore, which exhibits more lateral resistivity variation than observed in the two sidetrack sections (Figures 6.10 and 6.11). To interrogate the inversion results, data can be displayed in multiple formats. The simplest is to plot a slice along the wellbore (similar to a 1D inversion), a plan view, that shows lateral variation, and a view down the wellbore that combines the vertical and lateral distributions (Figure 6.12).

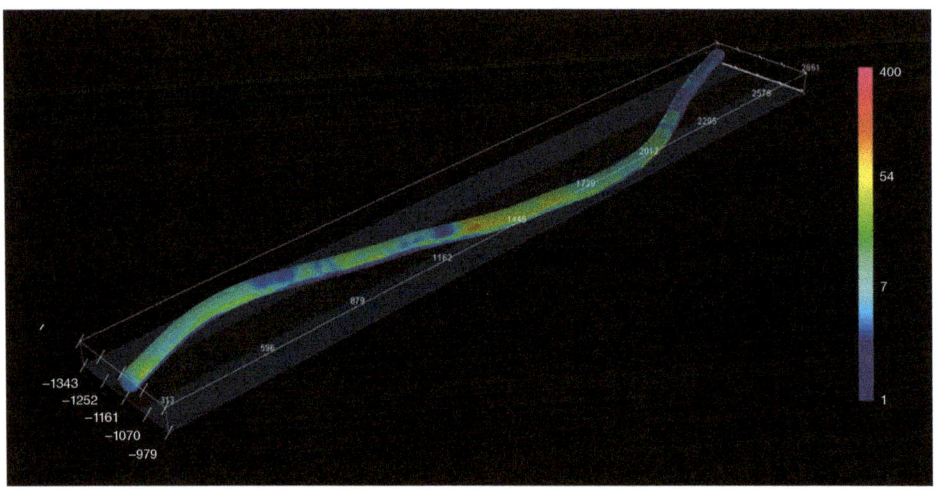

Figure 6.11 3D inversion results plotted for the first part of the main bore. The data has been cut off at a 25-m radius from the wellbore, representative of the tool sensitivity in this reservoir.

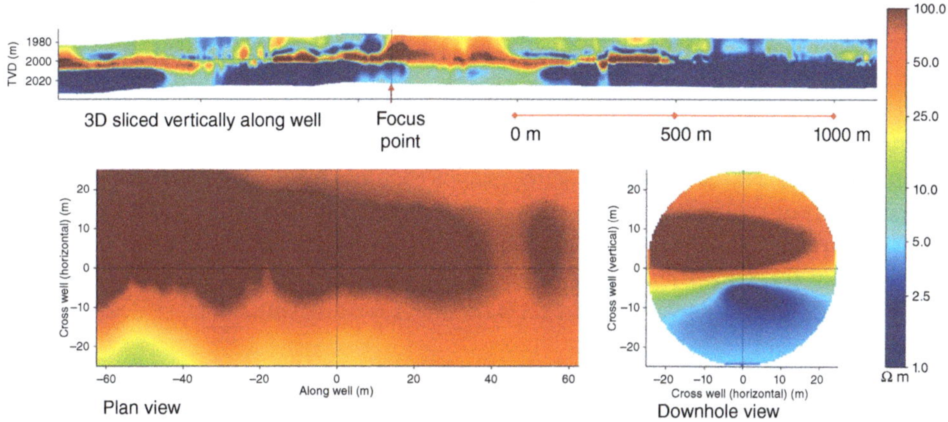

Figure 6.12 3D inversion results displayed at the focus point, at the beginning of the main high-resistivity geobody. Data has been cut off at a 25-m radius from the wellbore, representative of the tool sensitivity in this reservoir.

The 3D section (sliced vertically along the well) in Figure 6.12 shows the beginning of the main high-resistivity section. It is evident from the plan and downhole views that the high resistivity is predominantly distributed above and to the left of the wellbore. This matches the information from the ultra-deep azimuthal resistivity images (Figure 6.10).

Although the main high-resistivity zones are within the first 1500 m of the well, it is the more sporadic sands toward the toe of the well that exhibit the most interesting structures in the 3D results (Figure 6.13). For the 1D inversion (Figure 6.10), these appear to be sporadic, isolated small sand units. The associated ultra-deep azimuthal resistivity images (Figure 6.10) show that the sands exhibit considerable lateral variation, evident because

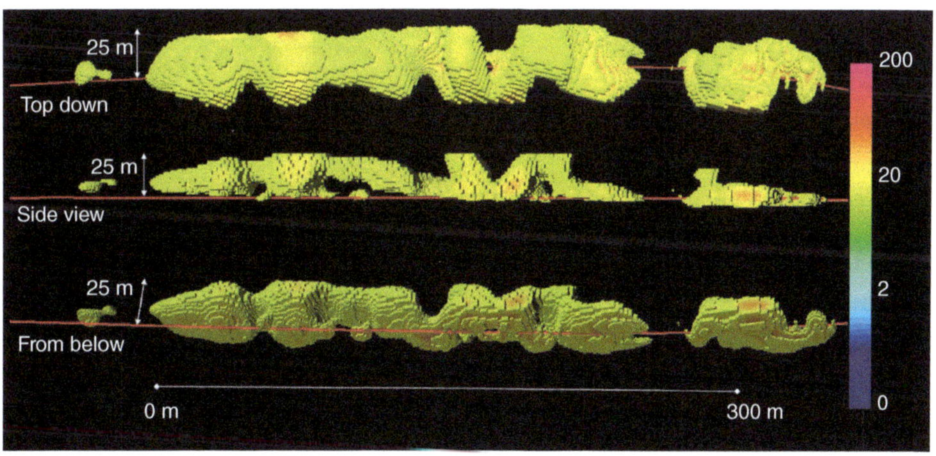

Figure 6.13 3D inversion results displayed toward the toe of the well. Points representing all data below a resistivity of 10 Ω m have been removed to expose the underlying 3D structure.

of the off-centered nature of the images for this section of the well. The lateral variation in resistivity distribution can result in a degradation of the 1D inversion because the 1D inversion algorithms assume that changes in the resistivity field only occur above and below the wellbore. This might be responsible for the apparent "flaring" of the 1D inversion results observed in many places along the well (Figure 6.10).

The structure represented by the 3D inversion is complex, exhibiting channel-type structures unevenly distributed across the well path. With such complex results, it is important to verify the structures are actually related to the geology and are not the result of inversion artifacts. Directly comparing the 3D inversion results to the azimuthal resistivity images where there are significant lateral changes shows the same distribution of resistivity around the wellbore (Figure 6.14).

Where the 3D inversion results exhibit high-resistivity zones distributed to the left or right of the wellbore, the same can be observed in the azimuthal image.

Dip picks from the density images exhibit a series of high-angle and low-angle planes representing the complex nature of the turbidite (Figure 6.15).

In general, dip results match the structures defined by both the 3D and 1D inversions, although because these density dip picks only exhibit changes extremely close to the wellbore, not all the major structures are defined.

The shallow and ultra-deep LWD measurements verify the 3D results, demonstrating the same distribution of formations around the borehole. This complicated turbidite depositional environment is a good test of the 3D inversion process.

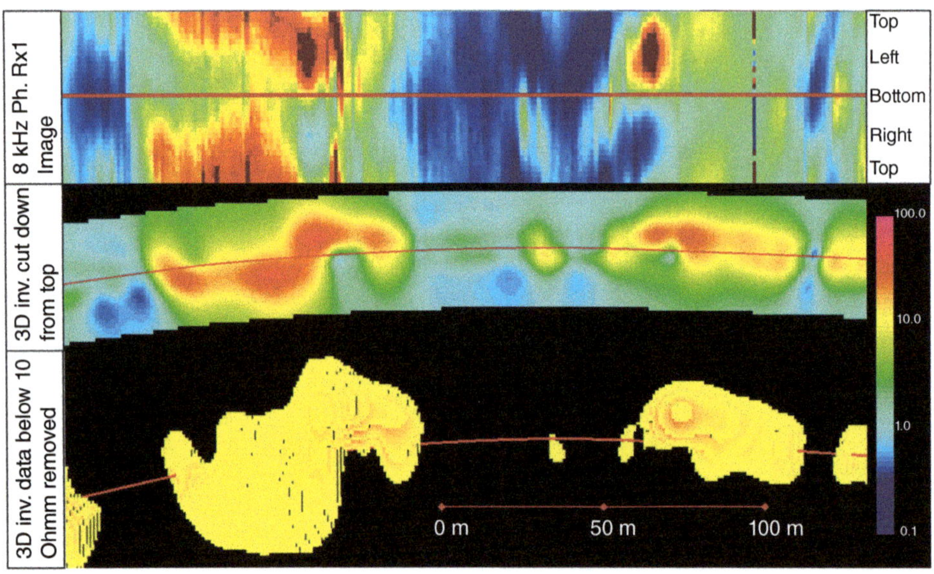

Figure 6.14 (Top) Azimuthal resistivity results displayed against a (middle) horizontal slice through the 3D inversion, and (bottom) a top-down view of the 3D results, where all resistivity below 10 Ω m has been removed.

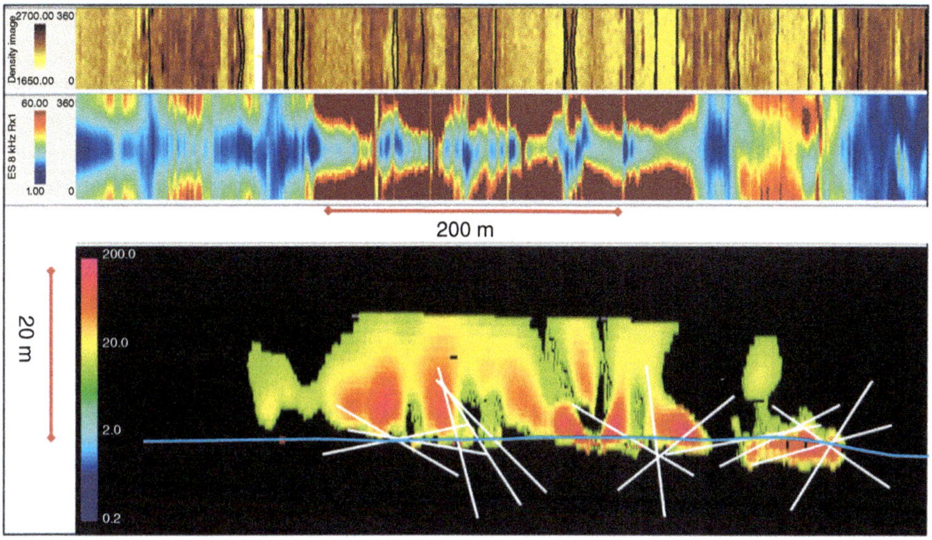

Figure 6.15 (Top) Azimuthal density image log; (middle) 8-kHz receiver-one azimuthal resistivity; (bottom) vertical slice along the well path through the 3D inversion results with superimposed density dip picks.

6.6 Conclusions

Where the geology is simple and there are no lateral variations in the position of formations, 1D inversions provide an excellent tool to assist in optimal well placement. However, the 3D nature of many reservoirs, resulting from the depositional environment and post-deposition structural deformation, requires a 3D inversion to provide a good representation of the reservoir. Ignoring lateral changes in the resistivity limits well placement strategies and has potential implications for production. For mature fields with a history of water injection, fluid boundaries can also have a significant impact; these often have a varied lateral distribution resulting from injection strategy and the influence of permeability, lithology, and structural controls.

The 3D inversion provides a model that represents the subsurface geology but is not a direct measurement. The quality of the inversion result can be directly compared to more direct measurements to demonstrate how close the model matches reality. Such comparisons are important for the verification of this new technology. Azimuthal resistivity images from the ultra-deep tool are not incorporated into the inversion and, because they have a similar depth of investigation and measure the same property, are good tools to investigate the quality of the 3D results. Verification can be expanded with the examination of shallow LWD measurements and, if available in sufficient detail, seismic data.

Synthetic examples can be used to demonstrate the expected responses of the 3D inversion to different geological scenarios, but an inversion of field data is the true test of the process. For the field example presented here, verification demonstrates that the 3D inversion results provide an excellent representation of the reservoir under investigation where the structure can be visualized as a true volume.

Acknowledgments

The authors thank Aker BP for supporting the ultra-deep resistivity LWD field trials and acknowledge partners ConocoPhillips and Lundin Norway for permission to publish. The authors also thank Halliburton and Computational Geosciences for permission to publish. Reviewed by Chandan Majumdar (Schlumberger) and Soumyait Mukherjee (IIT Bombay). Merryl Le Roux, Frank Otmar Weinreich, Joss Everett, Andrew Harrison, and the proofreading team (Wiley Blackwell) are thanked for their assistance. This chapter has been summarized in Mukherjee et al. (2023).

References

Bittar, M., Klein, J., Beste, R. et al. (2007). A new azimuthal deep-reading resistivity tool for geosteering and advanced formation evaluation. Presented at the SPE Annual Technical Conference and Exhibition, Anaheim, CA (11–14 November 2007). SPE-109971-MS. https://doi.org/10.2118/109971-PA.

Clegg, N., Djefel, B., Kunnas, J. et al. (2018). Identification and projected interception of oblique faults up to 85 ft away using ultra-deep azimuthal resistivity and geosignals: a case history from the Norwegian continental shelf. Presented at the Abu Dhabi International Petroleum Exhibition and Conference, Abu Dhabi, UAE (12–15 November 2018). SPE-192854-MS. https://doi.org/10.2118/192854-MS.

Clegg, N., Parker, T., Djefel, B. et al. (2019). The final piece of the puzzle: 3-D inversion of ultra-deep azimuthal resistivity LWD data. Presented at the SPWLA 60th Annual Logging Symposium, The Woodlands, TX (15–19 June 2019). SPWLA-2019-HHH.

Constable, M.V., Antonsen, F., Olsen, P.A. et al. (2012). Improving well placement and reservoir characterization with deep directional resistivity measurements. Presented at the SPE Annual Technical Conference and Exhibition, San Antonio, TX (8–10 October 2012). SPE-159621-MS. https://doi.org/10.2118/159621-MS.

Haber, E., Heldmann, S., and Modersitzki, J. (2007). An octree method for parametric image registration. *SIAM Journal on Scientific Computing* **29** (5): 2008–2023. https://doi.org/10.1137/060662605.

Mukherjee, S., Dasgupta, S., Majumdar, C. et al. (2023). Introduction to: Handbook of Petroleum Geoscience: Exploration, Characterization, and Exploitation of Hydrocarbon Reservoirs. In: *Handbook of Petroleum Geoscience: Exploration, Characterization, and Exploitation of Hydrocarbon Reservoirs*, Petroleum Industry Updates: Geoscientific and Economic Issues (ed. S. Mukherjee, S. Dasgupta, C. Majumdar, et al.). John Wiley & Sons, Hoboken, ISBN 978-1-119-68003-1.

Tilsley-Baker, R.M., Antonov, Y., Martakov, S. et al. (2013). Extra deep resistivity experience in Brazil geosteering operations. Presented at the SPE Annual Technical Conference and Exhibition, New Orleans, LA (30 September–2 October 2013). SPE-166309-MS. https://doi.org/10.2118/166309-MS.

Wilson, G., Marchant, D., Haber, E. et al. (2019). Real-time 3D inversion of ultra-deep resistivity logging-while-drilling data. *Society of Petroleum Engineers* https://doi.org/10.2118/196141-MS.

Wu, H.-H., Golla, C., Parker, T. et al. (2018). A new ultra-deep azimuthal electromagnetic LWD sensor for reservoir insight. Presented at the SPWLA 59th Annual Logging Symposium, London, UK (2–6 June 2018). SPWLA-2018-X.

7

Solving the Puzzle: Seven Effective Habits of Geosteering Team Members

Bronwyn Djefel

Epiroc, Kewdale, Western Australia, Australia
Author for correspondence: bronwyn.djefel@epiroc.com

Abstract

Do you like puzzles, such as four-dimensional ones influenced and affected by a variety of different systems and events over time? If so, you just might be a geologist. According to a quote from Louis Agassiz, "The world is the geologist's great puzzle box; he stands before it like the child to whom the separate pieces of his puzzle remain a mystery till he detects their relation and sees where they fit, and then his fragments grow at once into a connected picture beneath his hand."

Inherently, geology is a puzzle – just perform an internet search of "Death Valley sailing stones" to see how geology stumped people for many years until it was finally explained. The geosteering puzzle has another complication because it is not possible to physically observe the reservoirs being targeted many thousands of meters below the surface. Hence, geosteering team members should know how to piece together different datasets at different resolutions, apply their knowledge of depositional environments, choose the most likely from multiple interpretations, and determine whether the current well plan is optimal or if a change should be made. Decisions and their communication should occur within minutes during drilling; with rig rates reaching millions of dollars per day, the cost of a few minutes of nonproductive time can be enormous.

Keywords *Subsurface geology; drilling, 3D visualization; geological technique*

7.1 The Seven Effective Habits of Geosteering Team Members

What are the most effective habits geosteering specialists should possess to help maximize net-to-gross pay within a complex reservoir? There is much debate, but seven commonly identified habits include the following:

- Be tech-savvy
- Visualize in three dimensions (3D)

Handbook of Petroleum Geoscience: Exploration, Characterization, and Exploitation of Hydrocarbon Reservoirs, First Edition. Edited by Soumyajit Mukherjee, Swagato Dasgupta, Chandan Majumdar, Subhandip Mandal and Troyee Dasgupta.
© 2023 John Wiley & Sons Ltd. Published 2023 by John Wiley & Sons Ltd.

- Keep an open mind
- Be a 3D thinker
- Switch gears and focus
- Be polymathic
- Surf the seven Cs: conversational, comfortable, connected, capable, confident, collaborative, and calm

7.1.1 Be Tech-Savvy

Staying current with technological improvements, including hardware, software, and hosting applications, is challenging; however, it is extremely important for all geosteering team members. Service companies are developing deeper reading tools (Constable et al. 2012; Hartmann et al. 2014; Wu et al. 2018) where data is not interpreted but input into highly complex inversions (Farquharson and Oldenburg 1998; Chen et al. 2011). Inversions require further analysis and understanding of the acquisition methodology and sensitive volumes of data, which affect how the answer product should be interpreted.

Technology never rests and there are increasing amounts of data streamed in real time using a variety of systems that can be integrated into an understanding of current drilling activities, from bottom hole assemblies and bit performance to well and geological positioning in space. Only 40 years ago, hand-drawn paper logs were used to record data; however, fortunately, software has evolved and now provides visualization tools. More recently, various machine-learning and artificial intelligence have been applied to large datasets (Xu et al. 2019). However, with any new technology, there are potential risks that depend on the users' specific understanding of the input data and the variety of algorithms available as to whether the outputs make geological sense.

These types of analysis or applications are increasingly moving from traditional desktop or laptop computers to web portals hosted on cloud systems (Figure 7.1) or docker containers (Gutierrez et al. 2019; Yegulalp 2019). Suppliers of these applications want to simplify the interface, but some are configured for IT professionals rather than geosteering team members. Understanding how these systems work is important so that users can switch quickly to an updated alternate backup system, configure a new setup, or troubleshoot an unproductive service to prevent or limit nonproductive time.

7.1.2 Visualize in 3D

People "think" in different ways: accountants see everything in numbers, physicists, and mathematicians see interlocking theories and algorithms, and psychologists are more attuned to people's behaviors and underlying personalities. Each is using different sets of senses or abilities that are typically the foundation behind their career choice, conscious or not. Geosteerers habitually think in three dimensions and have a good understanding of perspective and space. This ability is extremely important because wells are drilled in increasingly complex depositional environments, using tools that provide measurements with a wide range of depths of investigation, from millimeters to over a hundred meters, and ranging from single measurements to azimuthal images and even 3D volumes (Clegg et al. 2019).

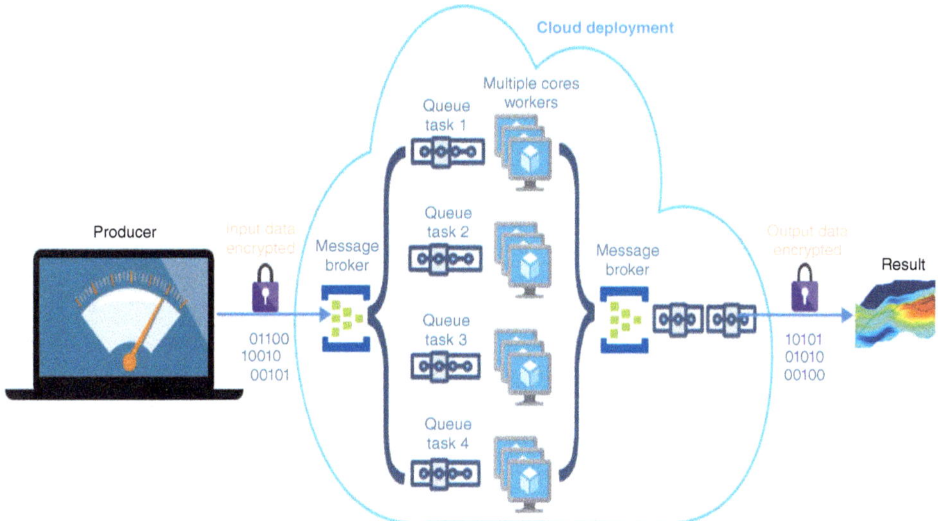

Figure 7.1 Cloud deployment of inversion processing (Gutierrez et al. 2019) is a step change from running everything on a laptop or desktop, involving connectivity and cloud management skills. *Source:* Gutierrez et al. 2019/International Petroleum Technology Conference.

Software helps visualize various datasets, but it can never replace how geologists internally analyze and integrate data in their minds. A good example of this is understanding how the propagation of electromagnetic fields varies depending on the formation properties in which they travel, resulting in varying depths of investigation. Another example is understanding how changes in input parameters to various processes affect the output, such as a computer-generated geological representation of the subsurface. The user for each case should have a 3D understanding of the geological environment to determine if the results are valid. If they are not, could they be explained by a slight variation of their own understanding? Without this expert feedback loop, a misrepresentative interpretation could be used.

7.1.3 Keep an Open Mind

Keeping an open mind, not just in geosteering, but in life, is an important but extremely challenging habit. A geosteering specialist works in a variety of disciplines and on a variety of teams, including internal service company teams and external operator teams. They provide knowledge and expertise gained from years of experience from a variety of geosteering operations and operator methodologies. It is important to the success of each geosteering operation that their approach is completely open and unbiased when learning about the operator's understanding of their reservoir and challenges. As a new team member, it can be easier to recognize existing bias within teams, and this should be discussed during planning phases.

It is helpful to remain aware of this potential barrier and to improve the understanding of operators' challenges and solutions by routinely taking steps to avoid it. Reminding each team member to take a step back, clearly examine all data and evidence, and thoroughly explore alternative scenarios can lead to an expanded set of possibilities, giving the team a broader understanding of the situation and allowing them to plan more effectively for actions and mitigation plans during the planning stages. Once drilling activities begin, they have more confidence in their abilities to make decisions based on their comprehensive shared understanding.

7.1.4 Be a 3D Thinker

Critical thinking is a common term, but 3D thinking is a concept where the brain and self are challenged to:

- See and recognize the truth in an opinion that is opposite to their own
- Truly think through the effects of their own decisions or actions
- Think not only of a solution to an existing challenge, but of a solution first and then determine what challenge it could solve
- Use both sides of the brain – the logical left as well as the more intuitive right

It is all about internal accountability, and it is difficult. The majority takes the easy route in "linear" thinking; however, without the inventions of Einstein, a 3D thinker, we would not be where we are today. Chuck Clayton authored an excellent article (2015) explaining this in detail and provided a warning that is inherent when geosteering and working within a diverse multidiscipline team balancing many scenarios, "*Warning!* This skill is great; however, being able to internalize the other side of an argument can occasionally lead to changing your position. This occurs when the facts and data, your logic, common sense, feelings, and emotions all show that your previous position was incorrect. This is powerful and maturing, but can be a bit frightening."

Great geosteerers can acknowledge they might be wrong, accept other assessments as valid, and work together on solutions.

7.1.5 Switch Gears and Focus

Agility is the skill to switch quickly between tasks or, more commonly, actions (in the sporting context) while priorities change (e.g. in football where a play has to be changed based on a split-second action by the opposition). Instinct, experience, and intuition can assist in minimizing hesitation, which can make the difference between success and failure. It has been proven that "switching" to a different task has productivity penalties because the brain must stop thinking of the previous task and start on the new one (Arrington et al. 2003). However, this penalty decreases if there are common links between the two tasks.

The geosteerer is typically trying to observe data trends that can either prove or disprove a geological interpretation, analyze how that interpretation can affect net-to-gross on the existing well path through to total depth, and effectively communicate recommended well path changes. Switching gears during tasks that should be continually repeated or updated is not something everyone is comfortable doing or is able to sustain. When new geosteerers are trained, it is interesting to observe during real-time exercises how they divide their time between monitoring data streams, interpreting data, communicating with various team members, writing reports, and exporting data. It is evident how they improve over time, and this habit of switching between tasks must be learned and developed, resulting in a competent, confident, and calm geosteerer.

7.1.6 Be Polymathic

This is an interesting term and is applicable to how a geosteerer can be most impactful during drilling operations. A polymath is "a person of great learning in several fields of study" (Dictionary.com). A polymathic person draws on this breadth of knowledge and applies unique solutions to complex problems. A geosteerer interacts with a significant number of related fields, including geology, geophysics, petro physics, petroleum engineering (all aspects of drilling and completing a well), electrical engineering and physics (tool design), mathematics (inversion, forward modeling, machine learning), computer science (software, cloud systems), reservoir engineering (fluid movement in reservoirs, completion design), and many more (Figure 7.2).

Building the habit to continuously learn in related fields, or even fields that might not seem directly related, assists a geosteerer in being able to effectively communicate within and across asset teams and helps ensure all team members have a consistent understanding across disciplines (Figure 7.3). Additionally, it helps when proposing geological explanations for interpretations that might not have been previously considered but which are supported by the data.

7.1.7 Surf the Seven Cs

Conversational, comfortable, connected, capable, confident, collaborative, and calm are traits or habits people are either born with or have developed over time that greatly assist in assimilating with new teams or team members. Geosteerers are ideally situated within the asset team of the operator. So, they constantly need to integrate themselves into a well-established team. For many, this is a daunting and uncomfortable experience that they would prefer to avoid.

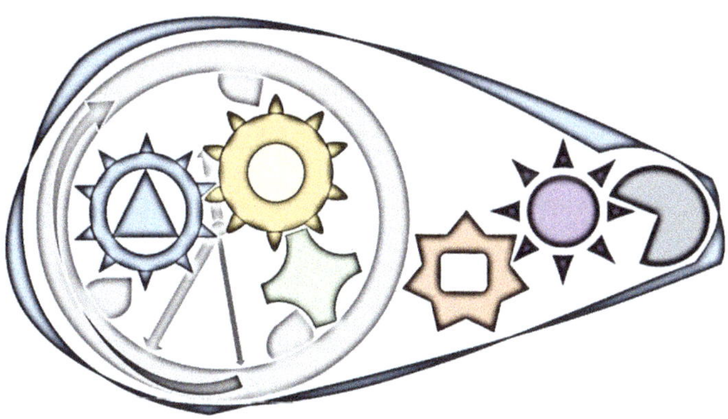

Figure 7.2 Some tasks can be performed in parallel, and some tasks affect other task completions. The majority of tasks are continuous during geosteering operations.

Figure 7.3 To become polymathic requires continuous learning over a broad range of subjects.

The summit of the seven Cs is becoming *confident* and *comfortable*, resulting from the development of the other five skills. Building *connections* during a career, either during work hours or industry-related networking events, is important in a job involving interactions with a variety of people. The geosteerer, the focal point of the service company directly sitting with the asset team, is the first point of information for anything occurring on the rig. It is important to remain in contact with various team members on the rig and in the office as well as helping operators connect with others within the organization. This is where being polymathic helps, and the geosteerer becomes the "go-to" person.

Being *capable* is having the desire to learn new things, assist with a task, or take ownership of a task. If someone is trusted to be capable, the tasks performed and delivered are of great value to the company and strengthen personal relationships. *Collaboration* is communicating effectively and efficiently what the geosteerer knows and thinks with others and, most importantly, listening to what the rest of the team knows and thinks before working together on a solution.

Don't persuade, defend, or interrupt. Be curious, be conversational, be real. And listen. Elizabeth Lesser.

Maintaining a sense of *calm* is important during a crisis. It enables the crisis to be examined clearly by keeping an open mind, employing 3D thinking, and using polymathic talents. The alternative can result in nonevidence-based scenarios that can descend into added confusion and crises as well as having a prevailing negative outlook. By keeping an open mind, it is possible to assess the situation to identify if it actually is a crisis, and manage it using various mitigation strategies, or if it stems from a misunderstanding or misperception. Although horizontal wells are increasingly common, there are many asset teams where this type of drilling is new, which is why conducting pre-well analyses and being co-located with the asset team during operations are so important. Further, being calm can encourage others to relax and release built-up tension.

Ever been stuck for something to say and been grateful that someone else was able to start a conversation that you could join? This is a skill that generally puts everyone at ease and helps people find connections and build relationships. Again, some people are natural *conversationalists* and others have to work at it. However, building this skill into a habit is beneficial during those awkward first meetings.

Finally, *comfortable* and *confident*: these are fairly self-explanatory – everyone likes a person who generates a feeling of comfort and who has confidence that is implicitly trusted. It is easier to talk to them and feel that their response is genuine. This can increase the operator's confidence to ask questions that might not be asked of someone else. Geosteerers become comfortable and confident in their own skin once they have mastered the skills needed to perform their job with ease and know who to contact if something occurs that requires more expertise. The seven Cs are the foundation of building trust with operators.

Habits or skills, these seven entities, when combined, contribute to a consummate geosteerer. Each of these is never complete and requires continuous learning, which is not always recognized within the industry, but it is inherently our own responsibility as humans to continue to develop and evolve.

Acknowledgments

The author acknowledges the many talented geosteerers that she has worked with for over 15 years who have inspired this chapter, especially Nigel Clegg, who epitomises the ultimate geosteerer, Beathe Hustoft, Jim Johnston and Mike Welch. Merryl Le Roux, Frank Otmar Weinreich, Joss Everett, Andrew Harrison, and the proofreading team (Wiley Blackwell) are thanked for their assistance. This chapter has been summarized in Mukherjee et al. (2023).

References

Arrington, C.M., Altmann, E.M., and Carr, T.H. (2003). Tasks of a feather flock together: Similarity effects in task switching. *Memory & Cognition* 31 (5): 781–789.

Chen, Y.H., Omeragic, D., Druskin, V. et al. (2011). 2.5D FD modeling of EM directional propagation tools in high-angle and horizontal wells. Presented at the SEG Annual Meeting, San Antonio, Texas (18–23 September 2011). SEG-2001-0422.

Clayton, C. (2015). The awesome power of 3-dimensional thinking. LinkedIN. https://www.linkedin.com/pulse/power-3-dimentional-thinking-chuck-clayton.

Clegg, N., Parker, T., Djefel, B. et al. (2019). The final piece of the puzzle: 3-D inversion of ultradeep azimuthal resistivity LWD data. Presented at the SPWLA 60th Annual Logging Symposium, The Woodlands, Texas (15–19 June 2019). SPWLA -2019-HHH.

Constable, M.V., Antonsen, F., Olsen P.A. et al. (2012). Improving well placement and reservoir characterization with deep directional resistivity measurements. Presented at the SPE Annual Technical Conference and Exhibition, San Antonio, Texas (8–10 October 2012). SPE-159621-MS.

Farquharson, C.G. and Oldenburg, D.W. (1998). Non-linear inversion using general measures of data misfit and model structure. *Geophysical Journal International* 134: 213–227.

Gutierrez E., Wu, H., Dong W. et al. (2019). Implementation of a scalable distributed cloud computing platform for fast real-time reservoir development. Presented at the International Petroleum Technology Conference, Beijing, China (26–28 March 2019). IPTC-19383-MS.

Hartmann, A., Vianna, A., Maurer, H.M. et al. (2014). Verification testing of a new extra-deep azimuthal resistivity measurement. Presented at the SPWLA 55th Annual Logging Symposium, Abu Dhabi, UAE (18–22 May 2014). SPWLA-2014-MM.

Mukherjee, S., Dasgupta, S., Majumdar, C. et al. (2023). Introduction to: Handbook of Petroleum Geoscience: Exploration, Characterization, and Exploitation of Hydrocarbon Reservoirs. In: *Handbook of Petroleum Geoscience: Exploration, Characterization, and Exploitation of Hydrocarbon Reservoirs*, Petroleum Industry Updates: Geoscientific and Economic Issues (ed. S. Mukherjee, S. Dasgupta, C. Majumdar, et al.). John Wiley & Sons, Hoboken, ISBN 978-1-119-68003-1.

Wu, H., Golla, C., Parker, T. et al. (2018). A new ultra-deep azimuthal electromagnetic LWD sensor for reservoir insight. Presented at the SPWLA 59th Annual Logging Symposium, London, UK (2–6 June 2018). SPWLA-2018-X.

Xu, C., Misra, S., Srinivasan, P. et al. (2019). When petrophysics meets big data: what can machine do? Presented at the SPE Middle East Oil and Gas Show and Conference, Manama, Bahrain (18–21 March 2019). SPE-195068-MS.

Yegulalp, S. (2019). What is docker? docker containers explained. InfoWorld from IDG. https://www.infoworld.com/article/3204171/what-is-docker-docker-containers-explained.html.

8

Driving Technology for Geosteering Decisions: Halliburton Geosteering

Bronwyn Djefel[1] *and Nigel Clegg*[2,*]

[1] *Epiroc, Kewdale, Western Australia, Australia*
[2] *Global Subject Matter Expert – Geosteering, Halliburton, Norway*
**Author for correspondence: Nigel.Clegg@Halliburton.com*

Abstract

Geosteering has been used for over 20 years in various forms, and technology development in both hardware and software has taken leaps and bounds over that time related to increasing challenges in operations and reservoirs. Remote operations where geosteerers are monitoring jobs in other countries, innovations such as wired pipe with the associated increased data handling, and complex inversions of ultra-deep data, all in real time, have changed how we operate and collaborate with our customers to help them meet their targets in reservoirs that were previously uneconomical. However, the right technology should be matched to the customer challenge. Hence, simple gamma geosteering is still relevant and in use today, but variants are also available such as azimuthal and spectral gamma. This article will track through time the technological developments of both hardware and software and look ahead to the future of geosteering with Halliburton.

Keywords *Geosteering; challenges; Halliburton; electromagnetic inversion*

8.1 Introduction

The major challenges that geosteering operations are conducted to overcome are summarized in Figure 8.1.

Arguably, the most critical piece of information to geosteer a well is knowing where the well is in 3-dimensional space and the tendencies of the bit as drilling continues through different formations. Understanding the bit/bottom hole assembly (BHA) tendencies enables geosteerers to plan the well ahead to optimize achievable geological targets and geoscientists to reduce uncertainties of reservoir and fluid boundaries by updating their

Handbook of Petroleum Geoscience: Exploration, Characterization, and Exploitation of Hydrocarbon Reservoirs, First Edition. Edited by Soumyajit Mukherjee, Swagato Dasgupta, Chandan Majumdar, Subhandip Mandal and Troyee Dasgupta.
© 2023 John Wiley & Sons Ltd. Published 2023 by John Wiley & Sons Ltd.

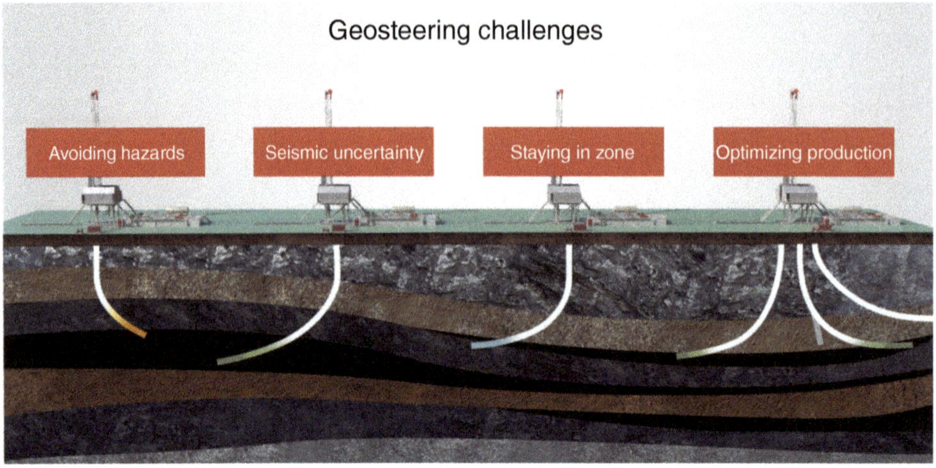

Figure 8.1 Common geosteering challenges. Avoiding hazards enables the successful completion of a well. Seismic uncertainty can lead to formations placed many meters vertically away from where they are expected to be. Staying in a zone is often the principal requirement as this is directly linked to production volumes and this ties in with optimizing production, where multiple zones may be targeted. So, these last two challenges have to be balanced. *Source:* Courtesy of Halliburton.

seismic and reservoir models. This enables optimization of full-field development while minimizing collision risks.

Halliburton started work on this challenge back in the 1920s when Sun Oil Co.'s vice president, J. N. Pew, set up a development team to devise tools and techniques for effective well surveying. The engineers had to create a system that could measure direction (north–south) and inclination at multiple points in the well so that drillers could more accurately calculate the bottom-hole location. The challenge at the time was that many dry holes were being drilled only a few 100 ft from producers; vertical holes were prevalent at this time and many geologists and drillers could not understand how this could be possible. Pews' hypothesis was that boreholes twisted and turned in a spiraling manner, and that formation boundaries themselves could cause deflections in inclination and direction.

The existing methods were only effective at defining the inclination at the base of the well and were a health and safety hazard due to associated burns. The most common method at this time was the "Acid Bottle Test". Hydrofluoric acid (a colorless fuming liquid) was poured into a glass bottle and lowered to the bottom of the borehole. It was left long enough for the acid to etch a line around the circumference of the glass bottle. Once it was brought to surface, the "inclination" of this line would infer the inclination of the bottom hole location.

Pews' team was successful in designing a directional survey unit, named SURWEL (see Figure 8.2), that could measure the inclination and direction at any point in the well. It used a gyroscope to measure borehole direction, with a bubble-level inclination unit and a camera placed above them. The unit was lowered to the bottom-hole location and the camera would take pictures of the gyro indicator superimposed over the position of the bubble. At surface, the film would be developed, and readings taken at intervals to define the well path from surface to bottom-hole location.

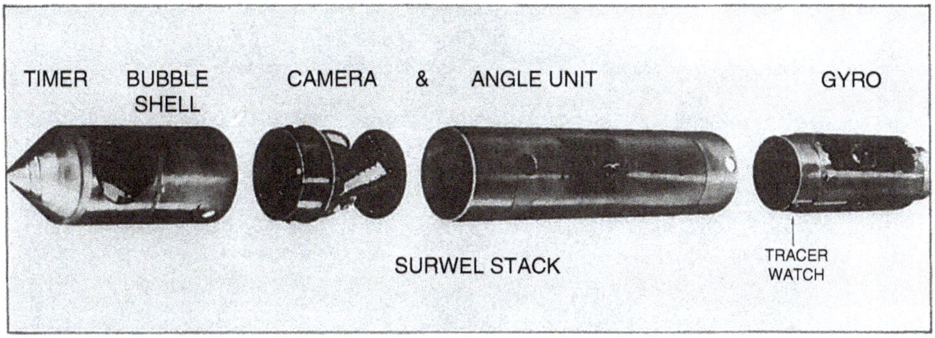

The 5 1/2"surwel components.

Figure 8.2 SURWEL 5 ½ in. tool developed by J. N. Pews' team in the 1920s. Using this device, it is possible to measure the inclination of a well at multiple depths and use these changes in inclination to calculate the trajectory of a well and, therefore, identify its position in the subsurface. *Source:* Courtesy of Halliburton.

This technology was developed due to an industry need to reduce the number of "dry" holes and was then very successful during a somewhat turbulent time in the USA drilling history when the courts became involved due to lease disputes (see Figure 8.3). Since the previous surveying was either not conducted or inaccurate, wells were drilled haphazardly and often close to lease lines. When they were surveyed, it was found that many wells infringed on lease lines and new laws were made governing well surveying which exist today.

Sperry-Sun continued development in those early days to improve the responsiveness of BHAs in order to orient the bit and improve directional drilling capabilities to hit various targets. Today, we are automating this process to reduce personnel at rig sites due to industry "de-manning" requirements and to improve the efficiencies and accuracy of placing horizontal wells. The iCruise® intelligent rotary steerable system (RSS) is designed with the latest technology that can be configured to drill a long lateral section fast, or to drill a vertical, curve, and lateral in a single run serving both mature field and unconventional markets (see Figure 8.4). Continuing the Sperry high-quality surveying tradition, the iCruise incorporates three distinct survey packages to measure the tool face 1000 times per second as well as an optional azimuthal gamma ray measurement ideal for geosteering applications.

It is paired with LOGIX™ Automated Drilling Director, which uses intricate physics-based models and machine learning to project the well path and avoid collisions, manage vibration and pressure parameters, and control steering (Demirer et al. 2019) by evaluating tool performance and characterizing formation tendencies in real time. The advanced algorithms enable modeling for a digital twin, which is automatically calibrated with data gathered from the downhole sensors in real time.

This system is also connected with geosteering software, RoxC®, to provide geosteerers the unique ability to optimize the well path ahead with reference to what the BHA downhole, in the specific reservoir, achieves. The inherent value is to enable directional drillers and geosteerers to focus more on the well delivery, meeting operator requirements than on making normal steering decisions that can be made by an automated, well informed continuously updating system.

(a)

Crooked Holes Give Unfair Drainage; May Make Surveying Compulsory

THE OIL AND GAS JOURNAL

Principles of Straight Hole Drilling Get Offic Recognition in Order of Texas Commission

By L. G. E. BIGNELL

Headlines from journal articles during the thirties.

(b)

Hurricane Mesa, during survey verification tests. Note rig platform and pipe running from rig to the lower left-hand corner.

Figure 8.3 On the left, examples of headlines in newspapers in the 1950s. On the right, the Sperry-Sun survey tool underwent testing at Hurricane Mesa to verify their accuracy at defining a well path. You can see the rig at the top and pipe running from the rig to the bottom left corner. *Source:* Courtesy of Halliburton.

The iCruise is the latest tool with continuous inclination and azimuthal surveying capability, developed with the capacity to increase the density of surveys, reduce the uncertainty of interpolation, and increase the positional accuracy of the well path. The first such tool Halliburton developed was the Gamma At Bit Inclination (GABI™) sensor, which is a sub

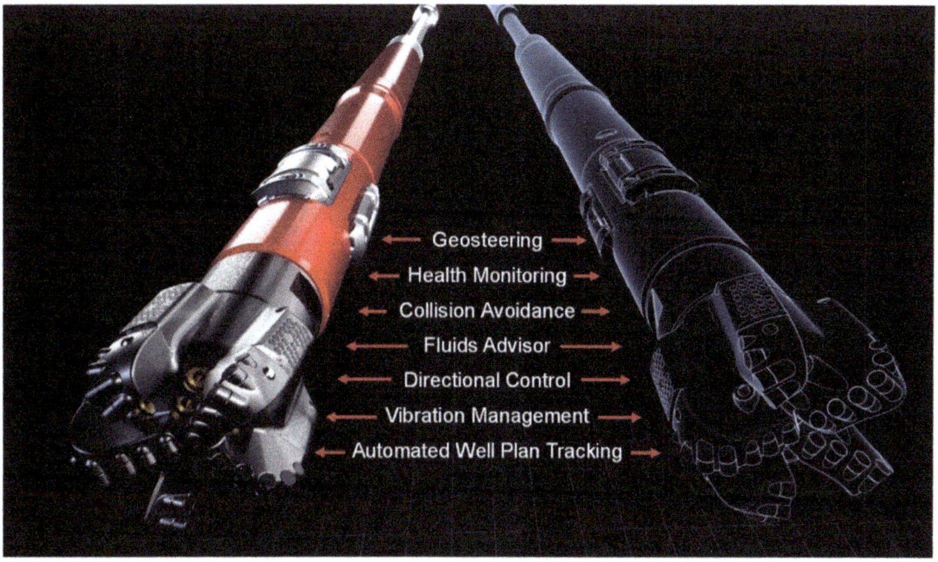

Figure 8.4 A digital twin model is built for iCruise RSS to simulate reservoir conditions for planning and use in autonomous directional drilling through LOGIX software. *Source:* Demirer et al. (2019), Society of Petroleum Engineers.

Figure 8.5 The GABI sensor provides azimuthal gamma ray and inclination surveys right behind the bit for accurate well placement. *Source:* Picture courtesy of Halliburton.

placed in between the bit and a motor assembly (see Figure 8.5). It incorporates continuous near-bit inclination measurements that provide immediate feedback on changes in the well trajectory, helping to minimize tortuosity, produce flatter longer horizontals, and reduce uncertainty in the well position. Having this information so close to the bit is essential when drilling in thin reservoir units and becomes critical when these thin reservoirs have variable dip in order to make quick interpretational decisions and adjustments to the well path to geosteer with high net to gross (NTG).

The GABI also includes four independent scintillation detectors, which provide measurements in four quadrants that produce a borehole image when sliding. Previously, azimuthal data would only be available while rotating with a motor; however, this innovation enables accurate well placement throughout the drilling run. Understanding the orientation of formation boundaries, relative to the borehole and true orientation, is paramount to make steering decisions and enhance reservoir understanding.

An example of thin reservoirs with dip uncertainty are the coal seams in the Eastern Australian gas fields, which vary from 1.5 to 4 m TVT with an average thickness of 2 m TVT. They are structurally complex with extensive faulting. Although internal seam character is fairly consistent, occasional splits develop with some internal laminations increasing and decreasing in gamma character. Geosteering is performed by using the realtime images from the GABI motor, which is located 3.5 m behind the bit to map these internal laminations. Due to the complex nature of these seams, geosteering decisions have to be made subtly in order to minimize tortuosity and avoid drilling out of the target seam (see Figure 8.6). Communication between the directional driller and geosteering specialist is the key to delivering well success. In total, 80 wells have been geosteered with the GABI motor in Australia, for a total footage of 90 500 m at an average of 96% in seam.

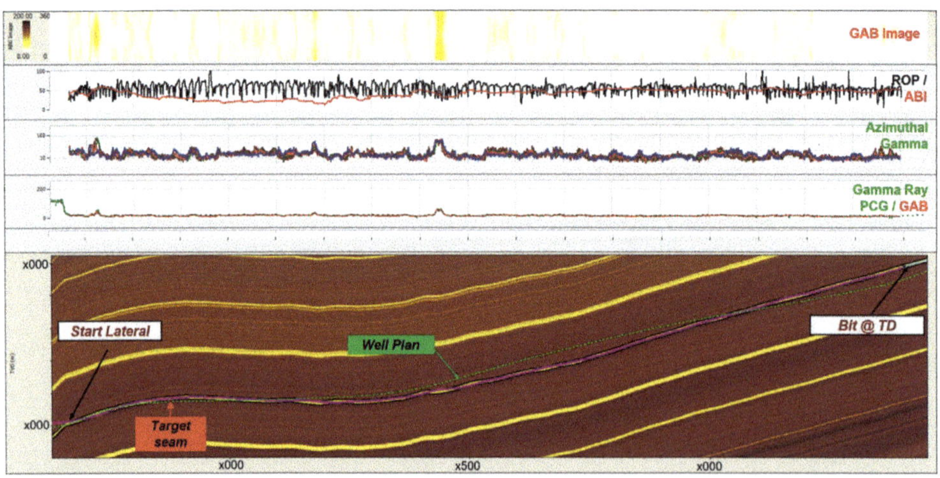

Figure 8.6 Example of GABI coal seam geosteering in thin zone with 100% in seam for 1739 m MD with complex dips (courtesy of Westside Corporation Limited). The original well plan is represented by the green dotted line, the geology has been updated during drilling to define the actual position of the target formation. The thin yellow zone shows the target zone based on its Gamma value (low gamma values are represented by yellow zones; high gamma values are marked as the brown zones). It is clear that for much of its length, the well plan would have been out of the target zone. The pink line denotes the actual well position.

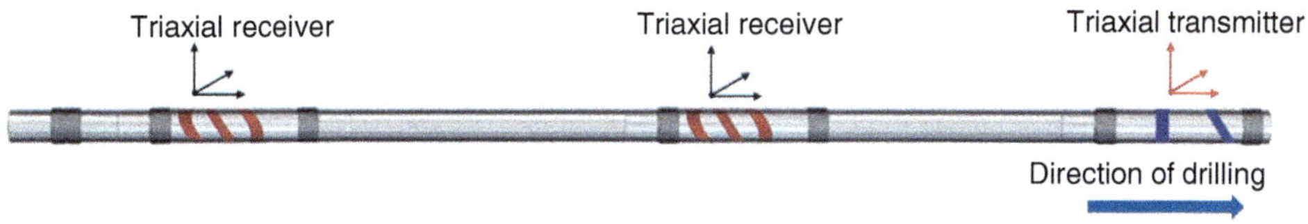

Figure 8.7 EarthStar ultra-deep resistivity tool configuration. The positions of the triaxial receivers and transmitted frequencies can be customized for the operator challenge. *Source:* Picture courtesy of Halliburton.

While some operators are challenged by thin reservoirs, others are challenged by mature fields with multiple stacked reservoirs with potential compartmentalization from either stratigraphic or structural means. These are further complicated with production and reservoir depletion. Understanding this movement and the structures is increasingly important to operators in order to maximize return on investment. In response to the market, Halliburton developed the EarthStar™ ultra-deep resistivity service for geostopping, geosteering, and geomapping up to 250 ft. distances from the wellbore. It takes advantage of the deep depth of investigation from induced electromagnetic fields using low-frequency transmissions and increased spacing of transmitter and receiver subs (see Figure 8.7) that are common to this type of tool. Halliburton also implemented design enhancements that increase the signal to noise ratio, such that very small signals at deeper depths in the reservoir are measured accurately.

The acquired data captures the nine component electromagnetic field responses at the various receivers and is inverted using either 1D or 3D methodology (Wu et al. 2018; Wilson et al. 2019) depending on the operator need. These inversions are resistivity models that best represent the formations and fluids in the surrounding volume. They are extremely useful in optimizing the landing of wells with seismic uncertainty, in mapping distances to top and base of the reservoir, as well as understanding complex geological structures (see Figure 8.8), providing an insight that is simply not available with the shallower measuring LWD and wireline tools.

Regardless of whether operators are drilling in thick or thin reservoirs, with simple dip orientations or more complex structural features requiring 3D geomapping analysis, the positional uncertainty associated with the wellbore position can impact geological interpretations as well as geosteering decisions. Positional uncertainties occur in both the horizontal and vertical planes, increase with measured depth and lateral lengths, and become a common concern for operators when geosteering.

In the past 10 years, increasing accuracy in defining well paths has been attained through both LWD measurement technology development and algorithm refinement of corrections. Increasing the frequency of surveying, using more than one sensor to verify a survey point, and correcting for misalignment of the sensor in the borehole sag correction (SAG) are all developments to assist operators in reducing their positional, and hence, geological interpretation uncertainties. Survey management is another area where the uncertainty around each survey can be reduced further, by applying various corrections for internal and external influences on the accelerometer and magnetometer readings. These include

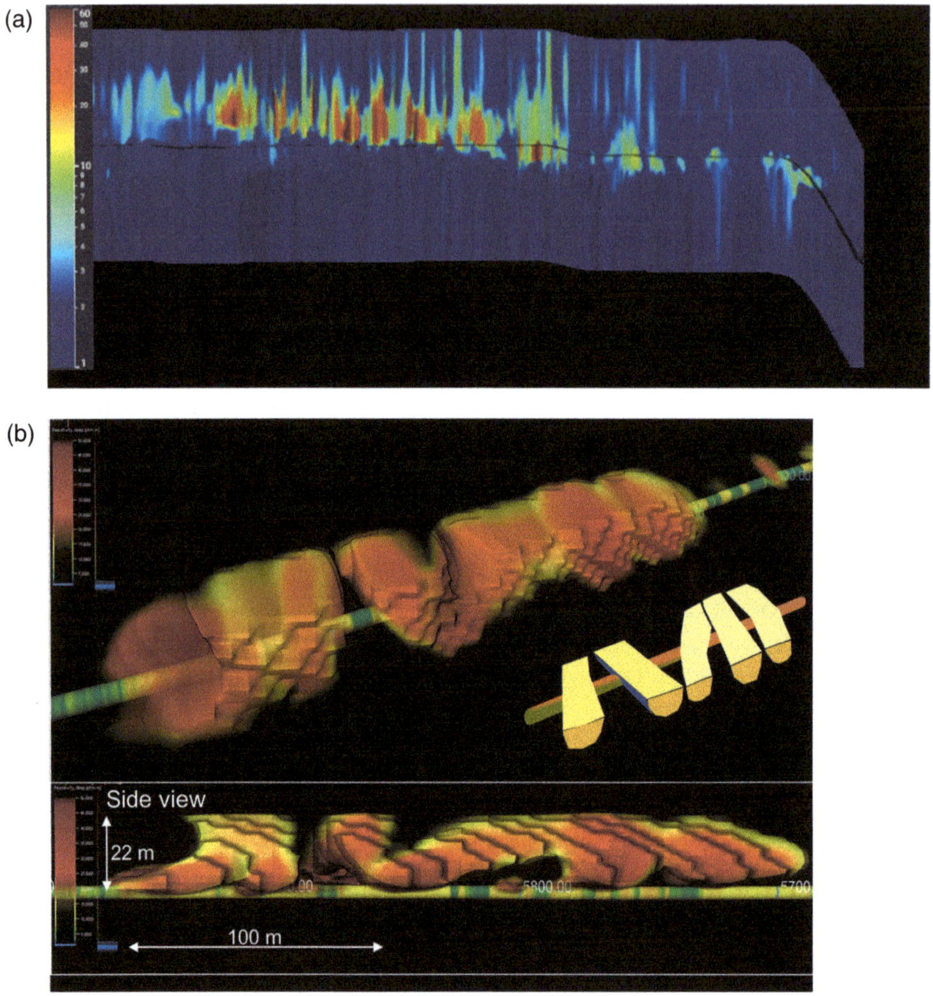

Figure 8.8 1D inversion canvas (top image) showing sparsely populated sand packages, and the corresponding 3D inversion (bottom images) clearly revealing channel sand packages and their direction relative to the wellbore (Wilson et al. 2019). The upper picture is a 1D inversion that represents changes in resistivity above and below the well path, changes to a well's inclination can be made to target zones identified on the canvas, or to avoid exiting the desired reservoir zones. Any changes in resistivity to either side of the well path can result in some distortion of the 1D canvas. Identifying resistivity boundaries distant from the well path allows changes to the well path to be made early, limiting the tortuosity of a well path. The lower image shows 3D inversion of the same dataset. The low resistivity data has been made transparent to reveal the 3D distribution of the target sands. Revealing the 3D reservoir structures allows geosteering decisions to be made to alter a well azimuth as well as inclination.

multi-station analysis (MSA) to minimize misalignment and sensor biases internal to the sensor package, in-field referencing (IFR) related to crustal magnetic offset, and interpolated in-field referencing (IIFR) where magnetic solar activity can cause surveys to become unreliable (also known as time-varying field corrections). Table 8.1 shows the TVD error at

Table 8.1 Standard TVD error for a 2000 m horizontal section (from Lawrence et al. 2010).

RESULTANT TVD ERROR FOR A 2000 m HORIZONTAL SECTION		
Error source (2 Sigma)	**TVD Error (2 Sigma)**	**TVD Error (%)**
Misalignment - MX/MY (0.12°)	±4.43 m	44.8%
Residual SAG (0.16°)	±6.14 m	30.9%
Gz bias (0.8 mG)	±1.68 m	24.0%
Root sum square	±7.63 m	

MX is the tool axial misalignment in the X axis and MY is the tool axial misalignment in the Y axis. Gz bias is related to the sensor accelerometer readings in the Z axis.

2 Sigma and the percentages of error attributed to each source for a 2000 m horizontal section. At +/− 7.63 m, this can exceed the accuracy that an operator requires if no additional corrections are made.

Another method to referencing this uncertainty is by recognizing known geological formations or fluid contacts on landing, and then zero the TVD uncertainty to be used as a reference point from there on. This is an acceptable industry practice (Lawrence et al. 2010) and focuses on understanding the local TVD error relating to a known geological point in space.

Halliburton applied a combination of survey management solutions in 2007, using a dual probe surveying system with improved accelerometer bias in an Australian well to reduce the TVD uncertainty. The dual probe solution used two survey sensors in the BHA at 90 ft. separation such that when the stand was drilled down and surveys were taken, the second survey sensor was located at approximately the same position as the previous first sensor survey station. Statistically, this results in lower associated uncertainty as measurements have repeatability and SAG corrections can be verified. Industry-first improvements were made in BHA analysis software that enabled the SAG correction to be improved from an industry standard SAG correction at the time of 0.08° at 1 Sigma to 0.04°. Making realtime corrections was then possible and the goal to stay within +/− 1 m TVD drilling window and a 1.5 m survey uncertainty window achievable within the lateral section (Lawrence et al. 2010). This is where the survey measurements of Directional Module (DM) and Pressure Case Directional (PCD) are plotted along with the At Bit Inclination (ABI™) and lower and upper TVD limits, improved TVD requirements, and the standard TVD calculations.

Timely and clear communication is critical for conducting geosteering jobs and Halliburton is at the forefront of developing Real Time Centers (RTCs) from inception in 1996 when Edgar Ortiz, Halliburton's now-retired Energy Services Group CEO, first introduced a vision for realtime operations; since then, Halliburton has built over 50 RTCs. Half of these were constructed for national and international oil companies and combine experienced service company personnel with operator teams throughout all phases of a geosteering project. Colocating all personnel leads to increased communication channels, building of team culture, and strengthening of relationships such that the normal barriers to good communication are broken down completely. It also leads to increased sharing of knowledge which naturally encourages cross-training of skills, mentoring, and career progression. The other centers are built as internal "hubs" for Halliburton's own service quality and operational excellence control. These are fully staffed by Halliburton experts and support operations within a country and/or region.

Today, many national and international oil companies continue to have their own geosteering remote operations centers (ROCs) and service company personnel colocate within these areas. With these types of setups, there is a need for data to be either in a central location so it can easily be transferred or visualized within multiple systems such that the entire asset team has access to the latest data for their interpretation and collaboration on decision making.

At Halliburton, we have developed tools, geosteering software, RTOCs and introduced new answer products that strengthen the communication and collaboration pathways. Our very first geosteering software, StrataSteer™, was introduced in the 1990s and incorporated a broad view of the planned horizontal well, the view of a single loaded offset well log profile, the drilling geological section with modeled, and realtime logs displayed across the top (see Figure 8.9). An innovative switch from the traditional vertical log interpretation, to viewing the logs horizontally naturally worked with the geological cross section for interpretation and aids collaboration with drilling departments.

There have been many iterations of the software through StrataSteer 3D and now our latest software RoxC, which has been driven primarily by the development of a new ultra-deep azimuthal resistivity service, EarthStar. The EarthStar itself makes millions of measurements capturing the reservoir response to induced electromagnetic fields. Inverting this data using either a 1D or 3D methodology requires high computing power and graphics capabilities that were built using the latest technology available (Gutierrez et al. 2019).

The RoxC web application provides rich 2D and 3D visualization without the need for the end user to install any software besides a web browser. Users can access RoxC using multiple devices such as mobile phones, tablets, and workstations, providing a seamless experience. By following a microservice architecture and using the RoxC high-performance computation platform based on open source technologies, users can scale the inversion-processing needs either on the Cloud, on-premise, or even using Edge devices in the field. RoxC takes advantage of the Cloud services' scalability by adjusting the amount of resources needed for the application deployment depending on the concurrent number of users.

As there are multiple representations that can explain the range of resistivity models that can fit the acquired data, new visualizations were developed including the five-card draw. The user selects a single slice of the 1D inversion result from the canvas and drops it to one of five vertical tracks arranged at the base of the display (see Figure 8.9b). Once dropped, the software displays the corresponding distribution curves for analysis; D5 (the lowest 5% of resistivity results), D50 (the median resistivity result), Davg (the average resistivity result), and D95 (the highest 5% of resistivity results) as per Figure 8.9b. The user is able to increase or decrease the MD location to see how the curves change in order to

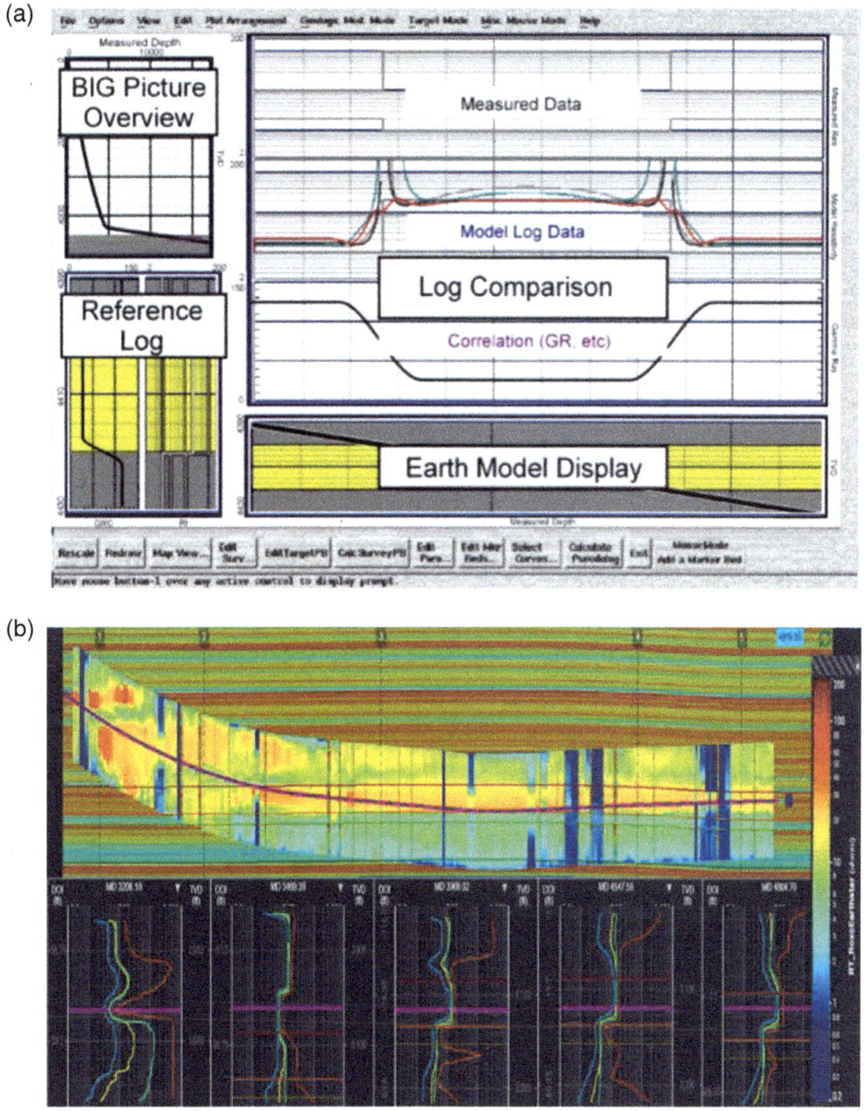

Figure 8.9 On the left, StrataSteer software was the first of its kind in the industry (courtesy of SPE and Jackson et al. 1998) and on the right, RoxC software, which is web based, incorporating cloud resources and new visualization tools. RoxC picture courtesy of Halliburton.

quantify uncertainty in both inverted resistivity and bed boundary positions relative to the well path (Wu et al. 2018).

As discussed earlier, the new 3D inversion requires additional visualization and analysis tools to discriminate the inverted 3D volume (see Figure 8.10). A new volume renderer was designed within RoxC with various tool sets that assist the user to analyze the data, including

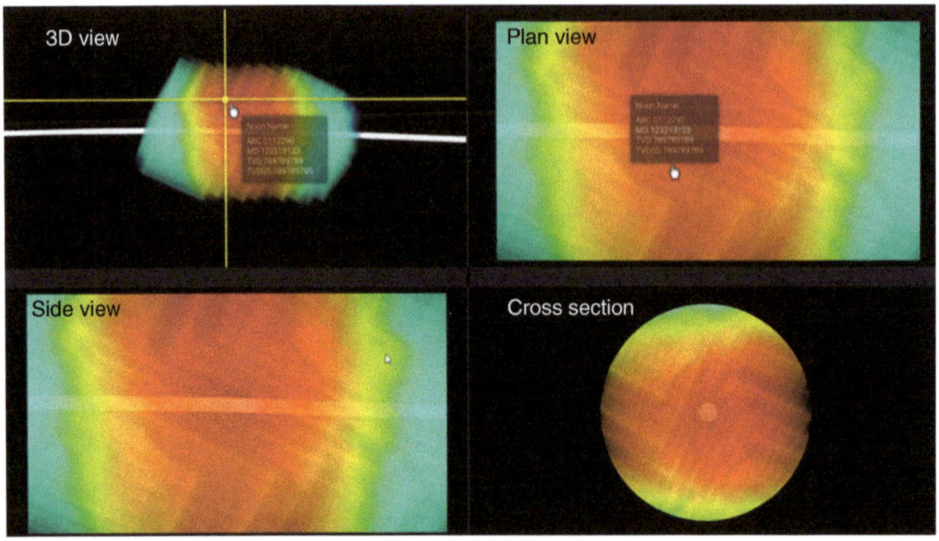

Figure 8.10 New 3D visualization within RoxC to enable analysis of the 3D-inverted reservoir volume. The top left image shows a small 3D segment with the well penetrating a red high resistivity zone. The cross hairs tool allows resistivity values within the volume to be interrogated. The top right image shows a plan view of the same segment. Any lateral changes in the resistivity will be displayed and allow changes in a well azimuth to be considered to target zones of interest that are identified to the sides of the well path. The bottom left image shows a side view, displaying changes in resistivity above and below the well path and allows changes in a wells TVD to be considered to optimize the vertical position of the well. The bottom right image shows a cross section along the well path, displaying vertical and lateral changes simultaneously.

the traditional color palette, but with improvements to apply transparency and add filters to resistivity zones of varying interest. A user can then focus solely on the high resistivity content, low or medium as per their interest, in understanding the 3D volume of the reservoir.

Other tools include being able to slice through the volume and stop at any point in 3D space for analysis. A vertical slice along the wellbore (varying as per the wellbore azimuth) is ideal for quality control with the 1D inversion results, the shallower data acquired by independent LWD tools and existing seismic data. A horizontal slice relative to the wellbore TVD (varying as per the wellbore TVD), provides the user with a quicklook understanding of geological and fluid features to the left and right, similar to the standard plan view.

Additional features in RoxC includes improved connectivity with earth modeling software with plugins enabled to send inversion results directly to operators' instances abiding by strict IT security authorization and authentication protocols. This reduces the need for geosteerers to export data, and operators to import data, enabling them to concentrate on interpretation of the data itself.

New developments are planned to follow the industry requirements for look-ahead services in order to optimize geo-stopping for drilling purposes to set casing prior to over- or under-pressured zones. This ensures the landing of the well is improved from today's service, which has already removed the need for pilot wells in the majority of cases. Automation or augmented geosteering tantalizes drilling departments, and machine learning techniques are progressing rapidly in developing inversions and predicting most likely geological projections. Other advances in visualization such as the tensor cloud from Landmark are enabling rapid updates of earth models and the application of this technology to reservoir modeling is exciting.

Acknowledgments

The authors thank Westside Corporation Ltd. and Aker BP for their contribution to field data. The authors also would like to acknowledge the support of the software and hardware development teams, the field verification team, and many others who contributed to the success of the development of the various technologies. Thanks also to the geosteering team members who work diligently with operators to deliver well placement success and to the management of Halliburton for their support of the various projects and encouragement to publish the work. Merryl Le Roux, Frank Otmar Weinreich, Joss Everett, Andrew Harrison, and the proofreading team (Wiley Blackwell) are thanked for their assistance. This chapter has been summarized in Mukherjee et al. (2023).

References

Demirer, N., Zalluhoglu, U., Marck, J. et al. (2019). Model predictive control method for autonomous directional drilling. Paper presented at the *SPE Annual Technical Conference and Exhibition* (30 September–2 October 2019). Paper Number: SPE-195917-MS.

Gutierrez E., Wu H., Dong W., and Efimov, I. (2019). Implementation of a scalable distributed cloud computing platform for fast real-time reservoir development. Paper presented at the International Petroleum Technology Conference, (26–28 March 2019). IPTC-19383-MS [Paper].

Jackson, C.E., Illfelder, H.M.J., and Pineda, G. (1998). Field implementation of geological steering techniques optimizes drilling in highly-deviated and horizontal wells. Paper presented at the *SPE International Conference on Horizontal Well Technology* (1–4 November 1998).

Lawrence, B., Mojsin, T., and Strachan, M. (2010). Improved accuracy of borehole positioning in horizontal wells. Paper presented at the *IADC/SPE Drilling Conference and Exhibition* (2–4 February 2010).

Mukherjee, S., Dasgupta, S., Majumdar, C. et al. (2023). Introduction to: Handbook of Petroleum Geoscience: Exploration, Characterization, and Exploitation of Hydrocarbon Reservoirs. In: *Handbook of Petroleum Geoscience: Exploration, Characterization, and Exploitation of Hydrocarbon Reservoirs*, Petroleum Industry Updates: Geoscientific and Economic Issues (ed. S. Mukherjee, S. Dasgupta, C. Majumdar, et al.). John Wiley & Sons, Hoboken, ISBN 978-1-119-68003-1.

Wilson, G., Marchant, D., Haber, E. et al. (2019). Real-time 3D inversion of ultra-deep resistivity logging-while-drilling data. Calgary, Alberta, Canada. SPE Annual Technical Conference and Exhibition (30 September–2 October 2019). SPE-196141-MS.

Wu, H., Golla, C, Parker, T. et al. (2018). A new ultra-deep azimuthal electromagnetic LWD sensor for reservoir insight. London, UK. SPWLA Annual Logging Symposium. SPWLA-2018-X (2–6 June 2018).

9

Rock Strength Estimation from Petrophysical Logs Through Core Data Calibration in Low Porosity and Low Permeability Carbonate Rocks

*Sankhajit Saha and Vikram Vishal**

Computational and Experimental Geomechanics Laboratory, Department of Earth Sciences, Indian Institute of Technology Bombay, Mumbai, Maharashtra, India

**Author for correspondence: drvikramvishal@gmail.com*

Abstract

Unconfined compressive strength (UCS) is one of the essential and fundamental parameters of geomechanical modeling and its applications. Laboratory-based core testing data provide most accurate information about the UCS values. However, it mainly provides point information, and the process is also time consuming and destructive. Thus, various empirical relations have been established based on petrophysical log and core calibration. Most of these published relations are developed for certain basins and specific geological characteristics. Thus, it is necessary to have a local calibration for UCS estimation. In this study, a log-derived UCS equation was rendered based on core testing results. Analysis has been carried out on the Proterozoic carbonate rocks from one of the oldest sedimentary basins in the central part of India. Petrophysical and mineralogical studies indicate that rocks are compact in nature with low porosity (0.14–4.01%), low permeability (0.01–0.46 mD) and contain high calcite content at 72%. Total seven core plugs were acquired and tested from four core samples in four different wells. Analysis indicates that compressional slowness-based exponential relation best fit to the data comparing to other published acoustic slowness-based empirical relations. This established relation shown here can be useful for compact, low-porosity, and low-permeable carbonate rocks having similar geological history.

Keywords *Uniaxial Compressive Strength (UCS); triaxial and uniaxial core test; carbonate rock*

9.1 Introduction

Evaluating rock strength is fundamental to the geomechanical study and in particular its applications in wellbore stability (Manshad 2014), sand production prediction (Javani et al. 2017), optimum mud weight design for drilling (Gholami et al. 2014). The uniaxial compression test (unconfined compression test) is one of the most important tests used

Handbook of Petroleum Geoscience: Exploration, Characterization, and Exploitation of Hydrocarbon Reservoirs, First Edition. Edited by Soumyajit Mukherjee, Swagato Dasgupta, Chandan Majumdar, Subhandip Mandal and Troyee Dasgupta.
© 2023 John Wiley & Sons Ltd. Published 2023 by John Wiley & Sons Ltd.

to measure rock strength and accurate measurement aids in better control when encountering drilling problems such as pipe sticking, tight hole, collapse, pack off, and sand production (Sarkar et al. 2012; Singh et al. 2013b; Vishal et al. 2015; Mahanta et al. 2018; 2020). Uniaxial compressive strength (UCS) is the most used parameter in mining engineering as well (Bieniawski 1976; Aladejare 2020; Pradhan et al. 2018; Singh et al. 2013a).

In order to determine UCS, direct and indirect techniques are used (Khaksar et al. 2009). In the direct approach, UCS is determined from the lab core tests. In indirect techniques, UCS is determined based on mathematical and empirical relationships. Khaksar et al. (2009) described best practices for rock strength estimation using both laboratory measurements and empirical methods. UCS is directly determined according to the American Society for Testing and Materials (ASTM D2938–95 2002), the International Society for Rock Mechanics (ISRM 2007) and other common standards. In general, core plug length vs. diameter should be 2 : 1 (Khaksar et al. 2009). Often, it is difficult to get sufficient samples with this ratio, and studies indicate that ratios within 1.6 : 1 have little effect on UCS value, when used with size correction (Agustawijaya 2007). Matthews and Clayton (1993) showed that there was no variation in UCS result for sample ratio between 2.0 and 2.5. For granular materials as per the ISRM standard, sample diameter needs to be 10 times to that of the largest grains (Fjaer 1992; Schön 2015). A few most common direct laboratory testing methods to estimate rock strength are (i) unconfined test conducted under zero confining pressure, which is cheaper and quicker, (ii) single-stage triaxial compressive test (SST) where testing is carried on multiple samples (ideally four to five plugs) at different confining pressures, and (iii) multistage triaxial compressive test (MST) when sample shortage is there, triaxial testing was carried out on single sample at different confining pressures (Bell 2005; Khaksar et al. 2009). Schön (2015) explained different parts of triaxial testing equipment (Figure 9.1) and

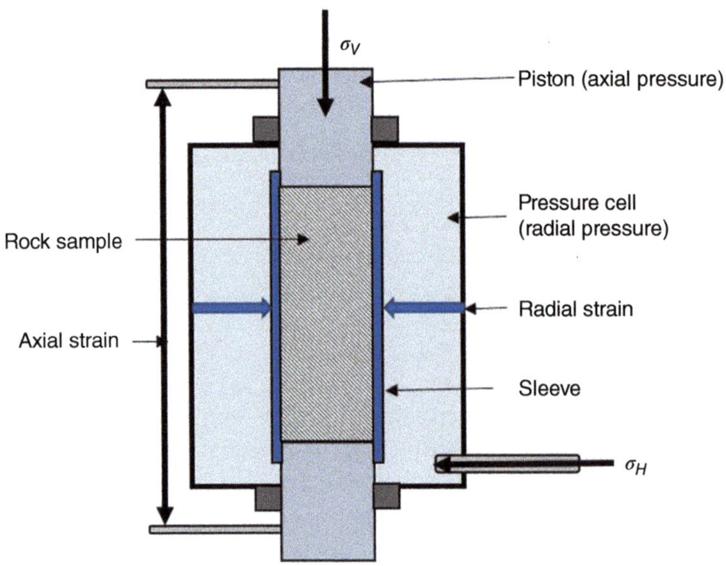

Figure 9.1 Triaxial pressure cell with cylindrical rock sample covered by sleeves. Axial and radial stress and strains are measured by corresponding stress and strain system.

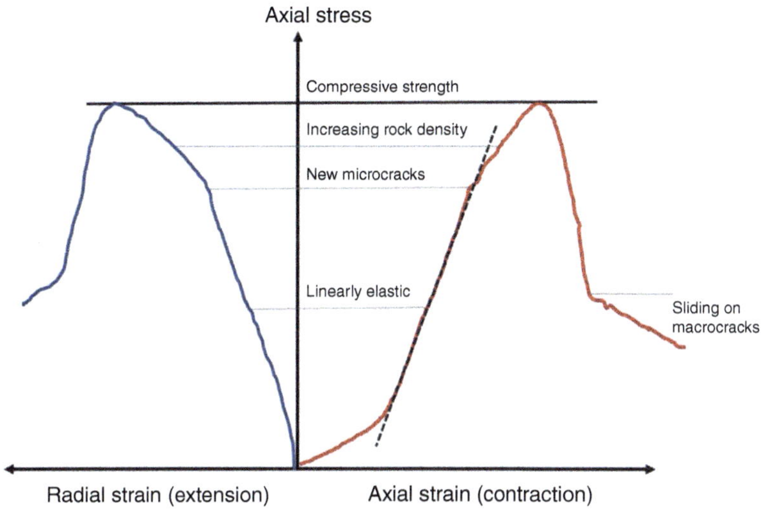

Figure 9.2 Schematic of stress-strain response curve indicates mechanical behavior during triaxial testing.

behavior of stress-strain curve (Figure 9.2) during each phase of testing. While in the earlier mentioned triaxial testing, horizontal stresses are uniform on a cylindrical plug sample, i.e. no intermediate radial stress component is considered, in a true triaxial test, intermediate stress is applied on cube-shaped samples (Haimson and Chang 2000; Schön 2015). Lab-based strength measurements are also affected by several physical properties e.g. water saturation, mineralogy, grain distribution (Agustawijaya 2007; Iskander et al. 2015), etc. Thus, it is also necessary to understand the mineralogical and structural properties (Chakraborty and Mukherjee 2020) of the core samples. X-ray diffraction (XRD) analysis serves the purpose to understand the mineralogical and structural composition of the plug samples (Wang 2016). In this study, the result of XRD analysis is used to understand the concentration of mineral(s) that may govern the rock strength.

Though the core test for rock strength is the most appropriate method, it does not lead to a continuous profile of rock strength along a wellbore. Laboratory direct measurement techniques are expensive and the results are sensitive to stress unloading (Chang et al. 2006). Unlike direct measurement approach, indirect methods are relatively simple and generally do not require any sample preparation. Chang et al. (2006) and Khaksar et al. (2009) showed a number of empirical relations based on petrophysical logs for different rock types (sandstone, shale, limestone, and dolomite). Table 9.1 shows some published empirical relation between UCS and rock physical properties for limestone and dolomite. Aladejare (2020) compares different published empirical relations with lab-based measurements.

All these empirical relations were core calibrated from certain basins. The ubiquitous application of these empirical relations for rock types under various geological settings may or may not define the actual strength of rocks. Thus, new relationships need to be established from core testing data for local scale studies. Ameen et al. (2009) calculated log-based rock mechanical properties in Arab-D carbonate reservoir, Ghawar field, and four hundred plug samples were used for calibration purpose.

Table 9.1 Empirical relationship between physical properties of rock and rock strength (UCS) for limestone and dolomite (Chang et al. 2006; Hadi 2018).

UCS (MPa)	Region where developed	General comments	Reference
$276\,(1-3\phi)^2$	Korobcheyev deposit, Russia	—	Rzhevsky (1971)
$(7682/\Delta t)^{1.82}/145$	—	—	Militzer and Stoll (1973)
$10^{(2.44+109.14/\Delta t)}/145$	—	—	Golubev and Robinovich (1976)
$174.8\exp(-9.3\,\phi)$		$R^2 = 0.68$	Farquhar (1994)
$13.256\,E_S{}^2 + 16.952\,E_s$			Lacy (1997)
$3.14\,E^{0.25}/\,\phi^{0.45}d_m{}^{0.5}$			Hatzor (1998)
$(Vp/0.0317) - 63.71$			Yasar and Erdogan (2004)
$13.8E^{0.51}$	—	Limestone with $10 < UCS < 300$ MPa	Chang et al. (2006)
$25.1E^{0.34}$	—	Dolomite with $60 < UCS < 100$ MPa	Chang et al. (2006)
$143.8\exp(-6.95\,\phi)$	Middle East	Representing low to moderate porosity ($0.05 < \phi < 0.2$) and high UCS ($30 < UCS < 150$ MPa)	Chang et al. (2006)
$135.9\exp(-4.8\,\phi)$		Representing low to moderate porosity $0 < \phi < 0.2$) and high UCS ($30 < UCS < 300$ MPa)	Chang et al. (2006)
$31031/(\Delta t^{1.5}\,d_m{}^{0.25})$			Prasad (2009)
$2.94\,(E^{0.834}/\phi^{0.088})$			Asef and Farrokhrouz (2010)
$11.05\,E_S{}^{0.66}$			Najibi (2015)
$3.67\,Vp^{2.14}$			Najibi (2015)
$81.386\exp(-6.5\,\phi)$			Faÿ-gomord (2016)

In this paper, a compressional slowness (Δt)-based empirical relation is established based on the available core mechanical testing data in a Proterozoic carbonate rock samples from central India. Both triaxial and uniaxial testing were carried out in core plug samples. Statistical methods were used to define the relationships between acoustic slowness and UCS values. Results of other published equations are also discussed in this paper.

9.2 Methodology

9.2.1 Core Inventory

Analyses were carried out in core samples acquired from Proterozoic carbonate reservoir from central India. The testing program for the samples retrieved involved both single-stage triaxial and uniaxial tests along with the determination of UCS for limestone samples from four wells. Total seven sets of core plugs were prepared for the testing. Three sets of samples were used for multistage triaxial testing and rest of the four sets of samples were considered for uniaxial testing (Table 9.1).

9.2.2 Lithology Description of Core Samples

Megascopic interpretation of core samples from a different depth indicates lithology varying from lime to silty-shale to siltstone. Deposition environment is shallow marine siliciclastic and carbonate (Sen and Mishra 2015). Alternate bands of shale were observed within limestone layers. Further, X-ray diffraction (XRD) analysis provided the mineralogical content. Relatively clean limestone zones comprised more than 70% calcite mineral with traces of clinochlore and illite. With increasing clay minerals, calcite concentration reduced to 42%. Helium porosity defined significantly low matrix porosity (0.14–4%). Air permeability shows the permeability range around 0.1–0.46 mD.

9.3 Results and Discussion

9.3.1 Core Testing Result

The main advantage of laboratory core testing is that the samples can be subjected to a whole suite of boundary conditions and stress constraints. This helps determine the point at which failure occurs where rock elastic properties can be determined. The triaxial compression tests were carried out to determine the cohesion and the friction angle based on multiple confining pressures using a high-pressure cell (Bell 2005). In the present study, the axial load was applied continuously between 0.5–1 MPa/s till failure of the sample. Testing of these samples was carried out in a hydraulic compression-testing machine using a 200-ton capacity load cell. Data for S3A sample was not used as it suffered damage during the testing. Figure 9.3 shows Mohr–Coulomb plot for triaxial testing of sample S1. The samples were tested in four different confining pressures. The plot shows calculated UCS as ~150 MPa. Cohesion (Co) appears around ~38.24 Mpa and internal friction angle (ϕ) ~36°. Based on the linear stress-strain behavior of the tested samples, the average Young's modulus and Poisson's ratio (PR) were calculated within the ranges of 13–59% of the failure stress uniaxial testing. Figure 9.4 illustrates stress and strain relation during uniaxial testing of sample S1. The compressional (P) and shear (S) wave velocities were also recorded by pulse transmission method using 0.1–1 MHz probes. Table 9.2 summarizes testing result of all the seven samples.

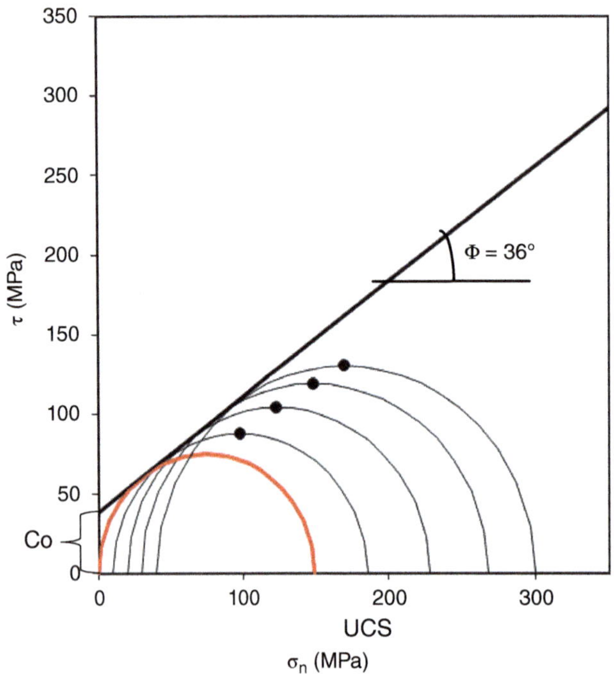

Figure 9.3 Mohr–Coulomb plot of multistage triaxial of sample S1. Based on plot-calculated UCS = ~150 MPa, Co = ~38 MPa and ϕ = ~36°. σ_n = Normal stress component, τ = Shear stress component.

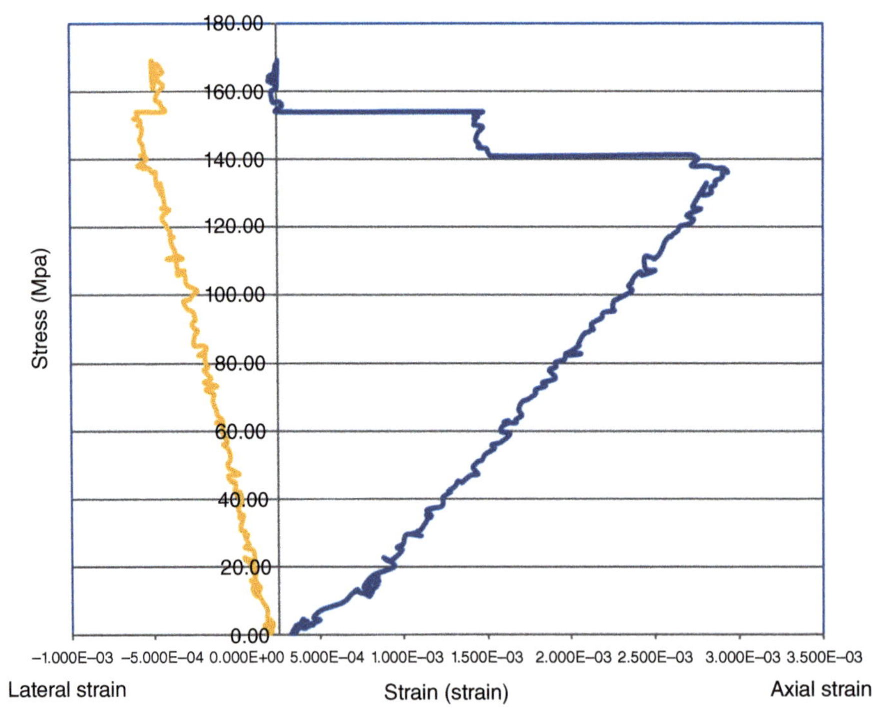

Figure 9.4 Stress and strain relation during uniaxial test of sample S1.

Table 9.2 Summarized rock mechanical testing result of five samples.

Sample ID	Lithology	P wave (m/s)	DTC (µs/ft)	S wave (m/s)	DTS (µs/ft)	UCS (Mpa)	Young's modulus (Gpa)	PR	Cohesion (Mpa)	Frictional angle (Degree)	Test type
S1	Limestone	6155	49.50	—	—	150	—	—	38.24	35.90	Triaxial
S2 A	Limestone	2685	90.30	—	—	95.48	—	—	24.8	35.25	Triaxial
S2 B	Limestone	2675	84.12	—	—	88.21	—	—	21.80	37.52	Triaxial
S2 C	Limestone	3878	78.50	2356	129.40	90.20	53.18	0.28	—	—	Uniaxial
S3 A	Limestone	5750	53.05	3153	96.69		—	—	—	—	Uniaxial
S3 B	Limestone	5778	52.72	3151	96.75	170	57.12	0.28	—	—	Uniaxial
S4	Limestone	5005	60.93	2541	119.98	190	56.84	0.28	—	—	Uniaxial

9.3.2 Core Calibration and Results

The UCS of rocks is calculated from the acoustic travel times using Eqs. (9.1 and 9.2) (Militzer and Stoll 1973; Golubev and Robinovich 1976, respectively) as these two methods are widely used for carbonate lithology:

$$UCS = \frac{\left(\frac{7682}{\Delta t}\right)^{1.82}}{145} \tag{9.1}$$

$$UCS = \frac{10^{2.44 + \frac{109.14}{\Delta t}}}{145} \tag{9.2}$$

Figure 9.5 shows comparison plot of UCS range derived using Militzer and Stoll (1973) (in blue), Golubev and Robinovich (1976) (in red), and core calibrated exponential relation (in green). In the figure, the green triangles indicate the UCS measured from the core tests. Plot shows that both the Militzer and Stoll (1973) (in blue) and the Golubev and Robinovich (1976) (in red) predictions do not seem to be well correlated to the measured values especially at lower slowness range (approximately <60 μs ft^{-1}). While the Golubev and Robinovich (1976) relation indicates an unrealistically high value of UCS, those determined by the standard Militzer and Stoll (1973) equation underpredict these values. Therefore, a modified equation had to be used instead to develop a better fit. The modified equation (Eq. 9.3), shown in the green curve is a better fit with the lab-derived UCS values. Table 9.3 indicates the core measured values along with those obtained from the standard equations. (Eqs. 9.1–9.3).

$$UCS = \frac{49490 \times e^{(-0.015 \times \Delta t)}}{145} \tag{9.3}$$

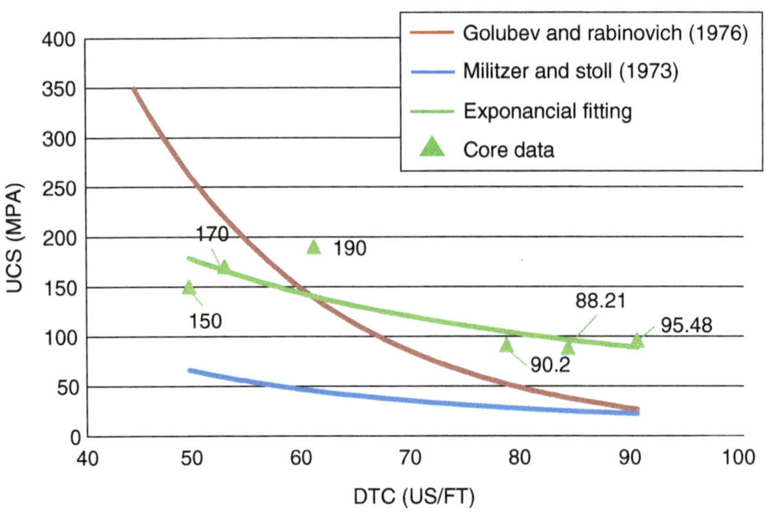

Figure 9.5 Plot shows DTC vs. UCS relations. Plot also indicates UCS value results based on Militzer and Stoll (1973) and Golubev and Robinovich (1976) relation showing significant deviation from actual measurement.

Table 9.3 Core-measured UCS value and calculated UCS value based on Militzer and Stoll 1973 and Golubev and Robinovich 1976 empirical relation based on core-measured acoustic slowness.

Sample ID	Δt ($\mu s\ ft^{-1}$)	UCS measured (MPa)	UCS Golubev and Robinovich (1976)(MPa)	UCS Militzer and Stoll (1973) (MPa)
S1	49.48	149.6	304.72	66.09
S3B	52.72	169	222.94	58.88
S4	60.93	192	117.29	45.24
S2A	90.33	95.45	30.64	22.09
S2B	84.10	88.14	37.65	25.16
S2C	78.49	90	46.62	28.53

DTC, Delay time of compressional wave (P wave); DTS, Delay time of Shear wave (S wave); P wave, Compressional wave velocity; PR, Poisson's ratio; S wave, Shear wave velocity.

Figure 9.6 shows the percentage variation histogram for all four samples considering three methods used to estimate the UCS. Analysis helps get an idea on accuracy of each method. As can be seen, the best-fitted regression equation works relatively better and the maximum variance is limited to 30% of the all core measured values. Militzer and Stoll (1973) and Golubev and Robinovich (1976) showed a variance of 50% of the true UCS, thereby reducing the reliability of the estimated properties in fast formation ($60\ \mu s\ ft^{-1} < \Delta t$). UCS values are calculated using the above-mentioned equations from the actual

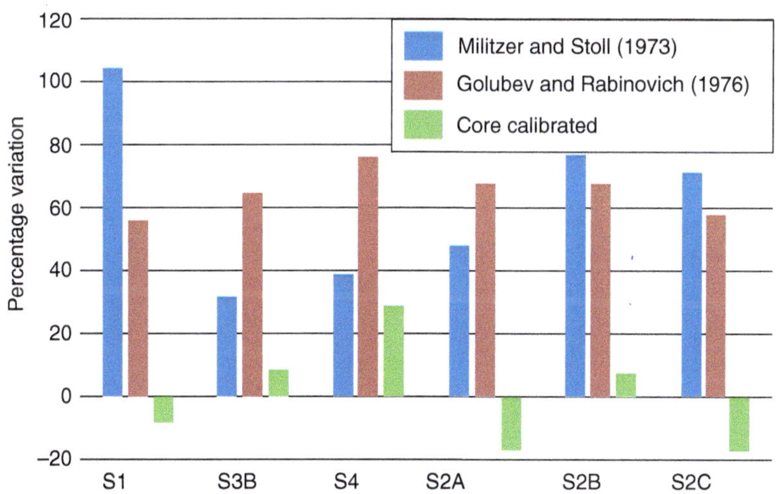

Figure 9.6 Plot shows percentage variation of calculated UCS value from measured value. It can be observed that Militzer and Stoll (1973) and Golubev and Robinovich (1976) based relation showing significant deviation from actual measurement for all the samples.

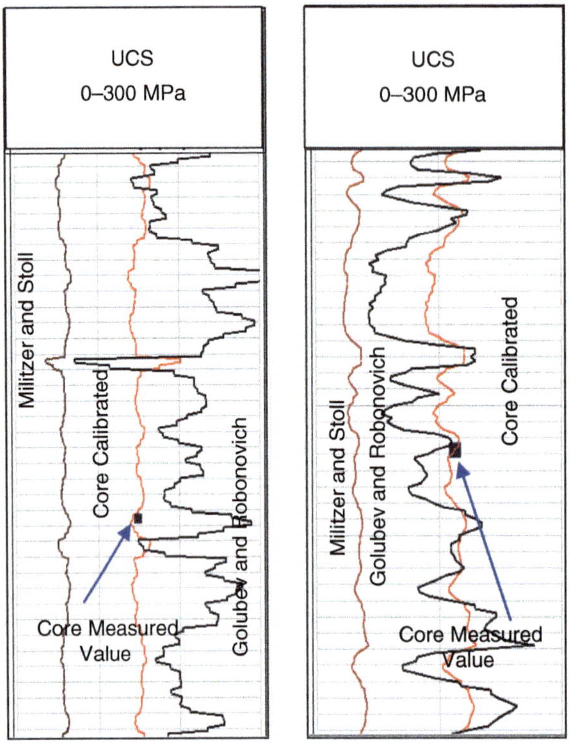

Figure 9.7 Continuous UCS value derived from acoustic log with core tested UCS in blue squares.

recorded acoustic log in the wells. Figure 9.7 shows UCS values derived from acoustic log in two wells. UCS derived from Militzer and Stoll (1973), Golubev and Robinovich (1976) and best-fitted regression curves are shown. Core-measured UCS values are presented as blue squares. From these two plots, we can further infer that best-fitted regression equation (Eq. 9.3) can be used for UCS calculation based on compressional slowness in lithology having high calcite content (>70%) and low matrix permeability.

9.4 Conclusion

Petrophysical log-based UCS calculation approach is a fast and cost-effective method and extensively used in the oil and gas industry.

To accomplish the objectives of this study, core testing data from different formations were utilized to calculate the rock strength along the wellbore. Various equations exist that relate petrophysical properties of rocks to rock strength (Chang et al. 2006). Although some equations may work reasonably well, rock strength variations with individual physical property measurements scatter considerably, indicating that most of the empirical equations are not significantly accurate to fit all data points from laboratory tests. Since rocks of the same lithology have different characterization due to different composition and depositional environments, it brings uncertainty to determine generic empirical equations for a certain rock type. Thus, most of the developed equations are presented for a specific region or rocks with specific characterization. This highlights the importance of local calibration for the equations before they are used to predict UCS for any stability analysis.

In this paper, a compressional slowness-based equation was established after calibrating core testing data in a low permeable Proterozoic carbonate formation. A total seven sets of core plugs were used for the calibration with the reservoir section. Mineralogical analysis done before shows high calcite content in the rock samples and the acoustic log-based exponential relation shows reasonably high matching with core-tested UCS value, unlike those with the commonly used empirical relations like Militzer and Stoll (1973) and Golubev and Robinovich (1976), that show significant deviation from the actual testing values. An uncertainty arises due to a smaller number of samples, which can be further modified using more sample test. However, the proposed regression equation can be useful for carbonate reservoir types having similar geological conditions.

Acknowledgments

Soumyajit Mukherjee invited, handled, and reviewed this article. Merryl Le Roux, Frank Otmar Weinreich, Joss Everett, Andrew Harrison, and the proofreading team (Wiley Blackwell) are thanked for their assistance. The article has been summarized in Mukherjee et al. (2023).

References

Agustawijaya, D.A. (2007). The uniaxial compressive strength of soft rock. *Civil Engineering Dimension* 9 (1): 9–14.

Aladejare, E.A. (2020). Evaluation of empirical estimation of uniaxial compressive strength of rock using measurements from index and physical test. *Journal of Rock Mechanics and Geotechnical Engineering* 12: 256–268.

Ameen, M.S., Smart, B.G.D., Somerville, J.M. et al. (2009). Predicting rock mechanical properties of carbonates from wireline logs (A case study: Arab-D reservoir, Ghawar field, Saudi Arabia). *Marine and Petroleum Geology* 26 (4): 430–444.

Asef, M.R. and Farrokhrouz, M. (2010). Governing parameters for approximation of carbonates UCS. *Electronic Journal of Geotechnical Engineering* 15: 1581–1592.

ASTM D2938-95 (2002). *Standard Test Method for Unconfined Compressive Strength of Intact Rock Core Specimens (Withdrawn 2005).* West Conshohocken, PA: ASTM International 1995.

Bell, F.G. (2005). Engineering geology |rock properties and their assessment. In: *Engineering Geology, Earth System and Environmental Science*, 566–580. Encyclopedia of Geology.

Bieniawski, Z. (1976). Rock mass classification in rock engineering. In: *Symposium Proceedings of Exploration for Rock Engineering* (ed. Z.T. Ieniawski), 97–106.

Chakraborty, M. and Mukherjee, S. (2020). Structural geological interpretations from unrolled images of drill cores. *Marine and Petroleum Geology* 115: 104241.

Chang, C., Zoback, M.D., and Khaksar, A. (2006). Empirical relations between rock strength and physical properties in sedimentary rocks. *Journal of Petroleum Science and Engineering* 51 (3–4): 223–237.

Farquhar, R.S.-W. (1994). porosity as a geomechanical indicator: an application of core and log data in rock mechanics. In: *European Petroleum Conference*, 481–489. London: Society of Petroleum Engineers.

Faÿ-gomord, O.D. (2016). Unraveling chalk microtextural properties from indentation tests. *Engineering Geology* 209: 30–43. https://doi.org/10.1016/j.enggeo.2016.05.005.

Fjaer, E.H. (1992). *Petroleum Related Rock Mechanics*. Amsterdam: Elsevier.

Gholami, R., Moradzadeh, A., Rasouli, V., and Hanachi, J. (2014). Practical application of failure criteria in determining safe mud weight windows in drilling operations. *Journal of Rock Mechanics and Geotechnical Engineering* 6: 13–25.

Golubev, A.A. and Robinovich, G.Y. (1976). Resultaty primeneia appartury akusticeskogo karotasa dlja predeleina proconstych svoistv gormych porod na mestorosdeniaach tverdychisjopaemych. *Prikl Geofiz Moskova* 73: 109–116.

Hadi, F. (2018). Geomechanical characterization and correlations to reduce uncertainty of carbonate reservoir analysis. Doctoral dissertation. Missouri University of Science and Technology, Geosciences and Geological and Petroleum Engineering. Missouri: Scholars' Mine a service of the Missouri S&T Library and Learning Resources. https://scholarsmine.mst.edu/doctoral_dissertations/2674.

Haimson, B. and Chang, C. (2000). A new true triaxial cell for testing mechanical properties of rock, and its use to determine rock strength and deformability of Westerly granite. *International Journal of Rock Mechanics and Mining Sciences* 285–296: https://doi.org/10.1016/S1365-1609(99)00106-9.

Hatzor, Y.P. (1998). A Microstructure-based Failure Criterion for Aminadav Dolomites. *International Journal of Rock Mechanics and Minings Sciences* 35 (6): 797–805. https://doi.org/10.1016/S0148-9062(98)00004-7.

Iskander, M., Omidvar, M., and Bless, S. (2015). Behaviour of granular media under high strain-rate loading. In: *Rapid penetration into Granular Media*, 11–63. Elsevier.

ISRM (2007). *The Complete ISRM Suggested Methods for Rock Characterization. Testing and Monitoring: 1974–2006*. International Society for Rock Mechanics, Commission on Testing Methods.

Javani, D., Aadnoy, B., Rastegarnia, M. et al. (2017). Failure criterion effect on solid production prediction and selection of completion solution. *Journal of Rock Mechanics and Geotechnical Engineering* 9: 1123–1130.

Khaksar, A., Tylor, P.G., Fang, Z. et al. (2009). Rock strength from core and logs: where we stand and ways to go. EUROPEC/EAGE Conference and Exhibition, Amsterdam, The Netherlands, (June 2009). SPE 121972.

Lacy, L.L. (1997). *Dynamic rock mechanics testing for optimized fracture designs*, vol. 58, 23–36. San Antonio, TX: SPE Annual Technical Conference and Exhibition https://doi.org/10.2118/38716-MS.

Mahanta, B., Singh, T.N., Ranjith, P.G., and Vishal, V. (2018). Experimental investigation of the influence of strain rate on strength; failure attributes and mechanism of Jhiri shale. *Journal of Natural Gas Science and Engineering* 58: 178–188.

Mahanta, B., Ranjith, P.G., Vishal, V., and Singh, T.N. (2020). Temperature-induced deformational responses and microstructural alteration of sandstone. *Journal of Petroleum Science and Engineering* 192: 107239.

Manshad, A.K. (2014). Analysis of vertical, horizontal and deviated wellbores stability by analytical and numerical methods. *Journal of Petroleum Exploration and Production Technology* 359–369.

Matthews, M.C. and Clayton, C.R.I. (1993). Influence of intact porosity on the engineering properties of a weak rock. In: *Geotechnical Engineering of Hard Soils – Soft Rocks* (ed. Anagnostopoulos et al.), 693–701. Rotterdam: Balkema.

Militzer, H. and Stoll, R. (1973). Einige Beiträige der Geophysik zur primäedatenerfassung im Bergbau Neue Bergbautechnik. *Leipzig* 3: 21–25.

Mukherjee, S., Dasgupta, S., Majumdar, C. et al. (2023). Introduction to: Handbook of Petroleum Geoscience: Exploration, Characterization, and Exploitation of Hydrocarbon Reservoirs. In: *Handbook of Petroleum Geoscience: Exploration, Characterization, and Exploitation of Hydrocarbon Reservoirs*, Petroleum Industry Updates: Geoscientific and Economic Issues (ed. S. Mukherjee, S. Dasgupta, C. Majumdar, et al.). John Wiley & Sons, Hoboken, ISBN 978-1-119-68003-1.

Najibi, A.R. (2015). Empirical relations between strength and static and dynamic elastic properties of Asmari and Sarvak limestones, two main oil reservoirs in Iran. *Journal of Petroleum Science and Engineering* 78–82: https://doi.org/10.1016/j.petrol.2014.12.010.

Pradhan, S.P., Vishal, V., and Singh, T.N. (2018). Finite element modelling of landslide prone slopes around Rudraprayag and Agastyamuni in Uttarakhand Himalayan terrain. *Natural Hazards* 94 (1): 181–200.

Prasad, U.D. (2009). Improved method for estimating the strength of carbonate rocks. In: *International Petroleum Technology Conference. IPTC*, 7–9. Doha, Qatar: International Petroleum Technology Conference.

Rzhevsky, V.N. (1971). *The Physics of Rocks*. Moscow: Mir Publisher.

Sarkar, K., Vishal, V., and Singh, T.N. (2012). An empirical correlation of index geomechanical parameters with the compressional wave velocity. *Geotechnical and Geological Engineering* 30 (2): 469–479.

Schön, J.H. (2015). Geomechanical properties. *Developments in Petroleum Science* 65: 269–300.

Sen, S. and Mishra, M. (2015). Geochemistry of Rohtas limestone from Vindhyan Supergroup, Central India: evidences of detrital input from felsic source. *Geochemistry International* 53 (12): 1107–1122.

Singh, R., Vishal, V., Singh, T.N., and Ranjith, P.G. (2013a). A comparative study of generalized regression neural network approach and adaptive neuro-fuzzy inference systems for prediction of unconfined compressive strength of rocks. *Neural Computing and Applications* 23 (2): 499–506.

Singh, T.N., Pradhan, S.P., and Vishal, V. (2013b). Stability of slopes in a fire-prone mine in Jharia Coalfield, India. *Arabian Journal of Geosciences* 2 (6): 419–427.

Vishal, V., Ranjith, P.G., and Singh, T.N. (2015). An experimental investigation on behavior of coal under fluid saturation, using acoustic emission. *Journal of Natural Gas Science and Engineering* 22: 428–436.

Wang, C.G. (2016). Slag processing. In: *The Utilization of Slag in Civil Infrastructure Construction*, 87–113. Elsevier Science.

Yasar, E. and Erdogan, Y. (2004). *International Journal of Rock Mechanics and Mining Sciences* 41 (5): 871–875.

10

Review on Organic Porosity in Shale Reservoirs

Deependra Pratap Singh[1,2], *Bodhisatwa Hazra*[1,2,*], *Vivek Singh*[1], *and Pradeep K. Singh*[1]

[1] *CSIR-Central Institute of Mining and Fuel Research, Dhanbad, Jharkhand, India*
[2] *Academy of Scientific and Innovative Research (AcSIR), Ghaziabad, Uttar Pradesh, India*

**Author for correspondence: bodhisatwa.hazra@gmail.com; bhazra@cimfr.nic.in*

Abstract

In recent years, with the commercial success of exploiting hydrocarbons from unconventional shale and coal reservoirs, porosity hosted by the organic matter within them has gained international attention. The organic porosity is presumably the more vital component of total porosity, as the hydrocarbons are primarily stored within them through adsorption. In this work, we review the information, analysis, and developments related to organic porosity in these strategically significant hydrocarbon reservoirs. Considering all factors, thermal maturity is the most critical factor that controls the formation and evolution of organic matter nano-porosity. Scanning electron microscopy (SEM) with argon ion milled surfaces has emerged as the more favored tool for understanding the porous structures within organic matter. However, one major limitation of SEM is its inability to image pores <5 nm in diameter. Correspondingly, microporosity, the most important site of gas storage in shale reservoirs, is not identifiable through SEM. Certain controversies exist regarding the usage of argon ion milling for preparing smooth samples for reliable SEM imaging. Notwithstanding, few researchers have highlighted its benefits. Other techniques, viz. gas adsorption has been used by researchers globally. Measurements made using N_2 and CO_2 at low temperatures and pressures have also revealed the existence of pores within organic matters, which allows gas to be stored. While both have been observed to map microporous structures, N_2 at low experimental temperatures of −197.3 °C lacks the thermal drive to access all microporous structures. Each technique has certain limitations, and for thorough assessment and documentation of organic porosity, a multifaceted approach comprising several techniques, beyond imaging and singular gas adsorption, is to be adopted.

Keywords *Porosity; Organic Matter; Gas Adsorption; Microscopy; Unconventional Reservoirs*

Handbook of Petroleum Geoscience: Exploration, Characterization, and Exploitation of Hydrocarbon Reservoirs, First Edition. Edited by Soumyajit Mukherjee, Swagato Dasgupta, Chandan Majumdar, Subhandip Mandal and Troyee Dasgupta.
© 2023 John Wiley & Sons Ltd. Published 2023 by John Wiley & Sons Ltd.

10.1 Introduction

One of the most critical and interesting aspects of geology is the presence of porous structures or porosity in rock formations (Mukherjee 2018a; Dasgupta and Mukherjee 2020). From its role in building stones to aquifers, from conventional oil and gas reservoir rocks to unconventional hydrocarbon reservoirs, the impact of porosity is significant, manifold, and consequently has received widespread attention (Anovitz and Cole 2015) including even in tectonic modeling (Mukherjee 2017, 2018b, 2018c). Among the sedimentary rocks, the main points of interest for commercial purposes, conventional sandstone and carbonate reservoirs are marked by coarser pores, which are in the magnitude of μm. In contrast, for unconventional reservoirs, viz. shale and coal bed methane (CBM), the pore sizes vary between nm and μm (Ougier-Simonin et al. 2016). For tight sandstone reservoirs, with exposure to generally high burial depths (>3.5 km) and concomitant increased pressure and temperature, porosity is reduced, brought forward by both compaction and cementation (Ehrenberg et al. 2009; Bowen et al. 2011). Further for tight sandstone reservoirs, the type and degree of diagenesis strongly affect the morphology, shape, distribution, and evolution of porous structures (Nelson 2009; Nabawy and David 2016). Previous researches have pointed out that the reduction in porosity of tight sandstone reservoirs due to greater depths has also been accompanied by presence of different types of mineral grain contacts and grain-to-grain sutured contacts (Higgs et al. 2007; Mozley et al. 2016). The first expression of compaction-reduction in response to burial depth was presented by Athy (1930), depicting an exponential decrease in porosity with increasing depth and compaction. The law presented by Athy (1930), Athy's law as it is widely known, is one of the most simplistic expressions of changing porosity with increasing depth, and consequently has been used in several studies (Baldwin and Butler 1985; Mello et al. 1994).

Although coming under the same category of unconventional hydrocarbon reservoirs as the tight sandstone horizons, pore systems in shale and CBM systems have received additional interest due to their variety, and the fact that both organic matter and inorganic matter within them control the pore system (Loucks et al. 2012; Jiao et al. 2014; Wood and Hazra 2017). Similar to sandstones, with increasing burial depths, the interparticle porosity and inorganic matter porosity in shales reduce due to compaction, cementation, etc. (Milliken and Day-Stirrat 2013; Mukherjee and Kumar 2018). The importance of inorganic porosity and its role in controlling the storage and movement of hydrocarbons in shale systems has been discussed (Ross and Bustin 2009; Curtis et al. 2012a). Porosity of clay minerals has also been studied and detailed by several researchers. For example, Cao et al. (2016), on their work on Chang-7 sandstones, observed a variety of pore shapes, viz. ink-bottle, slit-like, narrow slit-like, etc., present within clay minerals. Hazra et al. (2016) observed presence of microfractures and micro-cracks within an inorganic matter of Permian shales from India, which are coarser than those hosted by the organic matter. Previous researchers also documented the role of clay minerals on the gas-storage capacity of shales. Ross and Bustin (2009) documented that clay minerals, viz. illite have appreciable microporosity and can adsorb gases, while other clay minerals have significant mesoporosities. Earlier studies on natural clay minerals and soils have also indicated presence of microporous structures (Aringhieri 2004). For clay-rich shales, Ji et al. (2012) observed appreciable methane adsorption of capacity following the Langmuir monolayer coverage theory, although their results showed the highest methane sorption capacity for montmorillonite clays. However, their results also showed much higher sorption capacities for organic-rich shale members than the clay-rich members. They also observed that the methane adsorption capacity of clay-rich shales is mainly controlled by their specific surface areas (SSA). The mineral-SSA is a key factor that enumerates the interactive manner at solid–gas interfaces (Ji et al. 2012). Further, Cheng and Huang (2004) in their work on clays and organic matter in shaly-coal, oil-shale, and graphite, noted nearly-similar adsorption capacity of kaolinite and montmorillonite clays, compared to that of the Green River oil shale with a TOC content of 20.2 wt %. However, the adsorption capacity of the shaly coal and graphite was observed to be higher, indicating the role of organic richness and pore structure in the gas adsorption process. However, interestingly, the N_2-derived SSA for the above-mentioned clay minerals were observed to be much higher (~5–30 times) than the shaly coal and oil-shale samples. This and a few other results presented by other researchers (see Section 5) show exceptionally low N_2-SSA for organic-rich rocks (Hazra et al., 2021), and indicate that, unlike clays, the methane adsorption capacity of such rocks is not controlled by their SSA.

The clay minerals create a variable complex network within shales from microscopic and macroscopic scales. In fact, the more oriented the clay minerals are at microscales, the more anisotropic they are at macroscales (Sarout and Guéguen 2008a, 2008b). The clay sheets preferentially are aligned parallel to the bedding plane of the shales during the sedimentary processes (Ho et al. 1999), and this along with the orientation of the organics produces mechanical anisotropy within the shales (Vernik and Nur 1992; Vernik and Milovac 2011). If such a shale is deformed tectonically, deformation would intensify at the planes of anisotropy. In general, the bedding planes in organic-rich shales are not distinguishable ordinarily and often develop fissility, due to the bedding's parallel preferred orientation of the clay minerals (Blatt and Tracy 1996). Several researches showing the application of FE-SEM have documented the presence of elongated and/or flattened intercrystalline pores with clay mineral flakes oriented preferably. Interestingly, Schieber (2010) noted that for high-TOC shales, the pores within phyllosilicates are filled up by organic matter, while for the low-TOC shales, the same remains vacant providing a pathway for hydrocarbon flow.

While the inorganic matter-associated porosity in shale reservoirs is undoubtedly important, the organic mater hosted porosity in shale source-cum-reservoir rocks and in CBM systems has drawn more attention globally. Subsequent sections of this article will focus on the organic matter hosted porosity of the shale reservoirs.

10.2 Organic Matters and the Porosity Hosted by Them

Organic mater-hosted porosity has been commonly accepted as the more vital component of shale pore system and has been observed to be a critical element in successful shale gas reservoirs, viz. the Barnett, Woodford, Marcellus, etc., shale plays (Loucks et al. 2009, 2012; Milliken et al. 2013). Imaging techniques have greatly been used to document the nature and morphology of pores hosted by the organic components of shale rocks. Mann et al. (1994) first imaged organic matter pores in the Posidonia shales (Germany) using cryo-SEM. However, the fact that organic matter in shales hosts porosity and represents the most critical part of shale porosity gained popularity after Loucks et al. (2009) used argon-ion-milled samples for SEM analysis to document the presence of micropores in shale organic matter. Since then, this technique has been applied by several researchers to discuss the nature of the organic porosity of shale reservoirs. Organic matter in source rocks can generally be divided into two categories: (i) primary organic matter related to the depositional setting in which the rock was laid, and is of two broad categories – (a) non-extractable kerogen/maceral part, and (b) extractable or soluble part not related to hydrocarbon generation/expulsion; and (ii) secondary organic matter formed by alteration of primary components and is also composed of (a) extractable components (bitumen), and (b) non-extractable components (pyrobitumen) (Tissot and Welte 1984; Jarvie et al. 2007; Vandenbroucke and Largeau 2007). Microscopically, bitumen/pyrobitumen generally occurs as fracture filling or void filling materials, and are differentiated from kerogen, which typically shows their own morphologies (Jacob 1989; Mastalerz et al. 2018).

The mechanism of the generation of porosity in organic matter, in general, has been continuously searched by several researchers. A range of techniques, viz. imaging and scattering techniques and gas adsorption have been used in recent years to characterize the porosity of organic matter (Ross and Bustin 2009; Bernard et al. 2012; Mastalerz et al. 2012, 2017; Curtis et al. 2012b; Clarkson et al. 2013; Milliken et al. 2013; Labani et al. 2013; Bernard and Horsfield 2014; Romero-Sarmiento et al. 2014; Dong et al. 2015; Pommer and Milliken 2015; Holmes et al. 2017; Peng et al. 2017; Yang et al. 2017; Zhou et al. 2017; Sun et al. 2018, 2020; Hazra et al. 2018a, 2018b, 2018c, 2019a; Valentine et al. 2019; Vishal et al. 2019; Chandra and Vishal 2020, 2021; Wang et al. 2020; Chandra et al. 2020a, 2020b, 2022).

It has been commonly interpreted that the genesis and evolution of organic porosity are directly controlled by the thermal maturity levels of organic matter (Loucks et al. 2012). Researchers have argued that organic matter pores develop due to the expulsion of hydrocarbons from kerogen/macerals within the rock owing to increased maturity levels, thereby forming porous structures (Wang and Reed 2009). In their pioneering work on Barnett shales, Loucks et al. (2012) observed organic nanopore development to be strongly dependent on the thermal maturity levels of the samples (determined through vitrinite reflectance). They inferred that this is essentially caused due to conversion of kerogen to hydrocarbons and formation of pores within the organic matter in the process. Even more interestingly, they stressed upon the importance of presence of reactive and inert organic matter within the samples. For reactive or convertible organic components, as their hydrogen contents are higher, with increasing thermal maturity levels, hydrocarbons are generated and, consequently, pores are formed. On the other hand, for the inert organic matter with lower hydrogen contents, not many hydrocarbons are generated and, consequently, the formation of organic porosity in response to increased thermal maturity levels is little or insignificant (Loucks et al. 2012).

Liu et al. (2017) also discussed the importance of thermal maturity and organic matter type (reactive or inert). In one of the most fundamental works, using correlative optical-electron microscopy, they documented absence of any pores in the organic matter of the New Albany Shale (USA) at the early oil-window stage of maturity, while the samples at slightly higher thermal maturities (peak oil window) develop "occasional" mesoporous to macroporous structures. However, at a higher reflectance value of 1.41%, they observed a greater concentration of organic pores within the shale. They attributed these results to the generation and expulsion of greater amounts of hydrocarbons at higher thermal maturity levels. The increase in the volume of organic matter hosted porosity with increasing maturity compared to the low mature shales as observed by Liu et al. (2017) were essential for the solid bitumen present within the samples. For inertinites, cellular macropores (not dependent on thermal maturity level but on the inherent character) ranging between 200 and 10 000 nm occur within the shales across all thermal maturity levels. This also indicates the importance of the type of organic matter present, which would directly control the hydrocarbon generation potential, and consequently the formation of secondary porosity from the expulsion of hydrocarbons. For gas mature Barnett shales (USA), Bernard et al. (2012) also observed presence of nanoporous organic matter within samples collected from outcrops to depths of about ~2000 m. With increasing thermal maturity levels their results indicated that aromaticity of the kerogen increased, hydrogen indices dropped, generation of gaseous hydrocarbons due to secondary thermal cracking of oil/bitumen, and concomitant formation of secondary nanopores within the organic matter.

In another study on Eagle Ford shales (USA), Jennings and Antia (2013) observed lack of organic pores in samples with reflectance values <0.7%, while the organic pores occur within samples with vitrinite reflectance values greater than 1%. In fact, the organic matter hosted porosity was observed to be highest for samples with reflectance values between 1.4 and 1.6%. For samples with reflectance values >1.7%, they significantly observed organic matter hosted porosity to be less prominent and only a few such pores were observed. Similarly, Fishman et al. (2013) observed from the Eagle Ford shales poorly developed organic porosity within oil-window mature samples, while the same being well-developed within wet-gas window mature shales. The importance of thermal maturity and the type of organic matter present was also stressed by Curtis et al. (2012a). For the oil-window mature shales, they observed no organic porosity, while for samples with higher thermal maturities (vitrinite reflectance > 1.23%), a greater abundance of organic pores was observed. However, the importance of organic matter type was further stressed by their documentation presence of porous and nonporous organic matter adjacent to one another, for a sample with a reflectance value of 1.4%. Similar, porous and nonporous organic matter was also observed by Hazra et al. (2016) for Permian shale reservoirs (Lower Gondwana, India).

Using artificial maturation experiments as a tool, few researchers have tried to document the evolution of porosity in shale reservoirs. With such samples through anhydrous pyrolysis, Chen and Xiao observed an increase in nanoporosity of organic matter from the oil-window maturity stage to simulated vitrinite reflectance values of 3.5%. Jiang and Daigle (2019) artificially matured (pyrolyzed) two samples each from Eagle Ford Formation and Rock Mountains, USA, at four different temperatures (pyrolysis) of 110, 250, 450, ≥ 600 °C, and observed an increase in BET SSA, pore volumes, total porosity and permeability. They reasoned this to be caused due to progressive removal of hydrocarbons brought forward by elevated temperatures. The fact that organic matter hosted pore volume and overall microporosity increases with increasing maturity levels, especially beyond the oil window with the removal of hydrocarbons was also noted by Guo et al. (2017) during artificial maturation of type II kerogen-bearing Yanchang Shale. Similarly, Hu et al. (2015) during artificial maturation of Woodford shale to five distinct thermal maturity levels through hydrous pyrolysis noted an abundant increase in mesoporosity (up to 50 nm pore diameter) to mature to maximum oil-generation capacity. Finer mesopores (2–6 nm diameter) increase in volume for the sample when the oil is cracked to gas.

All the above points discussed put forward the importance of thermal maturity in controlling the evolution of porous structures within organic matter. While the type of organic matter is very important (as the certain organic matter may be more reactive or inert depending upon its hydrogen content), the importance of thermal maturity has been well documented by several researchers. Although the concept is well accepted now, several researchers have also come up with counter-observations (see next sections), which indicate that the evolution of organic porosity may be tricky and not straightforward in predicting.

10.3 Use of Scanning Electron Microscopy and Argon Ion Milling

A reason for the adoption of SEM as a tool to unravel the organic porosity of unconventional reservoirs is the emergence and utilization of techniques such as focused ion beam and argon ion milling of samples to prepare flat, smooth sample surfaces that allows exhaustive and precise imaging of microstructure in rocks. The need for better documentation of microstructures and pore systems has been the most defying task for petrophysicists and organic geochemists to explore certain and steadfast findings for the best interpretation of hydrocarbon accumulation and flow pattern in shale horizons (Loucks et al. 2009, 2012; Passey et al. 2010). It is thus imperative that the preparation of samples for SEM studies is precise so that important deductions can be made. Argon-ion milling has increasingly received more attention for the preparation of ultra-smooth polished surfaces, allowing going ahead with high magnification SEM, with high spatial resolution, to overcome the irregularities popped out by conventional techniques (Loucks et al. 2009; Wang and Reed 2009). During the conventional grinding of material, even following the utmost care, there is always a possibility of the creation of artificial pores and fractures by smearing and scratching out soft materials like clays and organic materials that once imaged could yield inaccurate porosity. Furthermore, such artificial pores are created along the edges of softer and more resilient phases like quartz and feldspar. To avoid these issues, argon-ion milling for polishing has been used by many researchers to iron out the artificial pores and their associated impact on sample observation and to go further analyze/characterize the true mineral texture and pore structure/morphology (Erdman and Drenzek 2013). The ion milling process aims to flatten the sample surface (± 5 nm) comparatively by spattering or ejecting particles from the target surface with controllable accelerated ions either through the broad ion beam (BIB) or focused ion beam (FIB) milling machines (Klaver et al. 2012). These instruments are designed in such a way that their voltage beam, ion-source, current, ion-gun angle and stage settings, e.g. rotation, rpm, and sample cooling can be maneuvered. The effect of milling rate on target largely depends on the above settings, as well as on the characteristics of the sample, viz. crystalline, noncrystalline, heterogeneity, thermal conductivity, etc.

The advantage of ion milling, especially BIB, is that it allows precise polishing of ductile materials such as clays and organic matters, without introducing any deformation of their structures (little deformation would be induced). In comparison to conventional polishing using water, ion milling represents a water-free polishing mechanism, which is especially significant when considering shale samples having a variety of clay minerals (selling clays, etc). Ion milling also removes any residual debris from the sample surface, thereby facilitating precise and correct imaging of pores and estimation of porosity through SEM at extremely high resolutions (Grobe et al. 2017).

Concerns have been raised regarding the potential impact of milling (FIB and argon ion) on porosity interpretations through SEM observations (Slatt and O'Brien 2011). Although ion milling does produce smooth surfaces that allow precise imaging (Desbois et al. 2009), several deleterious impacts of it on organic matter properties have been reported. Phaneuf (2005) first reported that owing to ion milling certain structural-fabric damages are induced within samples, causing destruction of porous structures within the mineral matter. This also implies that if ion milling reduces the non-organic pores, then in a way it may amplify the importance of organic pores of shales (Katz and Arango 2018). The technique of ion milling can itself cause certain modifications within the sample. Bassim et al. (2012), in their fundamental work on techniques for FIB milling of soft materials, observed that original coal samples and their milled counterparts had little spectral differences, as interpreted that carbonaceous materials are generally stable and strong under the influence of ion beams. It was opined that the organic microstructures present within samples are generally not altered within ion-milled samples, but the method of sample preparation can alter the nature of organic matter porosity. Grobe et al. (2017) through their extensive research documented that broad ion milling of coal samples helps in polishing organic matter without any mechanical damage. The broad ion milled surfaces were observed to be marked by enhanced surface fatness, absence of scratches and other residual debris (that often are encountered when water polishing is employed). For four coal samples with distinct thermal maturity levels having distinct vitrinite reflectance values (from conventional polishing methods), Grobe et al. (2017) observed an increase in reflectance of the samples when prepared through broad ion milling. The highest reflectance increase for BIB milled specimen in comparison to the conventional polishing specimen was observed for the graphite sample (5.29–10.04%). Many authors have used the rise of reflectance values owing to milling as an evidence of organic structure alteration (Sanei and Ardakani 2016; Mastalerz and Schieber 2017; Katz and Arango 2018). Using conventional grinding-polishing to prepare samples, followed by BIB polishing of the same sample, and then again repeating conventional polishing, Grobe et al. (2017) discarded the logic citing that had milling enhanced reflectance, the magnitude would have been more for immature or early mature samples, than graphite. They inferred that the rise in reflectance is essentially caused due to better sample surface polishing and flatness brought forward by milling (Figure 10.1).

Several authors stressed that ion milling can induce changes in organic matter properties. Klaver et al. (2016) investigated and documented that high vacuum conditions employed during sample preparation and polishing may induce devolatilization of labile organic compounds from organic matter, thereby forming secondary cracks, and leading to erroneous interpretations. A similar view was also put forward by Sanei and Ardakani (2016). In fact, the works of Sanei and Ardakani (2016) and Katz and Arango (2018) raised severe doubts about the usage of milling for sample polishing and its potential impact on thermal maturity interpretations of organic matter porosity. Using both cryogenic-broad ion beam milling (C-BIB) and FIB milling systems, Sanei and Ardakani (2016) showed a rise in reflectance values of macerals. While a 10–53% rise in reflectance values were observed when the

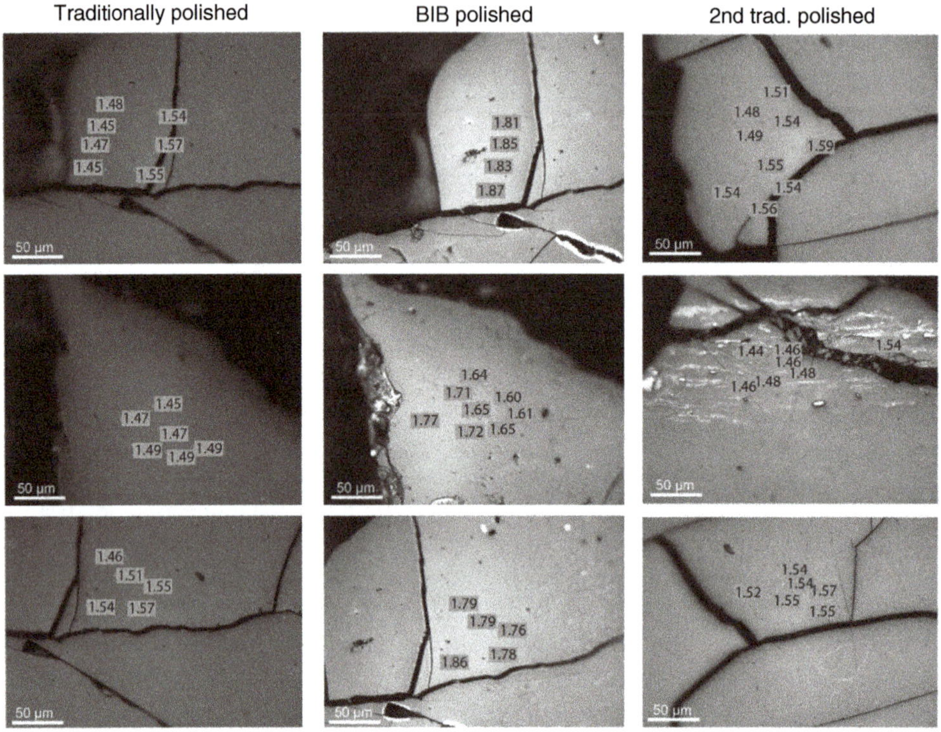

Figure 10.1 Coal sample having original reflectance of 1.52% (traditionally polished-grinded), showing increased reflectance of 1.70% (after BIB polishing). After a second conventional polishing, the reflectance values again dropped to 1.52%. *Source:* Grobe et al. (2017), Elsevier.

former was used, the application of the latter raised the reflectance values by almost ten times. The authors explained that this increase in reflectance in macerals was essentially caused due to ion milling induced thermal alteration and heat transfer to the specimen surface during milling. The heat transfer eventually induces the aromatization of organic materials and a concomitant rise in reflectivity. This lead the authors to further suggest that milling might create certain nanoporosity within organic matter due to devolatilization from increased heat transfer to the surface. Katz and Arango (2018) further documented the changes induced by milling and considered the importance of the organic composition of samples (prepared through hydrous pyrolysis) besides the reflectance measurements. For the oil-window mature sample, they observed a marked increase in reflectance for the ion milled sample split compared to the original sample split. Interestingly the solid bitumen content increased to 70% for the ion-milled split, while the same was nil in the original sample. This means that ion milling temperature on the surface of the samples rises and thereby induces pseudomorphic conversion and/or replacement of kerogen to solid bitumen. Note that Katz and Arango (2018) did not elaborate on the number of counts used to determine the maceral composition. For wet gas window mature samples, the authors observed changes in organic composition and increased reflectance in the ion milled sample-split, compared to the original sample. However, the quantum of change for the wet gas window mature sample was lower than that of the immature sample. This led the authors to suggest that the changes in composition and maturity of ion milled sample split from the gas window would be negligible.

Leaving aside the changes in organic composition, the inference on reflectance for ion milled samples presented by Katz and Arango (2018) is in exact contradiction to the result and idea presented by Grobe et al. (2017). Increased reflectance of macerals due to BIB compared to those prepared by conventional polishing was also noted by Valentine et al. (2019) in the order of liptinites> solid bitumen> vitrinites> inertinites. However, they did not observe any conversion or transformation of primary kerogen to solid bitumen, unlike that noted by Katz and Arango (2018). Valentine et al. (2019) considered coal and shale samples both across different thermal maturity intervals for their study to examine the impact of BIB on thermal maturity assessment. Further, the reflectance measurements were conducted on a large number of grains, across the different maceral types as mentioned above. Echoing the interpretations of Grobe et al. (2017), but with a different take, Valentine et al. (2019) inferred that kerogen modifies due to BIB, and the impact would have been larger for immature samples, which was not observed by them. They opined that the increased reflectance of different organic matter is caused due to increased surface flatness and smoothness resulting from BIB milling, and not due to thermal alteration. To validate their observations, they used atomic force microscopy (AFM) to show an increase in surface flatness and decrease in surface roughness brought forward by BIB milling, by comparing the same positions before and after milling. Significantly they inferred that the increased smoothness induced by BIB milling enabled the incident light to reflect from the sample surface at near-normal angles (specular reflectance) to the detector, unlike for mechanically polished samples with lower surface flatness, which allows a greater portion of reflectance to be diffuse reflectance (Figures 10.2 and 10.3).

Thus, different controversies exist about the potential impact of ion milling on organic microstructures. However, ion milling in producing flat surfaces for reliable imaging is also advantageous. Thorough documentation, using a wide range of samples [as shown by Valentine et al. (2019)], and corroborating the reflectance measurements from milled surfaces

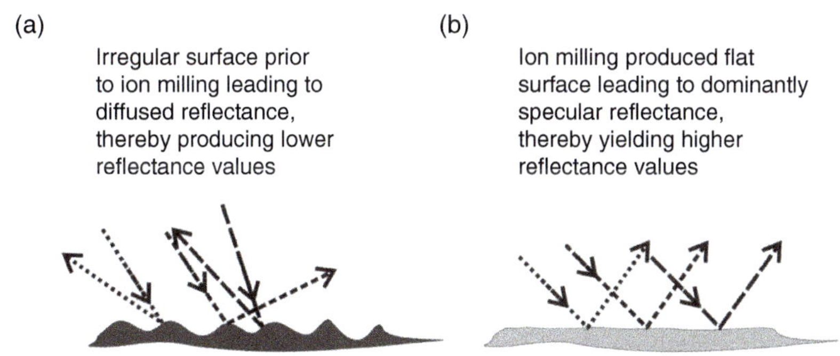

Figure 10.2 Schematic diagram showing the benefits of ion milling in producing samples with flat surfaces, thereby allowing specular reflectance of incident light rays and enhancing their reflectance values. *Source:* Diagram prepared based on the concept of Valentine et al. (2019)/with permission of Elsevier.

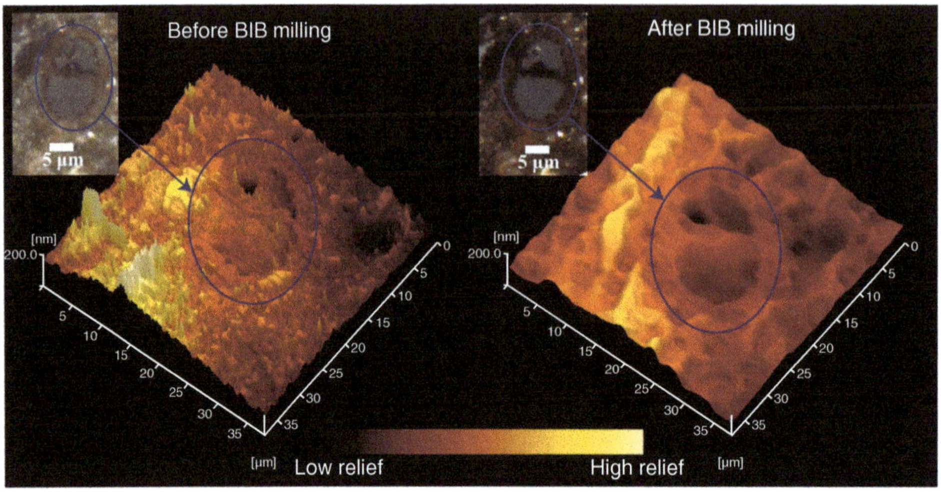

Figure 10.3 Before-milling and after-milling topographic features of a void-filling solid bitumen, as revealed by AFM analysis. An increase in solid bitumen reflectance from 0.39 to 0.55% was observed for before- and after-milled surfaces, respectively, owing to the increased surface flatness (as shown by AFM images) from ion milling [*Source:* Valentine et al. (2019), Elsevier.

with other maturity proxies, viz. Rock-Eval T_{max}, etc., should be adopted in the future. While reducing ion energy for cutting samples is one option, research focusing on the changes in measurements at different current levels across different sample types and maturities is needed in the future.

The major limitation of SEM examination of organic matter in shales or coals is that it cannot map smaller pores, which constitute a significant fraction of organic matter-hosted components (Loucks et al., 2009). Few researchers have detailed and put forward the importance of this fundamental while evaluating the organic porosity. For example, Klaver et al. (2015) observed that a major part of pores <5 nm in diameter was undercounted through SEM for the Haynesville shales. Liu et al. (2017) only considered pores >20 nm diameter. Imaging pores greater than that was found to be problematic through SEM. Several researchers using a combination of different techniques have discussed that a significant fraction of organic pores in gas shale and CBM reservoirs have widths <2 nm (Katz and Arango, 2018; Inan et al., 2018). For example, Yang et al. (2017) deduced abundant presence of pores <5 nm within the Longmaxi Shales (China). Pores within organics are generally smaller (Murphy et al. 2012): <1 nm in diameter for successful gas shale plays such as New Albany Shale (Strąpoć et al. 2010). It is imperative to understand that the <2 nm-sized pores significantly control the gas storage capacity of the above-mentioned reservoirs. In other words, pores not visible through SEM are the ones which have a stronger impact on the gas storage capacity of shale and coal reservoirs (Inan et al. 2018). Hence, many of the deductions on the impact of thermal maturity on the evolution of nanoporosity within organic matter are questionable and misleading and should be verified with other techniques such as gas adsorption, X-ray, neutron scattering, etc. Another factor that should be considered by researchers before applying SEM results for evaluating the organic porosity of reservoirs is the fact that SEM is generally conducted on small sample parts, and only represents a small part of the entire sample. More likely than not, the small portion of the sample being seen under SEM would not be representative of the entire sample or horizon. Thus, although SEM should be used for understanding the type of porous structures present within samples, using it to estimate porosity percentage may be erroneous, as it represents a localized presentation of the entire sample. Other techniques, viz. gas adsorption use crushed representative sample-part from a larger sample mass.

The last ten years have also seen the application of AFM for the characterization of organic matter in shales and coals. Javadpour et al. (2012) used different techniques, viz. phase-imaging, interactive-force spectroscopy, and friction mapping, of AFM to characterize organic materials in the shale system. Yang et al. (2017) in their landmark work introduced for the first time AFM coupled with infrared spectroscopy (AFM-IR) to assess both chemical and mechanical in-homogeneity of organic matter in unconventional reservoir rocks. For artificially matured New Albany shale through hydrous pyrolysis, using this unique combined feature, they showed that with increasing maturity as organic matter is aromatized, their stiffness also increased.

10.4 Primary Organic Porosity or Secondary Organic Porosity

As discussed in the preceding sections, the influence of thermal maturity on the evolution of organic porosity is undoubted. However, several works have also highlighted the lack of any relationship between the two. For example, Milliken et al. (2013) highlighted the importance of organic richness rather than thermal maturity levels in controlling organic porosity. Several researches focused only on porosity in the solid bitumen fraction (Hackley and Cardott 2016; Mastalerz et al. 2018), but not in the kerogen present within the samples. While some have opined that kerogen should not actually develop microporosity, as it is their structured nature that gets aromatized during hydrocarbon generation (Hackley and Cardott 2016; Inan et al. 2018). However, this is in strong contradiction to the previous landmark studies on gas adsorption capacity of coal samples, whereby the impact of kerogen type and thermal maturity levels have been highlighted (Crosdale et al. 1998; Laxminarayana and Crosdale 1999). Similarly, the importance of structured organic matter type (maceral) on the gas adsorption capacity of Permian shales and coals from India has been highlighted by Varma et al. (2014, 2015), Hazra et al. (2015). Figure 10.4a shows SEM image of a fusain lithotype (fossilized charcoal; typical type IV organic matter) within a Permian coal sample (Raniganj basin, India) showing presence of macropores, readily identifiable under the SEM. Figure 10.4b, on the other hand, shows the SEM image of a vitrain lithotype (typical type III organic matter) within a Permian coal sample (Raniganj basin, India) showing presence of pores, which are not clearly discernible under the SEM, even at higher magnifications. It is entirely possible that the porosity hosted by structured organic matter (as can be seen for the vitrain sample, Figure 10.4b) are extremely small (nanopores) missed in imaging by SEM, but can be identified by other porosity evaluation techniques. On the other hand, the porous structures of bitumens and type IV kerogen, being slightly larger, are imaged to a large extent by SEM and thus have received additional importance. It is thus understandable that interpretations of organic porosity should be validated across different probe types.

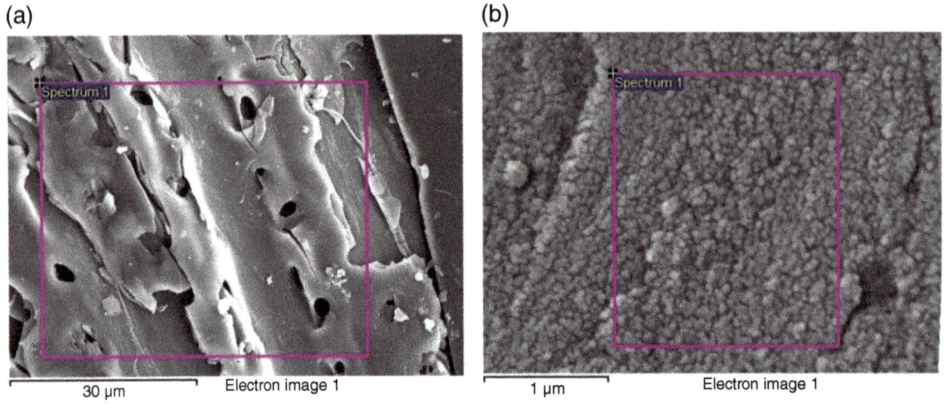

Figure 10.4 Scanning electron microscopy (SEM) image of fusain (a) and vitrain (b) lithotypes in Permian coals from India.

Löhr et al. (2015) questioned the evolution of porosity in organic matter. They observed presence of porous organic matter within kerogen and AOM of thermally immature (<0.4% reflectance) shales. Absence of such pores in mature samples (0.5–1.1% reflectance) further complicates the concept of the evolution of organic porosity. Similar to the immature samples, porosity was observed within the organic matter of the highly mature samples placed in the gas window. Even more interestingly, when the authors attempted to remove bitumen from the oil-window mature samples through solvent extraction, they observed an abundant presence of porosity. This led them to infer that the lack of porosity as observed under SEM within the oil-window mature samples is essentially caused due to infilling of porous structures by bitumen or oil. When solvent extraction is employed to remove the bitumen and/or oil, the pores open up and are observable under SEM. What this means is this inference fundamentally and possibly rightfully challenges the inferences by other researchers. It means that the organic matters have porosity even at lower maturity levels and are not seen because of the filling of the pores by oil, and once the oil is expelled, they become visible. That is not all pores are essentially formed due to kerogen conversion (secondary porosity), but are inherently present within the organic matter.

10.5 Low-Pressure Gas Adsorption for Measuring Organic Porosity

Subcritical gas adsorption techniques using N_2 and CO_2 at low pressures have emerged as another vital, more acceptable technique to evaluate the pore size spectrum of unconventional hydrocarbon reservoirs (Ross and Bustin 2009; Clarkson et al. 2013; Kuila and Prasad 2013; Mastalerz et al. 2013; Yang et al. 2014; Psarras et al. 2017; Hazra et al. 2018a, 2018b, 2018c, 2019a, 2019b). Usually, nitrogen gas is used for adsorption at low experimental temperatures of −197.3 °C (with relative pressure varying between >0 and 1 atm). It is now fairly well established that nitrogen accesses pore sizes within ~1.7 to <200 nm (Kuila and Prasad 2013). Thus, one major limitation poised by N_2 gas adsorption at low temperatures and pressures is its inability to access micropores [<2 nm; IUPAC classification; Sing et al. (1985)] and all macropores (>200 nm diameter). The details of the procedure of low-pressure gas adsorption (LPGA) using N_2, and the resultant adsorption isotherm can be found in Brunauer et al. (1940) and Sing et al. (1985). Because of its inability to access micropores, researchers have preferentially used CO_2 gas adsorption at 0 °C for micropore evaluation of coal and shale reservoirs (Clarkson and Bustin 1996; Ross and Bustin 2009; Mastalerz et al. 2012). Consequently, CO_2 gas adsorption can be convincingly used to map the micropore volumes of thermally mature organic matter, which cannot be achieved by N_2 gas adsorption or SEM (Hazra et al. 2019a).

While there is still no internationally accepted standard on the crush size for conducting LPGA in shales and coals, very recently, few authors have tried focusing on how crush sizes can alter interpretations of the porosity of organic-rich rocks. For experiments conducted on a Longmaxi shale (China) crushed to different sizes (ranging from 4 mm to 58 μm), Han et al. observed increased BET SSA and other pore volumes with lowering crushed-sample sizes, and based on their results, suggested 113 μm be used for LPGA experiments. On the other hand, Mastalerz et al. (2017) through their research opined 75 μm to be used as crush size for producing reliable data from LPGA analysis. In contrast, Hazra et al. (2018c) noted modification and probable damage of certain mesoporous and microporous structures at samples crushed below 212–75 μm sizes. Thus some gaps still exist in what optimum crush sizes should be used for conducting LPGA analysis. Less amusingly, this would be dependent on the type of mineral matter present (ductile clays or brittle quartz, etc.). The possible impact of crushing on pore structures of organic matter is also to be studied more comprehensively, possibly by using a range of high-TOC/carbonaceous shales and coals, across different thermal maturity levels. Similar to the impact of crush-size, the importance of degassing temperature and duration on shale pore attributes has also received attention (Singh et al., 2021).

Very recently, the analytical limitations of using N_2 for determining finer pore structures (micropores) in organic-rich shales have been highlighted by Hazra et al. (2019b, 2020, 2021). In these works, the authors highlighted that for several Indian Permian shales with TOC content >10 wt %, in spite of being marked by higher thermal maturity levels, the BET SSAs and BJH pore volumes were much lower compared to other shales with lower TOC and thermal maturity levels. These results presented by the authors contradict those presented by several authors (e.g. Ross and Bustin 2009) stressing that the surface areas and pore volumes determined from N_2-LPGA are generally governed by TOC and thermal maturity. It has been discussed by a few researchers that at low operating temperatures of −197.3 °C for N_2-LPGA, N_2 lacks much kinetic energy and, consequently, underestimates micropores (Ross and Bustin 2009; Hazra et al. 2019b, 2020). For samples inherently marked by mesoporous nature, the mesoporosity may be correctly represented by N_2. Although results from N_2 and CO_2 should not be compared mutually, due to the abovementioned limitation of N_2, often disparities are observed in the surface area measurements from the two gases, especially for coals (see Gan et al. 1972; Mahajan 1991; Okolo et al. 2015). Thus one work examining on a wide range of shales to coals with variable TOC contents evaluated for porosity using N_2 and CO_2 LPGA techniques may document the underestimation of nanoporosity by N_2. This also implies that fractal dimensions calculated from N_2-LPGA are also likely to be under-represented (Hazra et al. 2020).

10.6 Conclusions and Perspectives

While the evolution of organic matter porosity as a function of thermal maturity is undoubted, few researchers have pointed out additional factors that may play critical roles on the same. SEM aided with ion milling has helped in landmark progress in understanding the evolution of organic porosity. However, some controversies still exist about whether ion milling improves the surface flatness of samples, or whether it induces alteration of organic matter in samples. While some researchers have suggested that structured kerogen does not show any porosity, gas adsorption data on coals containing a majority of structured organic matter have shown adsorption capacities even greater than shales. Similarly, often it has been noted that high-TOC carbonaceous shales and coals show smaller SSA when analyzed using N_2, but show higher SSA when analyzed using CO_2. It essentially happens due to the microporous character of organic matter, which can be understood through CO_2 adsorption. A major limitation of SEM is also its inability to image pores <5 nm in diameter, and, consequently, the more important gas storage sites – the micropores – remain unidentified. It is thus imperative that a combination of techniques be used to map the porous structures in the organic matter of shales and coals.

Acknowledgments

The Director, CSIR-Central Institute of Mining and Fuel Research, is acknowledged for giving B. Hazra the CSIR-CIMFR in-house research project number: MLP-93/2019–2020). D. P. Singh and V. Singh thankfully acknowledge the Academy of Scientific and Innovative Research (AcSIR), (Ghaziabad, India) for being enrolled as research scholars. B. Hazra thankfully acknowledges AcSIR for being recognized as an Assistant Professor. Soumyajit Mukherjee (IIT Bombay) did a detailed review of the manuscript and is thankfully acknowledged for the vital suggestions. Merryl Le Roux, Frank Otmar Weinreich, Joss Everett, Andrew Harrison, and the proofreading team (Wiley Blackwell) are thanked for their assistance. The article has been summarized in Mukherjee et al. (2023).

References

Anovitz, L.M. and Cole, D.R. (2015). Characterization and analysis of porosity and pore structures. *Rev. Mineral. Geochem.* 80 (1): 61–164.

Aringhieri, R. (2004). Nanoporosity characteristics of some natural clay minerals and soils. *Clay Clay Miner.* 52: 700–704.

Athy, L.F. (1930). Density, porosity, and compaction of sedimentary rocks. *AAPG Bull.* 14: 1–24.

Baldwin, B. and Butler, C.O. (1985). Compaction curves. *AAPG Bull.* 69: 622–626.

Bassim, N.D., De Gregorio, B.T., Kilcoyne, A.L.D. et al. (2012). Minimizing damage during FIB sample preparation of soft material. *J. Microsc.* 245: 288–301.

Bernard, S. and Horsfield, B. (2014). Thermal maturation of gas shale systems. *Annu. Rev. Earth Planet. Sci.* 42: 635–651.

Bernard, S., Horsfield, B., Schulz, H.-M. et al. (2012). Geochemical evolution of organic-rich shales with increasing maturity: a STXM and TEM study of the Posidonia Shale (lower Toarcian, northern Germany). *Mar. Pet. Geol.* 31: 70–89.

Blatt, H. and Tracy, R.J. (1996). *Petrology: Igneous, Sedimentary and Metamorphic*, 2nde.

Bowen, B.B., Ochoa, R.I., Wilkens, N.D. et al. (2011). Depositional and diagenetic variability within the Cambrian Mount Simon sandstone: implications for carbon dioxide sequestration. *Environ. Geosci.* 18: 69–89. https://doi.org/10.1306/eg.07271010012.

Brunauer, S., Deming, L.S., Deming, W.S., and Teller, E. (1940). On a theory of the van der Waals adsorption of gases. *J. Am. Chem. Soc.* 62: 1723–1732.

Cao, Z., Liu, G., Zhan, H. et al. (2016). Pore structure characterization of Chang-7 tight sandstone using MICP combined with N 2 GA techniques and its geological control factors. *Sci. Rep.* 6 (1): 1–13.

Chandra, D. and Vishal, V. (2020). A comparison of nano-scale pore attributes of Barakar formation gas shales from Raniganj and Wardha Basin, India using low pressure sorption and FEG-SEM analysis. *J. Nat. Gas Sci. Engn.* 81: 103453.

Chandra, D., Vishal, V., Debbarma, A. et al. (2020a). Role of composition and depth on pore attributes of Barakar formation gas shales of Ib Valley, India using a combination of low pressure sorption and image analysis. *Energy Fuel* 34 (7): 8085–8098.

Chandra, D., Vishal, V., Bahadur, J., and Sen, D. (2020b). A novel approach to identify accessible and inaccessible pores in gas shales using combined low-pressure sorption and SAXS/SANS analysis. *Int. J. Coal Geol.* 228: 103556.

Chandra, D. and Vishal, V. (2021). A critical review on pore to continuum scale imaging techniques for enhanced shale gas recovery. *Earth-Science Reviews* 217: p.103638.

Chandra, D., Vishal, V., Bahadur, J. et al. (2022). Nano-scale physicochemical attributes and their impact on pore heterogeneity in shale. *Fuel* 314: p.123070.

Cheng, A.L. and Huang, W.L. (2004). Selective adsorption of hydrocarbon gases on clays and organic matter. *Org. Geochem.* 35: 413–423.

Clarkson, C.R. and Bustin, R.M. (1996). Variation in micropore capacity and size distribution with composition in bituminous coal of the Western Canadian Basin. *Fuel* 75: 1483–1498.

Clarkson, C.R., Solano, N., Bustin, R.M. et al. (2013). Pore structure characterization of north American shale gas reservoirs using USANS/SANS, gas adsorption, and mercury intrusion. *Fuel* 103: 606–616.

Crosdale, P.J., Beamish, B.B., and Valix, M. (1998). Coalbed methane sorption related to coal composition. *Int. J. Coal Geol.* 35 (1–4): 147–158.

Curtis, M.E., Cardott, B.J., Sondergeld, C.H., and Rai, C.S. (2012a). Development of organic porosity in the Woodford Shale with increasing thermal maturity. *Int. J. Coal Geol.* 103: 26–31.

Curtis, M.E., Sondergeld, C.H., Ambrose, R.J., and Rai, C.S. (2012b). Microstructural investigation of gas shales in two and three dimensions using nanometer-scale resolution imaging. *AAPG Bull.* 96: 665–677.

Dasgupta, T. and Mukherjee, S. (2020). *Sediment Compaction and Applications in Petroleum Geoscience*, Series: Advances in Oil and Gas Exploration & Production. Springer ISSN: 2509-372X.

Desbois, G., Urai, J.L., and Kukla, P.A. (2009). Morphology of the pore space in claystones–evidence from BIB/FIB ion beam sectioning and cryo-SEM observations. *eEarth Discussions* 4: 1–19.

Dong, T., Harris, N.B., Ayranci, K. et al. (2015). Porosity characteristics of the Devonian Horn River shale, Canada: insights from lithofacies classification and shale composition. *Int. J. Coal Geol.* 141–142: 74–90.

Ehrenberg, S.N., Nadeau, P.H., and Steen, Ø. (2009). Petroleum reservoir porosity versus depth: influence of geological age. *AAPG Bull.* 93 (10): 1281–1296. https://doi.org/10.1306/06120908163.

Erdman, N. and Drenzek, N. (2013). Integrated preparation and imaging techniques for the microstructural and geochemical characterization of shale by scanning electron microscopy. *AAPG Memoir* 102: 7–14.

Fishman, N., Guthrie, J., and Honarpour, M. (2013). The stratigraphic distribution of hydrocarbon storage and its effect on producible hydrocarbons in the Eagle Ford Formation, South Texas. In *Unconventional Resources Technology Conference (1540–1545) Society of Exploration Geophysicists*, American Association of Petroleum Geologists, Society of Petroleum Engineers. https://doi.org/10.1190/urtec2013-157.

Gan, H., Nandi, S.P., and Walker, P.L. (1972). Nature of the porosity in American coals. *Fuel* 51: 272–277.

Grobe, A., Schmatz, J., Littke, R. et al. (2017). Enhanced surface flatness of vitrinite particles by broad ion beam polishing and implications for reflectance measurements. *Int. J. Coal Geol.* 180: 113–121.

Guo, H., Jia, W., Zeng, J., and He, R. (2017). Evolution of organic matter and nanometer-scale pores in an artificially matured shale undergoing two distinct types of pyrolysis: a study of the Yanchang Shale with type II kerogen. *Org. Geochem.* 105: 56–66.

Hackley, P.C. and Cardott, B.J. (2016). Application of organic petrography in North American shale petroleum systems: a review. *Int. J. Coal Geol.* 163: 8–51.

Hazra, B., Varma, A.K., Bandopadhya, A.K. et al. (2015). Petrographic insights of organic matter conversion of Raniganj basin shales, India. *Int. J. Coal Geol.* 150–151: 193–209.

Hazra, B., Varma, A.K., Bandopadhyay, A.K. et al. (2016). FTIR, XRF, XRD and SEM characteristics of Permian shales, India. *J. Nat. Gas Sci. Eng.* 32: 239–255.

Hazra, B., Wood, D.A., Vishal, V. et al. (2018a). Porosity controls and fractal disposition of organic-rich Permian shales using low- pressure adsorption techniques. *Fuel* 220: 837–848.

Hazra, B., Wood, D.A., Kumar, S. et al. (2018b). Fractal disposition and porosity characterization of lower Permian Raniganj Basin Shales, India. *J. Nat. Gas Sci. Eng.* 59: 452–465.

Hazra, B., Wood, D.A., Vishal, V., and Singh, A.K. (2018c). Pore-characteristics of distinct thermally mature shales: influence of particle sizes on low pressure CO_2 and N_2 adsorption. *Energy Fuel* 32 (8): 8175–8186.

Hazra, B., Chandra, D., Singh, A.K. et al. (2019a). Comparative pore structural attributes and fractal dimensions of lower Permian organic-matter-bearing sediments of two Indian basins: inferences from nitrogen gas adsorption. *Energy Sources A: Recovery Util. Environ. Eff.* 41 (24): 2975–2988.

Hazra, B., Wood, D.A., Mani, D. et al. (2019b). Organic and inorganic porosity, and controls of hydrocarbon storage in shales. In: *Evaluation of Shale Source Rocks and Reservoirs*, 107–138. Springer Cham.

Hazra, B., Wood, D.A., Singh, P.K. et al. (2020). Source rock properties and pore structural framework of the gas-prone lower Permian shales in the Jharia basin, India. *Arab. J. Geosci.* 13 (13): 1–18.

Hazra, B., Vishal, V., and Singh, D.P. (2021). Applicability of Low-Pressure CO_2 and N_2 Adsorption in Determining Pore Attributes of Organic-Rich Shales and Coals. *Energy Fuels* 35: 456–464.

Higgs, K.E., Zwingmann, H., Reyes, A.G., and Funnell, R.H. (2007). Diagenesis, porosity evolution, and petroleum emplacement in tight gas reservoirs, Taranaki basin, New Zealand. *J. Sediment. Res.* 77: 1003–1025.

Ho, N.C., Peacor, D.R., and van der Pluijm, B.A. (1999). Preferred orientation of phyllosilicates in Gulf Coast mudstones and relation to the smectite-toillite transition. *Clay Clay Miner.* 47: 495–504.

Holmes, R., Rupp, E., Vishal, V., and Wilcox, J. (2017). Selection of shale preparation protocol and outgas procedures for applications in low-pressure analysis. *Energy Fuel* 31 (9): 9043–9051.

Hu, H., Zhang, T., Wiggins-Camacho, J.D. et al. (2015). Experimental investigation of changes in methane adsorption of bitumen-free Woodford Shale with thermal maturation induced by hydrous pyrolysis. *Mar. Pet. Geol.* 59: 114–128.

İnan, S., Al Badairy, H., İnan, T., and Al Zahrani, A. (2018). Formation and occurrence of organic matter-hosted porosity in shales. *Int. J. Coal Geol.* 199: 39–51.

Jacob, H. (1989). Classification, structure, genesis and practical importance of natural solid bitumen ("migrabitumen"). *Int. J. Coal Geol.* 11: 65–79.

Jarvie, D.M., Hill, R.J., Ruble, T.E., and Pollastro, R.M. (2007). Unconventional shale-gas systems: the Mississippian Barnett Shale of north-Central Texas as one model for thermogenic shale-gas assessment. *AAPG Bull.* 91 (4): 475–500.

Javadpour, F., Moravvej Farshi, M., and Amrein, M. (2012). Atomic-force microscopy: a new tool for gas-shale characterization. *J. Can. Pet. Technol.* 51 (04): 236–243.

Jennings, D.S. and Antia, J. (2013). Petrographic characterization of the Eagle Ford Shale, South Texas: mineralogy, common constituents, and distribution of nanometer-scale pore types. In: *Electron Microscopy of Shale Hydrocarbon Reservoirs*, vol. 102 (ed. W. Camp, E. Diaz and B. Wawak), 101–113. AAPG Memoir.

Ji, L., Zhang, T., Milliken, K.L. et al. (2012). Experimental investigation of main controls to methane adsorption in clay-rich rocks. *Appl. Geochem.* 27: 2533–2545.

Jiang, H. and Daigle, H. (2019). Effects of fluids removal and thermal maturation on permeability and pore structure of organic-rich shale: results from pyrolysis experiments. *J. Pet. Sci. Eng.* 183: 106453.

Jiao, K., Yao, S., Liu, C. et al. (2014). The characterization and quantitative analysis of nanopores in unconventional gas reservoirs utilizing FESEM–FIB and image processing: an example from the lower Silurian Longmaxi Shale, upper Yangtze region, China. *Int. J. Coal Geol.* 128-129: 1–11.

Katz, B.J. and Arango, I. (2018). Organic porosity: a geochemist's view of the current state of understanding. *Org. Geochem.* 123: 1–16.

Klaver, J., Desbois, G., Urai, J.L., and Littke, R. (2012). BIB-SEM study of the pore space morphology in early mature Posidonia Shale from the Hils area, Germany. *Int. J. Coal Geol.* 103: 12–25.

Klaver, J., Desbois, G., Littke, R., and Urai, J.L. (2015). BIB-SEM pore characterization of pore space morphology and distribution in postmature to overmature samples from the Haynesville and Bossier shales. *Mar. Petrl. Geol.* 59: 451–466.

Klaver, J., Desbois, G., Littke, R., and Urai, J.L. (2016). BIB-SEM pore characterization of mature and post mature Posidonia Shale samples from the Hils area, Germany. *Int. J. Coal Geol.* 158: 78–89.

Kuila, U. and Prasad, M. (2013). Specific surface area and pore-size distribution in clays and shales. *Geophys. Prospect.* 61: 341–362.

Labani, M.M., Rezaee, R., Saeedi, A., and Hinai, A.A. (2013). Evaluation of pore size spectrum of gas shale reservoirs using low pressure nitrogen adsorption, gas expansion and mercury porosimetry: a case study from the Perth and Canning Basins, Western Australia. *J. Pet. Sci. Eng.* 112: 7–16.

Laxminarayana, C. and Crosdale, P.J. (1999). Role of coal type and rank on methane sorption characteristics of Bowen Basin, Australia coals. *Int. J. Coal Geol.* 40 (4): 309–325.

Liu, B., Schieber, J., and Mastalerz, M. (2017). Combined SEM and reflected light petrography of organic matter in the New Albany Shale (Devonian-Mississippian) in the Illinois Basin: a perspective on organic pore development with thermal maturation. *Int. J. Coal Geol.* 184: 57–72.

Löhr, S.C., Baruch, E.T., Hall, P.A., and Kennedy, M.J. (2015). Is organic pore development in gas shales influenced by the primary porosity and structure of thermally immature organic matter? *Org. Geochem.* 87: 119–132.

Loucks, R.G., Reed, R.M., Ruppel, S.C., and Jarvie, D.M. (2009). Morphology, genesis, and distribution of nanometer-scale pores in siliceous mudstones of the Mississippian Barnett Shale. *J. Sediment. Res.* 79: 848–861.

Loucks, R.G., Reed, R.M., Ruppel, S.C., and Hammes, U. (2012). Spectrum of pore types and networks in mudrocks and a descriptive classification for matrix-related mudrock pores. *AAPG Bull.* 96: 1071–1098.

Mahajan, O.P. (1991). CO_2 surface area of coals: the 25 year paradox. *Carbon* 29: 735–742.

Mann, U., Neisel, J.D., Burchard, W.G. et al. (1994). Fluid-rock interfaces as revealed by cryo-scanning electron microscopy. *First Break* 12: 131–136.

Mastalerz, M. and Schieber, J. (2017). Effect of ion milling on the perceived maturity of shale samples: implications for organic petrography and SEM analysis. *Int. J. Coal Geol.* 183: 110–119.

Mastalerz, M., He, L., Melnichenko, Y.B., and Rupp, J.A. (2012). Porosity of coal and shale: insights from gas adsorption and SANS/USANS techniques. *Energy Fuel* 26 (8): 5109–5120.

Mastalerz, M., Schimmelmann, A., Drobniak, A., and Chen, Y. (2013). Porosity of Devonian and Mississippian New Albany Shale across a maturation gradient: insights from organic petrology, gas adsorption, and mercury intrusion. *AAPG Bull.* 97: 1621–1643.

Mastalerz, M., Hampton, L., Drobniak, A., and Loope, H. (2017). Significance of analytical particle size in low-pressure N_2 and CO_2 adsorption of coal and shale. *Int. J. Coal Geol.* 178: 122–131.

Mastalerz, M., Drobniak, A., and Stankiewicz, A.B. (2018). Origin, properties, and implications of solid bitumen in source rock reservoirs: a review. *Int. J. Coal Geol.* 195: 14–36.

Mello, U.T., Karner, G.D., and Anderson, R.N. (1994). A physical explanation for the positioning of the depth to the top of overpressure in shale dominated sequences in the Gulf Coast basin, United States. *J. Geophys. Res.* 99: 2775–2789.

Milliken, K.L. and Day-Stirrat, R.J. (2013). Cementation in mudrocks: brief review with examples from cratonic basin mudrocks. In: *Critical Assessment of Shale Resource Plays*, AAPG Tulsa, vol. 103 (ed. J.-Y. Chatellier and D.M. Jarvie), 133–150. USA: Oklahoma.

Milliken, K.L., Rudnicki, M., Awwiller, D.N., and Zhang, T. (2013). Organic matter-hosted pore system, Marcellus Formation (Devonian), Pennsylvania. *AAPG Bull.* 97: 177–200.

Mozley, P.S., Heath, J.E., Dewers, T.A., and Bauer, S.J. (2016). Origin and heterogeneity of pore sizes in the Mount Simon sandstone and Eau Claire formation: implications for multiphase fluid flow. *Geosphere* 12 (4): 1341–1361.

Mukherjee, S. (2017). Airy's isostatic model: a proposal for a realistic case. *Arab. J. Geosci.* 10: 268.

Mukherjee, S. (2018a). Locating center of gravity in geological contexts. *Int. J. Earth Sci.* 107: 1935–1939.

Mukherjee, S. (2018b). Moment of inertia for rock blocks subject to bookshelf faulting with geologically plausible density distributions. *J. Earth Syst. Sci.* 127: 80.

Mukherjee, S. (2018c). Locating center of pressure in 2D geological situations. *The Journal of Indian Geophysical Union* 22, 49–51.

Mukherjee, S. and Kumar, N. (2018). A first-order model for temperature rise for uniform and differential compression of sediments in basins. *Int. J. Earth Sci.* 107: 2999–3004.

Mukherjee, S., Dasgupta, S., Majumdar, C. et al. (2023). Introduction to: Handbook of Petroleum Geoscience: Exploration, Characterization, and Exploitation of Hydrocarbon Reservoirs. In: *Handbook of Petroleum Geoscience: Exploration, Characterization, and Exploitation of Hydrocarbon Reservoirs*, Petroleum Industry Updates: Geoscientific and Economic Issues (ed. S. Mukherjee, S. Dasgupta, C. Majumdar, et al.). John Wiley & Sons, Hoboken, ISBN 978-1-119-68003-1.

Murphy, M., Daniels, J., Cole, D., et al. (2012). Pore distribution in the Ordovician Shale of the Utica/Point pleasant Sub-Basin. AAPG Geologists Annual Convention and Exhibition, Long Beach, CA, April 22–25, Search and Discovery #50605.

Nabawy, B.S. and David, C. (2016). X-ray ct scanning imaging for the nubia sandstone as a tool for characterizing its capillary properties. *Geosci. J.* 20 (5): 1–14.

Nelson, P.H. (2009). Pore-throat sizes in sandstones, tight sandstones, and shales. *AAPG Bull.* 93 (3): 329–340.

Okolo, G.N., Everson, R.C., Neomagus, H.W.J.P. et al. (2015). Comparing the porosity and surface areas of coal as measured by gas adsorption, mercury intrusion and SAXS techniques. *Fuel* 141: 293–304.

Ougier-Simonin, A., Renard, F., Boehm, C., and Vidal-Gilbert, S. (2016). Microfracturing and microporosity in shales. *Earth Sci. Rev.* 162: 198–226.

Passey, Q.R., Bohacs, K.M., Esch, W.L., et al. (2010). From oil-prone rock to gas-producing shale reservoir – geological and petrophysical characterization of unconventional shale-gas reservoirs. *International Oil and Gas Conference and Exhibition in China*, Beijing, China. Society of Petroleum Engineers pp. 29, SPE 131350.

Peng, C., Zou, C., Yang, Y. et al. (2017). Fractal analysis of high rank coal from southeast Qinshui basin by using gas adsorption and mercury porosimetry. *J. Pet. Sci. Eng.* 156: 235–249.

Phaneuf, M.W. (2005). FIB for material science applications: a review. In: *Introduction to Focus Ion Beams: Instrumentation, Theory, Techniques, and Practice* (ed. L.A. Giannuzi and F.A. Stevie), 143–172. New York: Springer.

Psarras, P., Holmes, R., Vishal, V., and Wilcox, J. (2017). Methane and CO_2 adsorption capacities of kerogen in the Eagle Ford shale from molecular simulation. *Acc. Chem. Res.* 50 (8): 1818–1828.

Romero-Sarmiento, M.F., Rouzaud, J.N., Bernard, S. et al. (2014). Evolution of Barnett Shale organic carbon structure and nanostructure with increasing maturation. *Org. Geochem.* 71: 7–16.

Ross, D.J.K. and Bustin, R.M. (2009). The importance of shale composition and pore structure upon gas storage potential of shale gas reservoirs. *Mar. Pet. Geol.* 26: 916–927.

Sanei, H. and Ardakani, O.H. (2016). Alteration of organic matter by ion milling. *Int. J. Coal Geol.* 163: 123–131.

Sarout, J. and Guéguen, Y. (2008a). Anisotropy of elastic wave velocities in deformed shales: part 1-experimental results. *Geophysics* 73: D75–D89.

Sarout, J. and Guéguen, Y. (2008b). Anisotropy of elastic wave velocities in deformed shales: part 2-modeling results. *Geophysics* 73: D91–D103.

Schieber, J. (2010). Common themes in the formation and preservation of intrinsic porosity in shales and mudstones-illustrated with examples across the Phanerozoic. In SPE Unconventional Gas Conference, Society of Petroleum Engineers.

Sing, K.S.W., Everett, D.H., Haul, R.A.W. et al. (1985). Reporting physisorption data for gas/solid systems with special reference to the determination of surface area and porosity. *Pure Appl. Chem.* 57: 603–619.

Singh, D.P., Chandra, D., Vishal, V. et al. (2021). Impact of degassing time and temperature on the estimation of pore attributes in shale. *Energy & Fuels* 35 (19): pp. 15628–15641.

Slatt, R.M. and O'Brien, N.R. (2011). Pore types in the Barnett and Woodford gas shales: contribution to understanding gas storage and migration pathways in fine-grained rocks. *AAPG Bull.* 95: 2017–2030.

Strąpoć, D., Mastalerz, M., Schimmelmann, A. et al. (2010). Geochemical constraints on the origin and volume of gas in the New Albany shale (Devonian-Mississippian), eastern Illinois basin. *AAPG Bull.* 94: 1713–1740.

Sun, M., Yu, B., Hu, Q. et al. (2018). Pore structure characterization of organic-rich Niutitang shale from China: small angle neutron scattering (SANS) study. *Int. J. Coal Geol.* 186: 115–125.

Sun, M., Zhang, L., Hu, Q. et al. (2020). Multiscale connectivity characterization of marine shales in southern China by fluid intrusion, small-angle neutron scattering (SANS), and FIB-SEM. *Mar. Pet. Geol.* 112: 104101.

Tissot, B. and Welte, D.H. (1984). *Petroleum Formation and Occurrence*, 2e, 699. Berlin: Springer.

Valentine, B.J., Hackley, P.C., Hatcherian, J., and Yu, J.J. (2019). Reflectance increase from broad beam ion milling of coals and organic-rich shales due to increased surface flatness. *Int. J. Coal Geol.* 201: 86–101.

Vandenbroucke, M. and Largeau, C. (2007). Kerogen origin, evolution and structure. *Org. Geochem.* 38: 719–833.

Varma, A.K., Hazra, B., Samad, S.K. et al. (2014). Methane sorption dynamics and hydrocarbon generation of shale samples from West Bokaro and Raniganj basins, India. *Journal of Natural Gas Science and Engineering* 21: pp.1138–1147.

Varma, A.K., Biswal, S., Hazra, B. et al. (2015). Petrographic characteristics and methane sorption dynamics of coal and shaly-coal samples from Ib Valley Basin, Odisha, India. *International Journal of Coal Geology* 141: pp.51–62.

Vernik, L. and Milovac, J. (2011). Rock physics of organic shales. *Lead. Edge* 30: 318–323.

Vernik, L. and Nur, A. (1992). Ultrasonic velocity and anisotropy of hydrocarbon source rocks. *Geophysics* 57: 727–735.

Vishal, V., Chandra, D., Bahadur, J. et al. (2019). Interpreting pore dimensions in gas shales using a combination of SEM imaging, small-angle neutron scattering, and low-pressure gas adsorption. *Energy Fuel* 33 (6): 4835–4848.

Wang, F.P. and Reed, R.M. (2009). Pore Networks and Fluid Flow in Gas Shales. *SPE Annual Technical Conference and Exhibition*, New Orleans, Louisiana, Society of Petroleum Engineers, pp. 8, SPE 124253.

Wang, Y., Zhu, Y., Zhang, R. et al. (2020). SANS coupled with fluid invasion approaches for characterization of overall nanopore structure and mesopore connectivity of organic-rich marine shales in China. *Int. J. Coal Geol.* 217: 103343.

Wood, D.A. and Hazra, B. (2017). Characterization of organic-rich shales for Petroleum Exploration & Exploitation: a review- part 1: bulk properties, multi-scale geometry and gas adsorption. *J. Earth Sci.* 28 (5): 739–757.

Yang, F., Ning, Z., and Liu, H. (2014). Fractal characteristics of shales from a shale gas reservoir in the Sichuan Basin, China. *Fuel* 115: 378–384.

Yang, J., Hatcherian, J., Hackley, P.C., and Pomerantz, A.E. (2017). Nanoscale geochemical and geomechanical characterization of organic matter in shale. *Nat. Commun.* 8 (1): 1–9.

Zhou, S., Liu, D., Cai, Y. et al. (2017). Multi-scale fractal characterizations of lignite, subbituminous and high-volatile bituminous coals pores by mercury intrusion porosimetry. *J. Nat. Gas Sci. Engn.* 44: 338–350.

11

Experimental Understanding of Pore Structure and Wettability of the Unconventional Reservoir

Md Golam Kibria and Qinhong Hu*

Department of Earth and Environmental Sciences, The University of Texas at Arlington, Arlington, TX, USA

**Author for correspondence: kibriadu@gmail.com*

Abstract

Hydrocarbon movement through shale reservoirs is controlled by two main factors: pore size distribution (geometry), and pore connectivity (topology). One of the critical obstacles to sustainable shale gas/oil development is the low recovery factor, which we can investigate using several experimental methods. To understand shale reservoir nanopore structure and wettability characteristics of the USA shale, in this chapter, we examine pore structure, wetting characteristics, imbibition behavior, and edge-accessible porosity with the following complimentary tests: field emission-scanning electron microscopy (FE-SEM), mercury intrusion porosimetry (MIP), contact angle measurement of various fluids, fluid imbibition into initially dry shale, and edge-accessible pore connectivity from vacuum saturation-high pressure impregnation with tracers-containing *n*-decane. The wettability and imbibition tests use hydrophilic (DI water) and hydrophobic (*n*-decane) fluids. Experimental results indicate that different basin samples have diverse geologic and reservoir characteristics (e.g. total organic content, porosity, and permeability). The wettability and imbibition tests use different types of wetting fluids (DI water, brine, *n*-decane, acetone, and 20% isopropyl alcohol with DI water). At the time of the imbibition test, most of the shale showed a late-time imbibition slope close to ¼ for DI water, and the ½ slope for *n*-decane and acetone (a zwitter fluid) shows an imbibition slope of ~1. Most of the cause's high tortuosity value from MIP analysis and spontaneous water imbibition indicated that most of the shale matrices have low pore connectivity. MIP analysis of different shales from various basins represents that the median pore throat diameter for rock samples is 3.1–10.5 nm, and almost 50–85% of pore throats by volume are smaller than 100 nm. The matrix pore connectivity between organic nanopores and the stimulated fracture network appears to be controlled by hydrocarbon recovery. Comparing the pore-size distribution of shales results obtained from different methods helps understand and quantify the pore structure and its emergent fluid migration behavior in tight shale, which is critical to sustainable shale resource development.

Keywords *Pore structure; FE-SEM; MIP; wettability; imbibition*

Handbook of Petroleum Geoscience: Exploration, Characterization, and Exploitation of Hydrocarbon Reservoirs, First Edition. Edited by Soumyajit Mukherjee, Swagato Dasgupta, Chandan Majumdar, Subhandip Mandal and Troyee Dasgupta.
© 2023 John Wiley & Sons Ltd. Published 2023 by John Wiley & Sons Ltd.

11.1 Introduction

Unconventional hydrocarbon production has been highly successful in North America in recent years. The "shale gas revolution" in North America has had a significant influence on worldwide energy markets as well as in the United States. Shale gas production from China also increased dramatically within the last couple of years from the Sichuan basin (Wei et al. 2014; Hu et al. 2017; Yang et al. 2017). The success in unconventional hydrocarbon extraction rapidly grew with the additional experimental investigation on shale reservoir rock. Unconventional gas refers to shale gas trapped inside fine-grained sedimentary rocks called mudrocks (mudstone or shale). In general, mudrocks are rich in organic matter; thermally matured mudrocks offer shale gas with high combustion value and environmental friendliness (Curtis et al. 2012). Shale gas is an abundant natural resource, with reserves of up to 7299 trillion cubic meters of world shale gas resources (EIA 2017).

The shale matrix porosity is relatively low (5–10%). Its permeability is of the extremely small order of nano-Darcy, with highly heterogeneous (mineralogy, organic content, natural fractures, porosity, and other properties vary from place to place). Unlike conventional natural gas reservoirs, gas shale is incredibly complex in its mineral composition (Curtis et al. 2012), and it is characterized by low porosity, extremely low permeability, and a complicated pore structure (Curtis 2002; Sondergeld et al. 2016; Hinai et al. 2014).

Routine petrophysical testing is the most useful approach to study the properties of organic shale, which can measure resistivity, porosity, permeability, saturation, and is also an indispensable bridge between logs response and logs interpretation (Ross and Bustin 2009; Loucks et al. 2009; Bhattacharya and Carr 2019). Compared with conventional gas reservoirs, shale gas ones feature weakly sensitive geophysical parameters, largely increasing the difficulty of using geophysical well logging techniques to identify shale gas. For this reason, most of the causes used a combination method of gamma-ray (GR), resistivity, neutron porosity, density (DEN), litho-density log (LDL), interval transit time, and induction electrical log (IEL) to identify shale gas reservoirs (Loucks et al. 2009; Wei et al. 2014). Recently, some experimental techniques have been applied to characterize pore systems in different gas-producing shale formations. Several common methods such as mercury intrusion porosimetry (MIP), low-pressure gas physisorption, and small-angle neutron scattering/ultra small angle neutron scattering (SANS/USANS) used to identify pore throat, porosity, and pore structure (Loucks et al. 2009; Klaver et al. 2012; Pommer and Milliken 2015; King et al. 2015). Comparing all other methods, in most of the cases, MIP technique is one of the most efficient and cost-effective ways to characterize mesopore and macropore structure for porous solids such as shales (Webb 2001; Gao and Hu 2013; Vishal et al. 2019; Döner et al. 2021). As a widely accepted and well-established method, MIP is an effective method to investigate both the geometrical (e.g. porosity, pore-throat distribution, and pore volume) (Webb 2001) and topological (e.g. permeability and tortuosity) characteristics (Katz and Thompson 1986, 1987; Gao and Hu 2013; Labani et al. 2013; Hu et al. 2015a).

Besides pore size, type, and distribution, pore connectivity is another crucial factor in controlling shale oil or gas flow and production. High-resolution imaging methods, such as field emission scanning electron microscopy (FE-SEM), CT-scan, and transmission electron microscopy (TEM) used to investigate pore shape and pore connection (Clarkson et al.

2013; Kuila and Prasad 2013;Hinai et al. 2014; Peng et al. 2015; Pommer and Milliken 2015; Testamanti and Rezaee 2017). 3D pore structure and connection have been examined using focused-ion-beam (FIB)-SEM (Curtis et al. 2012; Dewers et al. 2012; Bernard et al. 2013; Kelly et al. 2016), and locally connected pore networks with different sizes of connected pore clusters were found.

Nonintrusive fluid detection methods, such as X-ray computerized tomography (X-CT), can simulate the 3D distribution of the PSD, fracture morphology, and fluid transport (Klaver et al. 2012; Blunt et al. 2013; Liu et al. 2018). Nuclear magnetic resonance (NMR) is only susceptible to hydrogen fluids in pores of saturated rocks and has been broadly used in petrophysical experiments and geophysical logging (Dunn et al. 2002; Tan et al. 2015). As a nondestructive and relatively new method, the NMR core analysis method has been used for the petrophysical characterization of sandstone, coal, and shales (Westphal et al. 2005; Josh et al. 2012; Chi and Heidari 2014; Zhao et al. 2017; Sondergeld et al. 2016; Sun et al. 2018). NMR new 1H probe technology is used for transverse relaxation time (T_2) analysis, which provides pore structure analysis for porous medium (Yao et al. 2010; Zheng et al. 2018), and allows studies of the transport of hydrogen-containing fluids in hydrocarbon reservoirs (Liu et al. 2017; Sun et al. 2018). Low-pressure N_2 and CO_2 physisorption is also used for characterizing mesopore and micropore structure more effectively, such as pore volume, specific surface area, and pore size distribution (PSD) in tight reservoirs (e.g. Clarkson et al. 2013; Kuila and Prasad 2013; Mastalerz et al. 2013; Hazra et al. 2018), and N_2 and CO_2 physisorption data has been shown to indicate that the porous materials could exhibit fractal geometries (Yao et al. 2010; Yang et al. 2014). These techniques, when combined with MICP, examine the pore size scale of shales because of the flexible access of the probe molecules (N_2) at measurement temperatures and pressures.

Small-angle and ultra-small-angle neutron scattering (SANS/USANS) techniques have been utilized to investigate the pore structure of coals and shales (Radlinski et al., 2004; Melnichenko et al., 2009; Mares et al., 2009). The combination of SANS and USANS methods is beneficial for investigating the wide spectrum of pore sizes, for the capacity to differentiate closed versus open porosity, and the ability to examine pore structure at sensible (reservoir) pressure–temperature (P–T) conditions and modifications in pore structure at varying P–T conditions (Zhang et al. 2020).

The wettability of rock is a crucial petrophysical parameter that can influence rock-fluid properties (e.g. relative permeability and capillary pressure), which is essential for realizing the hydrocarbon production performance (Anderson et al. 1987; Lan et al. 2015; Wu et al. 2017). Precise assessment of rock wettability is essential for production optimization, proper investigation of fracturing fluid loss, and choice of appropriate fracturing fluids and additives (Javaheri et al. 2017). For decades, several quantitative and qualitative methods have been used to evaluate rock wettability, such as contact angle measurement (Yassin et al. 2017), spontaneous imbibition (SI) (Chaturvedi et al. 2009; Gao and Hu 2016; Yang et al. 2018), and displacement capillary pressure (Jones and Roszelle 1978).

Nano chemical tracer investigates how pores are connected and how far nano tracer travels through the pores. Several researchers used nano tracer-bearing wettability fluids (API brine or *n*-decane) to assess the association of tracers with mineral or organic matter phases (Peng et al. 2012; Hu et al. 2017; Kibria et al. 2018), using a delicate and microscale elemental laser ablation followed by LA-ICP-MS mapping technique. For LAICP-MS tracer mapping two molecular tracers in *n*-decane fluid with the estimated sizes of 1.39 nm × 0.29 nm × 0.18 nm for 1-iododecane and (1.27 nm × 0.92 nm × 0.78 nm) for trichlorooxobis (triphenylphosphine) rhenium, much less penetration was observed for larger molecules of organic rhenium in shales with median pore-throat sizes of several nanometers by Hu et al. (2017). This LA-ICP-MS method can produce 2-D and 3-D maps of chemical distributions in rock at a spatial resolution of microns and a concentration limit of low mg kg^{-1} (Hu and Mao 2012; Peng et al. 2012).

In this chapter, we discussed several complementary tests that are currently used to investigate the pore structure, pore connectivity, and wettability characteristics of shale all over the world, including FE-SEM, MIP, contact angle measurement, SI into initially dry shale, and tracer diffusion into fluid-saturated in shale followed by the elemental mapping of laser ablation inductively coupled plasma-mass spectrometry (LA-ICP-MS). In this chapter, we incorporated the study of pore geometry and wettability-related fluid–matrix interactions in the shale. This will deliver an enhanced insight into both the reasons for the low hydrocarbon recovery rate and the causes of the enormous loss of fracturing fluid. This chapter describes shale's knowledge gap in the pore geometry and wettability relationship and offers a systematic combined characterization of pore geometry and wettability relationships in unconventional reservoirs.

11.2 Experimental Methods

11.2.1 Field Emission-Scanning Electron Microscopy (FE-SEM)

The mini-cube or plug samples (10 mm width, 5 mm height, and 1.5 mm length) were used for SEM characterization under higher resolution. Before the SEM experiment, the sample's bottom was ground using an abrasive to shorten the piece by ~0.5–1 mm. The surface was then further polished using a finer abrasive before using argon ion-beam milling to create a smooth surface utilizing the ion milling instrument with an acceleration voltage of 3–6 kV. The broad-ion-milling accomplished excellent polishing after four hours of milling for the whole bottom surface of the sample. After that, the sample was coated with carbon conductive tapes at a thickness of 10 nm to provide a conductive surface. The samples were then imaged using a Hitachi S-4800 II FE-SEM instrument with a 1.0 nm resolution. The lowest pixel resolution directly depends on the SEM instrument's acceleration voltage. This method can directly observe the microscopic pore types, morphology, and distribution (Loucks et al. 2009; Milliken et al. 2013). All high-resolution FE-SEM images are generally processed using different types of image software, such as ImageJ (Abrámoff et al. 2004), which has a function of applying quantitatively analyzed porosity in gray-scale SEM images due to its high accuracy and efficiency (Curtis et al. 2012; Bai et al. 2013).

11.2.2 Mercury Injection Porosimetry (MIP) Analysis

MIP approach provides cost-effective ways to determine pore structure characteristics such as bulk density, grain density, total pore area, porosity, pore-throat size distribution, permeability, and tortuosity (Gao and Hu 2013; Hu et al. 2017). Non-wetting liquid mercury will only invade pores once an external pressure is applied, with the diameter of the pore throats being inversely proportional to the applied pressure (Gao and Hu 2013). Utilizing a mercury intrusion porosimeter (AutoPore IV 9510 or AutoPore IV 9500 by Micromeritics Instrument Corporation or Pascal Mercury Porosimeters by Thermo Fisher Scientific), liquid mercury was incrementally injected with an increasing hydrostatic pressure to occupy the pore spaces of increasingly smaller pore-throat sizes in shale. Considering the high surface energy and non-wetting characteristics of mercury and assuming that the pores are cylindrical with a circular opening in cross-section, the pore-throat diameter at a certainly applied pressure is calculated based on the Young-Laplace Equation (Gan et al. 1972; Chalmers et al. 2012). Here we describe how an AutoPore IV 9510 MIP instrument to use for sample analysis. Before the MIP tests, a cube-sized (1 cm cube and smooth surface reduce conformance effect) shale sample was placed in an oven at 60 °C for at least two days to remove moisture. Then, the samples were held in a desiccator with a relative humidity of less than 10% at room temperature (23 °C) for cooling. At the start of the MIP test, the connected pore spaces in a sample are evacuated until 50 μmHg pressure (at 6.7 Pa or 99.993% vacuum). Then, the sample is subjected to low-pressure conditions, followed by a high-pressure test up to 413 MPa or 60 000 psi, which corresponds to a pore throat diameter of 2.8 nm using the corrected Washburn equation to consider variable contact angle and surface tension in nm-sized pore spaces (Wang et al. 2016). A penetrometer (narrow-bore sample holder), used for samples with less than 5% porosity, recorded the largest pore throat diameter at 50 μm under low-pressure MIP analysis with 5 psi filling pressure. An equilibration time of 10 seconds was selected as pressures proceeded from one level to another. A mercury volume change was set to be <0.1 μl for each step; initial evacuation and low-pressure mercury injection took approximately 30 minutes. For the high-pressure test, the equilibration time was 30 seconds with a total analysis time of around 3 hours at about 80 pressure points between 30 and 60 000 psi for both intrusion and extrusion tests. The surface tension and contact angle of mercury are adopted as 485 dyne cm^{-1} and 130°, respectively (Hu et al. 2015b; Kibria et al. 2018).

11.2.3 Wettability and Contact Angle Measurement

Wettability can be straightforwardly assessed by measuring the contact angle of different fluids on a solid surface. Wettability is essential to understanding the primary, secondary, and tertiary recovery processes of oil and gas because it explains the type of fluid that prefers to spread onto, or adhere to, the rock surface of the producing formation (Teklu et al. 2017). Fluid contact angle tests are conducted on an optical contact angle and interface tension meter (SL200KB, KINO Industry) under ambient laboratory conditions (Figure 11.1). The actual angle makes a difference in the micro- and nanometer scale when it touches the solid shale rock (the air-liquid interface may be creating this situation). Three reasons can explain this characteristic: (i) most hydrophilic areas filled with water at low contact angles and high surface energy conditions (Checco et al. 2003). (ii) In many solid surfaces, water molecules create a thin layer due to the disjoining pressure (the force per unit surface area) established by a long-distance van der Waals adhesion force (Derjaguin and Churaev 1974). The thickness of the liquid layer controls the size of this disjoining pressure, leading to constant thin films that alter the droplet's shape from a sphere to a precursor layer.

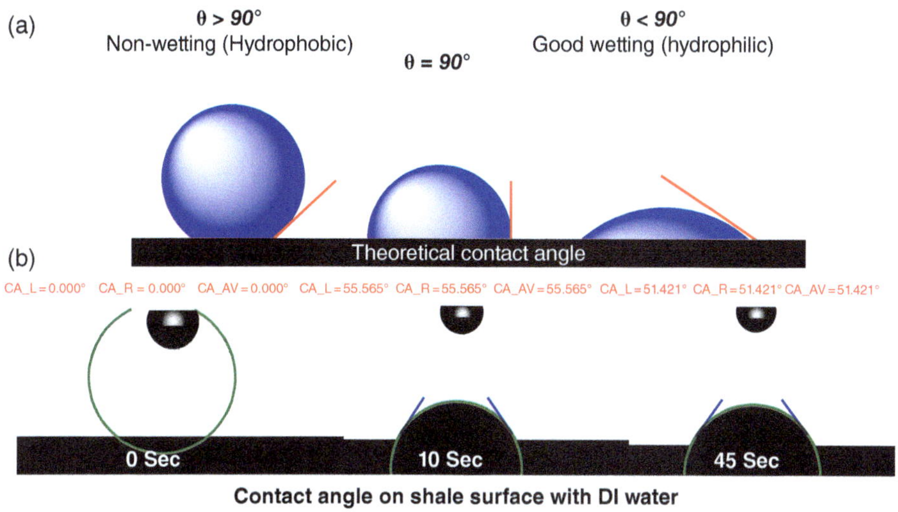

Figure 11.1 The illustration of contact angle (a) theoretical contact angle for hydrophobic and hydrophilic surface and (b) DI water droplet contact angle tests on shale sample (CA: contact angle; L: left; R: right; AV: average).

(iii) The role of surface tension becomes less significant compared to the line tension at the three-phase (liquid, air, and solid) contact line when liquid drop becomes smaller than the standard drop size (Marmur 1997).

The contact angle measurements require the solid surface to be smooth because surface roughness significantly affects the wetting angle (Singh 2016). To minimize the effect of heterogeneity of a rock surface, fluid contact angles for each sample are measured on several different random spots on the flat surface. Thus, the measured contact angles generally reflect the averaged wettability characteristics. Contact angle values were obtained for various polar and nonpolar fluids, including deionized water (DI water), API brine (8% NaCl and 2% $CaCl_2$ by weight; Crowe (1969)) (representative formation water), 20% IPA (isopropyl alcohol, C_3H_7OH) in DI water, and *n*-decane (a representative oil-phase fluid) (Yang et al. 2017). A droplet of each fluid at 2 μl was placed onto the shale sample surface, and an image for the air-liquid-solid interface was captured to calculate the contact angle using CAST™ 2.0 software.

11.2.4 Spontaneous Imbibition (SI)

SI is a capillary force-controlled process, during which a wetting fluid displaces a nonwetting one in a porous media under the influence of capillary suction (e.g. Hu et al. 2001; Takahashi and Kovscek 2010; Ghanbari and Dehghanpour 2015). Furthermore, the imbibition process helps assess pore connectivity for a given sample using the slope of log-imbibed fluid mass versus log-time (Hu et al. 2012), aided by the pore network model approach (Ewing and Horton 2002). Except for the top and bottom surfaces (transverse to the lamination), the other four sides of the cubic samples are coated with quick-cure epoxy to minimize vapor adsorption into the sample during the SI tests.

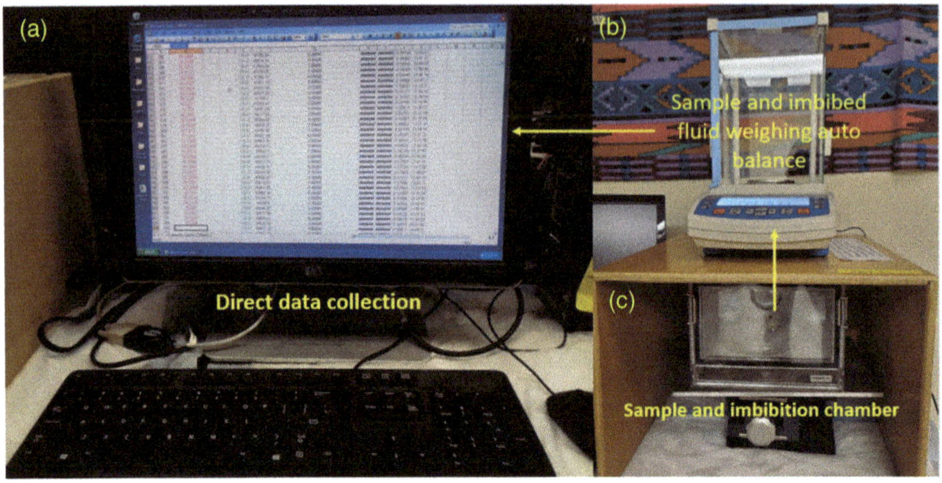

Figure 11.2 Schematic of spontaneous imbibition setup for air-liquid system, (a) data acquisition system, (b) electrical balance, (c) core sample, e.g. Cube immersed in liquid.

The apparatus for SI is illustrated in Figure 11.2, using a semi-microbalance at 0.01 mg resolution (Model AUW220D, Shimadzu, Kyoto, Japan). Hu et al. (2001) detailed the steps of conducting the imbibition tests and analyzing the resultant data. In general, DI water and *n*-decane were used as the imbibing liquids to displace air from pore spaces. Changes in sample weight over time were automatically recorded, and the fluid mass imbibed by the samples was calculated after adjusting for the buoyancy change and evaporation losses (Hunt 2001). The imbibition test was run for approximately three to four hours for *n*-decane and eight to nine hours for DI water before the sample was detached from the fluid contact.

When imbibition occurs vertically with negligible gravitational force, the SI process in a liquid–air system is expressed by handy's equation (Handy 1960): With the assumption of well-connected pore space for a porous medium, which shows a square-root-of-time relationship. A plot of cumulative imbibition mass/height vs. imbibition time on log-log coordinates shows Fickian diffusion-type behavior. The percentage of fluid absorption increases steadily with $t^{1/2}$ in the initial stage and then tends to level off near fluid saturation. Without an assumption of pore connectivity, Hu et al. (2012) reported three types of imbibition slopes for different rocks in such log-log plots, with values at 0.5, 0.26, and 0.26 transitioning to 0.5. Theoretical studies with percolation theory suggest that higher slope values represent good connectivity in a porous media, while lower slope values are a sign of low pore connectivity (Stauffer et al. 1994; Hu et al. 2012).

11.2.5 Vacuum Saturation and High-Pressure Impregnation

This test quantifies the edge-accessible connected pore distribution. The detailed test setup is illustrated in Figure 11.3. Only pores connected to the sample edge are accessible to an invasion of fluid or gas; such a distribution is related to pore connectivity (Hu et al. 2015b). Saturated diffusion with tracer-containing API brine (an aqueous phase) in the shale

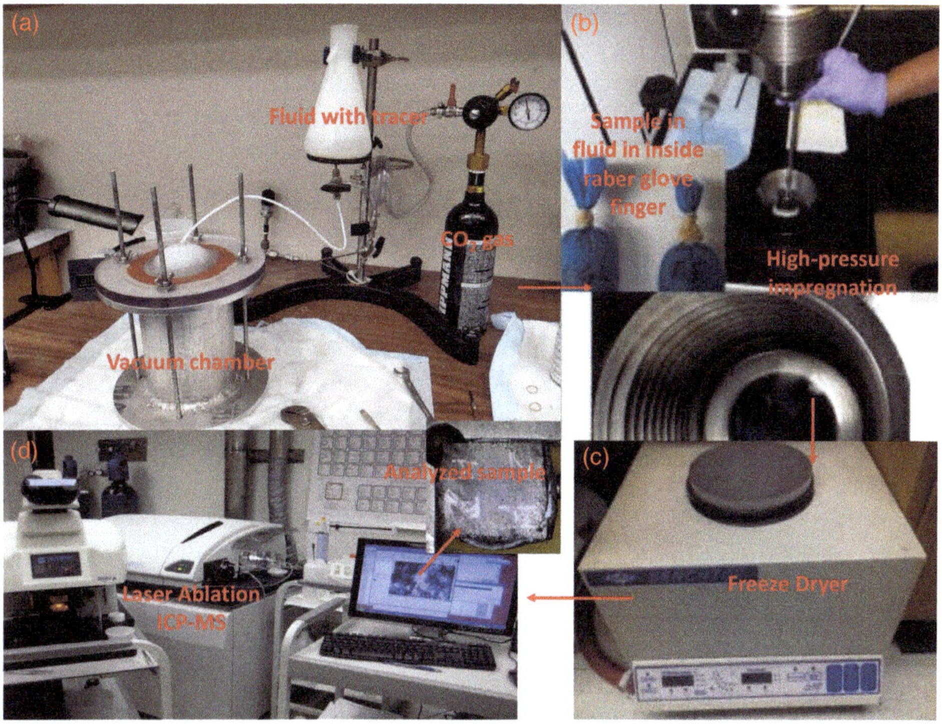

Figure 11.3 Representation ofvacuum saturation set up and high-pressure impregnation workflow – (a) vacuum setup with vacuum chamber, fluid with tracer container and CO_2 gas cylinder. (b) Sample in tracer fluid inside raber glove finger for high pressure impregnation. (c) Freeze dryer for kept tracer in place within samples. (d) Laser ablation ICP-MS for analyze tracer in sample.

investigates water-wet pores' spatial continuity (Hu and Mao 2012; Kibria et al. 2018). With the concentration gradient as a driving force, saturated diffusion tests are also used to quantify different tracer diffusion behaviors in the connected hydrophilic pores. In this test, hydrophilic fluid such as *n*-decane used for analyzing edge accessible pores, considering its wetting behavior on specific shale. The *n*-decane fluid includes two organic-phase molecular tracers with >99% pure tracer element-containing reagents: such as 1-iododecane {$CH_3(CH_2)_9I$, here termed organic-I in this chapter} and trichlorooxobis (triphenylphosphine) rhenium (V) {$[(C_6H_5)_3P]_2ReOC_{13}$, here termed organic-Re in this chapter}. Hu et al. (2017) reported that the molecular dimensions of organic-Re are 1.27 nm length $(L) \times 0.92$ nm width $(W) \times 0.78$ nm height (H), while organic-I has dimensions of 1.39 nm $(L) \times 0.29$ nm $(W) \times 0.18$ nm (H).

The vacuum saturation setup consists of a sample chamber connected to a vacuum pump (SD-90, Varian Inc., Palo Alto, CA), a compressed gas (CO_2) cylinder with 0–60 psi single-gauge CO_2 regulator (Chudnow Manufacturing, Oceanside, NY), and a reservoir of saturating fluid. The goal is to evacuate the sample chamber and, thus, the pores connected to the sample's edge. This is achieved by producing a vacuum to 0.05 Torr (1 Torr = 1 mm Hg = 133.3 Pa), equivalent to a 99.993% vacuum, as indicated from 0.01 to 20 Torr Pressure/Vacuum Meter for Pirani-Type Sensor (Cole-Parmer, Vernon Hills, IL) after

100 minutes, and staying at this pressure for nearly two hours. Flushing with CO_2 was conducted for three cycles at approximately one hour for each cycle, with vacuum pulling for one to two hours between cycles; this is followed by further evacuation of the sample at 0.04 Torr for 40 hours. Then, while the sample was still under a vacuum, a saturating fluid was introduced into the chamber until the sample was immersed so that the fluid would occupy the evacuated pore spaces connected to the sample edge. After immersion, pressurized CO_2 at 50 psi was applied to the chamber for 12 hours, further driving the tracer-containing fluid into the pore spaces. Following vacuum saturation, the saturated sample is further subjected to the pressure of 50 000–60 000 psi for one hour by placing the saturated sample inside a wrapped glove finger in the high-pressure chamber of the MIP instrument. Instead of pressurizing mercury into the sample during an MIP analysis, the tracer-containing *n*-decane is impregnated into the connected pore spaces under high hydrostatic pressure. The saturated samples were then weighed, immediately frozen in liquid nitrogen, freeze-dried at −50.5 °C and 0.054 Torr, and stored at low relative humidity before being mapped with Laser Ablation-Inductively Coupled Plasma-Mass Spectrometry (LA-ICP-MS) (Hu et al. 2015a, 2017). It used a 213-nm laser for the laser ablation system (UP-213, New Wave; Freemont, CA) to produce sub-micron depth holes in the shale sample. The laser ablates the shale into vaporized elements analyzed with ICP-MS. First, the top and bottom of the sample (reservoir) face are identified based on tracers' presence using a laser ablation spot grid. After that, the sample is cut, under dry conditions, in the middle from the top face crosswise relative to the imbibition direction. Then, LA-ICP-MS is performed, with a grid of spot analyses on the sawed inner face. LA-ICP-MS tracer mapping analysis is also conducted to map the tracer concentration at a spatial distribution to generate 2D/3D maps (including top, bottom, and interior surfaces). The detection limit of the LA-ICP-MS technique can reach as low as mg kg^{-1} level, and data reduction follows Peng et al. (2012).

11.3 Results and Discussion

11.3.1 Field Emission-Scanning Electron Microscopy (FE-SEM)

Investigated pore size and pore connectivity in a microscopic observation (~1000 μm^3) using field emission-scanning electron microscopy (FE-SEM) is more comfortable to evaluate shales compositional heterogeneity and pore connectivity (Loucks et al. 2009; Curtis et al. 2012). Loucks et al. (2009) introduce Ar-ion milling and field-emission SEM imaging for the first time to identify the governing nano-sized organic pores in Barnett Shale (similar to Woodford Shale). According to the classification system for shale pores proposed by Slatt and O'Brien (2011), shale pores can be separated, such as organic pores, fossil clastic pores, and interparticle flocculation pores, mineral grain intraparticle pores, and microfractures, fecal pellet pores, are observed in shales using FE-SEM. Reed et al., observed the morphology, size, and location of macro-meso-micropore in the Barnett shale reservoir. Shale reservoir spaces were divided into three types: intraparticle, interparticle, and organic matter pores (Loucks et al. 2009). Interparticle pores, e.g. pores between grains, micro-cracks developed between minerals and organic matter, or clay-related pores, are widely

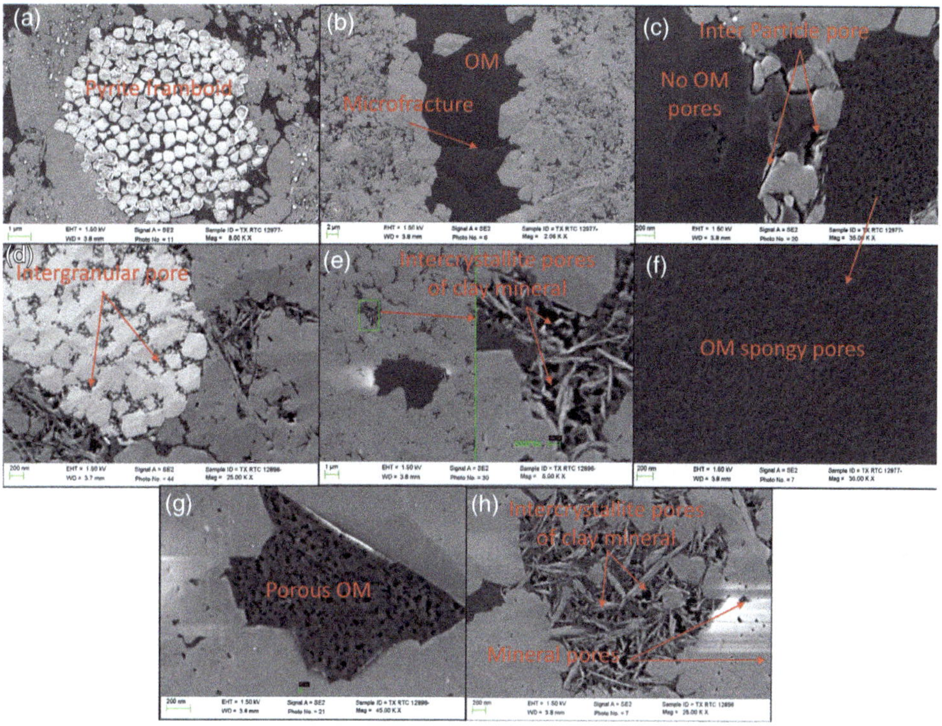

Figure 11.4 FE-SEM images of (a) pyrite framboid, (b) microfracture developed in the organic matter (OM), (c) inter particle in between mineral grains and OM without pores. (d) Intergranular pores within pyrite framboid, (e) intercrystallite pores of clay minerals, (f) OM spongy pores, (g) porous organic matter, (h) intercrystallite pores of clay minerals and mineral pore (intraparticle pore).

distributed in the rock matrix. Intraparticle pores, such as intercrystallite pores in pyrite framboids or cleavage pores in mica, are also easy to identify if there is a high proportion of pyrite and muscovite in the samples. OM pores are developed in the organic matter (either kerogen or bitumen) of the shales. OM pores are directly dependent on the basin of origin, such as the Barnett Shale from Texas has massively large OM pores. Liu et al. (2017) found few large organic pores (>500 nm) are observed in the Bakken samples. Based on the classification of different kinds of OM pores (Loucks et al. 2009), in Figure 11.4, we found several primary OM pores (Figure 11.4g), a few large bubbles pore with a large space of organic matter without pore (Figure 11.4c), and to a much larger extent small spongy pores (<50 nm) which are randomly distributed in the organic matter. These nanometer-scale pores could contribute a considerable amount to the total porosity; despite being so small, the large quantity of organic matter in these Woodford shales plays an important role. A similar type of sponge pore structure shale structure researchers such as Upper Bakken Shale from USA (Liu et al. 2017).

In Figure 11.4, as an example, we explained the storage space of Woodford shale from Texas, which includes pores and fractures. In the example picture, the pores are mainly intercrystallite pores of clay mineral, intergranular pores, OM sponge pores, and fracture. Pores formed influenced by overburden pressure and diagenesis are intergranular pores.

The intergranular pores between different minerals are with a pore size of 1–5 μm (Liu et al. 2017). The shapes of intergranular pores are mainly angular to subangular in shape. In Figure 11.4a and d, within the pyrite framboid, there are a lot of intergranular pores. These intergranular pores are usually well connected and can form an effective pore network. Therefore, they are favorable to the migration and preservation of oil and gas (Loucks et al. 2009). Microfractures are created by differential stress, which are small cracks in rock (Anders et al. 2014; Lai et al. 2017). Many factors control the development of fractures, among which tectonic stress, mineral composition and content, the abundance of organic matter, and sedimentary microfacies are the main factors (Zhang et al. 2017; Wang et al. 2018).

Additionally, fractures significantly improve the seepage ability and provide the necessary sites for the accumulation of a hydrocarbon shale reservoir (Zhang and Cao 2016). The width of the microfracture shown in Figure 11.4h is approximately 3.26 μm. The pores that develop in organic matter are often called organic matter pores. Organic matter pores are developed in the clay layer and have a size of 500 nm to less (Figure 11.4f and g). Prior research suggested that organic matter pores begin to appear when thermal maturation reaches a level of approximately 0.8% (Curtis et al. 2012). Intercrystallite pores along clay layers are widely developed due to the abundance of clay minerals in examples SEM picture (Figure 11.4e and h). Illite is the main clay mineral. So, this kind of pore is mainly interlaminar and mostly occurs as slit-shaped, generally linear and parallel to other pores. The pore sizes range from several nanometers to several hundred nanometers. All the clay minerals act as destructions to intergranular porosity, enhancing the mesopores and micropore porosities, and reducing the pore throat size (Huang and Zhao 2017), which are the main causes of the complicated relationship between clay mineral content and porosity (Zhu et al. 2016).

11.3.2 Mercury Injection Pressure (MIP) Analysis

The MIP experiments are based on centimeter-sized samples and deliver pore structure and topology-related data on a macroscopic scale. Figure 11.5 shows that the median pore throat diameter is 3.7–5.4 nm, and more than 70–80% of pores by volume are less than 100 nm in pore throat diameter for Woodford shale from Texas, USA. If any chemicals have the same size as small pore throat diameters, the steric interruption result can be significant (Hu and Wang 2003). Nelson (2009) showed that most concerned hydrocarbon species in oil/gas exploration have a molecular diameter of 0.5–10 nm. Therefore, hydrocarbon flow happens within a 5–100 nm pore size porous network.

As shown as arrows in Figure 11.5a, at least one major inflection point indicative of connected pore-throat networks is observed from the plot of intrusion pressure on the x-axis and log differential intrusion on the y-axis, which represent the increased intrusion of mercury into a particular pore-throat network when its capillary pressure is exceeded. These inflection points are used with the modified Washburn equation (Wang et al. 2016) to obtain the pore-throat size distribution and permeability and tortuosity for each connected pore-throat network (Hu et al. 2017; Kibria et al. 2018).

Representative Figure 11.5b shows that 20–30% pore spaces are 100 nm–50 μm of pore-throat size for five Woodford Shale samples. 100 nm to 50 μm size pores play essential roles as the primary conduits for mass transport and linking nanopore regions in samples from

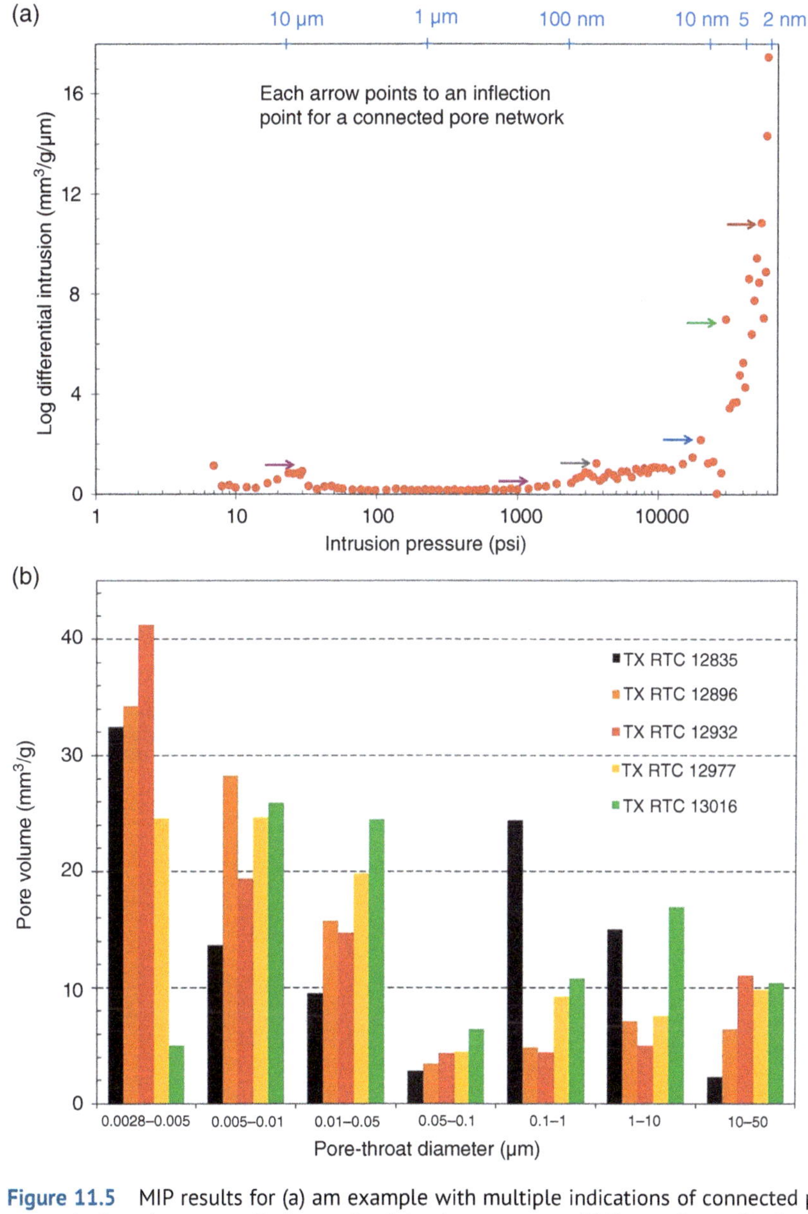

Figure 11.5 MIP results for (a) am example with multiple indications of connected pore networks from incrementally increasing intrusion pressures. (b) Pore volume distribution in diverse pore-throat regions for the Woodford Shale in five different depthssamples. *Source:* From Kibria et al. (2018)/with permission of Elsevier.

West Texas Woodford Shale. Using Katz and Thompson (1986) permeability calculation equation, Woodford samples matrix permeability is estimated in the range of 0.51–1.98 nD (Table 11.1). Hu et al., investigated the pore structure of several shale samples from New Mexico, USA, and the studied sample displays an average pore-throat size in the range of 1–10 μm (Table 11.1). Yang et al. (2017) showed that Chinese Longmaxi shales have extremely low porosities range between 2 and 6% with an average value of 4% and the

Table 11.1 MIP analysis-derived pore structure properties for the different US Shale samples.

Sample ID	Bulk density (g cm^{-3})	Apparent (skeletal) density (g cm^{-3})	Total pore area (m^2 g^{-1})	Porosity (%)	2.8–100 nm pore-throat network: Median pore-throat diameter (area) (nm)	Permeability	"Tortuosity" (D0/De)	L_e/L (square root of [tortuosity * total porosity])	References
Barnett 2167	2.52	2.58	6.34	2.61	6.1	2.17 nD	2081	6.05	Hu et al. (2015)
Barnett 2185	2.64	2.67	2.45	1.38	12.6	2.82 nD	5372	5.62	
Barette 2185	2.46	2.54	7.51	3.26	7.1	2.48 nD	3769	11.6	
TX RTC 12835	2.3	2.4	8.878	4.0497	3.7	2.16 nD	2107	9.23	Kibria et al. (2018)
TX RTC 12896	2.4605	2.4802	2.088	0.7956	3.9	0.54 nD	3930	5.59	
TX RTC 12932	2.3981	2.4296	3.582	1.2942	3.7	0.77 nD	6995	9.52	
TX RTC 12977	2.3821	2.4121	2.674	1.246	4.2	3.77 nD	2956	6.06	
TX RTC 13016	2.765	2.7885	0.98	0.85429	5.4	1.51 nD	2737	4.83	
Paddock 5931 D	2.589	2.898	0.228	6.75	0.0815	61.1 mD	38.6	68.2	
Abo 8956 D	2.293	2.728	0.641	15.9	0.137	9.08 mD	0.98	50.3	
Cisco Fm, NM	2.452–2.570	2.583–2.765	0.056–0.289	5.08–7.04	0.0967–0.467	24.6–26.2 mD	0.0464–10.4	15.4–292	
Spraberry, TX	2.29–2.32	2.51–2.53	7.33–9.69	8.60–9.04	23.9–62.3	192–622 nD	99.4–418	2.92–6.15	
Wolfcamp, TX	2.43–2.61	2.50–2.57	1.51–8.62	0.87–3.48	5.9–18.0	10–82 nD	2546–4515	6.15–10.1	

pore-throat diameters for the connected pore networks are predominantly in the nm size range. The Longmaxi shales interconnected pores with pore-throat sizes of 3–10 nm generally contribute to more than half of the total pore volume (Wei et al. 2014; Yang et al. 2017).

Effective tortuosity τ estimated applying Hager's (1998) equation on MIP data generated from an MIP machine. Effective tortuosity for Woodford samples is within the range of 2000–7000 which has no unit (Table 11.1). Effective tortuosity for Spraberry shale, TX samples is within the range of 99.4–418, a productive shale compared to Woodford. Tortuosity numbers described the intricate passageway level for fluid migration within a porous media (Epstein 1989; Hu and Wang 2003). Based on Hu et al. (2015b), it is easy to calculate the L_e/L ratios. Here the effective length of the pathway L_e denotes geometrical tortuosity, and L is the straight line's length between two endpoints of the channel. L_e/L ratios vary within 5.0–9.7 for different depths of Woodford Shale samples (Table 11.1). Tight shale formations show comparably large tortuosity values, which suggests that fluid within the rock will take a convoluted path to migrate from one location to another. In these Woodford Shale, the geometrical tortuosity L_e/L value of 4.83 for TX RTC 13016 sample represents that 1 cm straight distance will be replaced by 4.83 cm convoluted path. These tortuosity results imply that tight shale nanopores are rarely linked with each other. Sparsely connected nanopores take much more time to get a connection to migrate fluid.

11.3.3 Wettability and Contact Angle Measurement

Wettability demonstrates that the porous material solid surface has the propensity to wet by one fluid phase. Reservoir rock's wettability classifies as water wet, oil-wet, and intermediate wet based on the contact angle between fluid and rock surface (Craig 1971). The water-wetting phase has an attraction toward the water-wet rock surface. The oil-wet surfaces like oil to spread onto the surface rather than water affection. Intermediately wet surfaces have not a leading fondness for water or oil. Within the mixed wet reservoir rock, water- and oil-wet surfaces may coexist (Salathiel 1973; Takahashi and Kovscek 2010). Studying the wettability of shale surfaces is essential to understanding the gas flow and the flow back behavior of fracking fluids in shale formations (Fernø et al. 2011). Different reservoir properties include capillary pressure, relative permeability, hydrocarbon movement, and ultimate recovery related to wettability. The reservoir properties can be permanently damaged due to an incorrect assessment of wettability behavior (Morrow 1990; Sulucarnain et al. 2012; Singh 2016). A growing number of researchers have started to experimentally investigate various shales' wettability characteristics (Borysenko et al. 2009; Odusina and Sigal 2011; Makhanov et al. 2013; Gao and Hu 2016; Hu et al. 2017; Kibria et al. 2018).

Table 11.2 shows the contact angle measurements obtained from several different shale formations from the US shale basins using DI water, API brine, 20% IPA in DI water, and n-decane. In the table of Woodford shale, the contact angles of DI water range from 16.7° to 49.7° at 30 seconds after the droplet is in contact with the sample surface (Table 11.2). Figure 11.6 shows responses of Contact angle measurement of a representing samples from Woodford shale TX-RTC12896 where a vertical line is drawn at 30 seconds after the droplet tests. As shown in Table 11.2, these shale samples are slightly intermediate-wet; only the sample with lower contact angle shows a characteristic of being close to water wet; this sample also has the highest illite/smectite content, with clay minerals mostly being

Table 11.2 Air–liquid contact angles and imbibition slope of major oil- and gas-producing shale formation.

Formation and location	DI water	*n*-decane	API brine	20% IPA	Reference
	Contact angle (in degree) and imbibition slope				
Horn River, Canada	29.9–82.1	6.03–<3			
Wufeng-Longmaxi, Chongquing, China	2.8–20.7	<3			
Woodford, TX	16.7–49.7 0.207–0.331	5.77–9.42 0.352–0.591	70.9–81	16.8–61.2	Kibria et al. (2018)
Wolfcamp A, NM	45.1–67.6	<3			
Marcellus, PA	11.8–52.7	<3			
Yeso, NM, USA	85.7–9.3 0.164–0.180	6.63–6.86	55.2–88.1	13.2–14.3	
Abo Fm, NM, USA	10.6–62.5 0.068–0.103	6.26–<3	18.7–80.5	17.4–27.4	
Cisco Fm, NM, USA	36.2–42.1 0.100–0.147	4.55–5.72	41.7–60.8	8.22–35.0	
Haynesville, TX	7.4–58.3 0.288–0.545	<3			Wang et al. (2020)

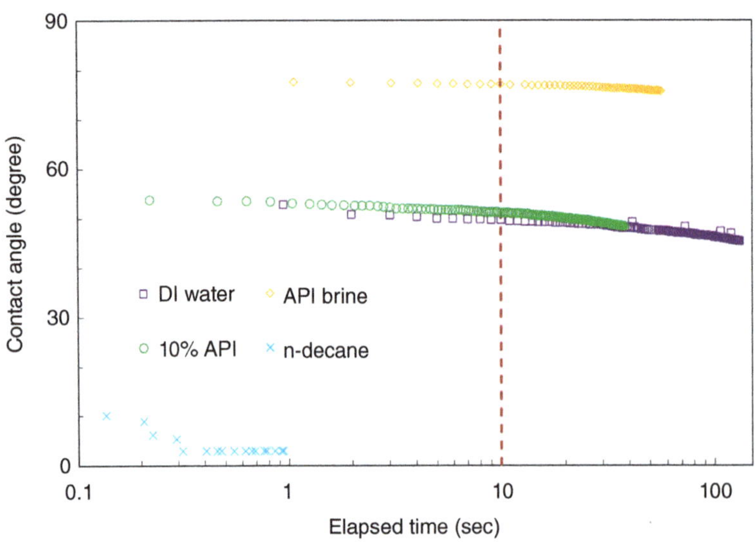

Figure 11.6 Contact angle measurement of a representing samples from Woodford shale TX-RTC12896; a vertical line is drawn at 30 seconds after the droplet tests.

water-wetting. In Table 11.2, samples from Cisco Fm, NM, and TX showed a moderate contact angle (41.7°–60.8°) for API brine, which means non-wetting for API brine. On the other hand, some shale, such as Woodford shale from TX, is non-wetting to API brine with a contact angle between 70.9°–81°.

As an oil-wetting fluid, *n*-decane shows a much lower contract angle (quickly changing from 5.77° to 9.42° at milli-second to <3°, which is the detection limit) throughout the Woodford shale, regardless of the mineralogy. This observation showed similar experimental results on shale from various formations around the world such as Longmaxi shale from China, suggesting that kerogen-rich shales tend to have a higher affinity for oil than for water (Table 11.2). Loucks et al. (2009) reported that shales contain oil wetting organic matter (kerogen) and water-wetting clay minerals. In the Woodford shale matrix, water-wetting clay minerals and oil-wetting organic matter (kerogen) grains may be dispersed in the rock which creates low-water- and API brine-connected wettability but well-connected oil-wet (*n*-decane) wettability.

Being a zwitter fluid, IPA can access both water- and oil-wetting pores, which was used to prepare a mixed liquid with DI water to observe the intermediate wettability behavior. For most cases such as Yeso Fm and Abo Fm from NM, USA showed contact angles 13.2 –143 and 17.4 –27.4 , respectively (Table 11.2).

11.3.4 Spontaneous Imbibition (SI)

The spontaneous imbibition process produced ½ slope for late-time imbibition [d log (liquid mass uptake)/d log (duration)] for well connected porous media with the contact of wetting fluid. Water-wetting Berea sandstone (benchmark reservoir rock in the hydrocarbon industry, which is Upper Devonian age from Eastern Kentucky) steadily exhibits a ½ slope, which indicates that this rock has well-connected pore spaces for water (Hu et al. 2012). In general, initially dried shale samples continuously show a late-time imbibition slope of approximately ¼ for DI water imbibition (which represent transitional wetting characteristics). Table 11.2 presented different shale sample SI results. Figure 11.7 serves as an exemplary DI water and n-decane imbibition figure for Woodford TX RTC 13016 sample. In general, low connectivity porous media show a characteristics slope of ¼ (Hu et al. 2015b; Gao and Hu 2016). Woodford Shale observed an imbibition slope close to ¼ for DI water imbibed samples, consistent with their wetting characteristics from contact angle measurements.

In Figure 11.7a, the first 20 seconds of imbibition data show a steep and noisy initial phase; this is because of the settling of the sample from being initially contacted with DI water. For some samples, during a subsequent short intermediate-time (within one minute), the stable slope of 0.50 ± 0.012 (the ½ slope for classical Fickian behavior) is also observed. The different researchers reported that fluid migration up to the sidewall region caused better connectivity (Hu et al. 2015b; Gao and Hu 2016; Singh 2016). Hu et al. (2015a) showed that when porous surface area gains moisture increasingly in a Fickian manner, it continuously shows ½ slope for SI. Within one minute's duration, the entire external surface is wetted, and the imbibition slopc gains ¼ value, which maintains this slop for the experimental period of eight to nine hours. Haynesville shale showed 0.288–0.545 imbibition slope for DI water imbibition (Wang et al. 2020).

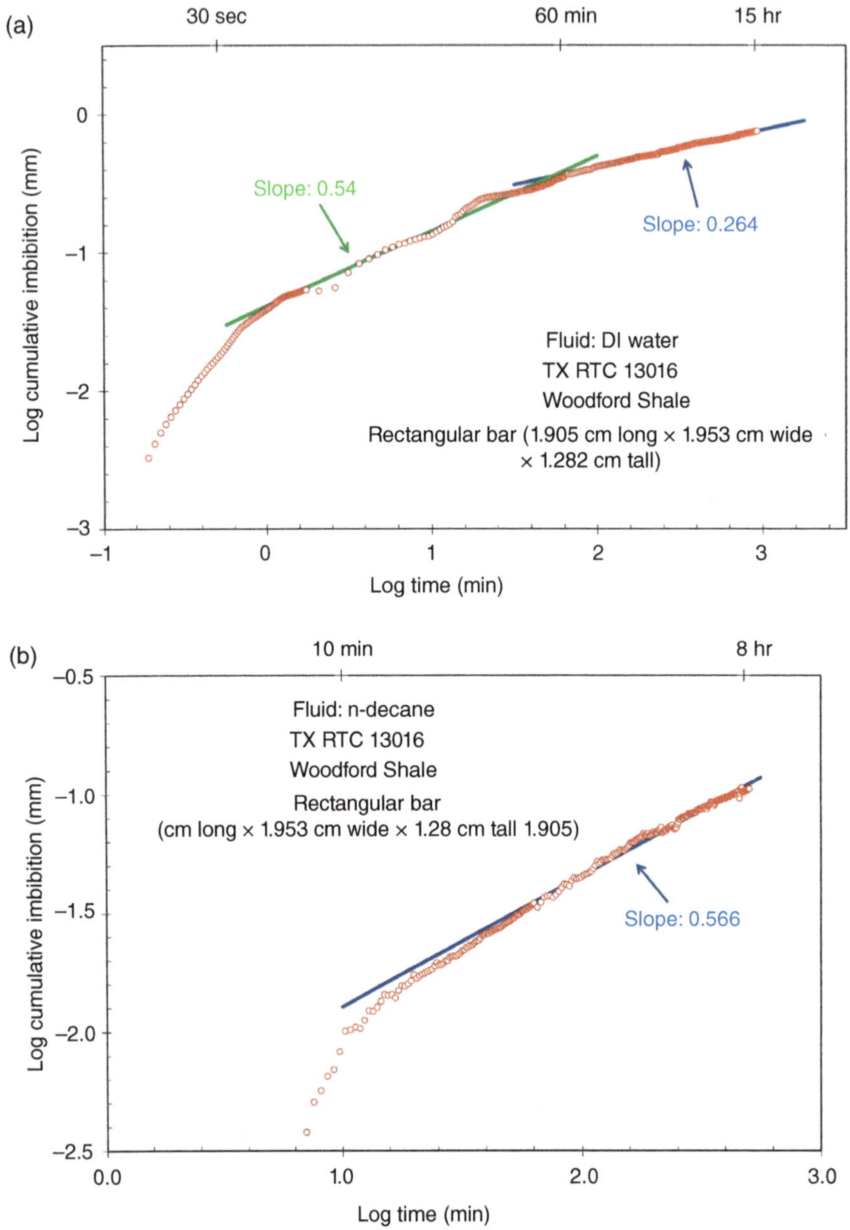

Figure 11.7 Log DI imbibition (mm) vs. Log time (min) for sample TX-RTC 13016: (a) DI water; (b) *n*-decane.

Table 11.2 shows the imbibition characteristics of *n*-decane fluid for several US shale, and Figure 11.7b represents a typical *n*-decane imbibition plot. Four samples out of five exhibit ½ slopes, which indicates that Woodford Shale is a good oil wetting rock, consistent with the contact angle measurement. The sample shows a lower imbibition slope at 0.352, though the sample has the same contact angle as other shales. This sample has a lower clay (10% illite/smectite) content (Table 11.2). The imbibition curve of n-decane displays a relatively large amount of n-decane uptake during the first 10–15 minutes, which represents samples are more oil-wet than DI water wet (Figure 11.8b). Gao and Hu (2016) observed that imbibed n-decane within samples could not be removed under 60 °C for more than

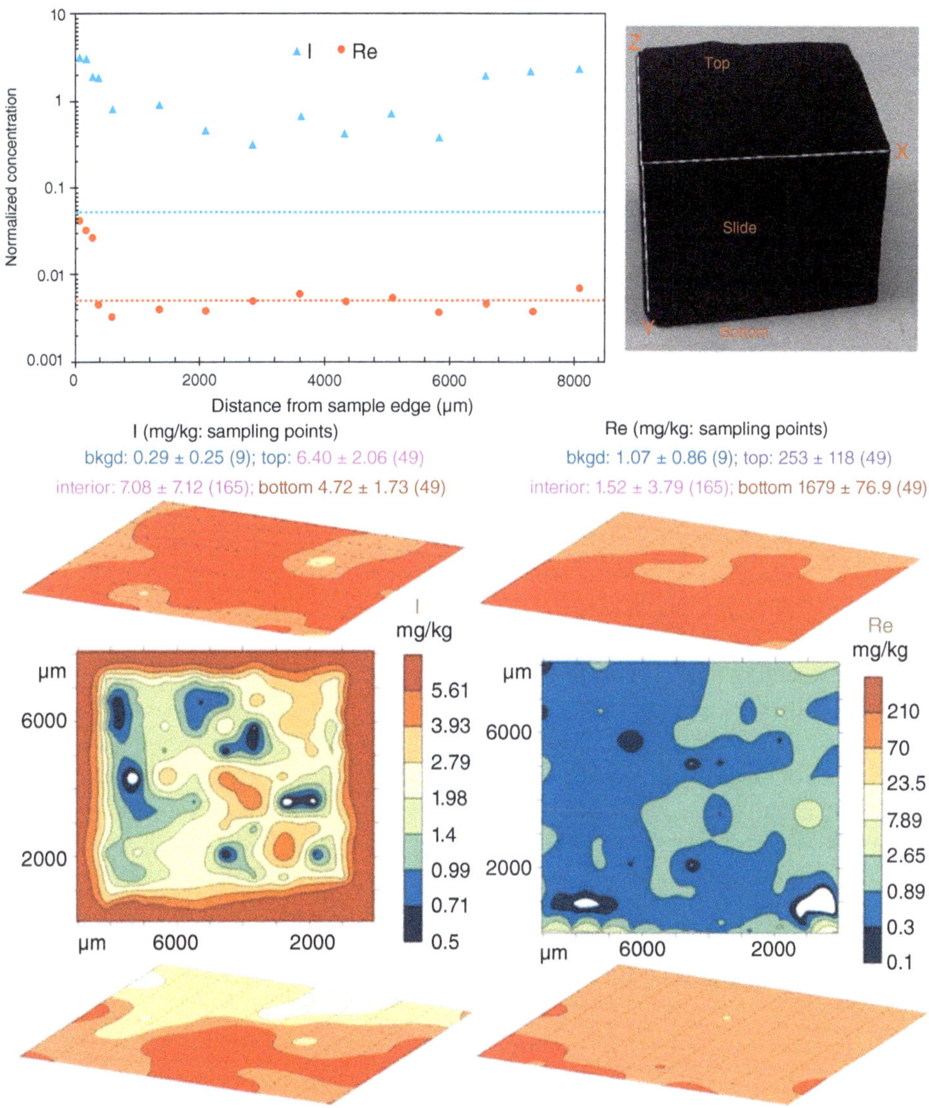

Figure 11.8 Profiles of organic-I and organic-Re tracers in the n-decane fluid after vacuum saturation-high pressure impregnation into sample TX RTC 12896. Laser spot size for all faces is 100 μm. The spacing between spots is 1500 μm for both top and bottom faces, and 100 μm for the first 300 μm which are then increased to 750 μm for the upward direction of the interior face. *Source:* From Kibria et al. (2018)/with permission of Elsevier.

two days. It was assumed that the water-wet pores associated with the clay minerals within the samples might become oil-wet to form an oil-wet pathway.

Different imbibition study shows that within shale formation, mineralogical composition changes significantly and systematically; these changes are sufficient to have a substantial influence on the mechanical properties of the shale, and, consequently, the potential for shale gas development. If any shale contains a modest amount (10–20%) of mixed-layer illite/smectite clays within the samples, and because it is an expandable component of smectite, it may be significantly challenging behavior for the imbibition.

11.3.5 Vacuum Saturation and High-Pressure Impregnation

To further study the intertwined relationship of wettability and connectivity, the edge assessable porosity distribution of connected hydrophobic pore spaces was interrogated with vacuum saturation, followed by high-pressure impregnation, of tracer-containing n-decane. As shown in Figure 11.8, the scatter plot on the top shows the normalized concentration (ratio to the averaged concentration from "bottom + top faces," with a total of 98 sampling spots; as the tracers come into the samples from all six faces of a cube) of tracers over the distance from sample edge (the sample height is 8500 μm) on the interior surface. Figure 11.8 at the bottom is mapped tracer distribution. Each vertical tracer panel consists of three (top, interior, and bottom) sample faces, with each dashed line on the top and bottom faces being 1000 μm apart. The LA-ICP-MS detection limits for I and Re tracer elements are 0.5 and 0.1 mg kg^{-1}, respectively; this is shown as the lowest value in scale bars. So, the color change precisely indicates the extent of tracer penetration (i.e. edge-accessible pore connectivity). The white zones characterize the areas of the tracer concentration below the detection limit.

Though Woodford Shale is oil-wet with good connectivity (from contact angle and imbibition tests), the hydrophobic pore network has a predominant pore-throat size of about 5 nm (from MICP analyses), which are most probably organic matter-hosted pores (from SEM images of shales with similar maturity, Loucks et al. (2009); Curtis et al. (2012)). An entanglement during molecular movement could surface when the ratio of pore: molecule diameter is less than 10 (Hu and Wang 2003), and this is exactly observed in our tracer-mapping results.

For organic-I tracing molecule with lower height and width (0.18–0.29 nm) than organic-Re (0.78–0.92 nm), it penetrates most pore spaces of the shale matrix within the zone (Zone A) of 0.5–1 mm from the sample edge, as seen by concentrations similar to these at "bottom + top faces." Deeper into the shale matrix from this well-connected Zone A, detected organic-I concentrations are lower and more sporadic, indicative of sparsely connected connectivity in Zone B. Furthermore, the larger height/width of organic-Re leads to molecular entanglement in pore networks at pore-throat sizes of only about 5 nm, as shown by its much lower presence in even Zone A. Such test design and the utility of wettability tracers of different molecular sizes help tease out the interlinked relationship of pore sizes, connectivity, and associated wettability of shale.

11.4 Conclusion

This chapter addresses experimental gaps regarding the pore structure and connectivity of several hydrocarbon-producing formations and the association with other fine-grained hydrocarbon reservoirs. In this chapter, the pore structure (connectivity) and wettability of several shale samples from Texas, Oklahoma, and New Mexico in the USA and Chinese Sichuan shale basin samples are studied with their coupled influences on tracer-containing fluid migration. Consistent with literature reports on many shale types, pore spaces in shales are predominantly at the nanometer scale with the permeability values in the nD range; pores with pore-throat sizes of 3–100 nm contribute to more than half of the total pore volume most of the producing shale formation. Intercrystallite pores in clay minerals depending on their clay mineral types, intergranular pores, dissolved pores, and microfractures are the main reservoir space types determined by FE-SEM observation in most of the shales all over the world. The contact angle measurement is a rapid method to describe the fluid wettability on a sample surface. SI could deliver not only pore connectivity evidence of the shale matrix but also reveal the wettability from the rate of the fluid movement of water against hydrocarbons.

Different shales have different types of wettability based on their maturation and mineral composition. In most of the cases, it was investigated that shale is mainly strongly oil-wetting and moderately strong water-wetting. Concentrations of non-sorbing and sorbing tracers decrease rapidly a few mm from the edge of the sample, which is associated with the predominant nanopore space and associated high tortuosity (highly restricted pathways), as well as the mixed wettability in Woodford and Barnette shale from the USA, and Longmaxi shale from China. In any shale, hydrocarbon recovery rate is related to its pore system, accessible porosity, mixed wettability, and intermediate to good pore connectivity.

Acknowledgments

Soumyajit Mukherjee (IIT Bombay) invited, handled, and reviewed this article. Merryl Le Roux, Frank Otmar Weinreich, Joss Everett, Andrew Harrison, and the proofreading team (Wiley Blackwell) are thanked for their assistance. The article has been summarized in Mukherjee et al. (2023).

References

Abràmoff, M.D., Magalhães, P.J., and Ram, S.J. (2004). Image processing with ImageJ. *Biophotonics international* 11 (7): 36–42.

Anders, M.H., Laubach, S.E., and Scholz, C.H. (2014). Microfractures: a review. *Journal of Structural Geology* 69: 377–394.

Anderson, T.F., Kruger, J., and Raiswell, R. (1987). CS-Fe relationships and the isotopic composition of pyrite in the New Albany Shale of the Illinois Basin, USA. *Geochimica et Cosmochimica Acta* 51 (10): 2795–2805.

Bai, B., Elgmati, M., Zhang, H., and Wei, M. (2013). Rock characterization of Fayetteville shale gas plays. *Fuel* 105: 645–652.

Bernard, S., Wirth, R., Schreiber, A. et al. (2013). FIB-SEM and TEM investigations of an organic-rich shale maturation series from the lower Toarcian Posidonia Shale, Germany: nanoscale pore system and fluid-rock interactions. *AAPG Memoir* 102: 53–66.

Bhattacharya, S. and Carr, T.R. (2019). Integrated data-driven 3D shale lithofacies modeling of the Bakken Formation in the Williston basin, North Dakota, United States. *Journal of Petroleum Science and Engineering* 177: 1072–1086.

Blunt, M.J., Bijeljic, B., Dong, H. et al. (2013). Pore-scale imaging and modelling. *Advances in Water Resources* 51: 197–216.

Borysenko, A., Clennell, B., Sedev, R. et al. (2009). Experimental investigations of the wettability of clays and shales. *Journal of Geophysical Research: Solid Earth* 114 (B7).

Chalmers, G.R., Bustin, R.M., and Power, I.M. (2012). Characterization of gas shale pore systems by porosimetry, pycnometry, surface area, and field emission scanning electron microscopy/transmission electron microscopy image analyses: Examples from the Barnett, Woodford, Haynesville, Marcellus, and Doig units Characterization of Gas Shale Pore Systems. *AAPG Bulletin* 96 (6): 1099–1119.

Chaturvedi, T., Schembre, J.M., and Kovscek, A.R. (2009). Spontaneous imbibition and wettability characteristics of Powder River Basin coal. *International Journal of Coal Geology* 77 (1-2): 34–42.

Checco, A., Guenoun, P., and Daillant, J. (2003). Nonlinear dependence of the contact angle of nanodroplets on contact line curvature. *Physical Review Letters* 91 (18): 186101.

Chi, L. and Heidari, Z. (2014). Quantifying the impact of natural fractures and pore structure on NMR measurements in multiple-porosity systems. IPTC 2014: International Petroleum Technology Conference, Doha, Qatar (19 January 2014). European Association of Geoscientists & Engineers, pp. cp-395.

Clarkson, C.R., Solano, N., Bustin, R.M. et al. (2013). Pore structure characterization of North American shale gas reservoirs using USANS/SANS, gas adsorption, and mercury intrusion. *Fuel* 103: 606–616.

Craig, F.F. (1971). *The Reservoir Engineering Aspects of Waterflooding*, vol. 3, 12–44. New York: HL Doherty Memorial Fund of AIME.

Crowe, C.W. and Dow Chemical Co (1969). Method of lessening the inhibitory effects to fluid flow due to the presence of solid organic substances in a subterranean formation. US Patent 3,482,636.

Curtis, J.B. (2002). Fractured shale-gas systems. *AAPG Bulletin* 86 (11): 1921–1938.

Curtis, M.E., Sondergeld, C.H., Ambrose, R.J., and Rai, C.S. (2012). Microstructural investigation of gas shales in two- and three-dimensions using nanometer-scale resolution imaging Microstructure of Gas Shales. *AAPG Bulletin* 96 (4): 665–677.

Derjaguin, B.V. and Churaev, N.V. (1974). Structural component of disjoining pressure. *Journal of Colloid and Interface Science* 49 (2): 249–255.

Dewers, T.A., Heath, J., Ewy, R., and Duranti, L. (2012). Three-dimensional pore networks and transport properties of a shale gas formation determined from focused ion beam serial imaging. *International Journal of Oil, Gas and Coal Technology* 5 (2-3): 229–248.

Döner, Z., Hu, Q., Kamal, M. et al. (2021). Petrophysical characteristics of Silurian mudstones from central Taurides in southern Turkey. *Journal of Earth Science* 32 (4): 778–798.

Dunn, K.J., Bergman, D.J., and LaTorraca, G.A. (2002). *Nuclear Magnetic Resonance: Petrophysical and Logging Applications*. Elsevier.

EIA, U.S., 2017. Annual energy outlook 2015: with projections to 2040. Center for Strategic and International Studies, Washington, DC (14 September 2017).

Epstein, N. (1989). On tortuosity and the tortuosity factor in flow and diffusion through porous media. *Chemical engineering science* 44 (3): 777–779.

Ewing, R.P. and Horton, R. (2002). Diffusion in sparsely connected pore spaces: temporal and spatial scaling. *Water Resources Research* 38 (12): 21–21.

Fernø, M.A., Grønsdal, R., Åsheim, J. et al. (2011). Use of sulfate for water based enhanced oil recovery during spontaneous imbibition in chalk. *Energy and Fuels* 25 (4): 1697–1706.

Gan, H., Nandi, S.P., and Walker, P.L. Jr. (1972). Nature of the porosity in American coals. *Fuel* 51 (4): 272–277.

Gao, Z. and Hu, Q. (2013). Estimating permeability using median pore-throat radius obtained from mercury intrusion porosimetry. *Journal of Geophysics and Engineering* 10 (2): 025014.

Gao, Z. and Hu, Q. (2016). Wettability of Mississippian Barnett Shale samples at different depths: Investigations from directional spontaneous imbibition. *AAPG Bulletin* 100 (1): 101–114.

Ghanbari, E. and Dehghanpour, H. (2015). Impact of rock fabric on water imbibition and salt diffusion in gas shales. *International Journal of Coal Geology* 138: 55–67.

Hager, J. (1998). Steam drying of porous media. PhD Thesis, Lund University, Sweden.

Handy, L.L. (1960). Determination of effective capillary pressures for porous media from imbibition data. *Transactions of the AIME* 219 (01): 75–80.

Hazra, B., Wood, D.A., Vishal, V. et al. (2018). Porosity controls and fractal disposition of organic-rich Permian shales using low-pressure adsorption techniques. *Fuel* 220: 837–848.

Hinai, A., Rezaee, R., Esteban, L., and Labani, M. (2014). Comparisons of pore size distribution: a case from the Western Australian gas shale formations. *Journal of Unconventional Oil and Gas Resources* 8: 1–13.

Hu, Q. and Mao, X. (2012). Application of laser ablation-inductively coupled plasma-mass spectrometry to studies of chemical diffusion, sorption, and transport in natural rock. *Geochemical Journal* 46 (5): 459–475.

Hu, Q. and Wang, J.S. (2003). Aqueous-phase diffusion in unsaturated geologic media: A review. *Critical Reviews in Environmental Science and Technology* 33 (3): 275–297.

Hu, M.Q., Persoff, P., and Wang, J.S. (2001). Laboratory measurement of water imbibition into low permeability welded tuff. *Journal of Hydrology* 242 (1–2): 64–78.

Hu, Q., Ewing, R.P., and Dultz, S. (2012). Low pore connectivity in natural rock. *Journal of Contaminant Hydrology* 133: 76–83.

Hu, Q.H., Clarke, J.J., Baird, R. et al. (2015a). Correlation of compositional phases and wettability-based fluid-tracers in rock. Presentation at the International Symposium of the Society of Core Analysts held in St. John's, Newfoundland and Labrador, Canada (16–21 August 2015). SCA 2015-A062.

Hu, Q.H., Liu, X.G., Gao, Z.Y. et al. (2015b). Pore structure and tracer migration behavior of typical American and Chinese shales. *Petroleum Science* 12 (4): 651–663.

Hu, Q.H., Liu, H., Yang, R. et al. (2017). Applying molecular and nanoparticle tracers to study wettability and connectivity of longmaxi formation in Southern China. *Journal of Nanoscience and Nanotechnology* 17 (9): 6284–6295. https://doi.org/10.1166/jnn.2017.14428.

Huang, X. and Zhao, Y.P. (2017). Characterization of pore structure, gas adsorption, and spontaneous imbibition in shale gas reservoirs. *Journal of Petroleum Science and Engineering* 159: 197–204.

Hunt, A.G. (2001). Applications of percolation theory to porous media with distributed local conductances. *Advances in Water Resources* 24 (3-4): 279–307.

Javaheri, A., Dehghanpour, H., and Wood, J.M. (2017). Tight rock wettability and its relationship to other petrophysical properties: a Montney case study. *Journal of Earth Science* 28 (2): 381–390.

Jones, S.C. and Roszelle, W.O. (1978). Graphical techniques for determining relative permeability from displacement experiments. *Journal of Petroleum Technology* 30 (05): 807–817.

Josh, M., Esteban, L., DellePiane, C. et al. (2012). Laboratory characterization of shale properties. *Journal of Petroleum Science and Engineering* 88: 107–124.

Katz, A.J. and Thompson, A.H. (1986). Quantitative prediction of permeability in porous rock. *Physical Review B* 34 (11): 8179.

Katz, A.J. and Thompson, A.H. (1987). Prediction of rock electrical conductivity from mercury injection measurements. *Journal of Geophysical Research: Solid Earth* 92 (B1): 599–607.

Kelly, S., El-Sobky, H., Torres-Verdín, C., and Balhoff, M.T. (2016). Assessing the utility of FIB-SEM images for shale digital rock physics. *Advances in Water Resources* 95: 302–316.

Kibria, M.G., Hu, Q., Liu, H. et al. (2018). Pore structure, wettability, and spontaneous imbibition of Woodford Shale, Permian Basin, West Texas. *Marine and Petroleum Geology* 91: 735–748.

King, H.E. Jr., Eberle, A.P., Walters, C.C. et al. (2015). Pore architecture and connectivity in gas shale. *Energy and Fuels* 29 (3): 1375–1390.

Klaver, J., Desbois, G., Urai, J.L., and Littke, R. (2012). BIB-SEM study of the pore space morphology in early mature Posidonia Shale from the Hils area, Germany. *International Journal of Coal Geology* 103: 12–25.

Kuila, U. and Prasad, M. (2013). Specific surface area and pore-size distribution in clays and shales. *Geophysical Prospecting* 61: 341–362.

Labani, M.M., Rezaee, R., Saeedi, A., and Al Hinai, A. (2013). Evaluation of pore size spectrum of gas shale reservoirs using low pressure nitrogen adsorption, gas expansion and mercury porosimetry: A case study from the Perth and Canning Basins, Western Australia. *Journal of Petroleum Science and Engineering* 112: 7–16.

Lai, J., Wang, G., Fan, Z. et al. (2017). Fracture detection in oil-based drilling mud using a combination of borehole image and sonic logs. *Marine and Petroleum Geology* 84: 195–214.

Lan, Q., Xu, M., Binazadeh, M. et al. (2015). A comparative investigation of shale wettability: the significance of pore connectivity. *Journal of Natural Gas Science and Engineering* 27: 1174–1188.

Liu, Q., Wu, J.E., Lim, Z.Y. et al. (2017). Evaluation of the metabolic response of *Escherichia coli* to electrolysed water by 1H NMR spectroscopy. *LWT-Food Science and Technology* 79: 428–436.

Liu, Y., Yao, Y., Liu, D. et al. (2018). Shale pore size classification: an NMR fluid typing method. *Marine and Petroleum Geology* 96: 591–601.

Loucks, R.G., Reed, R.M., Ruppel, S.C., and Jarvie, D.M. (2009). Morphology, genesis, and distribution of nanometer-scale pores in siliceous mudstones of the Mississippian Barnett Shale. *Journal of Sedimentary Research* 79 (12): 848–861.

Makhanov, K., Dehghanpour, H., and Kuru, E. (2013). Measuring liquid uptake of organic shales: A workflow to estimate water loss during shut-in periods. SPE Unconventional Resources Conference Canada. Society of Petroleum Engineers, Calgary, Alberta, Canada (5–7 November 2013).

Mares, T.E., Radliński, A.P., Moore, T.A. et al. (2009). Assessing the potential for CO2 adsorption in a subbituminous coal, Huntly Coalfield, New Zealand, using small angle scattering techniques. *International Journal of Coal Geology* 77 (1–2): 54–68.

Marmur, A. (1997). Line tension and the intrinsic contact angle in solid–liquid–fluid systems. *Journal of Colloid and Interface Science* 186 (2): 462–466.

Mastalerz, M., Schimmelmann, A., Drobniak, A., and Chen, Y. (2013). Porosity of Devonian and Mississippian New Albany Shale across a maturation gradient: Insights from organic petrology, gas adsorption, and mercury intrusion. *AAPG Bulletin* 97 (10): 1621–1643.

Melnichenko, Y.B., Radlinski, A.P., Mastalerz, M. et al. (2009). Characterization of the CO_2 fluid adsorption in coal as a function of pressure using neutron scattering techniques (SANS and USANS). *International Journal of Coal Geology* 77 (1-2): 69–79.

Milliken, K.L., Rudnicki, M., Awwiller, D.N., and Zhang, T. (2013). Organic matter–hosted pore system, Marcellus formation (Devonian), Pennsylvania. *AAPG Bulletin* 97 (2): 177–200.

Morrow, N.R. (1990). Wettability and its effect on oil recovery. *Journal of Petroleum Technology* 42 (12): 1–476.

Mukherjee, S., Dasgupta, S., Majumdar, C. et al. (2023). Introduction to: Handbook of Petroleum Geoscience: Exploration, Characterization, and Exploitation of Hydrocarbon Reservoirs. In: *Handbook of Petroleum Geoscience: Exploration, Characterization, and Exploitation of Hydrocarbon Reservoirs*, Petroleum Industry Updates: Geoscientific and Economic Issues (ed. S. Mukherjee, S. Dasgupta, C. Majumdar, et al.). John Wiley & Sons, Hoboken, ISBN 978-1-119-68003-1.

Nelson, P.H. (2009). Pore-throat sizes in sandstones, tight sandstones, and shales. *AAPG Bulletin* 93 (3): 329–340.

Odusina, E. and Sigal, R.F. (2011). Laboratory NMR measurements on methane saturated Barnett Shale samples. *Petrophysics* 52 (01): 32–49.

Peng, S., Hu, Q., Ewing, R.P. et al. (2012). Quantitative 3-D elemental mapping by LA-ICP-MS of a basaltic clast from the Hanford 300 Area, Washington, USA. *Environmental Science & Technology* 46 (4): 2025–2032.

Peng, Y., Liu, J., Pan, Z., and Connell, L.D. (2015). A sequential model of shale gas transport under the influence of fully coupled multiple processes. *Journal of Natural Gas Science and Engineering* 27: 808–821.

Pommer, M. and Milliken, K. (2015). Pore types and pore-size distributions across thermal maturity, Eagle Ford Formation, southern Texas Pores across Thermal Maturity, Eagle Ford. *AAPG Bulletin* 99 (9): 1713–1744.

Radlinski, A.P., Mastalerz, M., Hinde, A.L. et al. (2004). Application of SAXS and SANS in evaluation of porosity, pore size distribution and surface area of coal. *International Journal of Coal Geology* 59 (3-4): 245–271.

Ross, D.J. and Bustin, R.M. (2009). The importance of shale composition and pore structure upon gas storage potential of shale gas reservoirs. *Marine and Petroleum Geology* 26 (6): 916–927.

Salathiel, R.A. (1973). Oil recovery by surface film drainage in mixed-wettability rocks. *Journal of Petroleum Technology* 25 (10): 1–216.

Singh, H. (2016). A critical review of water uptake by shales. *Journal of Natural Gas Science and Engineering* 34: 751–766.

Slatt, R.M. and O'Brien, N.R. (2011). Pore types in the Barnett and Woodford gas shales: Contribution to understanding gas storage and migration pathways in fine-grained rocks. *AAPG Bulletin* 95 (12): 2017–2030.

Sondergeld, C., Tinni, A., Rai, C., and Besov, A. (2016). NMR considerations in shale evaluation. SPWLA 57th Annual Logging Symposium. Society of Petrophysicists and Well-Log Analysts, Reykjavik, Iceland (25–29 June 2016).

Stauffer, D., Adler, J., and Aharony, A. (1994). Universality at the three-dimensional percolation threshold. *Journal of Physics A: Mathematical and General* 27 (13): L475.

Sulucarnain, I.D., Sondergeld, C.H., and Rai, C.S. (2012). An NMR study of shale wettability and effective surface relaxivity. SPE Canadian Unconventional Resources Conference. Society of Petroleum Engineers, Calgary, Alberta, Canada (30 October–1 November 2012).

Sun, X., Yao, Y., Liu, D., and Zhou, Y. (2018). Investigations of CO2-water wettability of coal: NMR relaxation method. *International Journal of Coal Geology* 188: 38–50.

Takahashi, S. and Kovscek, A.R. (2010). Spontaneous countercurrent imbibition and forced displacement characteristics of low-permeability, siliceous shale rocks. *Journal of Petroleum Science and Engineering* 71 (1–2): 47–55.

Tan, M., Mao, K., Song, X. et al. (2015). NMR petrophysical interpretation method of gas shale based on core NMR experiment. *Journal of Petroleum Science and Engineering* 136: 100–111.

Teklu, T.W., Abass, H.H., Hanashmooni, R. et al. (2017). Experimental investigation of acid imbibition on matrix and fractured carbonate rich shales. *Journal of Natural Gas Science and Engineering* 45: 706–725.

Testamanti, M.N. and Rezaee, R. (2017). Determination of NMR T2 cut-off for clay bound water in shales: a case study of Carynginia Formation, Perth Basin, Western Australia. *Journal of Petroleum Science and Engineering* 149: 497–503.

Vishal, V., Chandra, D., Bahadur, J. et al. (2019). Interpreting pore dimensions in gas shales using a combination of SEM imaging, small-angle neutron scattering, and low-pressure gas adsorption. *Energy and Fuels* 33 (6): 4835–4848.

Wang, Y., Zhu, Y., Liu, S., and Zhang, R. (2016). Pore characterization and its impact on methane adsorption capacity for organic-rich marine shales. *Fuel* 181: 227–237.

Wang, T., Tian, S., Li, G. et al. (2018). Molecular simulation of CO_2/CH_4 competitive adsorption on shale kerogen for CO_2 sequestration and enhanced gas recovery. *The Journal of Physical Chemistry C* 122 (30): 17009–17018.

Wang, C., Gao, H., Gao, Y., and Fan, H. (2020). Influence of pressure on spontaneous imbibition in tight sandstone reservoirs. *Energy & Fuels* 34 (8): 9275–9282.

Webb, P.A. (2001). *An Introduction to the Physical Characterization of Materials by Mercury Intrusion Porosimetry with Emphasis on Reduction and Presentation of Experimental Data.* Norcross, GA: Micromeritics Instrument Corp.

Wei, Y., Jianbo, W., Shuai, L. et al. (2014). Logging identification of the Longmaxi mud shale reservoir in the Jiaoshiba area, Sichuan Basin. *Natural Gas Industry B* 1 (2): 230–236.

Westphal, H., Surholt, I., Kiesl, C. et al. (2005). NMR measurements in carbonate rocks: problems and an approach to a solution. *Pure and Applied Geophysics* 162 (3): 549–570.

Wu, K., Chen, Z., Li, J. et al. (2017). Wettability effect on nanoconfined water flow. *Proceedings of the National Academy of Sciences of the United States of America* 114 (13): 3358–3363.

Yang, F., Ning, Z., and Liu, H. (2014). Fractal characteristics of shales from a shale gas reservoir in the Sichuan Basin, China. *Fuel* 115: 378–384.

Yang, R., Hao, F., He, S. et al. (2017). Experimental investigations on the geometry and connectivity of pore space in organic-rich Wufeng and Longmaxi shales. *Marine and Petroleum Geology* 84: 225–242.

Yang, R., Hu, Q., He, S. et al. (2018). Pore structure, wettability and tracer migration in four leading shale formations in the Middle Yangtze Platform, China. *Marine and Petroleum Geology* 89: 415–427.

Yao, Y., Liu, D., Che, Y. et al. (2010). Petrophysical characterization of coals by low-field nuclear magnetic resonance (NMR). *Fuel* 89 (7): 1371–1380.

Yassin, M.R., Begum, M., and Dehghanpour, H. (2017). Organic shale wettability and its relationship to other petrophysical properties: a Duvernay case study. *International Journal of Coal Geology* 169: 74–91.

Zhang, H. and Cao, D. (2016). Molecular simulation of displacement of shale gas by carbon dioxide at different geological depths. *Chemical Engineering Science* 156: 121–127.

Zhang, X., Lu, Y., Tang, J. et al. (2017). Experimental study on fracture initiation and propagation in shale using supercritical carbon dioxide fracturing. *Fuel* 190: 370–378.

Zhang, Y., Hu, Q., Barber, T.J., Bleuel, M., Anovitz, L.M. and Littrell, K., 2020. Quantifying fluid-wettable effective pore space in the Utica and Bakken oil shale formations. *Geophysical Research Letters*, 47(14): p.e2020GL087896.

Zhao, P., Wang, Z., Sun, Z. et al. (2017). Investigation on the pore structure and multifractal characteristics of tight oil reservoirs using NMR measurements: Permian Lucaogou Formation in Jimusaer Sag, Junggar Basin. *Marine and Petroleum Geology* 86: 1067–1081.

Zheng, S., Yao, Y., Liu, D. et al. (2018). Characterizations of full-scale pore size distribution, porosity and permeability of coals: A novel methodology by nuclear magnetic resonance and fractal analysis theory. *International Journal of Coal Geology* 196: 148–158.

Zhu, G.P., Yao, J., Sun, H. et al. (2016). The numerical simulation of thermal recovery based on hydraulic fracture heating technology in shale gas reservoir. *Journal of Natural Gas Science and Engineering* 28: 305–316.

12

Analysis of Pore Characteristics of Select Indian Shale Samples and Assessment of Pore Connectivity by Conformance Correction of Mercury Intrusion Porosimetry Results

Tuli Bakshi[1,2], *Venkata Yasaswy Turlapati*[2], *Vikram Vishal*[1,*], *B.K Prusty*[2], *and Khanindra Pathak*[2]

[1] *Computational and Experimental Geomechanics Laboratory, Department of Earth Sciences, Indian Institute of Technology Bombay, Mumbai, Maharashtra, India*
[2] *Clean Coal Technology Laboratory, Department of Mining Engineering, Indian Institute of Technology Kharagpur, Kharagpur, West Bengal, India*

**Author for correspondence: drvikramvishal@gmail.com*

Abstract

Recent research has speculated that the organic-rich shale deposits in the eastern part of India have the potential for future shale gas exploration. Limited research and published data in this domain pose a significant challenge in evaluating the gas capacity estimates and producibility of these shales. In this research, pore characteristics of a few subsurface shale samples were investigated in detail using the mercury injection method and low pressure N_2—CO_2 adsorption study. SEM images revealed the presence of organic matter pores, clay pores, interlayer clay pores, intragranular pores, and microcracks. Pore size distribution study combining CO_2—N_2 isotherm and mercury injection exhibited smooth curves that corroborate with each other and indicates lit-shaped pores in all the samples. Porosity values were calculated using the mercury intrusion capillary pressure (MICP) and pore connectivity was assessed for the studied samples based on MICP data. The pore volume percentage calculated from the pore-size data divulged a predominance of mesopores, which can be attributed to the high clay content. High organic matter content in all the shale samples causes dominance of micropore volumes among the total pore volume. Bulk volume and porosity were calculated, followed by conformance correction, by adhering to Bailey's method. The mercury injection plots suggested that the samples with lower tightness and higher connectivity have higher macropore volume percentages. On the other hand, the tighter samples are dominantly mesoporous. These findings support the existing researches being done on the pore characteristic studies on shale.

Keywords *Shale; mercury porosimetry; pore volume; conformance correction*

Handbook of Petroleum Geoscience: Exploration, Characterization, and Exploitation of Hydrocarbon Reservoirs, First Edition. Edited by Soumyajit Mukherjee, Swagato Dasgupta, Chandan Majumdar, Subhandip Mandal and Troyee Dasgupta.
© 2023 John Wiley & Sons Ltd. Published 2023 by John Wiley & Sons Ltd.

12.1 Introduction

The success of shale gas in the USA since the first decade of this century has inspired and redefined the global energy resources industry. The advent of cutting-edge technologies has been successfully tackling the complexities of heterogeneous shale reservoirs and reshaping the standard of exploration, production, and development of unconventional reservoirs. Over the past two decades, India's economic growth has been associated with a significant surge in energy consumption. The increasing energy demand has triggered a considerable burden on the country's limited resources (e.g. Varigonda 2015), and, hence, optimal usage of resources has become vital. In recent years, the demand for natural gas has increased significantly in India. It is estimated to reach 746 MSCMD by 2029–2030, which can be attributed to the development of substantial gas reserves and the adoption of cleaner energy solutions, and technological advancement. Due to the affordability and purity of natural gas, India has decided to increase its consumption from 6.5% at present to 15% in the next few years; to combat the issue of growing greenhouse gas emissions caused by the burning of higher hydrocarbons and coal. The role of shale gas will be instrumental in meeting this growing requirement for energy in India.

The primary component of any shale includes organic matter and clay minerals (e.g. Holmes et al. 2017; Bakshi et al. 2020). The various mineral grains, clay, and organic matter create a diverse pore geometry (Chandra and Vishal 2020; Kuila et al. 2012). Due to heterogeneities that can either be inherent (Mukherjee 2018) or tectonics-induced (Mukherjee and Kumar 2018), several types of pores exist in gas shale systems in both inorganic minerals and organic matter. Davudov et al. (2017) classified shale pores into accessible and inaccessible based on the leaky-tube model of Civan (2011). Interconnected pores are primarily accessible, and the different types of inaccessible pores include the ones, which are naturally isolated or induced isolated. Through the MICP test, connectivity of shale pores with sizes >3 nm can be measured if the samples are partially saturated with mercury (Chalmers et al. 2012; Davudov et al. 2018). Through their studies, King et al. (2015) have considered MICP as one of the best techniques to provide direct information on pore throats. Different attributes determined from MICP, like pore throat distribution (PTD), pore sizes, geometry, and surface area (SSA), largely govern the nature of pores in the shale matrix. The pore attributes determine the accessible porosity, pore connectivity, permeability, matrix diffusion rate, and reservoir quality (Ruppert et al. 2013; King et al. 2015; Topór 2020). Hu et al. (2015) thoroughly studied the pore connectivity of Barnett shale and reported poor connectivity in the analyzed shale samples. They further linked this to the overall low recovery and declining gas production.

Moghanloo (2015) emphasized the necessity of research on connected pores that highlights the scope of pore compressibility since decompression of rock and gas is the primary mechanism of gas production from shale. The closure volume is the required volume of mercury that would conform to the sample's surface roughness confined in the outer layer of grains. Thus, the closure volume depends on the grain and the sample size. In crushed samples, particle grains are positioned together in a sample holder, and artificial pore space is created between particles. The low entry pressure of this artificial pore space, which is treated as a closure volume, can be easily distinguished from that of the right pore space present inside the particles. Compressibility results from the relative volume change of the sample caused by the mercury imposing a stress field. Intrusion starts with surpassing the largest pore throat's capillary entry pressure and ends with filling all the pore spaces with mercury (Bailey 2009).

The Damodar Valley and the Upper Assam basin are the two crucial Indian basins with a significant prospect for shale gas exploration and exploitation. India has marked its position as a shale gas resource after identifying shale gas wells in the Damodar Valley basin and the Cambay basins. However, the assessment of the Damodar Valley and the Assam basin shales have been minimal, with very few published results on pore connectivity in the shale matrix. The present article fills up this knowledge gap.

The application of Bailey's (2009) method of calculating the total pore compressibility, combined with the MICP data, is a critical approach to developing an accurate understanding of pore characteristics. A power law regression is adopted, and the Hg porosimetry is performed to investigate the shale mesopore and macropore characteristics thoroughly. Accessible pore compressibility, pore connectivity, and the influence on permeability before and after the conformance corrections were estimated.

12.2 Methodology

Samples were collected from the Damodar Valley basin (Jharkhand) coalfields and the Upper Assam basin (Assam) coalfield in the eastern part of India (Figure 12.1). The sample name, corresponding coalfield, and depth are listed in Table 12.1.

The mercury intrusion study was done in *Quantachrome PoreMaster* situated in the CSIR-Central Glass & Ceramic Research Institute, Kolkata to study the porosity and pore size

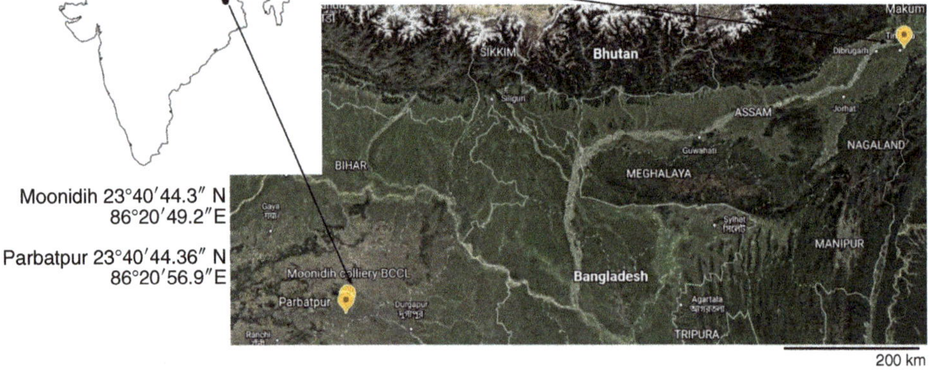

Figure 12.1 Sample location map.

Table 12.1 Location of the studied samples with depth and formation.

Sample name	Colliery	Formation	Depth (m)
MOON	Moonidih	Barakar	421
PB	Parbatpur		180
TKPDG	Makum	Tikak Parbat	63
TKPGS	Makum		150

distribution of the samples. The helium pycnometer determined the skeletal density of the same samples. The helium pycnometer was performed after crushing the samples in a 1000–2000 μm size range. Moisture and different contaminants were removed through the oven-drying process. The samples were further degassed for 30 minutes before the final analysis. Approximately 1–3 g of samples were considered for the study, and mercury was injected up to 40 000 psi.

Porosities were calculated from the Hg intrusion data. The difference between the bulk density (ρ_{bulk}) from the MICP and skeletal density (ρ_{grain}) provides the porosity values (Bustin et al. 2008; Tian et al. 2013). The Archimedes principle of displacement measures the bulk density of samples, and the grain density is obtained by helium pycnometer based on the principles of Boyle's law. Helium is used to measure shale grain density due to its inert nature and low kinetic diameter, which leads to greater access to the fine pore structure of shales (Bustin et al. 2008).

$$\varphi = 100 \times \left(1 - \rho_{\text{bulk}}/\rho_{\text{grain}}\right) \tag{12.1}$$

Generally, the mercury volume during the MICP test is rectified to consider conformance (Bailey 2009). Conformance is the amount of mercury that envelopes the sample's exterior portion and its shape before the intrusion and is a function of sample shape irregularities and its size. The method suggests that the mercury volume recorded until the intrusion/entry pressure should be subtracted from the total volume. A ratio of MICP-calculated porosity to total porosity (ϕ_{LPP}) was considered to evaluate accessible pore volume fraction.

12.3 Results and Discussions

12.3.1 Type of Pores

Apart from storing gas, the gas-shale reservoir pores also act as the gas migration pathway, imparting immense importance to production. Researchers have classified the shale reservoir pores into interparticle, intraparticle, and organic (Loucks et al. 2012). This article has investigated the nature of pores in the Damodar Valley and the Upper Assam shale. SEM images have displayed a wide variety of pores of different shapes and sizes. A large number of organic matter pores were also identified (Figure 12.2a, c, and d). The images depict a sporadic presence of framboid pyrites in all the samples, which create interparticle pores within

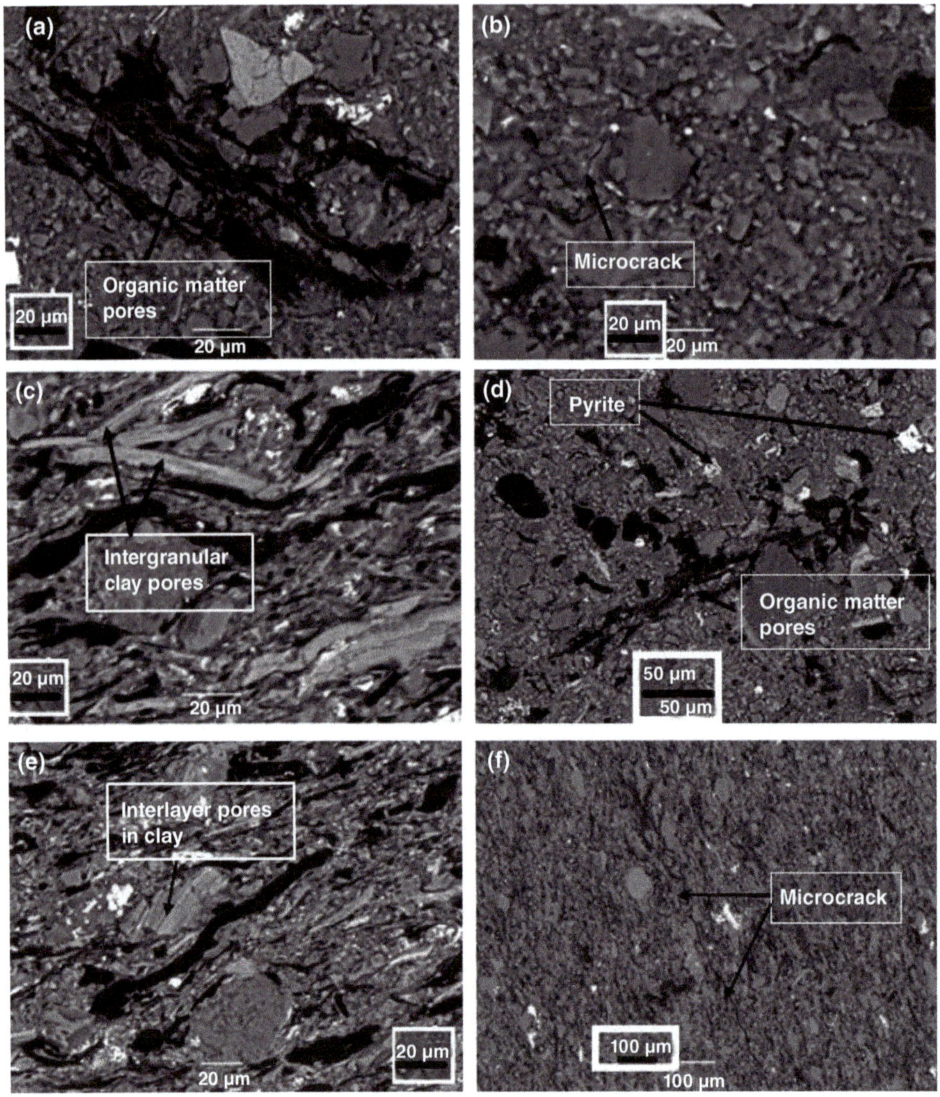

Figure 12.2 SEM images showing different types of pores in shale.

them. These pyrites are also associated with organic matter creating a pore network. Some interlayer pores were observed in the shale, formed by the interlayer of various clay sheets such as illite and chlorite (Figure 12.2c and e). Intraparticle pores found in quartz, feldspar, and other mineral particles, are not very well developed in the mineral assemblages, indicating poor connectivity. Microcracks were observed in the shale matrix, forming a three-dimensional pore network with intergranular pores (Figure 12.2b and f) and between the contacts of hard mineral grains (Figure 12.2b and e). Intergranular pores were detected between mineral grains.

12.3.2 Conformance Correction

When a rock sample is kept in the sample holder for mercury injection, mercury conforms to the rock surface in the initial pressure ranges. It is confined to the outer region of the sample. The entry pressure of mercury to the largest pore throats is known as intrusion pressure. Generally, an abrupt change of slope in the mercury saturation vs. injection pressure graph (Figure 12.3) corresponds to the entry pressure. In conventional reservoir rocks, intrusion and closure overlap at relatively low pressure and small volumes (Bailey 2009). However, in tight rocks with low macroporosity and permeability, this sudden change of slope occurs at very high pressure and toward a relatively larger volume. This behavior corresponds to the rock's bulk compression, making the process of identification of the conformance pressure challenging and yet crucial. Bailey (2009) suggested that the power-law functions concerning pressure best express tight formations' compressibility; therefore, the compression phase is the linear portion of the log-log plot generated from the incremental mercury volume change vs. pressure graph. Any nonconformity with the line could be either due to conformance on low pressure or intrusion at the high-pressure region (Davudov and Moghanloo 2018).

Pore Volume Compressibility is calculated by:

$$C_p = (1/PV)(dPV/dP) \tag{12.2}$$

Here, PV is the residual air-filled pore volume (Bailey 2009).

12.3.2.1 Identification of Conformance

The plot of mercury saturation and injection pressure (Figure 12.3) showcases the sudden change in slope. However, this change is identified at significantly higher pressures and cannot be termed as the conformance pressure. Further, it is also evident that before the slope change, the volume intrusion increases steadily with pressure. Using Bailey's method, the conformance pressure can be identified at the origin of the straight line in the log-log plot of injection pressure and pore volume compressibility (orange line in Figure 12.4). The

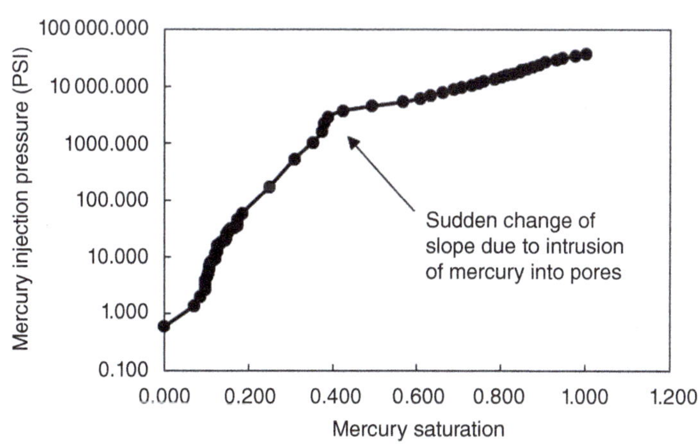

Figure 12.3 Hg injection vs. saturation graph pointing out the sudden change in slope.

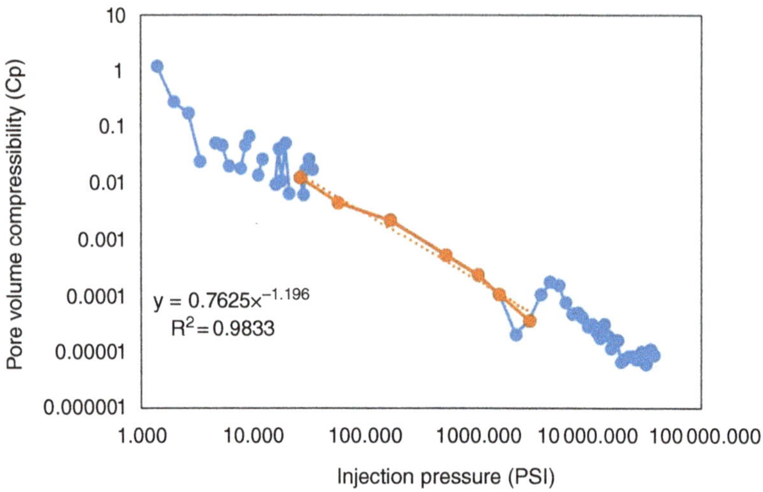

Figure 12.4 MICP curve showing conformance pressure, compression pressure, and injection pressure.

points above and below this straight line are initial pressure prior to conformance and pore intrusion pressures, respectively. It is assumed in this method that the grain compressibility is negligible. Hence, the entire bulk compression appearing in Figure 12.4 is regarded as pore compression, and theoretically, once the pores are filled with mercury, they bounce back in shape.

A comparative analysis of the results of MICP and helium porosity was conducted on the samples from all the Damodar Valley and the Upper Assam shale. Table 12.2 shows the MICP experimental data obtained for the Damodar Valley and the Upper Assam shale.

The MICP study was conducted to explore the porosity, pore size distribution, and pore connectivity. All the tests were performed after drying the samples in an oven at 108 °C. Pore volume is determined by the mercury left in the penetrometer stem. An increase in pressure causes a higher amount of mercury to move into the sample pores. Bailey (2009) highlighted that conformance corrections are required in MICP porosity calculations. In this study, an attempt has been made to understand the porosity both before and after the conformance correction. ρ_b is the bulk density determined through the MICP

Table 12.2 MICP experimental data.

Sample	Sample weight W_s (g)	Bulk volume (cm^3)	Total pore volume V_{Hg} (g cc^{-1})	Conformance-corrected pore volume V_{Hg} (g cc^{-1})
MOON	0.7076	0.3705	0.0066	0.0028
PB	0.8682	0.4134	0.0129	0.0076
TKP-G	1.0526	0.4314	0.0089	0.0052
TKP-DG	0.5456	0.491	0.0371	0.0073

analysis in g cm^{-3} at an instrument pressure 40 000 psi. W_s is the sample weight in g. Porosity is determined from MICP by using the formulae:

$$\Phi_{MICP} = \frac{V_{Hg}}{V_b} \tag{12.3}$$

Here, V_{Hg} is the pore volume at 40 000 psi. V_b is the bulk volume of the sample, which can be calculated by:

$$V_b = \frac{W_s}{\rho_b} \tag{12.4}$$

12.3.3 Pore Connectivity Evaluation

The shapes of mercury injection curves provide a basic understanding of pore connectivity and morphology (Vafaie et al. 2015). Figure 12.5 presents that MOON and TKPDG have attained ~100% mercury saturation ~100 psi, indicating accessibility of the pores or good connectivity. The 100% mercury intrusion implies the complete network of linked pore spaces is accessed by mercury. In PB and TKPG, ~40% pore network was invaded around 10 psi, and 100% saturation was realized around 10 000 psi, indicating a less interconnected nature of pores. These samples are tighter as100% saturation is attained >10 000 psi. The incremental Hg intrusion plots suggest that the samples with good connectivity have higher macropore volumes (Figure 12.7) than those of the remaining. Similar observations were reported in Garau shale, where samples with lower tightness were dominated by macropores (Vafaie et al. 2015).

12.3.4 Pore Size Distribution

The sample pore size distributions were evaluated by merging the mercury data with the low-pressure N_2, CO_2 dV/dlogr plots (Figure 12.6) (Bakshi et al. 2017, 2018, 2020) to get the total picture of micropores, mesopores, and part of macropores. The mercury data was observed to be less consistent with LPGA data in MOON and TKPDG, and the transition from LPGA data to mercury data was not smooth. This is because mercury intrusion

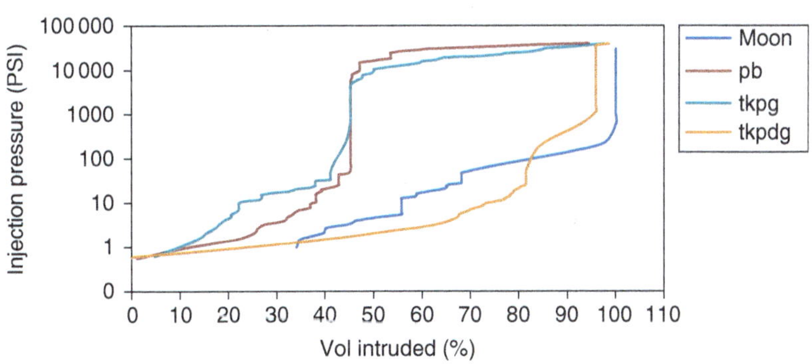

Figure 12.5 The pressure of injection vs. cumulative mercury saturation for shale samples.

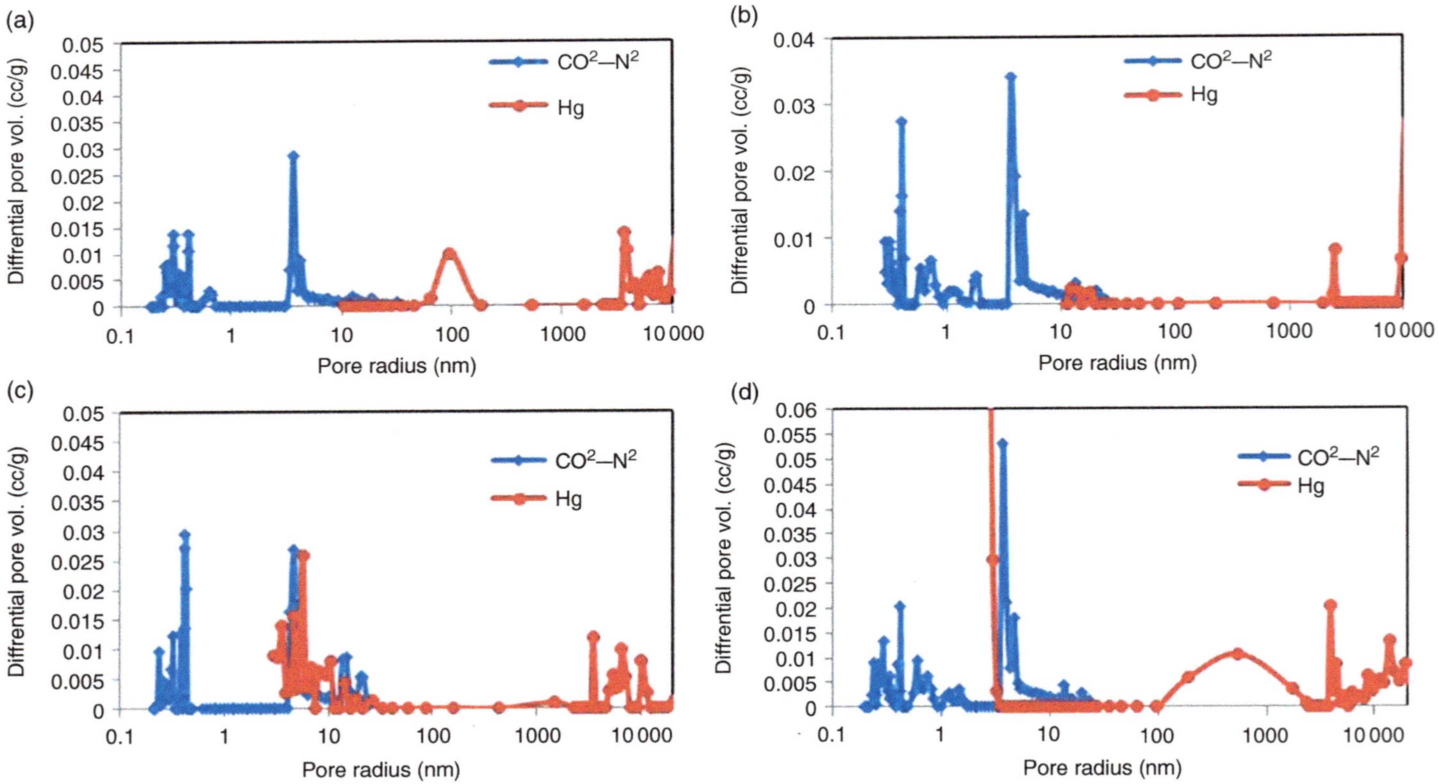

Figure 12.6 Pore volume distribution using N_2—CO_2—Hg d*V*/dlogr plots showing (a) MOON, (b) PB, (c) TKPG, (d) TKPDG.

measures only pore throats, whereas BJH theory pore size distributions in LPGA are obtained by measuring pore bodies (Clarkson et al. 2013; Kuila and Prasad 2013). In the case of slit-shaped pores, pore throats were the same as pore bodies yielding similar adsorption and Hg analysis results. Hence, a possibility exists that in PB and TKPG samples, the pores might be of slit-shaped geometry. However, for other samples (MOON, TKPDG), the pore size data from mercury intrusion and N_2—CO_2 adsorption isotherm were distinct. This deviation is due to the pore geometries other than that of the slit shape. According to these figures, the samples display a multimodal pore size distribution.

12.3.5 Contribution of Different Pore Sizes to the Total Porosity

The pore volume percentages V_{mic}, V_{me}, and V_{ma} depict pores of <2 nm, between 2 and 50 nm, and >50 nm, respectively. From the data of N_2—CO_2 and MICP pore size distribution, the volume percentage of micropores, mesopores, and macropores to the total pore volume has been calculated and presented in Figure 12.7. It was noted that the contribution of mesopore volume to the total pore volume is highest (except MOON), followed by micropore volume and macropore volume, respectively. A similar phenomenon where mesopore volume is a primary contributor to the total pore volume was reported in the gas-shale reservoirs from the Longmaxi Formation in the Sichuan basin, China (Tian et al. 2013; Xiong et al. 2015). The presence of an extensive amount of clay mineral results in the creation of high mesopore volume as clay is primarily associated with mesopores (Ross and Bustin 2009). A high amount of micropore volume than mesopore volume in the case of the MOON is due to the higher TOC content than the rest of the shale samples (Figure 12.6). Since organic matter is the principal contributor to the micropore surface area, high TOC concentrations in all the samples could have resulted in its high micropore volume. The reliability of these findings is affirmed since they conform with some notable work on gas shale in North American basins (Chalmers and Bustin 2007, 2008; Ross and Bustin 2009).

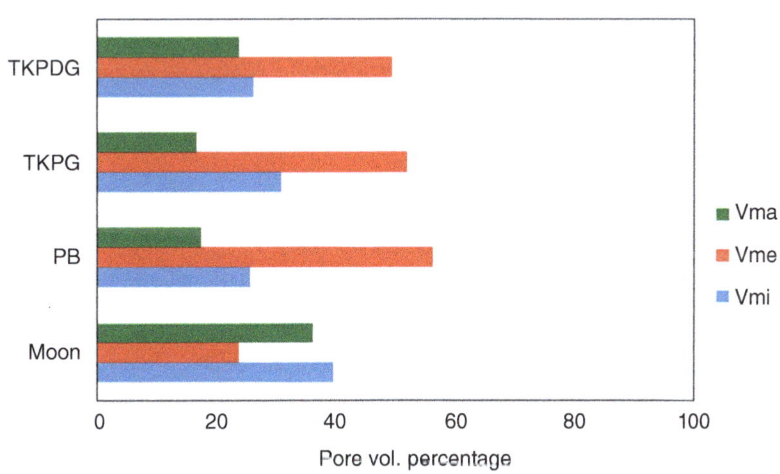

Figure 12.7 Distribution characteristics of whole pore volume in micro-, meso-, and macropore size range.

12.3.6 Permeability Calculated from the MICP Data

For an explicit understanding of matrix connectivity for the Gondwana and the Assam shale samples, the absolute permeability calculation was done based on MICP experimental data following an empirical model by Dastidar et al. (2007). This model is based on the weighted geometric mean of the pore throat radius distribution (R_p). The entire spectrum of pore throats was considered for this calculation. The R_p is calculated as:

$$R_p = \exp\left[\frac{\sum_{i=1}^{n} W_i \ln(R_i)}{\sum_{i=1}^{n} W_i}\right] \tag{12.5}$$

Here R_i, pore throat radius at ith capillary pressure, and W_i:R_i, weight at ith capillary pressure. W_i is the ratio of the incremental volume intruded into the sample (V_i) at a specific capillary pressure to the total volume intruded (V_{Hg}):

$$W_i = \frac{V_i}{V_{Hg}} \tag{12.6}$$

Dastidar et al.'s proposed equation for permeability calculations is:

$$K = 4073\, R_p^{1.64} \varphi^{3.06} \tag{12.7}$$

Table 12.3 lists the calculated permeability values for the studied shale samples. Results show that permeability is somewhat influenced by porosity. Permeability being a function of effective porosity, increases with increasing porosity. Figure 12.5 shows that the samples with the least connectivity have the least permeability and the ones with higher permeability have better connectivity.

12.4 Conclusions

The complex geology and genesis of shale result in a high degree of heterogeneity. Hence, the outcome and interpretation of some analyses may not be representative of other shales from the same basin or the same formation. However, the results can offer some understanding of the complex nature of the pore characteristics of these formations. The

Table 12.3 Porosity and permeability before and after Bailey correction.

Sample	Porosity before correction (%)	Porosity after correction (%)	Permeability before Bailey correction (mD)	Permeability after Bailey correction (mD)	Clay (%)	TOC (%)
MOON	1.78	0.75	0.007	0.0005	24.97	30.90
PB	3.12	1.80	0.005	0.001	40.39	12.12
TKPG	1.93	1.27	0.005	0.001	42.21	19.65
TKPDG	7.6	1.40	0.127	0.007	10.48	4.86

uncorrected mercury porosity of the tested shale samples varies from 1.7 to 7.6%, whereas the conformance corrected porosity varies from 0.75 to 1.4% in all the samples. The uncorrected permeability varies from 0.005 to 0.127 mD, while the permeability after Bailey correction varies from 0.0005 to 0.007 mD in all the samples. Hg injection curves corroborate with the calculated permeability values. Lower permeability is observed in the samples with low pore connectivity. Samples with higher macropore volume display increased pore connectivity and a higher value of permeability. The Hg d*V*/dlogr, when combined with CO_2—N_2 data trend, has been found to be nearly consistent with each other, suggesting the presence of slit-shaped pores. CO_2—N_2—Hg d*V*/dlogr indicates the dominance of mesopore for all shale samples, caused by a high amount of clay and organic matter.

Acknowledgments

The authors are extremely grateful to Electrosteel Castings Limited, Bharat Coking Coal Limited, and North Eastern Coalfields for providing samples for this work. The authors are indebted to Dr. K. Muraleedharan and Dr. Ashim Kumar Chakraborty for permitting us to do the sample analysis at the material characterization laboratory of CSIR-CGCRI Kolkata. The authors express gratitude to the reviewers for the valuable suggestions. Merryl Le Roux, Frank Otmar Weinreich, Joss Everett, Andrew Harrison, and the proofreading team (Wiley Blackwell) are thanked for their assistance. The article has been summarized in Mukherjee et al. (2023).

References

Bailey, S., 2009. Closure and compressibility corrections to capillary pressure data inshales. DWLS Workshop, Golden, CO (19 October 2009).

Bakshi, T., Prusty, B.K., Pathak, K. et al. (2017). Source rock characteristics and pore characterization of Indian shale. *Journal of Natural Gas Science and Engineering* 45: https://doi.org/10.1016/j.jngse.2017.06.010.

Bakshi, T., Prusty, B.K., Pathak, K., and Pal, S.K. (2018). Pore characteristics of Damodar valley shale and their effect on gas storage potential. *Journal of Petroleum Science and Engineering* https://doi.org/10.1016/j.petrol.2017.10.091.

Bakshi, T., Vishal, V., Prusty, B.K., and Pathak, K. (2020). Carbon dioxide and methane sorption characteristics of Damodar Valley and Upper Assam shale, India. *Energy and Climate Change* https://doi.org/10.1016/j.egycc.2020.100002.

Bustin, R., Bustin, A., Cui, A., Ross D., and Murthy Pathi, V. (2008). Impact of shale properties on pore structure and storage characteristics. In: *SPE paper 119892 Presented At The Society of Petroleum Engineers Shale Gas Production Conference in Fort Worth*, Texas (16–18 November 2008).

Chalmers, G.R.L. and Bustin, R.M. (2007). The organic matter distribution and methane capacity of the Lower Cretaceous strata of northeastern British Columbia. *Canada. International Journal of Coal Geology.* 70: 223–239.

Chalmers, G.R.L. and Bustin, R.M. (2008). Lower Cretaceous gas shales in northeastern British Columbia, part I: geological controls on methane sorption capacity. *Bulletin of Canadian Petroleum Geology*. 56: 1–21.

Chalmers, G.R., Bustin, R.M., and Power, I.M. (2012). Characterization of gas shale pore systems by porosimetry, pycnometry,surfacearea,andfield emission scanning electron microscopy/ transmission electron microscopy image analyses: examples from the Barnett, Woodford, Haynesville, Marcellus,andDoig units. *AAPG Bulletin*. https://doi.org/10.1306/10171111052.

Chandra, D. and Vishal, V. (2020). A comparison of nano-scale pore attributes of Barakar Formation gas shales from Raniganj and Wardha Basin, India using low pressure sorption and FEG-SEM analysis. *Journal of Natural Gas Science and Engineering* 81: 103453. https://doi.org/10.1016/j.jngse.2020.103453.

Civan, F. (2011). Porous media transport phenomena. In: *Porous Media Transport Phenomena*, 10–53. Hoboken, NJ: John Wiley & Sons, Inc. https://doi.org/10.1002/9781118086810.

Clarkson, C.R., Solano, N., Bustin, R.M. et al. (2013). Pore structure characterization of North American shale gas reservoirs using USANS/SANS, gas adsorption, and mercury intrusion. *Fuel*. https://doi.org/10.1016/j.fuel.2012.06.119.

Dastidar, R., Sondergeld, C.H., and Rai, C.S. (2007). An improved empirical permeability estimator from mercury injection for tight clastic rocks. *Petrophysics*..

Davudov, D. and Moghanloo, R.G. (2018). Impact of pore compressibility and connectivity loss on shale permeability. *International Journal of Coal Geology* 187: 98–113. https://doi.org/10.1016/j.coal.2018.01.008.

Davudov, D., Moghanloo, R. G., Lan, Y., & Vice, D. (2017). Investigation of shale pore compressibility impact on production with reservoir simulation. *Paper presented at the SPE Unconventional Resources Conference*, Calgary, Alberta, Canada (February 2017). doi: https://doi.org/10.2118/185059-MS.

Davudov, D., Moghanloo, R.G., and Lan, Y. (2018). Evaluation of accessible porosity using mercury injection capillary pressure data in shale samples. *Energy and Fuels* 32 (4): 4682–4694. https://doi.org/10.1021/acs.energyfuels.7b03930.

Holmes, R., Rupp, E.C., Vishal, V., and Wilcox, J. (2017). Selection of shale preparation protocol and outgas procedures for applications in low-pressure analysis. *Energy and Fuels*. https://doi.org/10.1021/acs.energyfuels.7b01297.

Hu, Q., Ewing, R.P., and Rowe, H.D. (2015). Low nanopore connectivity limits gas production in Barnett formation. *Journal of Geophysical Research: Solid Earth* 120 (12): 8073–8087. https://doi.org/10.1002/2015JB012103.

King, H.E., Eberle, A.P.R., Walters, C.C. et al. (2015). Pore architecture and connectivity in gas shale. *Energy and Fuels* 29 (3): 1375–1390. https://doi.org/10.1021/ef502402e.

Kuila, U. and Prasad, M. (2013). Specific surface area and pore-size distribution in clays and shales. *Geophysical Prospecting*. https://doi.org/10.1111/1365-2478.12028.

Kuila, U, Prasad, M., Derkowski, A., & McCarty, D. K. (2012). Compositional controls on mudrock pore-size distribution: an example from Niobrara formation. *Paper presented at the SPE Annual Technical Conference and Exhibition*, San Antonio, Texas, USA (October 2012).

Loucks, R.G., Reed, R.M., Ruppel, S.C., and Hammes, S. (2012). Spectrum of pore types and networks in mudrocks and a descriptive classification for matrix-related mudrock pores. *American Association of Petroleum Geologists Bulletin* 96 (6): 1071–1098.

Moghanloo, R.G., Yuan, B., Ingrahama, N. et al. (2015). Applying macroscopic material balance to evaluate interplay between dynamic drainage volume and well performance in tight formations. *Journal of Natural Gas Science and Engineering* 27: 466–478. https://doi.org/10.1016/j.jngse.2015.07.047.

Mukherjee, S. (2018). Moment of inertia for rock blocks subject to bookshelf faulting with geologically plausible density distributions. *Journal of Earth System Science* 127: 80. https://doi.org/10.1007/s12040-018-0978-4.

Mukherjee, S. and Kumar, N. (2018). A first-order model for temperature rise for uniform and differential compression of sediments in basins. *The International Journal of Earth Sciences* 107: 2999–3004. https://doi.org/10.1007/s00531-018-1634-6.

Mukherjee, S., Dasgupta, S., Majumdar, C. et al. (2023). Introduction to: Handbook of Petroleum Geoscience: Exploration, Characterization, and Exploitation of Hydrocarbon Reservoirs. In: *Handbook of Petroleum Geoscience: Exploration, Characterization, and Exploitation of Hydrocarbon Reservoirs*, Petroleum Industry Updates: Geoscientific and Economic Issues (ed. S. Mukherjee, S. Dasgupta, C. Majumdar, et al.). John Wiley & Sons, Hoboken, ISBN 978-1-119-68003-1.

Ross, D.J. and Bustin, M.R. (2009). The importance of shale composition and pore structure upon gas storage potential of shale gas reservoirs. *Marine and Petroleum Geology* 26 (6): 916–927.

Ruppert, L.F., Sakurovs, R., Blach, T.P. et al. (2013). A USANS/SANS study of the accessibility of pores in the Barnett shale to methane and water. *Energy and Fuels* 27 (2): 772–779. https://doi.org/10.1021/ef301859s.

Tian, H., Pan, L., Xiao, X. et al. (2013). A preliminary study on the pore characterization of Lower Silurian black shales in the Chuandong Thrust Fold Belt, southwestern China using low pressure N_2 adsorption and FE-SEM methods. *Marine and Petroleum Geology*. https://doi.org/10.1016/j.marpetgeo.2013.07.008.

Topór, T. (2020). An integrated workflow for MICP-based rock typing: A case study of a tight-gas sandstone reservoir in the Baltic Basin (Poland). *Nafta-Gaz* 76 (4): 219–229. https://doi.org/10.18668/ng.2020.04.01.

Vafaie, A., Habibnia, B., and Moallemi, S.A. (2015). Experimental investigation of the pore structure characteristics of the Garau gas shale formation in the Lurestan Basin, Iran. *Journal of Natural Gas Science and Engineering*. https://doi.org/10.1016/j.jngse.2015.06.029.

Variginda, K.C. (2015). India: Petroleum Policies and Geopolitics. In: *Peroleum Geosciences: Indian Contexts* (ed. S. Mukherjee), 271–294. Springer Geology. ISBN: 978-3-319-03119-4.

Xiong, J., Liu, X., and Liang, L. (2015). Experimental study on the pore structure characteristics of the Upper Ordovician Wufeng Formation shale in the southwest portion of the Sichuan Basin, China. *Journal of Natural Gas Science and Engineering*.

13

Geochemical Modeling of Diagenetic Reactions in the Eocene Sediment-Gravity-Flow Deposit Reservoirs Influenced by Salt Tectonics: The Espírito Santo Basin, Brazil

Marcos Antonio Klunk[1,], Sudipta Dasgupta[2], Mohuli Das[2], Renzo D'souza[2], Soyane Juceli Siqueira Xavier[3], and Paulo Roberto Wander[1]*

[1] *Department of Mechanical Engineering, University of Vale do Rio dos Sinos, São Leopoldo, Brazil*
[2] *Department of Earth Sciences, Indian Institute of Technology Bombay (IIT Bombay), Mumbai, Maharashtra, India*
[3] *Department of Geology, University of Vale do Rio dos Sinos, São Leopoldo, Brazil*

Author for correspondence: mklunk@gmail.com

Abstract

Diagenetic alteration in the Eocene turbidite reservoirs of Espírito Santo Basin has been envisaged through geochemical modeling using TOUGHREACT software. Extensive albitization has been encountered, accompanied by the dissolution of detrital plagioclase and K-feldspars. The diapiric impregnation of the succession by halokinesis and the resulting high salinity of pore water suggest that diagenesis was invigorated by the entry of brine from the salt diapirs. The circulation of high-temperature fluids from the diapirs through fractures is the main cause of the late diagenetic reactions, thereby influencing the reservoir quality. This behavior is accompanied by concomitantly growing authigenic quartz crystals as derived from the fluid-inclusion study. Based on the available geological and geochemical data, a series of computer simulations present the effects of fluid percolating through the reservoir and non-reservoir depositional elements deposited from sediment-gravity flows, along with the influence of this interaction with the salt domes. The simulations were carried out in batch mode to stabilize the water with sandstone and the results were then utilized in a series of 1D simulations. A nominal water flow of 10 cm y^{-1} was assumed in isothermal conditions at 100 °C. In this scenario, the precipitation of calcite in sandstone was considered to occur due to the presence of thermobaric flow. Its authigenic precipitation is strongly controlled by the dissolution of plagioclase. The simulations carried out with TOUGHREACT reveal the potential of the simulator to model the complex diagenetic cases of turbiditic reservoirs in an analog stratigraphic configuration. The simulations also demonstrate the need to dynamically integrate the diagenesis associated with the sandstones so that the obtained results can be consistent with the real diagenetic patterns.

Keywords *Diagenesis, Sandstone; 1D simulation; Espírito Santo basin; Geochemical modeling; Saturation index*

Handbook of Petroleum Geoscience: Exploration, Characterization, and Exploitation of Hydrocarbon Reservoirs, First Edition. Edited by Soumyajit Mukherjee, Swagato Dasgupta, Chandan Majumdar, Subhandip Mandal and Troyee Dasgupta.
© 2023 John Wiley & Sons Ltd. Published 2023 by John Wiley & Sons Ltd.

13.1 Introduction

Numerical modeling has gained substantial significance in the understanding of complex diagenetic processes, thereby aiding in the prediction of geochemical reactions in the oil reservoirs (Klunk et al. 2015; Wei et al. 2016). For the construction of geochemical models, the computational codes utilize the physicochemical properties of minerals (pH, Eh, temperature, salinity solutions, and porosity) and depositional environments in which these reactions occur (Szczepański 2018). Mathematical models simulate the interaction of aqueous solutions with minerals, providing a better understanding of the diagenetic reactions (Paraska et al. 2014; Klunk et al. 2020).

Kinetic and thermodynamic parameters that govern diagenetic reactions, as well as geological data from a reservoir of the Espírito Santo Basin, can be used to generate a model, which can predict the integrity of the petroleum system (Klunk et al. 2018). Therefore, the objectives of this work are to establish a geochemical model, evaluate the impact of geological changes and water-rock interaction, and predict the diagenetic reactions in a turbiditic reservoir of the Espírito Santo Basin.

This study aims to generate a tool which is capable of assisting the hydrocarbon exploration geologists to conceptualize the reservoir in detail (Lai et al. 2015). For this particular study, the TOUGHREACT geochemical modeling software in batch and 1D mode has been used (Klunk et al. 2021). The simulations allowed the recreation of the diagenetic reactions and thereby expansion of the understanding of the water-rock interaction in a geological environment. The results in batch mode (water-sandstone interaction) permitted the initial water to attain a steady state in approximately 35 000 years. Once the stability was gained, the same water was used for 1D simulations. These simulations revealed the phenomenon of albitization accompanied by the reduced precipitation of kaolinite and the dissolution of plagioclase (González-Acebrón et al. 2010). The albitization data produced in the simulations confirmed the abundance of albite. This approach provides a tool to estimate the intensity of albitization in the reservoir. This information can be successfully utilized during exploration and production operations in the hydrocarbon industry for estimating the porosity and permeability of the reservoir rock (Zhang et al. 2015).

13.2 Geological Context

The sandstones of the Espírito Santo Basin, diagnosed for this particular work, represent the turbiditic reservoirs of the Urucutuca Formation belonging to the Eocene Epoch (Mansurbeg et al. 2012). Most of these sandstones have a significant concentration of feldspars in close proximity to the salt domes which determines the reservoir structure and diagenetic evolution (Vandeginste et al. 2017).

13.3 Compositional Data

For the conceptual model of the reservoir, the following average composition of sandstones was used according to Table 13.1.

The chemical composition of the water used in geochemical modeling is shown in Table 13.2.

Table 13.1 Mineralogical composition of the sandstone.

Mineral	**Sandstone (% TVR[a])**
Quartz	45.10
K-feldspar	10.35
Plagioclase	15.14
Albite	4.36
Calcite	1.25
Dolomite	5.06
Kaolinite	8.33

[a] total volume of rock.

Table 13.2 Water composition used in the simulations.

Solutes	**Concentration (mol l^{-1})**
Na^+	2.055
Cl^-	3.123
K^+	0.151
Fe^{2+}	0.0002
HCO_3^-	0.029
H^+	$3.16e^{-7}$

The first simulations were carried out in batch mode, at a temperature of 100 °C in order to achieve a steady state between water and the sandstone. In the 1D modeling, the water composition at a temperature of 120 °C and the local interactions between the sandstones were simulated to recreate the diffusive flow processes, which are controlled by the concentration gradients (De Lucia et al. 2016).

The analysis resulted in consideration of the complex interactions between water and the reservoir rock, thereby creating a reactive transport system, generated by the chemical instability of the lithology (Li et al. 2020). The fluid flow is considered to be 10 cm year^{-1}, which is a generic value typically used in simulations of diagenetic systems.

The minerals used in the simulation include quartz, feldspars (plagioclase, K-feldspar, and albite), carbonates (calcite and dolomite) and kaolinite. The most important reaction of clay minerals in the sandstone diagenesis is the formation of illite. The occurrence of this mineral can remarkably affect the sandstones (Yuan et al. 2017).

13.4 Results and Discussions

13.4.1 Batch Mode Simulations

Such simulations were used to stabilize the formation of water considering the composition of the rocky reservoir (Kirste et al. 2014). The quantities of Na^+ and Cl^- used in the simulations prove the influence of the salt domes near the reservoir rock. The concentrations of other solutes come from the interaction of water with the geological environment (Kuo et al. 2011). The simulations (in batch mode) of the studied sandstones revealed that it took 35 000 years to attain a steady state (Figure 13.1). The water was initially saturated with a series of minerals for allowing the simulator to produce concentrations of various solutes in conditions close to equilibrium with the reservoir rocks.

13.4.2 1D Simulation

The results of these simulations were used to perform the 1D simulations. The simulation domain is used in lengths of 100 m. The fluid obtained by simulation in batch mode at 120 °C was used to simulate the conditions of entry of the most heated fluid, which is influenced by the salt dome in the reservoirs (Speight 2009). Figure 13.2 shows the saturation of the mineral over the domain (100 m) where a reaction front can be observed in the initial 20 m, where water mixes with the sandstone. Kaolinite is supersaturated due to the absence of

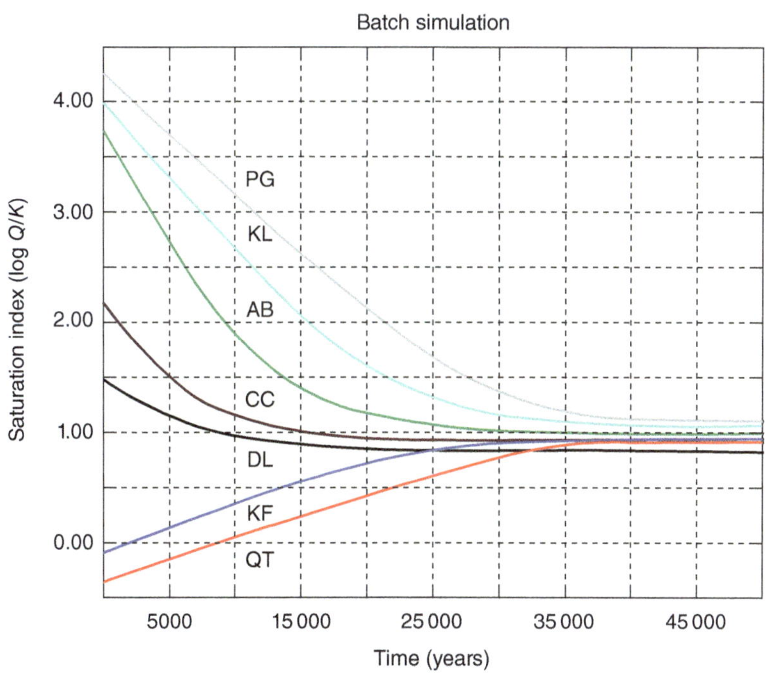

Figure 13.1 Batch mode simulations between formation water and sandstone. The minerals present are quartz (QT), albite (AB), plagioclase (PG), calcite (CC), dolomite (DL), kaolinite (KL), and K-feldspar (KF).

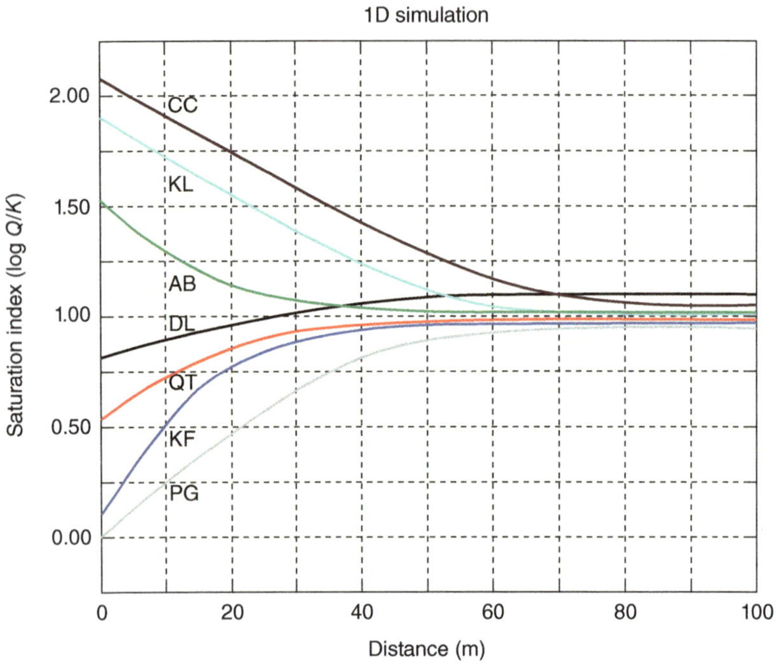

Figure 13.2 Simulations in 1D after the water reaches a steady state. The minerals present are quartz (QT), albite (AB), plagioclase (PG), calcite (CC), dolomite (DL), kaolinite (KL), and K-feldspar (KF).

clay transformation reactions and low sandstone reactivity. Calcite precipitation is more pronounced at the reaction front (before 80 m), which is, therefore, the fate of Ca^{2+} released by the dissolution of other minerals (Xiao et al. 2020).

Variations in the composition of water along the sandstone produce variations in the reactions and also in the rate of change of total porosity. Consequently, changes in porosity can vary significantly over time and space in the reservoir. This variation is reflected in the rates of precipitation and dissolution of K-feldspar, calcite, quartz, and kaolinite along the sandstone. Notably, the porosity is initially increased by the dissolution of feldspar, later being reduced by the precipitation of albite and dolomite, as well as quartz and calcite in small quantities (Gao et al. 2017). The simulations performed by TOUGHREACT displayed the potential of the simulator to model complex situations of reservoir diagenesis (Klunk et al. 2021).

13.5 Conclusions

The simulations performed with the TOUGHREACT geochemical modeling software were used to predict the diagenetic reactions observed in sandstones in the Espírito Santo Basin, Brazil. This work was necessary for the qualitative understanding of the diagenetic alteration of sandstones when exposed to the flow of saline water. The construction of a conceptual model allowed the configuration of geochemical models based on petrographic data.

The chosen scenario was the water-rock interaction between the salt dome and the reservoir with the percolation of fluids with high salinity and a temperature of 120 °C.

Using simulations in batch and 1D mode, it was possible to observe the behavior of minerals when subjected to different reaction conditions. Parameters such as saturation index, mineralization rate, rate of change in porosity, and change in total porosity allowed the understanding of diagenetic reactions in the geological environment.

The simulations in batch mode formation enabled the formation water to reach a steady state and later use it for the 1D simulations. We observed that the stability achieved by the sandstones was 35 000 years. The 1D simulations reveal that the albitization is accompanied by reduced kaolinite precipitation. This approach provides a tool to estimate the intensity of the albitization in relation to a saline fluid entry point in reservoirs. This and other qualitative observations provide empirically controlled rules that could be utilized in exploration and production operations in the oil industry.

The results also indicate changes in sandstones with loss of porosity, significant albitization, and small amounts of precipitation of calcite and kaolinite. The results of the simulations can be further used for the approximation of the total porosity variation. The ability to predict diagenetic reactions using geochemical modeling will become more reliable with additional simulations that can examine the effect of the CO_2 level on the water, which may be more complex than this study.

Acknowledgments

Soumyajit Mukherjee (IIT Bombay) invited and handled this article. Merryl Le Roux, Frank Otmar Weinreich, Joss Everett, Andrew Harrison, and the proofreading team (Wiley Blackwell) are thanked for their assistance. The article has been summarized in Mukherjee et al. (2023).

References

De Lucia, M., Kempka, T., Afanasyev, A. et al. (2016). Coupling of geochemical and multiphase flow processes for validation of the MUFITS reservoir simulator against TOUGH. *Energy Procedia* 97: 502–508.

Gao, Z., Feng, J., Cui, J. et al. (2017). Physical simulation and quantitative calculation of increased feldspar dissolution pores in deep reservoirs. *Petroleum Exploration and Development* 44 (3): 387–398.

González-Acebrón, L., Arribas, J., and Mas, R. (2010). Role of sandstone provenance in the diagenetic albitization of feldspars. *Sedimentary Geology* 229 (1–2): 53–63.

Kirste, D., Haese, R., Boreham, C., and Schacht, U. (2014). Evolution of formation water chemistry and geochemical modelling of the CO_2CRC Otway Site residual gas saturation test. *Energy Procedia* 63: 2894–2902.

Klunk, M.A., Damiani, L.H., Feller, G. et al. (2015). Geochemical modeling of diagenetic reactions in Snorre Field reservoir sandstones: a comparative study of computer codes. *Brazilian Journal of Geology* 45: 29–40.

Klunk, M.A., Dasgupta, S., and Conceiçao, R.V. (2018). Computerized geochemical modeling of burial diagenesis of the Eocene turbidite reservoir elements: Urucutuca Formation, Espírito Santo Basin, southeastern Brazil passive margin. *Journal of Palaeogeography* 7: 12–26.

Klunk, M.A., Dasgupta, S., Das, M. et al. (2021). Application of geochemical modelling software as a tool to predict the diagenetic reactions between the marine connate water and the salt dome in a petroleum system. *Journal of South American Earth Sciences* 109: 103272–103280.

Klunk, M.A., Shah, Z., Caetano, N.R. et al. (2020). CO_2 sequestration by magnesite mineralization through interaction between Mg-brine and CO_2: integrated laboratory experiments and computerized geochemical modelling. *International Journal of Environmental Studies* 77 (3): 492–509.

Kuo, C.-C., Huang, C.-S., and Yeh, H.-D. (2011). Transient analysis for fluid injection into a dome reservoir. *Advances in Water Resources* 34 (12): 1553–1562.

Lai, J., Wang, G., Ran, Y., and Zhou, Z. (2015). Predictive distribution of high-quality reservoirs of tight gas sandstones by linking diagenesis to depositional facies: evidence from Xu-2 sandstones in the Penglai area of the central Sichuan basin, China. *Journal of Natural Gas Science and Engineering* 23: 97–111.

Li, H., Zhang, L., Liu, L., and Shabani, A. (2020). Impact of rock mineralogy on reservoir souring: a geochemical modeling study. *Chemical Geology* 555: 119811.

Mansurbeg, H., De Ros, L.F., Morad, S. et al. (2012). Meteoric-water diagenesis in late Cretaceous canyon-fill turbidite reservoirs from the Espírito Santo Basin, eastern Brazil. *Marine and Petroleum Geology* 37 (1): 7–26.

Mukherjee, S., Dasgupta, S., Majumdar, C. et al. (2023). Introduction to: Handbook of Petroleum Geoscience: Exploration, Characterization, and Exploitation of Hydrocarbon Reservoirs. In: *Handbook of Petroleum Geoscience: Exploration, Characterization, and Exploitation of Hydrocarbon Reservoirs*, Petroleum Industry Updates: Geoscientific and Economic Issues (ed. S. Mukherjee, S. Dasgupta, C. Majumdar, et al.). John Wiley & Sons, Hoboken, ISBN 978-1-119-68003-1.

Paraska, D.W., Hipsey, M.R., and Salmon, S.U. (2014). Sediment diagenesis models: review of approaches, challenges and opportunities. *Environmental Modelling & Software* 61: 297–325.

Speight, J.G. (2009). Reservoirs and reservoir fluids. *Enhanced Recovery Methods for Heavy Oil and Tar Sands* 59–94.

Szczepański, P. (2018). Chemometric method for Donnan dialysis physicochemical model simplification. Prediction of: Transport, recovery, concentration, and desalination efficiency. *Desalination* 444: 6–12.

Vandeginste, V., Stehle, M.C., Jourdan, A.-L. et al. (2017). Diagenesis in salt dome roof strata: Barite – calcite assemblage in Jebel Madar, Oman. *Marine and Petroleum Geology* 86: 408–425.

Wei, W., Zhu, X., Meng, Y. et al. (2016). Porosity model and its application in tight gas sandstone reservoir in the southern part of West Depression, Liaohe Basin, China. *Journal of Petroleum Science and Engineering* 141: 24–37.

Xiao, M., Wu, S., Yuan, X. et al. (2020). Diagenesis effects on the conglomerate reservoir quality of the Baikouquan Formation, Junggar Basin, China. *Journal of Petroleum Science and Engineering* 195: 107599.

Yuan, G., Cao, Y., Zhang, Y., and Gluyas, J. (2017). Diagenesis and reservoir quality of sandstones with ancient "deep" incursion of meteoric freshwater – an example in the Nanpu Sag, Bohai Bay Basin, East China. *Marine and Petroleum Geology* 82: 444–464.

Zhang, Y., Pe-Piper, G., and Piper, D.J.W. (2015). How sandstone porosity and permeability vary with diagenetic minerals in the Scotian Basin, offshore eastern Canada: implications for reservoir quality. *Marine and Petroleum Geology* 63: 28–45.

14

Stratigraphic Boundary Detection Using UDWT and Edge-Detection on Well Log Data

Somali Roy[1,*], *Biplab Kumar Mukherjee*[2], *and Sudipta Dasgupta*[1]

[1] *Department of Earth Sciences, Indian Institute of Technology Bombay, Mumbai, Maharashtra, India*
[2] *LTI, Mind Space SEZ (Serene Properties), Navi Mumbai, Maharashtra, India*

**Author for correspondence: Email roysomali29@gmail.com*

Abstract

The success of the oil and gas industry depends on utilizing a good geological model. The traditional interpretation of a geological model requires the use of both seismic and well log data. Since seismic data is an indirect measure of the lithology, it can miss out on some important lithological boundaries due to absence of high-frequency data in the deeper section. Well log data, on the other hand, is a direct measure of the lithology and can be a good addition to demarcate lithological boundaries. This paper identifies the lithological boundaries using gamma-ray log based on two different techniques without much involvement from the interpreter. The workflow considers the amplitude spectrum of the seismic data to tie with the well-derived output. This data-driven automated process can help an interpreter draw and develop a geological model quickly and efficiently.

Keywords *UDWT; gamma-ray; stratigraphic boundary; wavelet transform; well log; Weiner filter; edge detection*

14.1 Introduction

Understanding lithological boundaries through geological and geophysical studies is of enormous importance in pure and applied geoscience. This is because such boundaries are the preferential place of deformation intensification and shear hear generation (e.g. Mukherjee and Biswas 2015; Mulchrone and Mukherjee 2015). Oil and gas exploration requires the acquisition of data like seismic (e.g. Misra and Mukherjee 2018), well log (e.g. Dasgupta et al. 2019), core (e.g. Chakraborty and Mukherjee 2020), and others. All these data are crucial to creating an efficient geological model of the subsurface. Geophysical measurements come in multirange scales. Scaling of these datasets plays a very

Handbook of Petroleum Geoscience: Exploration, Characterization, and Exploitation of Hydrocarbon Reservoirs, First Edition. Edited by Soumyajit Mukherjee, Swagato Dasgupta, Chandan Majumdar, Subhandip Mandal and Troyee Dasgupta.
© 2023 John Wiley & Sons Ltd. Published 2023 by John Wiley & Sons Ltd.

important role in final geological interpretation. Seismic and well logs being the most popular in the oil and gas industry need to be frequently correlated for lithological boundaries and understanding the facies variation. Gamma-ray values give us information about the lithology of the subsurface as the presence of natural radioactivity in formations concentrated in clay-rich minerals is used for identifying lithologies to distinguish between shales and non-shales and for correlating zones unless there is any other contamination present in depositional sediments.

Well-processed seismic data provides a fair resolution laterally and vertically and is a powerful tool to map lithology changes over a large area. Both vertical and horizontal resolution of seismic data depends on the wavelength of the signal at a certain depth (Alsadi 2017). Frequency content in the shallower section is high and provides high resolution, whereas resolution decreases with depth due to lower frequencies in the deeper seismic section. Whereas the vertical resolution of well log depends on multiple factors like logging speed, matrix density, source, etc. (Wahl 1983; Davies and Elliott 1996). Although a precise vertical resolution of the gamma-ray log is hard to decide but often takes greater than 30 cm (Diniz-Ferreira and Torres-Verdín 2012). Although well log is the direct measurement but is localized and not valid from a regional perspective. In order to get a regional perspective and to create a lithological model, we need to tie the well log data with regional seismic surfaces. This gives a perspective on the gross depositional environment and facies variation. This requires the identification of lithological boundaries using the rock properties of well log data. The first step for this is to identify the lithological boundaries using undecimated discrete wavelet transformation (UDWT). The next step is to detect higher-order lithological sequences through the edge-detection algorithm written in Octave.

Well log data are generally nonstationary (Honório et al. 2012). Hence, signal-processing tools are applied to analyze the logs to extract the subsurface information (Pan et al. 2008; Honório et al. 2012). Different linear and nonlinear statistical methods, e.g. arithmetic, geometric, harmonic means, median, etc., are available to upscale the data set. Several complex algorithms also have been developed by several researchers (e.g. Singh et al. 2016; Srivardhan 2016) for extracting lithological boundaries from the Gamma-ray logs.

A new workflow has been proposed to complete the objectives for the present study and create a V-shale (volume of shale) log for better correlation with the seismic maps and lithological boundaries.

14.2 Study Area

The study area rests in a Mahanadi shelf-slope basin, Odisha in the East Coast of India (Figure 14.1). Figure 14.2 shows the seismo-geological section where the well X6 intersected the Plio-Pleistocene to Miocene deltaic shallow marine channel complex sands. The well has been drilled till 4100 m. The zone of interest of the well in this study area is from 900 to 4100 m. The sediment input in this area mainly comes from the Ganga–Brahmaputra, Brahmani, and the Subarnarekha river system (Roy et al. 2015), which produces sediments in the magnitude of 7.10×10^9 kg year^{-1} (Subrahmanyam et al. 2008).

Figure 14.1 Well location in the Mahanadi basin.

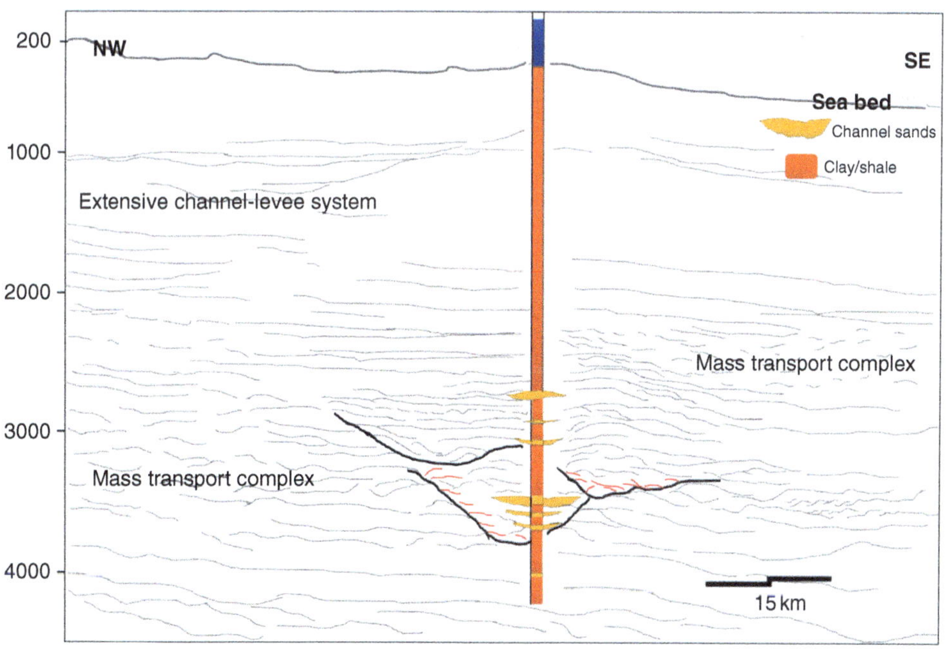

Figure 14.2 The study area in the outer shelf-slope region, where a well X6 has been drilled. The well encountered many sand bodies in the late Miocene age.

14.3 Theory

To analyze a signal, we should honor time and frequency. Till the 1970s, the most powerful mathematical tool for analyzing signals was the Fourier method (Gao and Yan 2010). In the Fourier transform, the signal is a superposition of sine and cosine waves with different frequencies (Bracewell 1989). In fact, with few practices of sine and cosine waves, it helps us discover the main characteristics of a signal. Well logs are generally considered nonstationary in nature (Honório et al. 2012) as the time period is not constant with baseline properties change with depth.

Fourier transformation yields coefficients for certain frequencies contributing to the average strength of that frequency in a signal. Fourier transformation uses sine and cosine functions, which vary uniformly up to infinity, hence not suitable to analyze the local behavior of a signal. So, applying Fourier transformation in well logs will not reveal much information (Gabor 1946). In 1946, Dennis Gabor adapted the short-time Fourier transform (STFT) to address this shortcoming (Gabor 1946). In computing STFT, the longer time signal is divided into shorter segments of equal length and then compute Fourier transform for each window. Hence, it provides some information about when and at what frequencies an event has occurred. The inadequacies of this theory lie in the size of the window with which the mapping of the signal is done. Once the window length is decided for one event, it remains the same for all the events. Hence, it cannot capture both the lower and higher frequencies in a single window.

Therefore, to remove these deficiencies, wavelet transformation is used. The fundamentals of the wavelet theory come from the Fourier analysis but with some important differences, e.g. wavelets are short functions and they can break a complicated signal into several components and process them individually using various window lengths, hence characterizing a signal efficiently (Guan et al. 2004).

Wavelet-based filtering methods such as continuous wavelet transform (CWT) and discrete wavelet transform (DWT) are very efficient in pattern recognition and processing nonstationary signals (Sharma et al. 2015). The wavelet transformation provides varying time as well as frequency resolution using a window of varying length which was missing in the Fourier transformation (Polikar 1999). The result of this transformation is a set of wavelet coefficients given by

$$W_{\psi}(a,b) = \frac{1}{\sqrt{|a|}} \int_{-\infty}^{+\infty} x(t)\overline{\psi}\frac{t-b}{a}\mathrm{d}t \tag{14.1}$$

Here ψ is a wavelet function and $\overline{\psi}$ is a complex conjugate of ψ. The variables a and b are the scale and phase of the signal ((Wickerhauser 1994; Boggess and Narcowich 2001).

UDWT transforms the original signal at each point and saves the detailed coefficient (high-frequency coefficients) and uses the approximation coefficients (low-frequency coefficients) for the next level (Wang et al. 2010). Hence, the outcomes at each level are of the same length as the original signal. This form is also known as stationary wavelet transform, which gives a better approximation than the conventional DWT.

Conventionally, a frequency filter (Fourier transformation) is used on the well logs to denoise the data, though the suitability of a particular filter remains uncertain depending

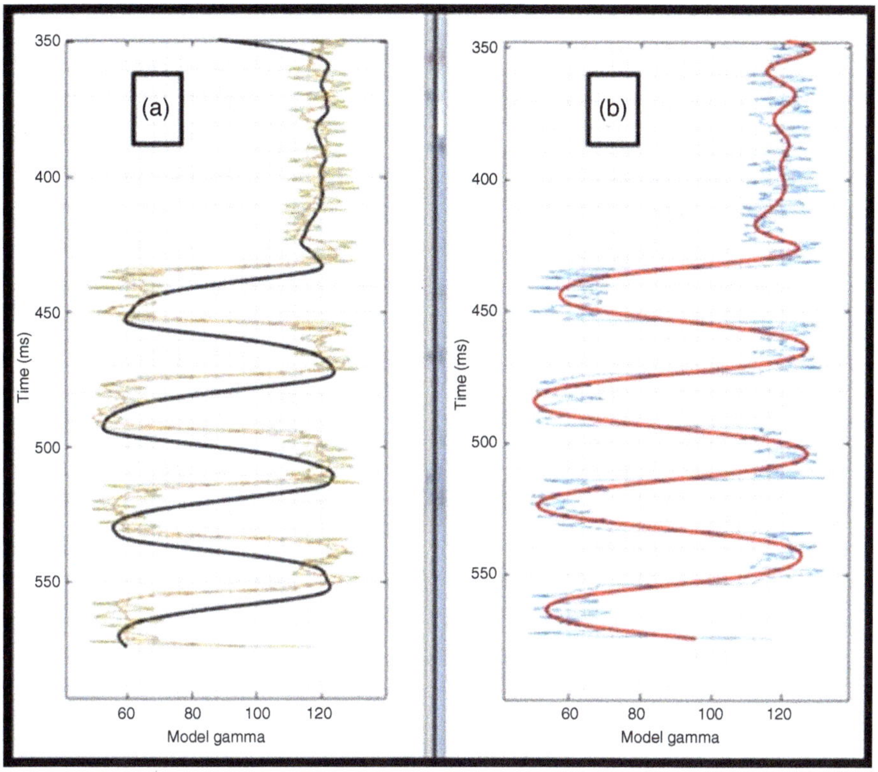

Figure 14.3 Upscaled well log overlaid on synthetic gamma log A. The black line is the upscaled synthetic gamma using a frequency filter B. The red line is the upscaled synthetic gamma using wavelet decomposition.

on the nature of noise associated with the signal (Zhou and Rabinowitz 1997). A bandpass filter is tested on a synthetic gamma log to understand the efficiency of the technique. The used synthetic log was prepared based on a hypothetical geological model of shale and sand of varying thickness. Figure 14.3 shows the difference in output after using the Butterworth filter applied for forward filtering after Fourier transform (2A) and wavelet decomposition filtering (2B), where the latter data is a better fit to the input log in comparison to the former one.

Apart from the bandpass filter, other popular denoising filters like UDWT, median filter, and moving average (coding done using GNU Octave Version 5.2.0) are also tested. Moving average and median filters are both sliding window-based conventional methods where the center value of the window is replaced by the mean or median value, respectively. Figure 14.4 shows that the filtered smooth curves (shown in red) created by median filter, moving average and UDWT overlaid on the actual synthetic curve. The filtered curve (median and moving average) is valid in the thicker section but provides unreliable results for thinner units. On the other hand, filtering through wavelet decomposition having multi-resolution property handles thick and thin beds equally well with little loss of details as shown in the rightmost column.

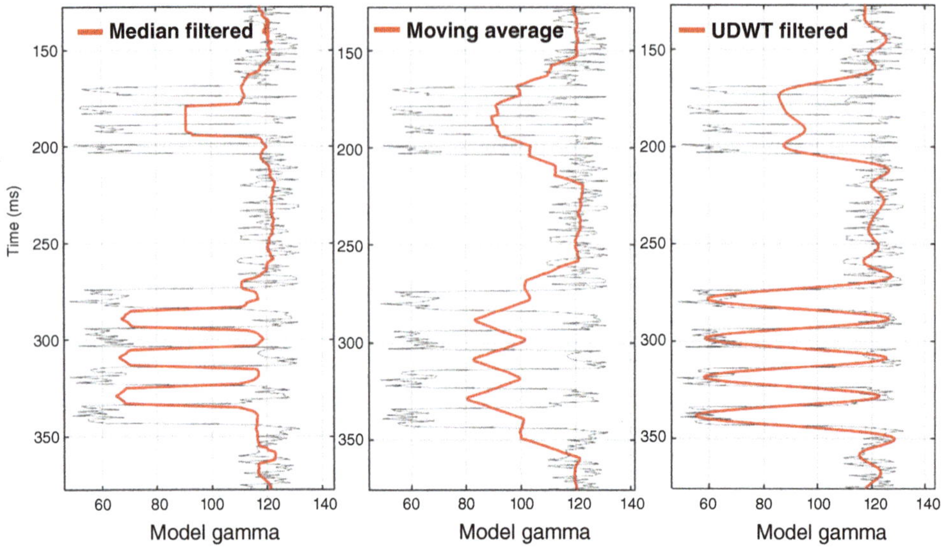

Figure 14.4 Comparative study between three different filtering techniques.

14.4 Methodology and Workflow

In the present study, the workflow starts with the identification of the major interfaces using UDWT by decomposing a gamma ray log into the coefficient of approximation (cA) and coefficient of details (cD) (Srivardhan 2016).

Firstly, the recorded well logs are converted from depth domain to time domain using the check-shot data of the well to run wavelet decomposition. Applying UDWT in the time domain enhances more confidence in our calculation of the frequency content. However, Pan et al. (2008) mentioned that well log in depth domain can also be used for this purpose.

Finally, pre-stacked migrated seismic data is extracted at the well location in order to match both the frequencies from the seismic and well log. For this, an amplitude spectrum has been developed, which is discussed in Section 14.5. The spectra show an envelop where the seismic data ranges between 5 and 85 Hz.

The wavelet transformation in this article was performed using GNU Octave, which is a scientific programming language. The output wavelet coefficients depend upon two important factors, which are the types of wavelets used and the level of decomposition applied. There are several wavelets, among which the Haar wavelet is the simplest with positive and negative pulses representing step function (Guan et al. 2004). Other wavelets, viz. Coiflet, Daubechies, and Morlet waves are popular because of their multiple applications (Guan et al. 2004; Sharma et al. 2015). In this specific workflow, we have tested various Daubechies wavelets: db2, db4, db6, and db8 (Figure 14.5). The Daubechies wavelets have been used by several authors (Pan et al. 2008; Srivardhan 2016) for this purpose. We found the "db6" wavelet among the six Daubechies vanishing moments to be the most suitable for our study. The higher the vanishing moments, the higher are the complexities of the wavelet.

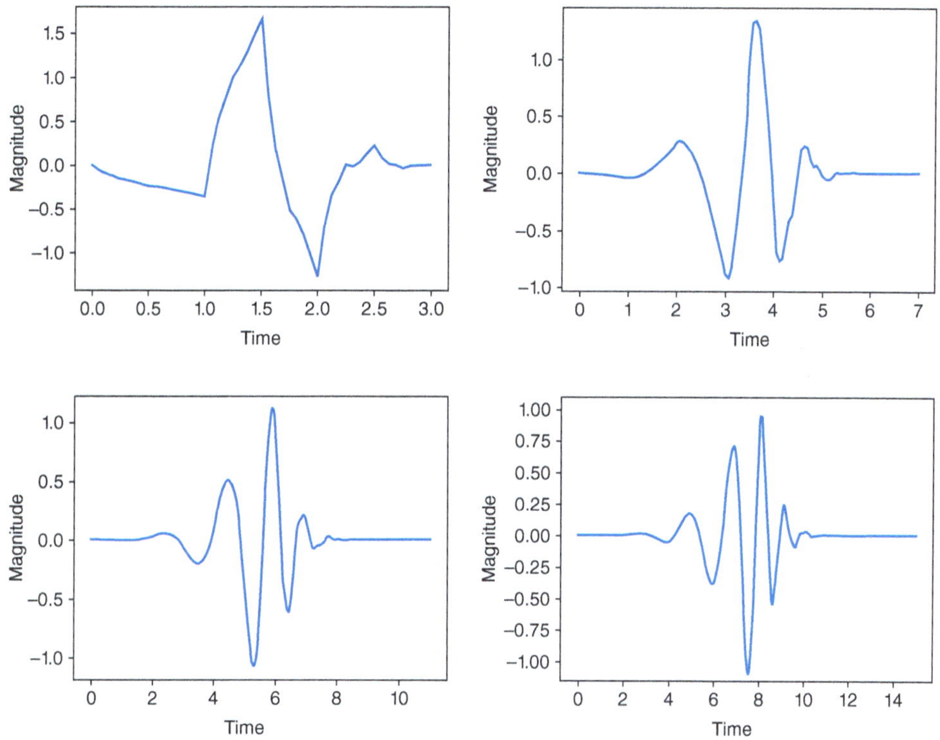

Figure 14.5 Daubechies wavelet for various vanishing moments.

UDWT method has been applied on recorded Gamma-ray log, as it is nonstationary in nature, to upscale and detect lithological boundaries (Srivardhan 2016). Figure 14.6 shows the workflow followed to upscale the gamma log and identify lithological interfaces. A bandpass filter and the Hilbert transform have been applied successively on the reconstructed detail (RcD) to identify the sharp interfaces (Huang et al. 1998).

For filtering purposes, the high-frequency component of the decomposed signal is removed using a threshold criterion based on seismic spectra.

Figure 14.7 shows the coefficients generated by using UDWT, which can either be filtered using a soft or a hard filter. A soft filter reduces the coefficient value by a constant, whereas a hard filter eliminates all the coefficients below a threshold value (Figure 14.8). The cD has further been filtered using the Weiner threshold algorithm, which is a popular linear filtering technique (Kaur et al. 2002). Weiner filter performs better in comparison to soft or hard thresholding. It minimizes the error between the predicted signal to the observed signal. This filter behaves in between soft and hard thresholding.

The final low-frequency sub-band (cA) is reconstructed to recreate the filtered Gamma log. The filtered log effectively preserves resolution as UDWT decomposes a gamma log into wavelet coefficients and not frequency components, which provide better-filtered match compared to the median and moving average filtering (Figure 14.4).

Oil and gas industries commonly use the relative gamma-ray deflection as a shale volume (VSH) indicator. VSH indicates the actual shale volume in the lithology, whereas a gamma-ray

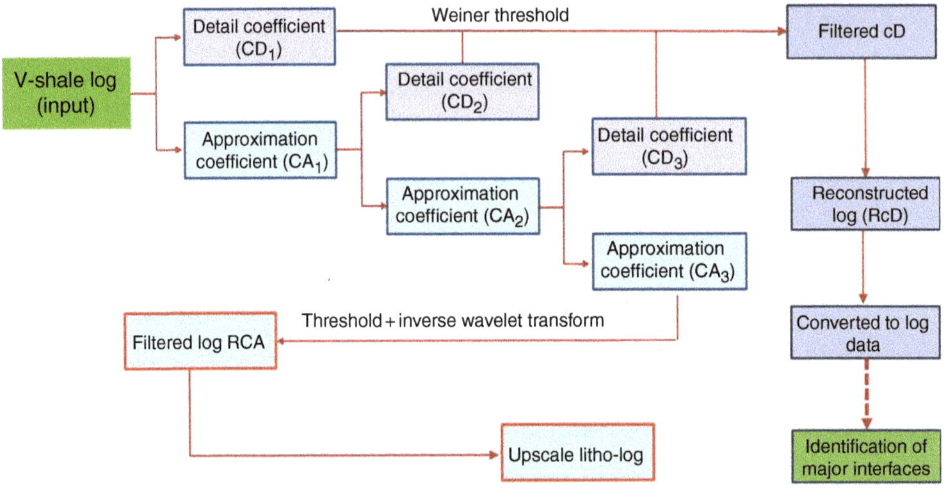

Figure 14.6 Workflow for detecting lithological boundaries used for gamma-ray log.

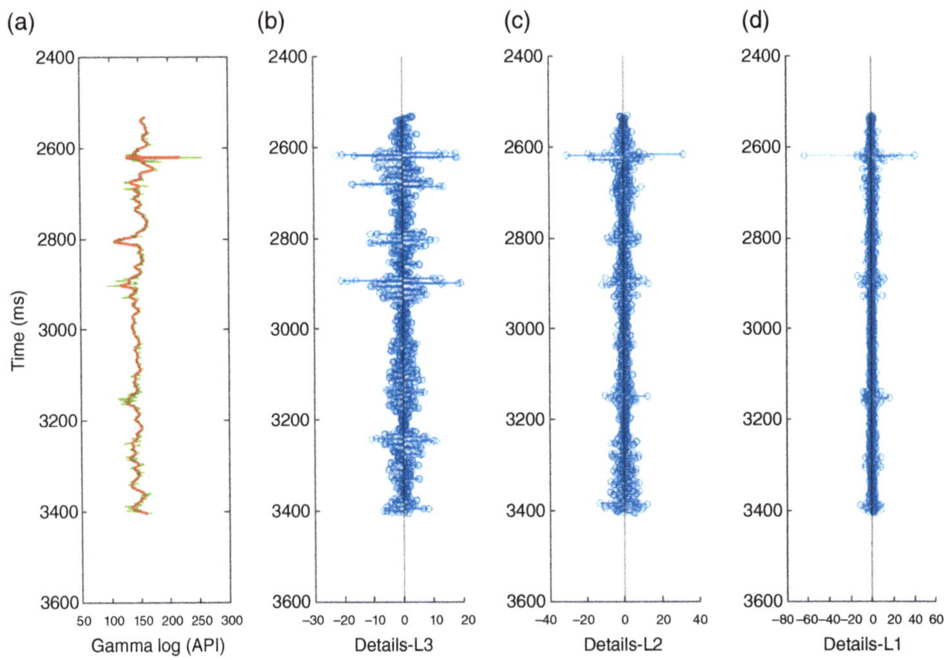

Figure 14.7 The output (leftmost column) after the undecimated wavelet transform. The green line is the recorded gamma-ray log overlaid by a red curve which is the filtered output of the gamma-ray. The right-hand side of the image contains the detailed coefficients.

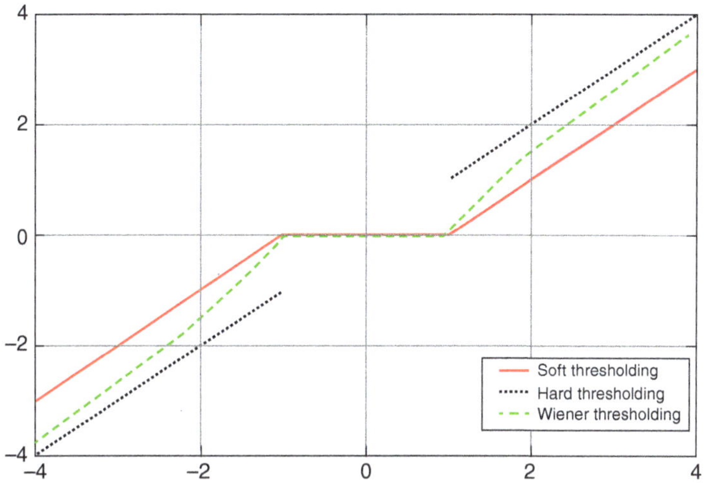

Figure 14.8 The output after the un-decimated wavelet transform. The green line is the recorded gamma-ray log overlaid by red curve which is the filtered output of the gamma-ray. The right-hand side of the image contains the detailed coefficients for each level.

log is an indirect measure between shales and non-shales. Hence, VSH log is used in this processing workflow for a high-frequency lithofacies' variation indicator. The simplest procedure to calculate VSH is to scale the gamma-ray log between its minimum and maximum values from 0 to 100% shale. Worldwide, the gamma-ray log is used in several linear and nonlinear empirical relations. However, in the present study, due to the absence of any radioactive mineral, GR is considered as good indicator of shale. Hence, the linear equation of the gamma ray has been used as a shale volume indicator.

$$\mathrm{IGR} = \frac{\mathrm{GR}_{\log} - \mathrm{GR}_{\min}}{\mathrm{GR}_{\max} - \mathrm{GR}_{min}} \tag{14.2}$$

Here, IGR, gamma-ray index; GR_{log}, gamma-ray reading of formation; GR_{min}, minimum gamma-ray (clean sand or carbonate), and GR_{max}, maximum gamma-ray (shale). Then, a generic edge detection algorithm is used in the newly created VSH log to delineate boundaries of higher order cycles of facies variation as shown in Figure 14.9.

The next step is to upscale the VSH log using various averaging methods like arithmetic, harmonic, geometric, and quadratic average. A comparative study has been provided in Section 14.5.

14.5 Results and Discussions

The two important steps in UDWT are the selection of wavelets and determining the level of decomposition, as discussed earlier.

Figure 14.10 shows the output of the Hilbert transform from RcD-filtered spectrum as shown in Figure 14.6. The cyan plot is the HT curve, which has multiple peaks representing a strong change in gamma-ray responses. As we understand that the gamma-ray log represents the lithological responses, these peaks in the HT curve denote the change in lithological changes.

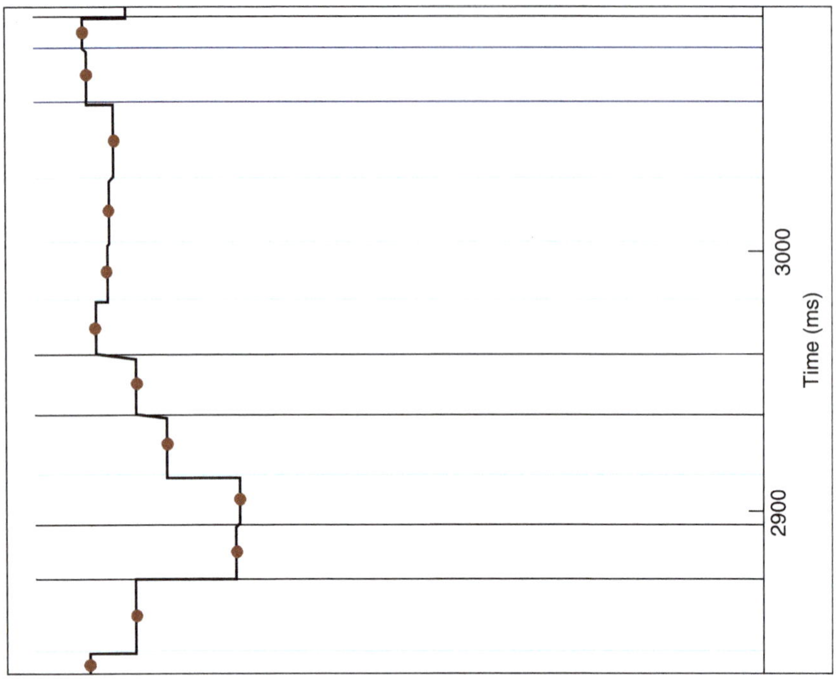

Figure 14.9 The black curve shows the output after the Edge-Detection algorithm on UDWT VSH log. The blocky curve signifies the interfaces across different litho-facies.

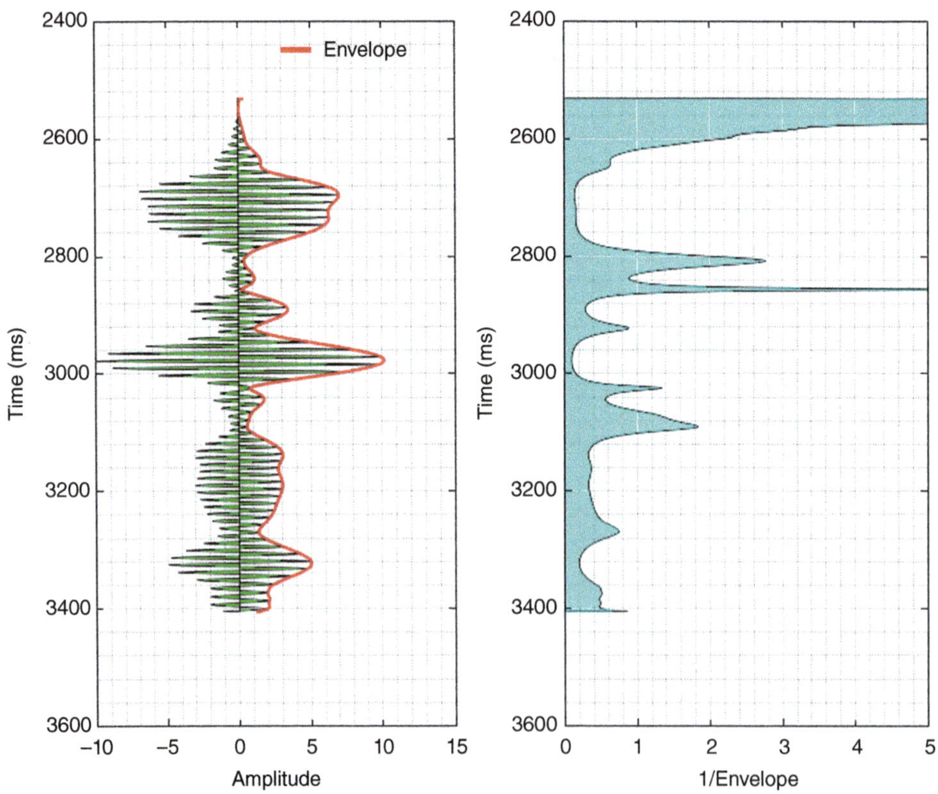

Figure 14.10 Left-hand side of the figure shows the response after the reconstructed detailed coefficient (RcD) is filtered with a bandpass filter. The red line is an envelop created by the Hilbert transform of the filtered RcD. The right-hand side of the figure shows the inverted value of the Hilbert curve indicating the strong interfaces.

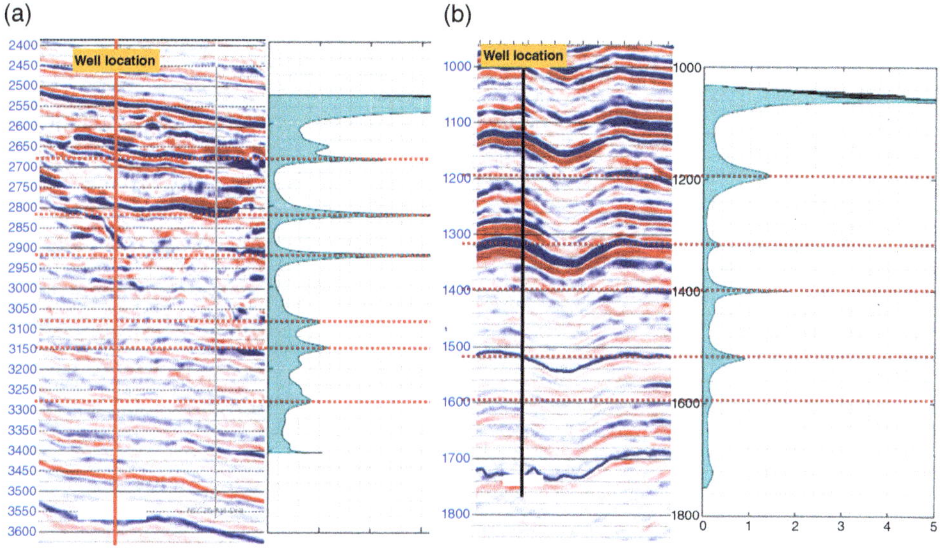

Figure 14.11 1D Interfaces represented by the HT peaks in cyan correlated with the PreSTM seismic section where blue is peak and red is trough.

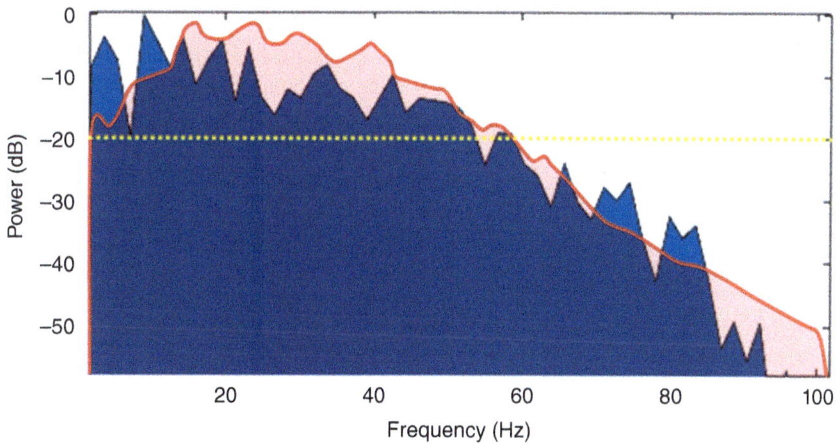

Figure 14.12 UDWT-filtered VSH log spectrum (blue area) underlaid on the seismic amplitude spectrum (red line) at the well location. The yellow dashed line is −20 dB line.

The seismic responses in Figure 14.11 show the major depositional interfaces, which corroborate largely with the HT curve. The red horizontal lines connect the seismic reflectors with HT peaks. Sometimes, we cannot rely on the HT responses at both the ends of the well log due to the edge effect.

The VSH curve is filtered using the UDWT method as discussed in Section 14.4. The level of filtering was restricted by the amplitude spectrum of the seismic of the same location as the output should be in synchronization with the seismic data. Figure 14.12 shows a

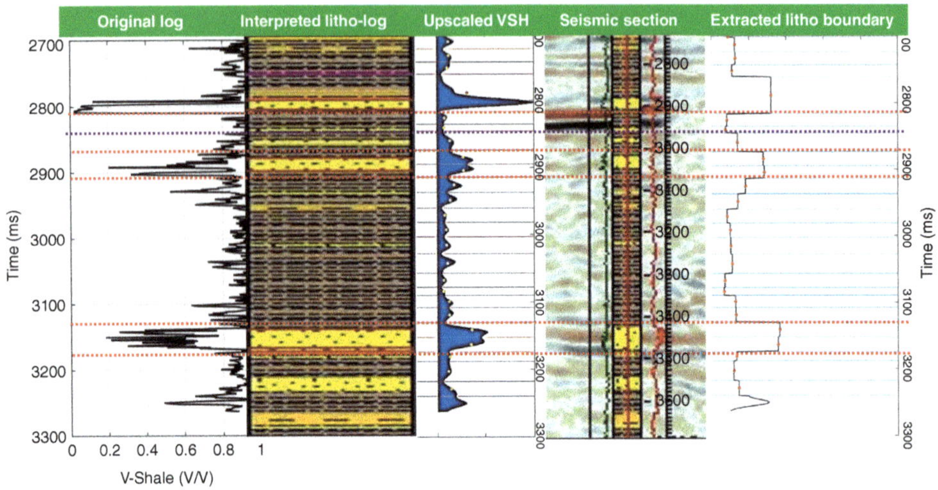

Figure 14.13 The extreme right shows the edge-detection output using the VSH log and compared with the strat column.

comparison between the seismic and filtered VSH amplitude spectrum. The extract litho-boundary output is combined with a seismic section overlaid by well litho-column, shown in Figure 14.13. There is a good match between the extracted litho-boundary (ELB) and the litho-column as shown by horizontal dotted red lines. These results give an idea of the lithological boundaries and eventually interpret the sequence boundary. Furthermore, combining these results with the upscaling of the well log data will create a good geological and rock property model. Figure 14.13 also shows the finer details than those in Figure 14.11.

14.6 Conclusions

The study shows that the wavelet transform is a better process than the Fourier-based filtering technique. The workflow developed provides considerable help in identifying major lithological boundaries and good match when compared to seismic section. Although the present process does not handle the data at the edge of the well log, which needs further study. The interfaces delineated by the UDWT process can be used for geological modeling. This process was improved by using an edge-detection algorithm over the UDWT-filtered VSH log, which helped mark the litho-facies variation between shale and non-shale components. The interfaces delineated by this process can further be used for facies modeling, which is beyond the scope of this paper.

Acknowledgments

This work has been reviewed by Soumyajit Mukherjee and Swagato Dasgupta. Merryl Le Roux, Frank Otmar Weinreich, Joss Everett, Andrew Harrison, and the proofreading team (Wiley Blackwell) are thanked for their assistance.

We are grateful to Reliance Industries Limited for providing this data.

References

Alsadi, H.N. (2017). The seismic reflection signal. In: *Seismic Hydrocarbon Exploration*, 2D and 3D Techniques, Seismic Waves (ed. I.H.N. Alsadi), 187. Springer International Publishing.

Boggess, A. and Narcowich, F.J. (2001). *A First Course in Wavelets with Fourier Analysis*. Hoboken, NY: Wiley.

Bracewell, R.N. (1989). The Fourier transform. *Scientific American* 260 (6): 86–95. Retrieved from https://www.jstor.org/stable/24987290.

Chakraborty, M. and Mukherjee, S. (2020). *Structural Geological Interpretations from Unrolled Images of Drill Cores*, vol. 115. Marine and Petroleum Geology.

Dasgupta, S., Dasgupta, T., Singh, V.N., and Mukherjee, S. (2019). *A Numerical Study of Epithermal Neutron Log Response and Application of Image Log for Porosity Determination* (ed. A. Billi and A. Fagereng), 237–258. Mumbai: Elsevier. ISSN: 2542-9000. ISBN: 9780128140482.

Davies, S.J. and Elliott, T. (1996). Spectral gamma ray characterization of high-resolution sequence stratigraphy: examples from upper carboniferous fluvio-deltaic systems, County Clare, Ireland. *Geological Society, London, Special Publications* 104 (1): 25–35.

Diniz-Ferreira, E.L. and Torres-Verdín, C. (2012). *Improved estimation of pore connectivity and permeability in deepwater carbonates with the construction of multi-layer static and dynamic petrophysical models, SPWLA 53rd Annual Logging Symposium*, OnePetro, Cartagena, Colombia (June 2012). Society of Petrophysicists and Well-Log Analysts.

Gabor, D. (1946). *Theory of communication Journal of the Institution of Electrical Engineers* 93 (26): 426–457.

Gao, R.X. and Yan, R. (2010). From Fourier Transform to Wavelet Transform: A Historical Perspective. *Springer, Boston, MA*. https://doi.org/10.1007/978-1-4419-1545-0_2.

Guan, L., Du, Y., and Li, L. (2004). *Wavelets in Petroleum Industry: Past*. Present and Future: Society of Petroleum Engineers.

Honório, B.C., Drummond, R.D., Vidal, A.C. et al. (2012). Well log denoising and geological enhancement based on discrete wavelet transform and hybrid thresholding. *Energy exploration and exploitation* 30 (3): 417–433.

Huang, N.E., Shen, Z., Long, S.R. et al. (1998). The empirical mode decomposition and the Hilbert spectrum for nonlinear and non-stationary time series analysis. *Proceedings of the Royal Society of London* 454 (1971): 904–995. https://doi.org/10.1098/rspa.1998.0193.

Kaur, L., Gupta, S., and Chauhan, R.C. (2002). Image denoising using wavelet thresholding. *ICVGIP* 2: 16–18.

Misra, A.A. and Mukherjee, S. (2018). *Atlas of Structural Geological Interpretation from Seismic Images*. Wiley Blackwell.

Mukherjee, S. and Biswas, R. (2015). Biviscous horizontal simple shear zones of concentric arcs (Taylor Couette flow) with incompressible Newtonian rheology. In: *Ductile Shear Zones: From Micro- to Macro-scales* (ed. S. Mukherjee and K.F. Mulchrone), 59–62. Wiley Blackwell ISBN: 978-1-118-84496-0.

Mulchrone, K.F. and Mukherjee, S. (2015). Shear senses and viscous dissipation of layered ductile simple shear zones. *Pure and Applied Geophysics* 172: 2635–2642.

Pan, S.-Y., Hsieh, B.-Z., Ming-Tar, L., and Lin, Z.-S. (2008). Identification of stratigraphic formation interfaces using wavelet and Fourier transforms. *Computers & Geosciences* 34 (1): 77–92.

Polikar, R. (1999). The story of wavelets. In: *Physics and Modern Topics in Mechanical and Electrical Engineering*, 192–197. World Scientific and Engineering Academy and Society.

Roy, S., Choudhuri, M., and Gupta, P. (2015). Rift grabens and crustal architecture of the offshore north 217 East Coast-Mahanadi Basin, Eastern Continental Margin of India. In: *Petroleum Geosciences: Indian Contexts* (ed. S. Mukherjee). Mumbai: Springer Geology.

Sharma, M., Gupta, R., and Singh, G. (2015). Application of wavelet – an advanced approach of transformation. *Advanced Research in Electrical and Electronic Engineering* 1: 28–34.

Singh, A., Maiti, S., and Tiwari, R.K. (2016). Modelling discontinuous well log signal to identify lithological boundaries via wavelet analysis: an example from KTB borehole data. *Journal of Earth System Science* 125 (4): 761–776.

Srivardhan, V. (2016). Stratigraphic correlation of wells using discrete wavelet wavelet transform with Fourier transform and multi-scale analysis. *Geomechanics and Geophysics for Geo-Energy and Geo-Resources* 2 (3): 137–150. https://doi.org/10.1007/s40948-016-0027-1.

Subrahmanyam, V., Subrahmanyam, A.S., Murty, G.P.S., and Murthy, K.S.R. (2008). Morphology and tectonics of Mahanadi basin, northeastern margin of India from geophysical studies. *Marine Geology* 253: 63–72.

Wahl, J.S. (1983). Gamma-ray logging. *Geophysics* 48 (11): 1536–1550.

Wang, X.-Y., Yang, H.-Y., and Fu, Z.-K. (2010). A new wavelet-based image denoising using undecimated discrete wavelet transform and least squares support vector machine. *Expert Systems With Applications* 37 (10): 7040–7049.

Wickerhauser, M.V. (1994). *Adapted Wavelet Analysis from Theory to Software*. New York: IEEE Press.

Zhou, Z. and Rabinowitz, P.D. (1997). A class of recursive weighting filters and their application to well logging data processing. In: *SEG Technical Program Expanded Abstracts 1997*, 301–304. Society of Exploration Geophysicists.

15

Source Rock Geochemistry for Shale Characterization

Devleena Mani[1,], Nihar Ranjan Kar[2], and M.S. Kalpana[1]*

[1] *Centre for Earth, Ocean and Atmospheric Sciences, University of Hyderabad, Hyderabad, Telangana, India*
[2] *Centre for Earth, Ocean and Atmospheric Sciences, University of Hyderabad, and CSIR-National Geophysical Research Institute, Hyderabad, Telangana, India*

*Author for correspondence: dtiwarisp@uohyd.ac.in

Abstract

Shale, a clastic sedimentary rock, acts as a major source rock for the conventional petroleum systems. Due to its fine grain-size, it can accumulate organic matter significantly. The thermal transformation of shale's organic content in the subsurface produces oil and natural gas. For the unconventional petroleum systems such as shale gas, the shale itself is the source and the reservoir rock. In order to quantify the generative potential of shale source rocks, mapping of sedimentary organic matter in terms of its richness, kerogen quality, and maturity is essential. Sedimentary biomarkers and their isotopic ratios measure the compositional changes associated with hydrocarbon generation, migration, accumulation, and production. This study presents the gas-generation potential of Upper Permian shales from the Damodar Valley in eastern India using biomarker and stable isotope proxies, along with an overview on the recent analytical developments in the geochemical characterization of sedimentary organic matter in a shale rock for successful evaluation of a shale gas play.

Keywords *Shale gas; organic matter; biomarkers; isotopes; Damodar Valley*

15.1 Introduction

Organic matter in sediments forms the source of gaseous hydrocarbons, primarily methane, in shale gas plays. The gaseous hydrocarbons are stored in the organic-rich, fine-grained matrix of the shale rocks. Production of organic matter and its deposition, preservation, and maturation define the quality and quantity of gas in shales. Sedimentary organic matter is derived from the biopolymers, lipids, carbohydrates, and proteins of once living organisms. Biopolymers transform into macromolecular geopolymer "kerogen" over time, upon burial in the subsurface. The resistance to the diagenetic conditions governs the

Handbook of Petroleum Geoscience: Exploration, Characterization, and Exploitation of Hydrocarbon Reservoirs, First Edition. Edited by Soumyajit Mukherjee, Swagato Dasgupta, Chandan Majumdar, Subhandip Mandal and Troyee Dasgupta.
© 2023 John Wiley & Sons Ltd. Published 2023 by John Wiley & Sons Ltd.

preservation of biomolecules, with lipids, cutins, and lignins having lower susceptibility for degradation compared to proteins (Hunt 1996). The mass production of organic matter occurs through photosynthesis by the autotrophic green plants, blue-green algae, and photosynthetic bacteria across the lands and oceans. Higher organic content in sediments and rocks is enhanced by processes such as higher organic productivity, nutrient supply and upwelling, reducing environments with low oxygen content and restricted water circulation, lack of bioturbation, very fine-grained sediment particles of less than 2 μm, and optimum sedimentation rates (Bjørlykke 2015).

The preservation of organic matter in sediments is a rare event and a fraction of <0.1% gets preserved, depending upon the geological and the redox conditions of sedimentation. Under the influence of diagenetic, catagenetic, and metagenetic processes, the organic matter thermally cracks into oil and gas. Diagenesis occurs at shallow depths, leading to the formation of biogenic gas and the temperature is less than 60 °C. The biopolymers evolve into kerogen. Thermal cracking of kerogen during catagenesis results in the formation of oil, condensates, and wet gases. At higher temperatures of 150–200 °C, the organic matter gets cracked to dry gas containing largely methane. This is followed by metagenesis, which is the last stage of evolution of sediments in the subsurface (Hunt 1996).

The gas systems have been classified into biogenic and thermogenic, depending upon their origin. Biogenic gases are diagenetic products of shallow subsurface microbial reactions, whereas the thermogenic gases result from the cracking of kerogen at high temperatures. The stable isotopic variations of elements such as carbon provide significant information on the origin and correlation of natural gas to its source (Magoon and Dow 1994; Peters et al. 2005).

15.2 Composition of Sedimentary Organic Matter

Sedimentary organic matter comprises largely of kerogen (80–90%) and a minor fraction of bitumen (10–20%). The dispersed, amorphous macromolecular organic matter insoluble in organic solvents is kerogen, whereas bitumen is the lipid content, synthesized by the earlier living organisms and gets inherited in the sediments with minimal structural changes in organic components during the diagenetic processes associated with sedimentation (Figure 15.1). Bitumen dissolves in organic solvents and constitutes the sedimentary biomarkers.

Shale source rocks are characterized by the quantity, type, and thermal maturity of the kerogen and the bituminous biomarkers present in them (Mani et al. 2014; Hazra et al. 2019; Mani 2019). Organic richness is measured by the total organic carbon content, described in weight percent (% TOC). Generally, the higher the TOC, the greater is the hydrocarbon generation potential. The type or quality of organic matter is dependent upon the type of kerogen, which has varying capacity for oil and gas generation. Kerogen is classified into four types, namely I, II, III, and IV, based on their carbon (C), oxygen (O), and hydrogen (H) contents (Van Krevelen 1961; Vandenbroucke 2003; Mani et al. 2015; Mani et al. 2017). The structural changes that take place in the kerogen and biomarker compounds with the temperature indicate the thermal maturity experienced by the source rocks over time, and, in turn, provide the qualitative and quantitative estimates of the gas generated. Some of the important conventional thresholds are used to define the shale's

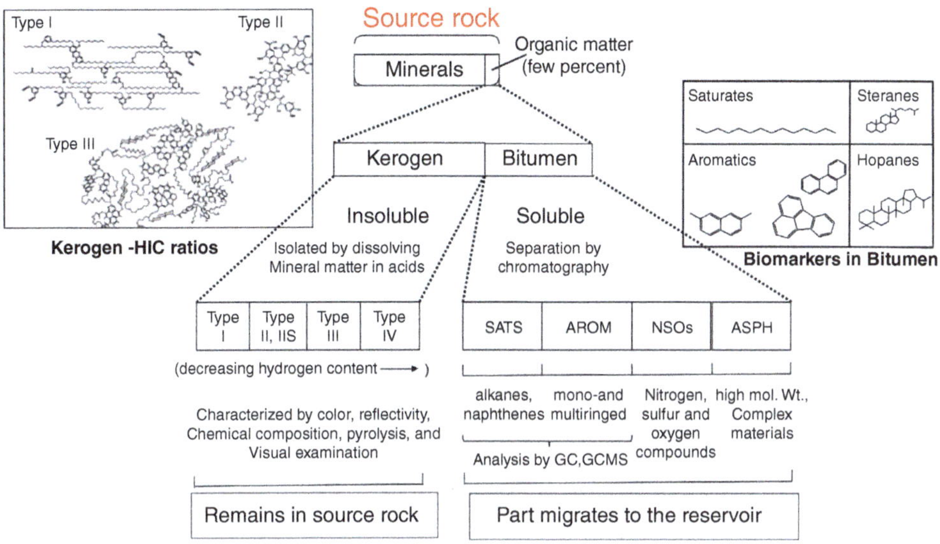

Figure 15.1 Composition of sedimentary organic matter.

generative potential. Based on organic matter quantity, a TOC with less than 0.5% is defined poor and that greater than 2% is defined very good. The hydrogen content on the organic matter indicates the kerogen type. Thus, organic matter with hydrogen index of 0–150 is gas-prone, whereas that greater than 300 is oil-prone. The thermal maturity of organic matter defines the oil and gas window. Parameters such as kerogen transformation ratios, TR (TR = ~0.1 – top of oil window, TR ~ 0.4 – bottom of oil window), Tmax and vitrinite reflectance (VR) are used to classify mature, immature, or overmature sedimentary organic matter.

The formation of shale gas requires higher thermal maturity, which is achieved by greater burial depths. Hence, the thermal maturity parameters are higher for gas shales compared to conventional oil- and gas-producing source shale rocks. In general, the productive gas shales have TOC greater than 3 wt% and possess higher hydrogen content (HI > 350 mg HC g^{-1}TOC). The kerogen is dominated by Type II, marine origin organic matter with thermal maturity greater than 1.10% Ro (Horsfield and Schulz 2012; Mani et al. 2015). Shale gas produced in the USA generally has a VR maturity of >1.4% Ro to produce dry gas (Whitelaw et al. 2019). Biomarkers and their stereoisomers along with the carbon isotopic compositions form important thermal maturity indicators for shale gas evaluation.

15.2.1 Shale Biomarkers

Biological markers are complex organic molecules containing carbon, hydrogen, and other elements such as nitrogen, sulfur, and oxygen occurring in sediments, rocks, and crude oils (Moldowan et al. 1986; Peters et al. 2005). These compounds originate from the lipid content of formerly living organisms and show no major structural change from their precursor or parent organic molecules. Select lipids and their hydrocarbon remnants, which are derived from earlier organisms, carry a high degree of taxonomic specificity. The uniqueness helps to reconstruct the past organismic diversity from the sedimentary inventory of fossil biological markers. Thus, biomarkers help in identifying the source, depositional environment, and thermal maturity of the organic matter and are particularly used in source rock correlation studies (Briggs and Summons 2014) (Table 15.1).

Being less susceptible toward microbial and diagenetic alterations, biomarkers like alkanes and aromatic compounds along with their stereoisomers archive reliably the thermal and depositional history and sources of organic matter contribution. The sources vary

Table 15.1 Biomarkers as indicators of source, redox, and type of organic matter in sediments.

Biomarkers	C range	Proxy information
***n*-alkanes**		
[a]CPI > 5	C_9–C_{21}	Marine, lacustrine algal source, C_{15}, C_{17}, and C_{19} dominant
	C_{25}–C_{37}	Terrestrial plant wax source, C_{27}, C_{29}, and C_{31} dominant
CPI < 1	C_{12}–C_{24}	Bacterial source: oxic, anoxic, marine, and lacustrine
	C_{20}–C_{32}	Saline or anoxic environments: carbonates and evaporites
Acyclic isoprenoids		
Pristane	C_{19}	Chlorophyll, and oxic and suboxic environments
Phytane	C_{20}	Chlorophyll, anoxic, and saline
Botrycoccane	C_{34}	Lacustrine, brackish
Sesquiterpenoids		
Cadalene, eudesmane	C_{15}	Terrestrial plants
Diterpenoids		
Abietane, pimarane, kaurane, retene	C_{19}–C_{20}	Higher plant resins
Tricyclic terpanes	C_{19}–C_{45}	Degradation products of cell wall lipids of bacteria & alga
Tetracyclic terpanes	C_{24}–C_{27}	Degradation of pentacyclic triterpenoids
Hopanes	C_{27}–C_{40}	Bacteria
Norhopanes	C_{27}–C_{28}	Anoxic marine
2- and 3-methyl hopanes	C_{28}–C_{36}	Carbonate rocks
Benzohopanoids	C_{32}–C_{35}	Carbonate environments
Hexahydrobenzohopenoids	C_{32}–C_{35}	Anoxic, carbonate-anhydrite depositions
Gammacerane	C_{30}	Hypersaline environment
Oleananes, lupanes	C_{30}	Products of late cretaceous and tertiary flowering plants
Bicadinane	C_{30}	Resins from gymnosperm tree
β-carotane	C_{40}	Arid, hypersaline environments
Steranes	C_{19}–C_{23}	Eukaryote organisms, plants, and animals
	C_{26}–C_{30}	
24-*n*-propylsteranes	C_{30}	Restricted to marine sediments
4-methylsteranes	C_{28}–C_{30}	Marine and lacustrine dinoflagellates
Dinosteranes	C_{30}	Marine, triassic, or younger rocks

Source: After Hazra et al. (2019).
[a] CPI = Carbon preference index.

according to their occurrence of terrestrially or marine-origin or that derived bacterially (Peters et al. 2005). An important proxy to identify the source and thermal maturity of organic matter is the odd and even carbon number predominance in a hydrocarbon compound (Tissot and Welte 1984). The carbon preference index (CPI) is the ratio of the sum of odd carbon-numbered alkanes to that of even alkanes in the range of nC_{24}–nC_{34}. CPI values <1 indicate marine input with low maturity; whereas a value >1 indicates the terrestrial matter input with high maturity. The redox-sensitive biomarkers such as pristine-phytane ratios (Pr/Ph) along with their correlation with C17 and C18 alkanes in a particular sedimentary rock are used for understanding the different types of depositional environments and the oxygenation conditions. Proxy ratio $\{P_{aq} = (C_{23} + C_{25})/(C_{23} + C_{25} + C_{29} + C_{31})\}$is used to know the type of macrophytes. Where the values <1 suggest non-emergent plant matter input, values from 0.1–0.4 suggest emergent macrophytes and values from 0.4–1 suggest submerged/floating macrophytes (Ficken et al. 2000).

Tetracyclic and pentacyclic terpenoids such as steranes and hopanes along with their stereoisomers of R/S and α/β configurations are important markers due to their structure complexities (Table 15.1). Naturally occurring hopanoids have thermodynamically less stable 17β(H),21β(H) stereochemistry. With subsequent diagenesis and maturation of organic matter, saturated hopanes with more stable 17α(H), 21β(H) isomers are formed. Thus, most mature source rocks contain 17α(H), 21β(H) isomer. Figure 15.2 illustrates

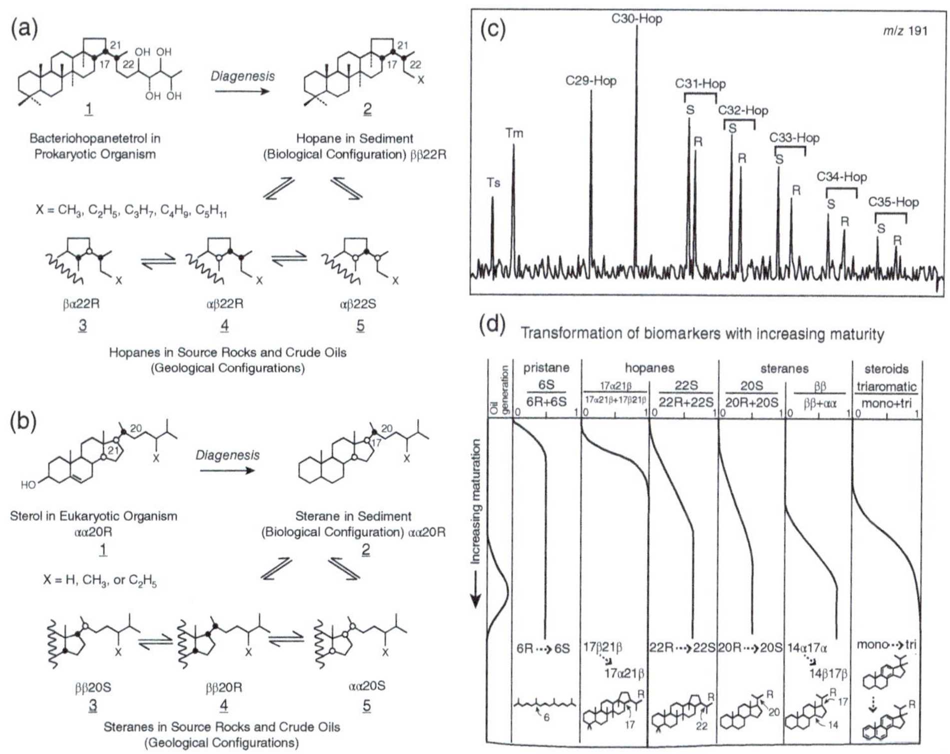

Figure 15.2 Biomarkers as indicators of increasing thermal maturation of sedimentary organic matter. (a) and (b) indicate the geological configuration of steranes and hopanes. (c) Mass spectra of hopane at characteristic m/z = 191. (d) Change in biomarker ratios as function of increasing maturation. *Source:* Adapted from Peters et al. (2005) and Gaines et al. (2009).

the changes in stereoisomers of different biomarkers in response to thermal stress associated with increasing burial.

25, 28, 30-trisnorhopane shows bacterial input with the anoxic marine condition of deposition (Grantham et al. 1980; Schouten et al. 2001). C35 17α,21β(H)-hopane suggests bacterial input and reduction to an anoxic depositional environment (Peters and Moldowan 1991; Koster et al. 1997). Norhopane (C29 hopane) originates from various sources and has a carbonate- or evaporate-type environment. 2-methylhopanes are sourced from a cyanobacterial population and show an enclosed basin-type depositional environment (Summons et al. 1999). 3β-methylhopanes show a methanotrophic bacterial source and a lacustrine environment (Collister et al. 1992; Summons and Jahnke 1992). Hexahydrobenzohopanes show a bacterial input with an anoxic carbonate-anhydrite environment (Connan and Dessort 1987). Ts (18α-22,29,30-trisnor-neohopane) and Tm (17α-22,29,30-trisnorhopane) can be used to know the thermal maturity of the source rock (Moldowan et al. 1986) and are represented by Ts/(Ts + Tm). C27–C29 steranes show algal and higher plant input and can have various kinds of depositional environments (Moldowan et al. 1985; Volkman 1986). 4-methylsteranes have a bacterial or dinoflagellate input and a lacustrine to marine-type environment of deposition (Brassell et al. 1985; Wolff et al. 1986). Diasteranes show an algal/higher plant input and clayey-type environment of deposition (Van Kaam-Peters et al. 1998). Dinosteranes show a dinoflagellates input source and a marine-type environment (Summons 1987; Goodwin et al. 1988). Aromatic markers such as polycyclic aromatic hydrocarbons (PAHs) and aromatic steranes are pervasive components of marine, lacustrine, and terrestrial sediments. Their relative abundance and ratios reflect the thermal maturity of sedimentary organic matter (Mani et al. 2017).

15.2.2 Carbon Isotopes

Stable carbon isotope ratios (^{13}C vs. ^{12}C) have widely been used to identify the sources of hydrocarbon gases (Mani et al. 2011). The fractionation of carbon leads to the enrichment of ^{12}C in biogenic carbon as compared with the inorganic carbon reservoir of the surficial environment (atmospheric carbon dioxide and dissolved marine carbonate) (Schidlowski 1987). This results in variations in stable ratios of different carbon-containing compounds (Figure 15.3).

The marine and terrestrial plants exhibit different $\delta^{13}C$ signatures, which is mainly attributed to the differences in the isotopic values of varied carbon sources (Mani et al. 2015). The gas systems have been classified broadly into biogenic (or microbial) and thermogenic. There also occur mixtures of the two gas types (Mani et al. 2015). The occurrence of thermogenic gas in the subsurface indicates the possibility or presence of source rock and provides information on the elements of the hydrocarbon system, which otherwise lacks microbially produced gases (Figure 15.3).

The stable carbon isotope ratios are affected by processes of migration, thermal maturation of hydrocarbons, and bacterial oxidation and are susceptible toward fractionation. Thus, care should be taken that the shales are not contaminated by weathering processes (Mani et al. 2011, 2015). The thermally desorbed light hydrocarbon gases upon pyrolysis of shales reflect the pristine signatures, which are derived from the kerogen cracking (Mani et al. 2015).

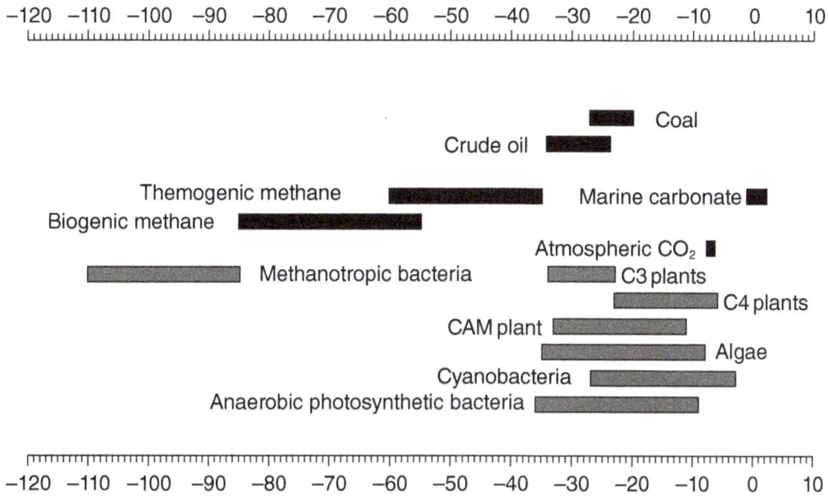

Figure 15.3 Variation of stable isotopes of carbon in nature. *Source:* Adapted from Vieth and Wilkes (2010).

15.3 Geochemical Characterization of Shales

Different geochemical methods provide different information about the source rock, which are integrated to give a reliable representation of its gas generation prospect. Among these, the rock eval pyrolysis along with kerogen degradation kinetics, biomarkers, and isotopic parameters are quite useful in shale characterization (Figure 15.4).

15.3.1 Bulk Organic Matter Analysis Using Rock-Eval Pyrolysis

Rock-Eval pyrolyzer is an open system, temperature-programmed instrument used to determine the volatile hydrocarbons (S_1), generative potential (S_2), total organic carbon (TOC), mineral carbon (Min_C), total carbon (TC), hydrogen index (HI), oxygen index (OI), and temperature at which maximum HC is generated by cracking of kerogen (T_{max}) in powdered shale samples (Mani et al. 2015, 2017; 2019). These parameters are used to determine the producing quality, capacity, and thermal maturity of the source rocks. The processes occur in two ovens, pyrolysis oven (inert environment, without the presence of oxygen) and oxidation oven (combustion, i.e. with the presence of oxygen). The flame ionization detector (FID) and the infrared (IR) detector detect the hydrocarbons and concentrations of CO and CO_2, respectively, liberated during pyrolysis and oxidation. Generally, the temperature program is set at 300–650 °C for the pyrolysis oven and from 300 to 850 °C for the oxidation oven, at the rate of 25 °C min^{-1} (Behar et al. 2001; Mani 2019).

Temperature and time are important factors for organic matter maturation and can be simulated to know the hydrocarbon potential of shales (Tissot and Espitalié 1975; Braun and Burnham 1987). Kerogen gets transformed into oil and gas via several n parallel reactions, each of which obeys the first-order kinetics and Arrhenius's law. The rate at which the reaction occurs can be represented as

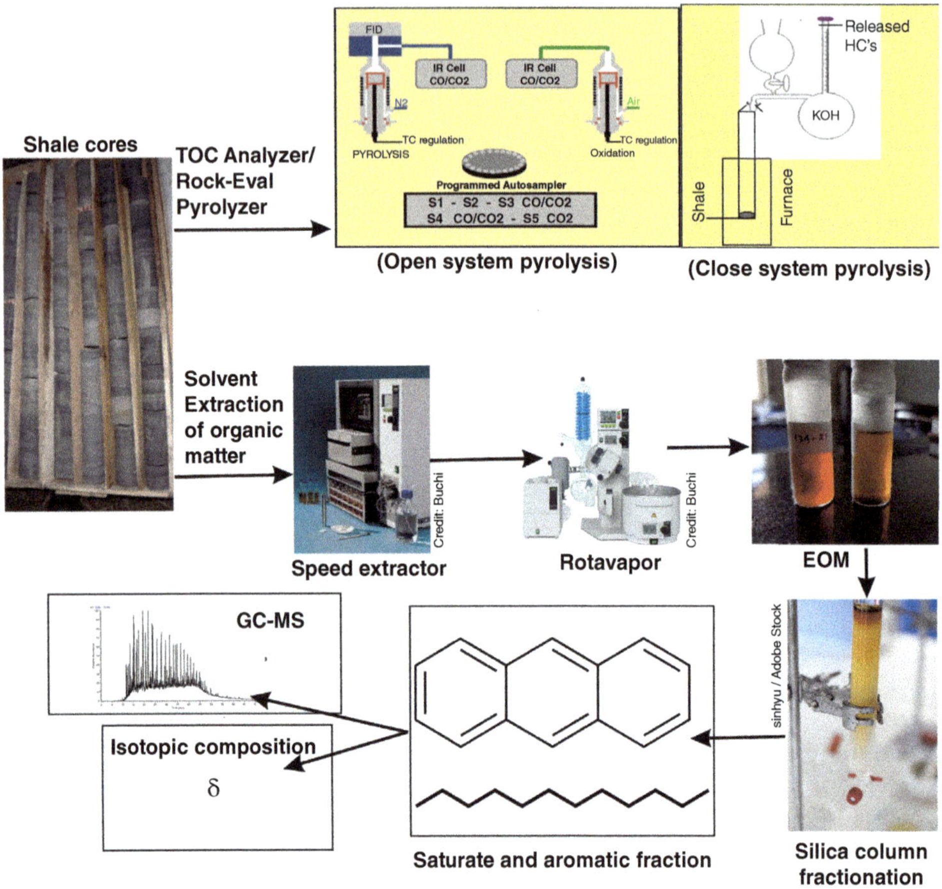

Figure 15.4 A schematic for characterization of kerogen and bitumen in shale samples using Rock Eval pyrolysis and organic matter extraction techniques. The rock eval schematic is from laboratory manual. *Source:* Vinci Technologies.

$$\mathrm{d}q_i/\mathrm{d}t = -k_i{}^*q_i \tag{15.1}$$

where, $\mathrm{d}q_i/\mathrm{d}t$ is the rate of hydrocarbon generation (coincides with the S_2 peak of the rock eval pyrolysis), $\mathrm{d}t$ is the time in second, q_i is the amount of pyrolysable organic matter by the reaction i, and k_i is the rate constant for the ith equation, which depends on the absolute temperature. The negative sign shows the decrease in organic matter concentration, being cracked to oil and natural gas as the reaction proceeds.

Being first-order law of kinetics, characterized by Arrhenius law, Eq. (15.1) can be represented as

$$k_i\,(T) = A\ \exp.(-E_i/RT) \tag{15.2}$$

where, A is the Arrhenius constant (Sec^{-1}), E_i is the activation energy of reaction i and is represented as J mol^{-1}, R is gas constant (0.00199 kcal mol^{-1}), T is absolute temperature in Kelvin.

Substituting k from Eq. (15.2) in Eq. (15.1),

$$dq_i/dt = A\ \exp.(-E_i/RT)^* q_i$$

where dq_i/dt shows the rate of hydrocarbon generation and q_i is the residual petroleum potential of the organic matters that were involved in the reaction i.

Kinetics studies provide quantitative information related to kerogens' generation potential, transformation ratios, and rates for production of hydrocarbons (Espitalie et al. 1987).

15.3.2 Biomarker Extraction and Analysis

Bitumen is soluble in organic solvents, hence can be extracted using techniques of ultrasonication, Soxhlet apparatus, or through speed extractors. After purification into different organic fractions, it is analyzed using GC–MS.

15.3.2.1 Solvent Extraction of Organic Matter

For the present study, the shale samples were extracted using a Buchi Speed Extractor E-914. Solvent extraction of organic compounds from the powdered shale samples was done at elevated temperature (100 °C) and pressure (100 bar) using 9 : 1 DCM-MeOH (dichloromethane and methanol) mixture. About 2–3 g of powdered shale was mixed with diatomaceous earth in extraction cell of capacity of 80 ml. 4 μl of 250 ppm n-hexatriacontane-d_{74} (n-$C_{36}D_{74}$) was added to the sample to know the percent recovery of the extracted organic matter. A three-step extraction program was run to obtain the sedimentary organic matter. The extracted organic matter (EOM) was concentrated to about 1 ml under a gentle stream of nitrogen. It was dissolved in ~3 μl of DCM: MeOH (93 : 7 v/v) per mg of EOM, and n-pentane was added in 40-fold excess volume of EOM + DCM : MeOH and centrifuged in order to remove asphaltene fraction. Sulfur was then removed from the obtained maltene fraction by adding activated (HCl-treated) copper.

15.3.2.2 Column Chromatography

The total lipid extract (TLE) was fractionated into saturated and aromatic fractions using silica gel column chromatography. A burette (30 × 2 cm id), filled with 10 cm silica gel 60 (70–230 mesh) was taken, above which 1.5 cm of alumina was added. The EOM mixed with alumina was placed on the top of the silica column. The aliphatic fraction was eluted using 100 ml of AR-grade hexane and the aromatic fraction was eluted using 150 ml of hexane: DCM (1 : 4) mixture. The solvent fractions containing the analytes were reduced to 0.5–1 ml under the gentle stream of N_2 gas. Internal standards, deuterated androstane, and deuterated tetracosane, $d_2C_{27}\alpha\alpha\alpha$ (20R)-cholestane and d_{10}-naphthalene were added to the alkane and aromatic fraction, respectively for the quantification of target analytes (Mani et al. 2017).

15.3.2.3 Gas Chromatography–Mass Spectrometry (GC–MS)

A Varian 3800 GC coupled to 320 Triple Quad Mass Spectrometer, equipped with DB-1 ms (100% dimethylpolysiloxane)-fused silica capillary column (60 m length × 0.25 mm inner diameter × 0.25 μm film thickness) was used to analyze the aliphatic fractions. Helium (99.999% purity) at a constant flow rate of 1.2 ml min^{-1} was used as a carrier gas. The temperature of the injector port was maintained at 300 °C. The temperatures of the MS transfer line and ion source were set at 310 and 200 °C, respectively. Electron ionization mode (70 eV) was set for fragmentation of analytes in the ion source. The GC column-oven temperature was increased from 35 °C with a hold time of 1 minute to 130 °C at 10 °C min^{-1} and then to 310 °C at the rate of 4 °C min^{-1}, after which it remained isothermal for 10 minutes. One μl of the sample was injected with the help of an autosampler in splitless mode. The mass spectrometer was operated in full scan mode, with a mass range (m/z) between 50 and 600 amu. Compound identification was performed by comparing the GC retention times, fragmentation pattern, and m/z peaks of molecular ions with that of biomarker standards from CHIRON AS, Norway, Norwegian Geochemical Standard – Jet Rock (NGS JR-1), and published mass spectra. For aliphatic hydrocarbon fraction, in particular, an alkane standard, S-4106-100-HX *n*-alkane C10-C40 (even + pristane and phytane) from CHIRON AS was used, and the *n*-alkanes were identified from m/z 57 chromatograms. The pentacyclic triterpenes were identified from the characteristic m/z 191 along with the 370, 398, 412, and 426 molecular ions for C27, C29, C30, and C31 homologs, respectively. Sterane distributions were quantified by measuring peak areas in the m/z 217 chromatograms (Mani et al. 2017).

For the aromatic markers, the GC–MS oven was programmed at an initial temperature of 60 °C with a hold of 1 minute, and then ramped in three stages – initially to 175 °C at a rate of 6 °C min^{-1} with a hold of 4 minutes, followed by 235 °C at the rate of 3 °C min^{-1} with a hold of 4 minutes, and, finally, to 300 °C at the rate of 5 °C min^{-1}, with a hold of 15 minutes. GC–MS was operated in full scan mode, scanning from m/z 50–600 amu. The methyl, dimethyl, and trimethyl naphthalenes were assigned from the m/z 142, 156, and 170 chromatograms, respectively; and phenanthrene, dibenzothiophene, and methyl and dimethyl phenanthrenes from m/z 178, 184, 192, and 206 chromatograms, respectively. The individual peaks were identified by comparison with literature and using the full mass spectra library (NIST). The quantification was achieved by comparing the peak areas of the target compounds with that of a known concentration of internal standard anthracene-d10. Method blanks were performed to check for the presence of any trace organic contaminants from the laboratory and were found to be negligible. The recovery of deuterated hexatriacontane added prior to extraction in the powdered shale samples ranged from 85 to 92%. The EOM yield varied between 4 and 96 mg g^{-1} dry weight of the respective shale sample (Mani et al. 2017).

15.3.3 Isotope Ratio Mass Spectrometer (IRMS)

The ratios of stable isotopes of lighter elements such as carbon ($^{13}C/^{12}C$) and oxygen ($^{18}O/^{16}O$) are determined using an isotope ratio mass spectrometer (IRMS). Continuous flow-IRMS are IRMS instruments coupled online to sample preparatory lines. Commonly used peripherals include the (i) gas chromatography–combustion–IRMS (GC–C–IRMS), used for determining the compound-specific isotope ratio of individual hydrocarbon compounds, (ii) gas bench–IRMS (GB–IRMS), used for the C and O isotope ratio measurement in carbonates, and (iii) elemental analyzer–IRMS (EA–IRMS), for bulk isotopic compositions of the coexisting organic matter in rocks.

In the present study, the bulk carbon isotope (δ^{13}Corg) analysis of the kerogen concentrate was performed using a thermo finnigan flash-elemental analyzer connected to Delta plusXP IRMS. Powdered shale samples were taken in tin containers, and placed inside the auto-sampler drum, where they were purged with a continuous flow of helium and later dropped at preset intervals into a vertical quartz tube, which was maintained at 1020 °C (combustion reactor). The samples were flash-combusted in the helium stream, temporarily enriched with pure oxygen. The gas mixture formed in this process was directed to the chromatographic column (Porapak PQS). The isotopic composition of purified carbon dioxide (CO_2) was measured using IRMS. The Flash EA-IRMS was calibrated using international standards, NBS-20 and graphite, and the precision was found to be within $\pm 0.3‰$ of the reported value (Mani et al. 2015).

Coupled techniques of a gas chromatograph and the mass spectrometer, along with the introduction of a sample combustion interface into the gas chromatograph-IRMS helps in the reliable determination of compound-specific isotope ratios. An Agilent 6890 GC coupled to a Finnigan-Delta PlusXPIRMS via a GC combustion III interface was used for the determination of carbon-specific isotope ratios of the desorbed gases. The gas chromatographically separated products of the gas sample mixture, carried in the stream of helium, was eluted at the output of the gas chromatograph. It was then passed through an oxidation/reduction reactor and was introduced into the ion source of the mass spectrometer for the ratio determination. The instrument was calibrated using Vienna Pee Dee Belemnite (VPDB) and natural gas standard (NGS) to know the variations in results.

15.4 Generative Potential of Permian Shales from Jharia Coal Field, Damodar Valley

15.4.1 Generalized Geology and Stratigraphy

The Damodar Valley forms a part of the Satpura-Damodar belt of the Indian Gondwana region and comprises series of sub-basins, formed as a result of the deposition of sediments in river valleys (Chandra 1992). The Damodar Valley basin takes its name from the Damodar River, which flows through the Raniganj, Jharia, and Bokaro sub-basins, and is one of the most extensive and the best-developed subdivisions of the Gondwana system (Figure 15.5).

The region is rich in varied mineral resources, particularly coal and mica. Jharia sub-basin has a NW-SE linear trend and is found between the Bokaro and Raniganj sub-basin toward the eastern part of the Damodar Valley. In Neoproterozoic basement, the Phanerozoic sedimentation was initiated with the deposition of Late Carboniferous Gondwana sediments. In Jharia, the oldest Gondwana unit is the Talchir. Overlying Precambrian metamorphics, it is successively followed by the Barakar, Barren Measure, and Raniganj Formations. The lowest member of Talchir is a boulder-bed, which is succeeded by greenish shales and sandstones. The Barakar Formation consists of white- to fawn-colored sandstones and grits with occasional conglomerates and beds of shale. It contains much carbonaceous matter in the form of streaks, lenticles, and seams of coal and is chiefly coal-bearing (Chandra 1992). The Barren Measure, which intervenes between the Barakar and Raniganj

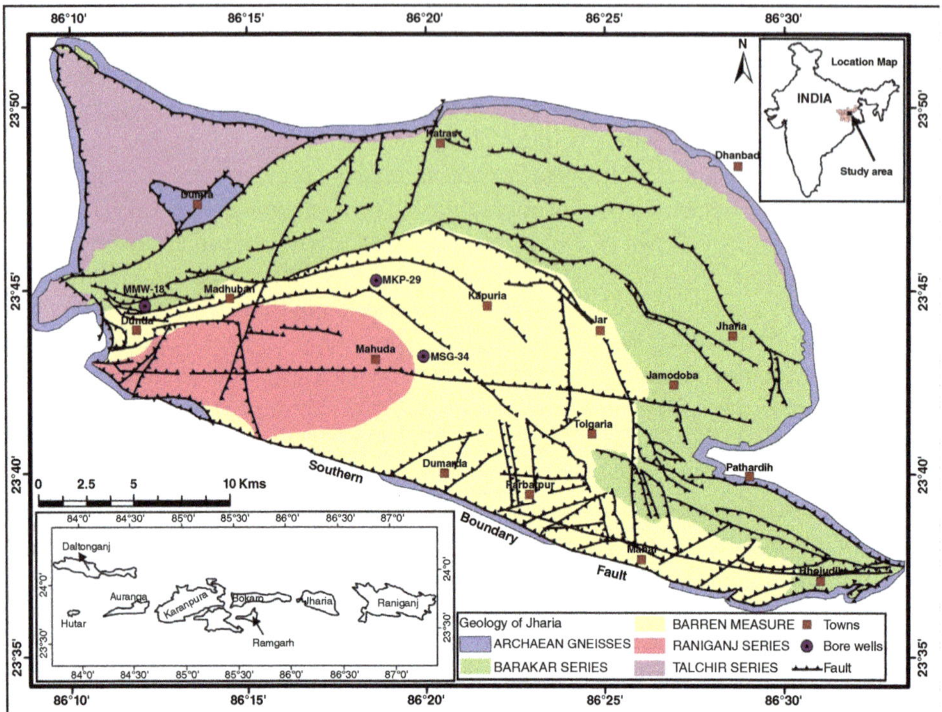

Age	Formation	Thickness (m)
Jurassic or Tertiary Lower Jurassic		
Upper Permian	Raniganj	800
Middle Permian	Barren Measure	730
Lower Permian	Barakar	1250
Upper Carboniferous	Talchir	245
	unconformity	
Archean		

Figure 15.5 Geological map and generalised stratigraphy of Damodar Valley. *Source:* After Mani et al. (2015).

Formations, is entirely barren of coal seams. The Formation consists of sandstones and shales. The shales are, in general, carbonaceous with clay-ironstone nodules, which are sideritic at depth, but when oxidized near the surface, become limonitic (Chandra 1992). Ironstone shale is found in this Formation. It is overlain by the Raniganj series, having a thick succession of fine-grained feldspathic sandstones, shales and coal. Raniganj series shows a regressive phase of coal-bearing Formation in transition to the lower Barren measure Formation (Table 15.1). Late Permian shales from Raniganj, Barren Measure, and Barakar Formations of the Jharia sub-basin are considered for the present study. Subsurface-drilled shale cores were taken from three blocks of the Jharia sub-basin, namely West Mohuda, Singra, and Kapuria (Figure 15.5). The shales were washed with Milli-Q water to remove the drilling mud and air-dried at room temperature. Homogeneously powdered samples were used for the bulk organic geochemical analysis using Rock-Eval pyrolyzer, extraction of organic matter, and carbon isotope studies, following analytical procedures described previously (Section 15.3).

15.4.2 Results and Discussions

The Permian shales from the Jharia sub-basin are rich in organic matter and thermally mature, constituting predominantly the type III kerogen. The bulk organic geochemical data comprising TOC (%), kerogen type, and thermal maturity of the studied formations is shown in Table 15.2.

The pyrolysis of organic matter in Rock-Eval pyrolyzer provides an important thermal maturity parameter, Tmax, which is the temperature at which the maximum amount of hydrocarbons gets liberated during the cracking of kerogen. A Tmax temperature <435 °C shows immature samples, i.e. the rock has not experienced high temperature exposure for the production of oil and gas. A Tmax temperature between 435 and 455 °C shows the oil window stage for the Type-II kerogen, whereas that between 435 and 465 °C shows the oil window stage for the Type-III kerogen. Temperature >455 °C shows the gas window stage for Type-II kerogen and >465 °C shows the gas window stage for Type-III kerogen (Tissot et al. 1987). The Tmax of Jharia shales indicates that shales are in the mature to post-mature stage of gas generation. The maturity of organic matter with depth recorded in the form of Tmax is shown in Figure 15.6.

Oil and gas get generated by the thermal cracking of kerogen. This cracking of kerogen depends on the different bond activation energies, which are required to be broken (Braun and Burnham 1987). Simpler kerogen like Type 1 possesses a very homogeneous distribution of organic matter. So, it requires lower activation energy ranges to break the bonds. Type II or III possesses increasingly heterogeneous organic matter. As different ranges of bonds require different activation energy for their breaking, the activation energy for such

Table 15.2 Organic richness, kerogen type, and thermal maturity of shales from Raniganj, Barren Measure, and Barakar Formations of Jharia Coal Field.

Formation	TOC (%)	Kerogen type	Thermal maturity
Raniganj	2.9–23.1	Mixed Types II–III	Mature (Oil window)
Barren measure	4.4–4.5	Type III	Mature (Oil window-gas condensate zone)
Barakar	5.9–16.4	Type III	Post mature (Condensate wet gas and dry gas zone)

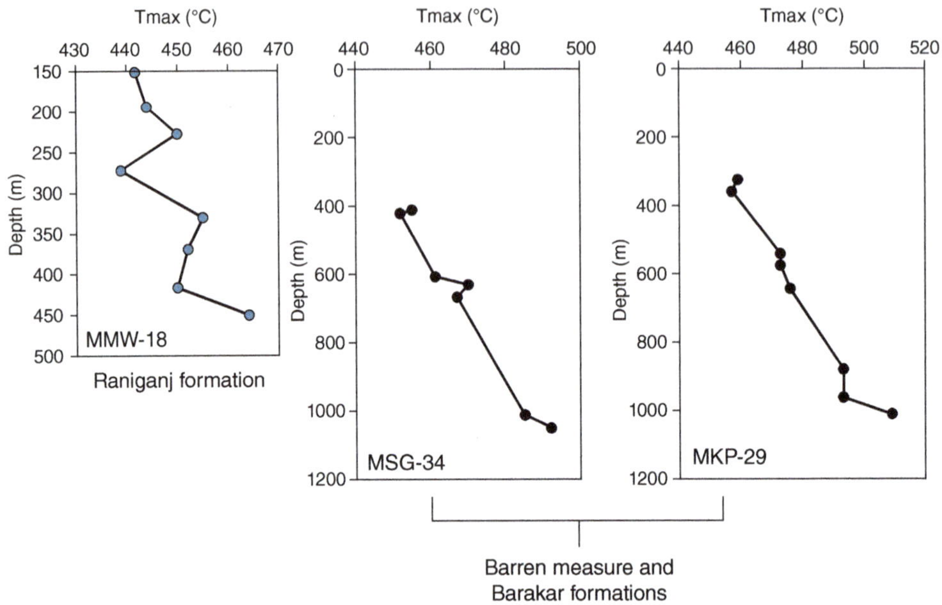

Figure 15.6 Variation of Tmax with depth in shale cores of Jharia basin.

kerogen types will be spread in a broader range (Tissot and Espitalie 1975). The activation energies obtained for the kerogen from Jharia shales are shown in Figure 15.7. The broad spread of activation energy suggests a heterogenous Type III kerogen in most samples.

The saturated biomarkers extracted from the shales comprise n-alkanes, acyclic isoprenoids – pristane and phytane, hopanes, and steranes. A representative chromatogram showing the presence of these compound classes in Raniganj Formation is shown in Figure 15.3. The alkanes are dominated by higher carbon numbers of C23–C27, indicating a terrestrial source to the organic matter in shales (Figure 15.3). The CPI and odd to even preference index (OEP) are greater >1 for most of the samples, corroborating the terrestrial input and high thermal maturity (Scalan and Smith 1970). A higher abundance of the Pristane (Pr) compared to phytane (Py) indicates a peat swamp-type depositional environment, primarily in an oxidizing condition (Figure 15.8).

The samples are characterized by high values of Ts/Tm ratios and C29 sterane C20S/(20S + 20R) and $[\alpha\beta\beta/(\alpha\beta\beta + \alpha\alpha\alpha)]$ ratios, which are indicators of increased thermal maturity.

The aromatic markers found in Permian shales are derived from the transformation and thermal degradation of steroid and terpenoid molecules. Phenanthrenes are mainly derived from steroids and terpenoids from alkylnaphthalenes (Mani et al. 2017). The α-substituted isomers of the polynuclear aromatic hydrocarbons are more sterically hindered compared to the β-substituted isomers, which makes them thermodynamically less stable. Thus, the ratios of α to β isomers are maturity-dependent, and their concentrations are applied to calculate temperature-sensitive maturity parameters such as methylphenanthrene indices (MPI), and dibenzothiophene (DBT) ratios and di, tri, and tetra methylnaphthalene ratios (Figure 15.9).

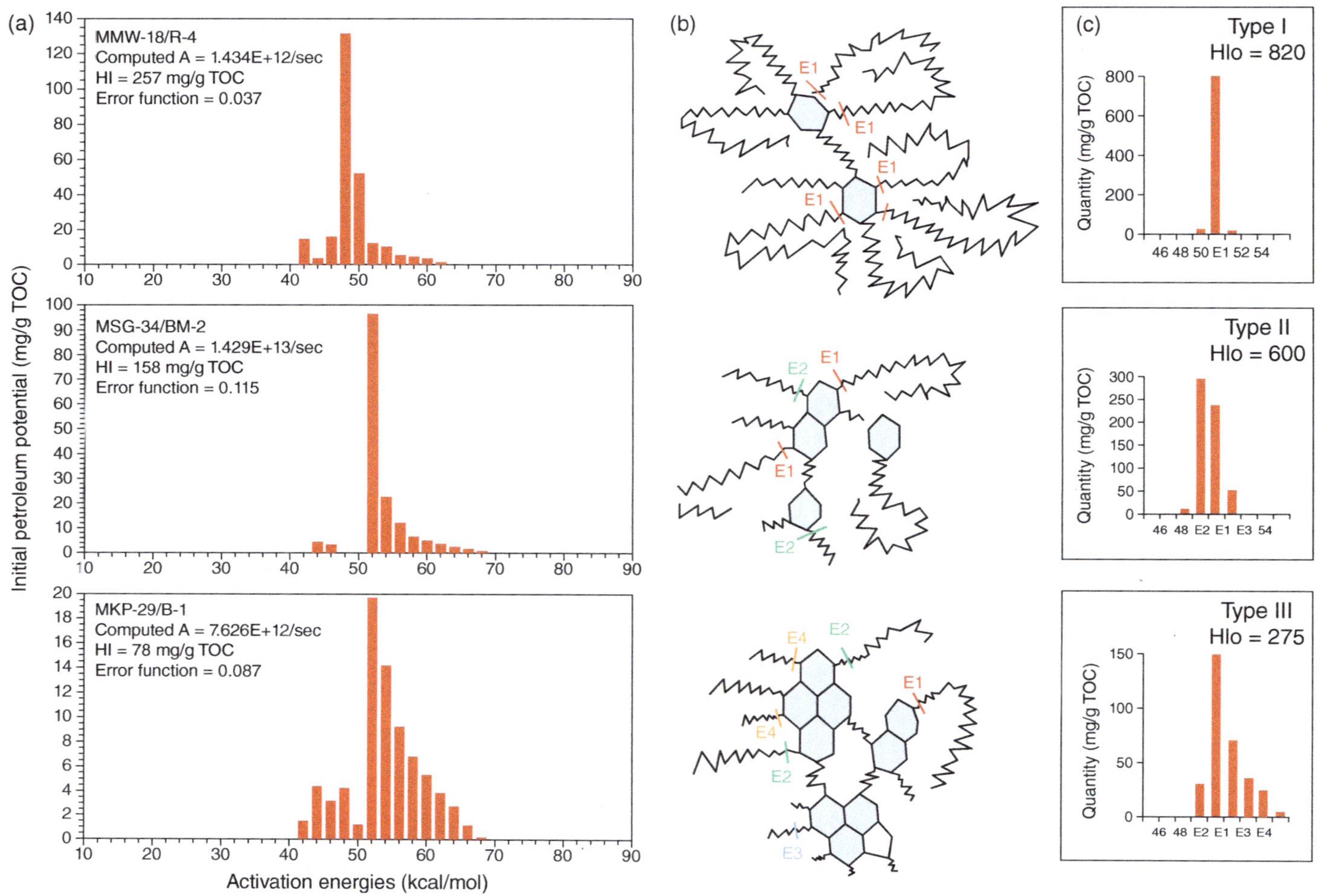

Figure 15.7 Activation energies of Permian shales from Jharia (a) Broad activation energies indicating Type III heterogenous kerogen; (b and c) show comparison of different bonds showing activation energies and initial generation potential (HIo) in the three main Kerogen types. *Source:* Tissot and Welte (1984)/Springer Nature.

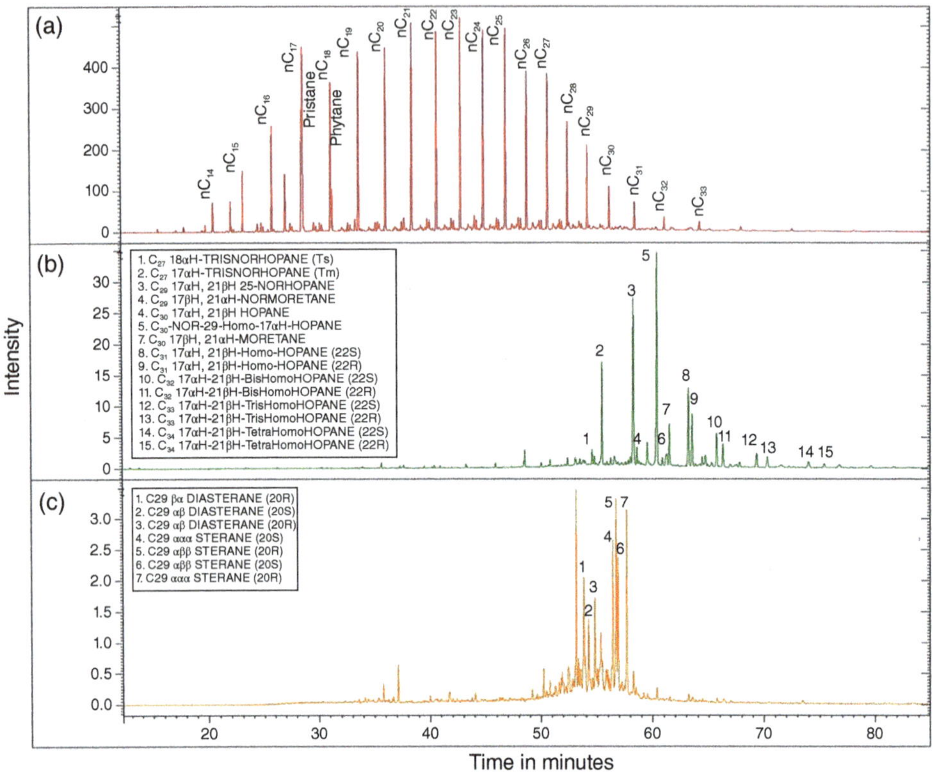

Figure 15.8 Mass chromatogram of saturate biomarkers extracted from Raniganj shales. (a) Alkanes (m/z = 57). (b) Hopanes (m/z = 191). (c) Sterane (m/z = 217).

Diagenetic processes lead to the formation of monoaromatic A rings and C rings in steranes. Both these rings are found in immature sediments, but only the C-ring monoaromatics are prominently present in mature source rocks. These kinds of triaromatic steroids do not occur in recent immature sediments because they are formed from monoaromatic steroids with increasing depth of burial as a result of increasing thermal maturity. Triaromatic steroids are observed in some Raniganj Formations in the sample formation (Figure 15.9).

The bulk- and compound-specific carbon isotope ratios were measured for the kerogen and also the desorbed gas from shales. The δ^{13}C ratios range for, −22.7 to −30.8%, indicating a dominant contribution from a terrestrial source. The desorbed gases from the shale samples constitute dominantly methane, along with lesser proportions of ethane and propane. The carbon isotopic composition of methane lies in thermogenic ranges and indicates its contribution in shales from thermal degradation of kerogen (Table 15.3).

Thus, the composition of sedimentary biomarkers and the stable carbon isotope ratios of kerogen concentrates and gases in the studied shales from the Jharia coalfield suggest it to be organic-rich and thermally mature, with predominantly Type III kerogen. The desorbed gaseous hydrocarbon indicates its origin from deep thermal sources rather than bacterial or near surface sources.

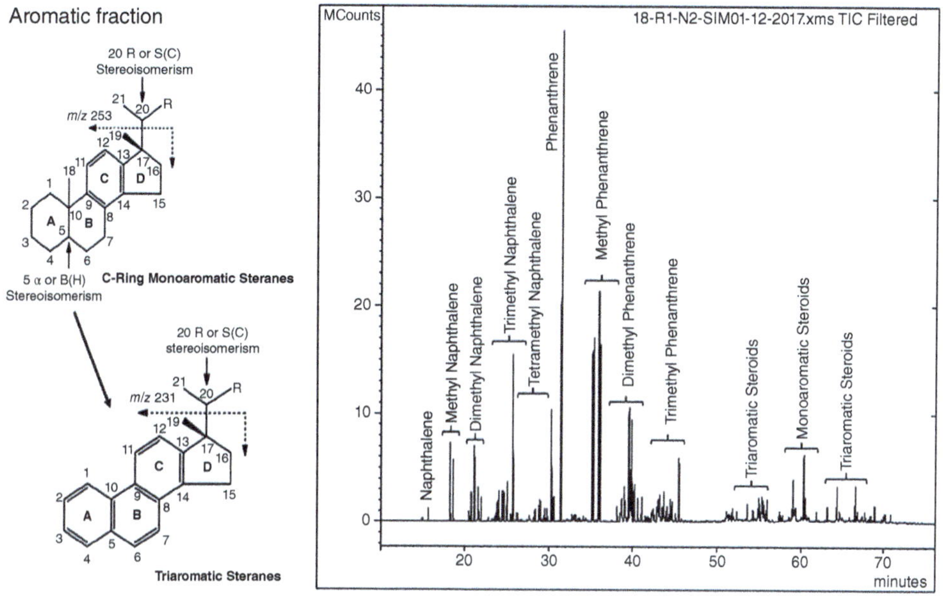

Figure 15.9 Alkyl naphthalenes and mono and tri aromatic steranes in the Raniganj Shales from Jharia sub-basin.

Table 15.3 Stable carbon isotope ratio of bulk organic matter and desorbed gases from Jharia shales.

Sample ID	Bulk $\delta^{13}C$ ratio (‰)	Sample	$\delta^{13}C_1$ (‰)	$\delta^{13}C_2$ (%)	$\delta^{13}C_3$ (%)
MKP-29-BM-2	−30.8	MSG-34 BM-II	−40.4	—	−28.9
MKP-29-B-3	−25.0	MSG-34 BM-II	−42.5	—	−27.0
MMW-18-R4	−24.3	MSG-34 BM-II	−43.0	—	−26.9
MSG-34-BM-1	−24.6	MSG-34 BM-II	−44.6	−29.7	−29.2
MSG-34-B-VA	−22.7	MSG-34 BM-II	−44.8	−30.3	−27.0

B = Barakar; BM = Barren Measure; R = Raniganj.

15.5 Recent Approaches to Shale Characterisation

A major part of organic matter in sedimentary rocks occurs in the form of kerogen, which is the macromolecular precursor material for hydrocarbons. The quantity, quality, and properties of kerogen in shales are important for resource evaluation and also in understanding the processes by which gas shales store, retain, and expel natural gas. Shale gas reservoirs pose numerous challenges in terms of their reservoir characterization. The kerogen and bituminous compounds studied using organic and isotope geochemistry provide the source, type, and thermal maturity of the organic matter. However, understanding the geochemical evolution of organics within shales with increasing thermal maturity is essential to have close estimates of gas generation. The intrinsically complex kerogen is composed of an amorphous porous carbon skeleton and its pore size distribution spans from a few Å to the μm range (Bousige et al. 2016). The microporosities exhibited by kerogen, besides the known meso- and macroporosity, is largely responsible for their ultralow permeability when compared with that of conventional reservoirs (Bousige et al. 2016). The nanoscale geochemical heterogeneities, which develop as a response to maturation need high-resolution imaging techniques for kerogen characterization (Bernard and Horsfield 2014) along with organic and isotopic measurements in order to define the transport mechanisms in such matrix. It has been observed that previously proposed model equations based on structural parameters of kerogen underestimate the thermal maturity while overestimating the hydrocarbon generation potential (Agrawal and Sharma 2019). This is because the kerogen samples, which were used to reconstruct the old models were mostly derived from immature shales having VRs of less than one (Agrawal and Sharma 2019). Realistic molecular models of kerogen with its maturity series provide a detailed picture of kerogen's nanostructure (Bousige et al. 2016). Imaging techniques like focused ion beam-scanning electron microscopy (FIB-SEM), scanning transmission X-ray microscopy (STXM)-based X-ray absorption near edge structure (XANES) spectra (Bernard and Horsfield 2014) (STEM) provide the spatial variability in chemistry and structure of thermally mature organic-rich shales at the sub-micrometer scale. The corroboration of thermal maturity biomarkers and stable isotopic ratios with the molecular models and imaging techniques essentially improves the understanding of the geochemical nature of shale gas systems and provides reliable estimation of reserves.

Acknowledgments

DM is thankful to Soumyajit Mukherjee (SM), IIT Bombay for invitation to write this article. An anonymous reviewer and SM provided review comments. DM and NRK are thankful for the financial support from UoH-IoE (MHRD (F11/9/2019-U3(A)). NRK acknowledges UGC, New Delhi for the Junior Research Fellowship. MSK is grateful to Director, CSIR-NGRI for the permission for this publication. Merryl Le Roux, Frank Otmar Weinreich, Joss Everett, Andrew Harrison, and the proofreading team (Wiley Blackwell) are thanked for their assistance. The article has been summarized in Mukherjee et al. (2023).

References

Agrawal, V. and Sharma, S. (2019). Improved kerogen models for determining thermal maturity and hydrocarbon potential of shale. *Scientific Reports* 8 (1): 1–8.

Behar, F., Beaumont, V., Penteado, H.L., and De, B. (2001). Rock-Eval 6 technology: performances and developments oil & gas science and technology. *Oil & Gas Science and Technology* 56 (2): 111–134.

Bernard, S. and Horsfield, B. (2014). Thermal maturation of gas shale systems. *Annual Review of Earth and Planetary Sciences* 42: 635–651.

Bjørlykke, K. (2015). Introduction to petroleum geology chapter 1. In: *Petroleum Geoscience: From Sedimentary Environments to Rock Physics* (ed. K. Bjørlykke). Berlin: Springer-Verlag Berlin Heidelberg https://doi.org/10.1007/978-3-642-34132-8_1.

Bousige, C., Ghimbeu, C.M., Vix-Guterl, C. et al. (2016). Realistic molecular model of kerogen's nanostructure. *Nature Materials* https://doi.org/10.1038/NMAT4541.

Brassell, S.C., Eglinton, G., and Fu, J.M. (1985). Biological marker compounds as indicators of the depositional history of the Maoming oil shale. *Organic Geochemistry* 10: 927–941.

Braun, R.L. and Burnham, A.K. (1987). Analysis of chemical reaction kinetics using a distribution of activation energies and simpler models. *Energy & Fuels* 1: 153–161.

Briggs, D. and Summons, R.E. (2014). Ancient biomolecules: their origins, fossilization, and role in revealing the history of life. *BioEssays* 36 (5).

Chandra, D. (1992). *Jharia Coalfield*Mineral Resources of India, 5, 149. Bangalore: Geological Society of India.

Collister, J., Summons, R.E., Lichtfouse, E., and Hayes, J.M. (1992). An isotopic biogeochemical study of the Green River oil shale. *Organic Geochemistry* 19: 265–276.

Connan, J. and Dessort, D. (1987). Novel family of hexacyclic hopanoid alkanes (C32–C35) occurring in sediments and oils from anoxic paleoenvironments. *Organic Geochemistry* 11: 103–113.

Espitalie, J., Marquis, F., and Sage, L. (1987). Organic geochemistry of the Paris Basin. In: *Petroleum Geology of North West Europe* (ed. J. Brooks and K. Glennie), 71–86. London: Graham and Totman http://www.uky.edu/KGS/coalcoal-macerals.php#.

Ficken, K.J., Li, B., Swain, D.L., and Eglinton, G. (2000). An n-alkane proxy for the sedimentaryinput of submerged/floating freshwater aquatic macrophytes. *Organic Geochemistry* 31: 745–749.

Gaines, S.M., Eglinton, G., and Rullkötter, J. (2009). *Echoes of Life: What Fossil Molecules Reveal about Earth History*, 355. Oxford: Oxford University Press. ISBN: 978-0-19-517619-3.

Goodwin, N.S., Mann, A.L., and Patience, R.L. (1988). Structure and significance of C30 4-methylsteranes in lacustrine shales and oils. *Organic Geochemistry* 12: 495–506.

Grantham, P.J., Posthuma, J., and DeGroot, K. (1980). Variation and significance of the C27 and C28 triterpane content of a North Sea Core and various North Sea crude oils. In: *Advances in Organic Geochemistry 1979* (ed. A.G. Douglas and J.R. Maxwell), 29–38. Oxford, UK: Pergamon Press.

Hazra, B., Wood, D.A., Mani, D. et al. (2019). Source-Rock Evaluation Using the Rock-Eval Technique. In: *Evaluation of Shale Source Rocks and Reservoirs. Petroleum Engineering*. Cham: Springer International Publishing https://doi.org/10.1007/978-3-030-13042-8_3.

Horsfield, B. and Schulz, H.M. (2012). Shale gas exploration and exploitation. *Marine and Petroleum Geology* 31 (1): 1–2.

Hunt, J.M. (1996). *Petroleum Geology and Geochemistry*, 617. San Francisco, CA: W. H. Freeman and Company.

Koster, J., Van Kaam-Peters, H.M.E., Koopmans, M.P. et al. (1997). Sulphurisation of homohopanoids: effects on carbon number distribution, speciation, and 22S/22R epimer ratios. *Geochimica et Cosmochimica Acta* 61: 2431–2452.

Magoon, L.B. and Dow, W.G. (1994). The petroleum system: from source to trap. *AAPG Memoir* 60: 655.

Mani, D. (2019). Characterizing the source rocks in petroleum systems using organic and stable isotope geochemistry: an overview. *The Journal of Indian Geophysical Union* 23 (1): 10–27.

Mani, D., Patil, D.J., and Dayal, A.M. (2011). Stable carbon isotope geochemistry of near surface adsorbed alkane gases in Saurashtra basin. *Chemical Geology* 280 (1–2): 144–153.

Mani, D., Dayal, A.M., Patil, D.J. et al. (2014). Gas potentiality of Proterozoic and Phanerozoic shales from NW Himalaya, India: inferences from new pyrolysis data. *International Journal of Coal Geology* 128–129: 81–95.

Mani, D., Patil, D.J., Dayal, A.M., and Prasad, B.N. (2015). Thermal maturity, source-rock potential and kinetics of hydrocarbon generation in Permian shales of Damodar Valley basin, eastern India. *Marine and Petroleum Geology* 66: 1056–1072.

Mani, D., Kalpana, M.S., Patil, D.J., and Dayal, A.M. (2017). Organic matter in gas shales: origin, evolution and characterization. In: *Shale Gas: Exploration, Environmental and Economic Impacts*, vol. 3, 25–52. Elsevier ISBN 9780128095737.

Moldowan, J.M., Seifert, W.K., and Gallegos, E.J. (1985). Relationship between petroleum composition and depositional environment of petroleum source rocks. *American Association of Petroleum Geologists Bulletin* 69: 1255–1268.

Moldowan, J.M., Sundararaman, P., and Schoell, M. (1986). Sensitivity of biomarker properties to depositional environment and/or source input in the lower Toarcian of S.W. Germany. *Organic Geochemistry* 10: 915–926.

Mukherjee, S., Dasgupta, S., Majumdar, C. et al. (2023). Introduction to: Handbook of Petroleum Geoscience: Exploration, Characterization, and Exploitation of Hydrocarbon Reservoirs. In: *Handbook of Petroleum Geoscience: Exploration, Characterization, and Exploitation of Hydrocarbon Reservoirs*, Petroleum Industry Updates: Geoscientific and Economic Issues (ed. S. Mukherjee, S. Dasgupta, C. Majumdar, et al.). John Wiley & Sons, Hoboken, ISBN 978-1-119-68003-1.

Peters, K.E. and Moldowan, J.M. (1991). Effects of source, thermal maturity, and biodegradation on the distribution and isomerization of homohopanes in petroleum. *Organic Geochemistry* 17: 47–61.

Peters, K.E., Walters, C.C., and Moldowan, J.M. (2005). *The Biomarker Guide, Volume 2: Biomarkers and Isotopes in the Petroleum Exploration and Earth History*. Cambridge, UK: Cambridge University Press.

Scalan, R.S. and Smith, J.E. (1970). An improved measure of the odd-to-even predominance in the normal alkanes of sediment extracts and petroleum. *Geochimica et Cosmochimica Acta* 34: 611–620.

Schidlowski, M. (1987). Application of stable carbon isotopes to early biochemical evolution on earth. *Annual Review of Earth and Planetary Sciences* 15: 47–72.

Schouten, S., de Loureiro, M.R.B., SinningheDamste, J.S., and de Leeuw, J.W. (2001). Molecular biogeochemistry of Monterey sediments, Naples Beach, California. I: distribution of hydrocarbons and organic sulfur compounds. In: *The Monterey Formation: From Rocks to Molecules* (ed. C.M. Isaacs and J. Rullkotter), 150–174. New York: Columbia University Press.

Summons, R.E. (1987). Branched alkanes from ancient and modern sediments: isomer discrimination by GC/MS with multiple reaction monitoring. *Organic Geochemistry* 11: 281–289.

Summons, R.E. and Jahnke, L.L. (1992). Hopenes and hopanes methylated in ring a: correlation of the hopanoids from extant methylotrophic bacteria with their fossil analogues. In: *Biological Markers in Sediments and Petroleum* (ed. J.M. Moldowan, P. Albrecht and R.P. Philp), 182–200. Englewood Cliffs, NJ: Prentice-Hall.

Summons, R.E., Jahnke, L.L., Hope, J.M., and Logan, G.A. (1999). 2-Methylhopanoids as biomarkers for cyanobacterial oxygenic photosynthesis. *Nature* 400: 554–557.

Tissot, B.P. and Espitalie, J. (1975). Thermal evolution of organic matter in sediments: application of a mathematical simulation. Petroleum potential of sedimentary basins and reconstructing the thermal history of sediments. *Oil & Gas Science and Technology* 30 (5): 743–778.

Tissot, B.P. and Welte, D.H. (1984). *Petroleum Formation and Occurrence*. New York: Spring-Verlag.

Tissot, B., Pelet, R., and Ungerer, P. (1987). Thermal history of sedimentary basins maturation indices, and kinetics of oil and gas generation. *AAPG Bulletin* 71 (12): 1445–1466.

Van Kaam-Peters, H.M.E., Heidy, M.E., Kostter, J. et al. (1998). The effect of clay minerals on diasterane/sterane ratios. *Geochimica et Cosmochimica Acta* 62: 2923–2929.

Van Krevelen, D.W. (1961). *Coal: Typology-Chemistry-Physics-Constitution*. Amsterdam: Elsevier Science.

Vandenbroucke, M. (2003). Kerogen: from Types to Models of Chemical Structure. *Oil & Gas Science and Technology* 58 (2): 243–269.

Vieth, A. and Wilkes, H. (2010). Stable isotopes in understanding origin and degradation processes of petroleum. In: *Handbook of Hydrocarbon and Lipid Microbiology* (ed. K.N. Timmis). Berlin, Heidelberg: Springer https://doi.org/10.1007/978-3-540-77587-4_5.

Volkman, J.K. (1986). A review of sterol markers for marine and terrigenous organic matter. *Organic Geochemistry* 9: 83–99.

Whitelaw, P., Uguna, C.N., Stevens, L.A. et al. (2019). Shale gas reserve evaluation by laboratory pyrolysis and gas holding capacity consistent with field data. *Nature Communications* 10: 3659. https://doi.org/10.1038/s41467-019-11653-4.

Wolff, G.A., Lamb, N.A., and Maxwell, J.R. (1986). The origin and fate of 4-methyl steroid hydrocarbons 1. 4-methyl sterenes. *Geochimica et Cosmochimica Acta* 50: 335–342.

16

A GIS-Based Approach to Explore the Possibility of N–S Gondwana Rift in the South-Eastern Part of India

Subhobroto Mazumder, Ravi Prakash, and D.S. Mitra*

Remote Sensing and Geomatics, KDM Institute of Petroleum Exploration, ONGC, Dehradun, Uttarakhand, India
**Author for correspondence: subhom007@gmail.com*

Abstract

During Early Permian to Middle Triassic, the Gondwana assembly of plates rifted several times into many intracratonic rift basins, some of which occur in the central and eastern parts of India. The present study is a GIS-based approach to explore the possibility of the existence of an N–S-oriented Early Permian-Middle Jurassic rift aligned along the south-eastern margin of the Indian Plate extending from the Cauvery Basin in the south, up to the southern part of the KG Basin in the north. Multisource data derived from analysis of data-sets corresponding to different verticals of geology and geophysics have been integrated and analytical methods implemented to derive evidence of the premised rift and prepare a geological model. Furthermore, a GIS-based tectonic reconstruction with Eastern Antarctica has been used to validate the rift model. The sediments of the premised rift may act as an additional target for hydrocarbon exploration in these basins, thus redefining its domains in both time and space.

Keywords *Gondwana Rift; GIS-based tectonic reconstruction; Permian; Cauvery Basin; Palar Basin; Pennar Basin*

16.1 Introduction

The end of Neo-Proterozoic was characterized by a progressive agglomeration of crustal blocks referred to as Pangea that existed for ~120 Ma from the mid-Carboniferous up to Lower Liassic (de Lamotte et al. 2015), i.e. about 195 Ma ago (Figure 16.1a). Disintegration and breakup of Pangea initiated from the latest Carboniferous and during the Early Permian intracratonic rift systems were established in numerous places in the Pangean assembly (Langhi and Borel 2004; de Lamotte et al. 2015). These rifts originated due to failure of the attenuated crust along pre-existing zones of weakness inherited from Precambrian structural fabric (Chakraborty and Ghosh 2005) and were subsequently filled with

Handbook of Petroleum Geoscience: Exploration, Characterization, and Exploitation of Hydrocarbon Reservoirs, First Edition. Edited by Soumyajit Mukherjee, Swagato Dasgupta, Chandan Majumdar, Subhandip Mandal and Troyee Dasgupta.
© 2023 John Wiley & Sons Ltd. Published 2023 by John Wiley & Sons Ltd.

(a)

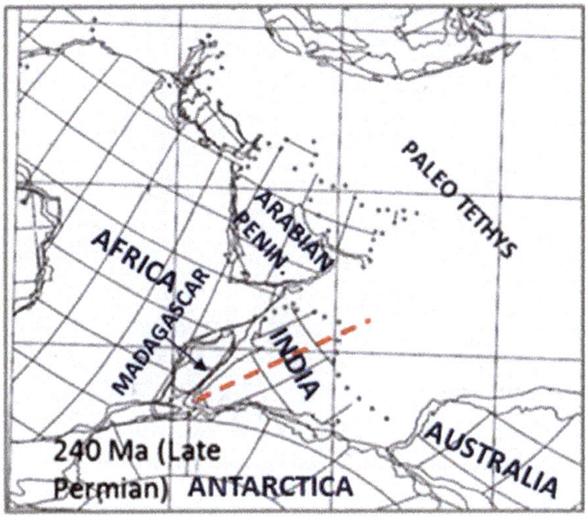

(b)

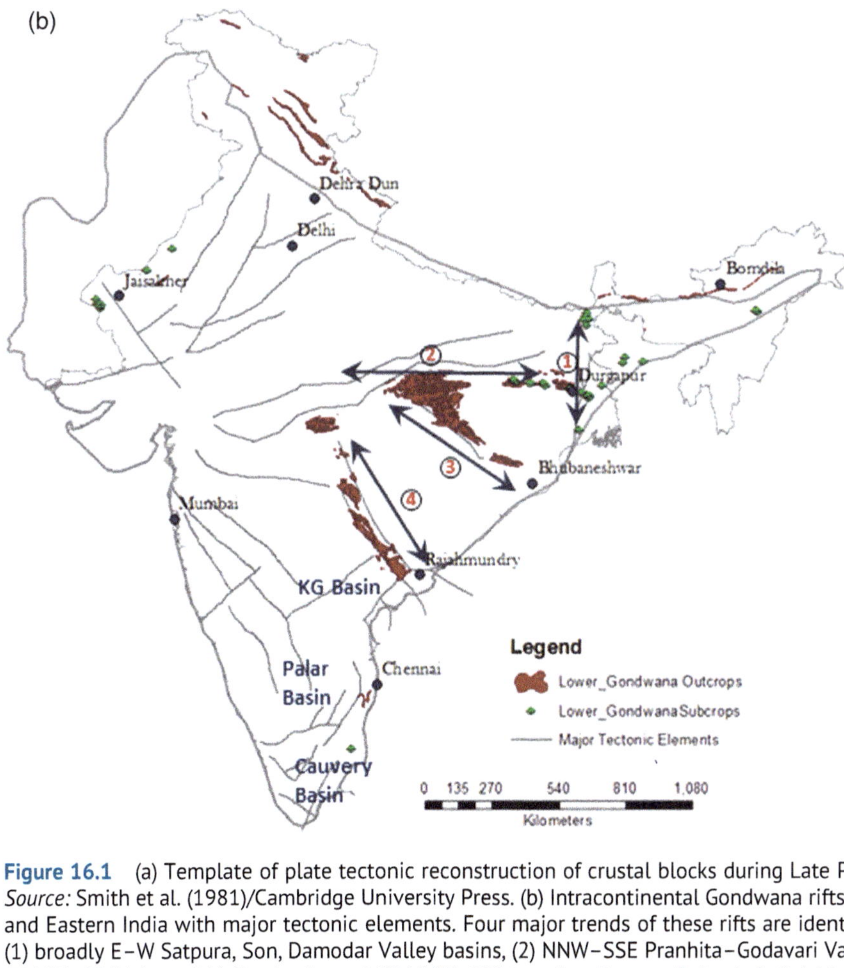

Figure 16.1 (a) Template of plate tectonic reconstruction of crustal blocks during Late Permian. *Source:* Smith et al. (1981)/Cambridge University Press. (b) Intracontinental Gondwana rifts in Central and Eastern India with major tectonic elements. Four major trends of these rifts are identified: (1) broadly E–W Satpura, Son, Damodar Valley basins, (2) NNW–SSE Pranhita–Godavari Valley basin, (3) NW–SE Mahanadi Valley basin, and (4) NNW–SSE trending Purnea–Rajmahal–Galsi basin.

a thick sequence of continental clastic sediments with minor marine intercalations in an early phase. Such exposures of fault-bounded Gondwana sediments oriented broadly E–W, N–S, and NW–SE occur mainly in the central and eastern parts of India (Figure 16.1b), and had been studied in detail by a number of workers (Casshyap 1979; Casshyap and Tewari 1988; Tewari 1998; Bhattacharya et al. 2012; Bhattacharya and Bhattacharya 2015).

The present study is a GIS-based approach to the premise and delineates an additional N–S-oriented Early Permian-Mid Jurassic rift (East Coast Gondwana Rift [ECGR]) aligned along the eastern coast of India extending from the Cauvery Basin up to the southern part of the Krishna-Godavari (KG) Basin.

The south-eastern margin of the Indian plate was attached to parts of the Enderby Basin, the Princess Elizabeth Trough, and the Davis Sea Basin of East Antarctica (Gaina et al. 2007) during Carboniferous-Permian. This formed the eastern part of Pangea until they rifted apart in the Late Jurassic-Early Cretaceous that caused the evolution of the KG Basin as well as its adjacent Cauvery Basin (Padhy et al. 2015; Nagendra and Reddy 2017). In both these basins, continental stretching resulting from NW–SE extension initiated passive margin rifting (see Misra and Mukherjee (2015) for review on similar issues) leading to NE–SW striking series of horst and graben (ridges and basins) morphology (Figure 16.2). These structures are parallel to the Indian coast in the KG Basin but turn oblique (15–20° anticlockwise) to it further southward. Stratigraphically, in the KG Basin, the initial sediments encountered post-initiation of India-Antarctica rifting have been established to be base of Tithonian (Raju et al. 1994) or early Oxfordian (Aswal et al. 2001). Similarly, in the Cauvery Basin, the corresponding sediments encountered in subcrops have a Tithonian to Tithonian-Berriasian age connotation (Prasad and Phor 2009). In accordance, any older sediments observed in these basins can be considered to be pre-rift or infra-rift (rift associated with India-Antarctica separation) sediments of Gondwana affinity that were deposited in an older Carboniferous-Permian-Triassic rift. This proposed rift is older than these NE–SW-oriented horst and graben structures and is aligned at a high angle to them suggesting a basin opening that is much earlier than both KG and Cauvery.

Subsequent to the separation from Antarctica, the northern part of the eastern coastal margin of Greater India displayed rift characteristics. The southern part of it beyond the KG Basin showed evidences of shearing (Subrahmanyam and Chand 2006) along a ridge-transform system (Chand et al. 2001) implying strike-slip tectonism. This strike-slip may have initiated along a pre-existing older Precambrian lineament called the Proto-Boundary Fault (Chari et al. 1995) and was eventually accommodated by a 95 km southward displacement of Sri Lanka from its initial position adjacent to India. This formed the Gulf of Mannar or Mannar Basin in between them (Figure 16.2) (Premarathne et al. 2015; Desa et al. 2018) in two phases with the first phase initiating around 160–165 Ma (upper-middle Jurassic) (Ratnayake et al. 2018). This was associated with the counterclockwise rotation of Sri Lanka with the northern part of the island acting as a pivot (Premarathne et al. 2015) and the development of NE–SW and NW–SE discontinuities. This second phase was more prominent and resulted in further counterclockwise rotation of Sri Lanka, more widening of the Mannar basin and the initiation of the Cauvery and Lanka Basins (Baillie et al. 2004; Kularathna et al. 2015). This anticlockwise rotation of Sri Lanka also resulted in Cretaceous marine entry into the Mannar Basin, which was earlier than that in adjacent areas in the Cauvery Basin (Premarathne et al. 2015). Evidences of marine

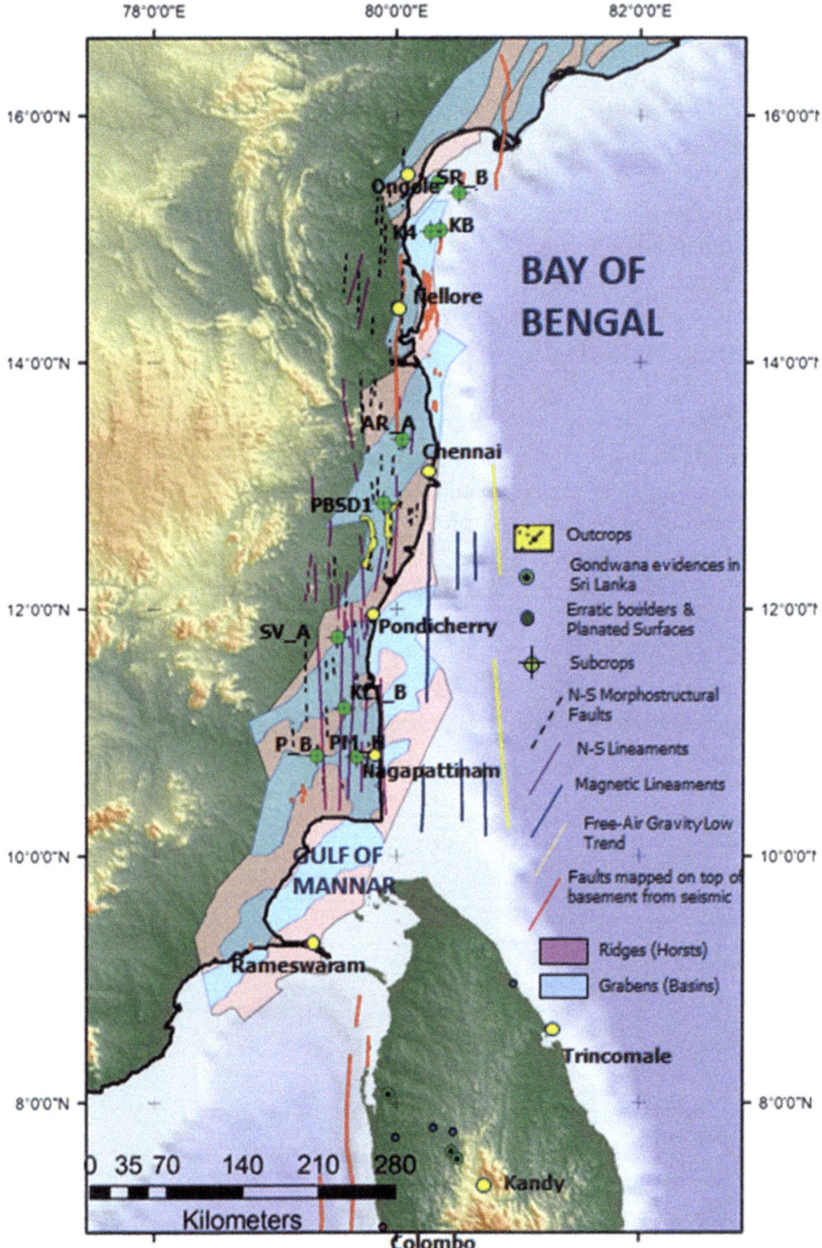

Figure 16.2 Horsts and Grabens along the south-eastern margin of India developed due to Antarctica-India separation in Late Jurassic-Early Cretaceous. Outcrops and subcrops of Permian-Triassic-Middle Jurassic sediments are encountered in these areas forming a broad N–S alignment. Also, evidences of Permian sedimentation have been observed in Sri Lanka that formed a contiguous southward landmass to these areas. N–S oriented morphostructural faults, regional lineaments, faults from seismic data mapped at basement level, gravity trends, and magnetic lineations may serve as evidence of the N–S rift bounding those sediments.

influence in sediments in these areas earlier than that may be unrelated to or prior to the India-Antarctica-rifting event or post-/syn-dating any Lower Gondwana rifting. The anticlockwise rotation of India and Sri Lanka, however, continued until ~96 Ma (Cenomanian, Late Cretaceous) (Chari et al. 1995), whereas the separation of the conjoined continental blocks of East Antarctica and West Australia from Greater India continued as the new seafloor developed between them (Gibbons et al. 2013) until the blocks rifted apart from each other around 85 Ma (Santonian, Late Cretaceous) and Australia and Antarctica reached their respective present positions.

16.2 Methodology and Analysis

The present study involves GIS-based extraction of geological and geophysical (G&G) evidences of the premised rift like outcrops, subcrops, tectonic elements, and seismic and gravity-magnetic data from existing maps/data volumes. Additionally, surface features like faults derived from drainage-based morphostructural analysis and lineaments from different image data sets have been used to substantiate the rift model and indicate its geological alignment and disposition. Subsequently, these data sets have been incorporated into a GIS platform to build up a pattern and derive a model of the proposed rift. Furthermore, an attempt has been made to rotate the rift back to a Late Carboniferous–Early Permian position, again in a GIS environment and integrate it with evidences of rifting reported in contiguous parts in East Antarctica associated with the KG and Cauvery basins (Figure 16.3).

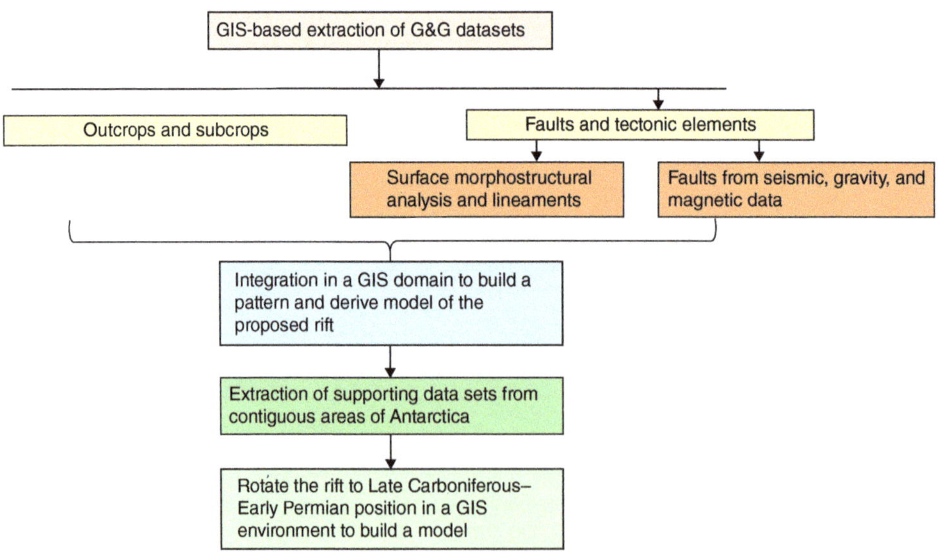

Figure 16.3 Flowchart defining the workflow of the study.

16.2.1 Outcrops and Subcrops

One of the primary data sets that suggest the existence of pre-Cretaceous sediments in the KG-Cauvery Basin area adjacent to the south-eastern coast of India are the isolated outcrops of Ongur Formation in the Palar Basin. The Palar Basin occurs as a small intracratonic rift basin situated between the larger KG and Cauvery sedimentary Basins and shares a tectonic history broadly similar to these adjacent larger basins. Lithologically, the Ongur Formation is the oldest sequence in this basin overlying the pre-Cambrian basement and is constituted of boulder beds and greenish shales believed to be of fluvioglacial origin (Rangaraju et al. 1993). However, the major characteristic that makes this sequence significant from the point of view of this article is that the palynoflora recovered from this sequence indicates an early Permian age of the formation in an otherwise Early Cretaceous-Late Jurassic basin. These assemblages are comparable to those recorded from the Talchir Stage of Rewa and Manendragarh Gondwanas with a similar marine influence (Vairavan 1993). Additionally, the Ongur Formation is linearly disposed with N–S structural trend (Figure 16.2) within a basin, which has its axis oriented broadly in a NE to NNE alignment (Mazumder et al. 2013b). Petrographically, sandstones within Ongur Formation are texturally and mineralogically immature with subangular to subrounded well to poorly sorted framework grains, subarkosic with almost fresh feldspars and rock fragments of gneiss, granite, schist, recrystallized chert, phyllite, and volcanic fragments (Prakash et al. 2018). These petrographic characteristics connote short transportation, rapid deposition, and a slope-derived origin, all of which may indicate a deposition in a rift setting and synrift mechanism of sediment dispersal.

Other than outcrop data, subcrop samples derived from shallow and deep wells in the Palar Basin have been analyzed to locate correlative litho-units in the northern part of the Palar Basin. In shallow borehole PBSD-1 (Figure 16.2), drill–core samples were observed to be coal-bearing and on palynostratigraphic analysis indicated a Late Permian age. Reworked Permian and Triassic palynomorphs have also been reported in the drill-core samples of PBSD-1 (Tripathi and Vijaya 1997). Further northward, biostratigraphic studies in deep well AR-A (Figure 16.2) drilled by ONGC in 2014 to the depth of 2400 m and encountering pre-Cambrian basement indicates the oldest sediments to be of Middle Jurassic (Bajocian-Bathonian) age suggesting older pre-rift/infra-rift sediments (Basavaraju et al. 2016). Sedimentological studies of conventional cores recovered from well AR-A further depict that these sediments are ill-sorted sandstones with poor textural maturity consisting of feldspar and lithic fragments. These sediment characters also signify a probable rift-related deposition (Davison and Underhill 2013; Zaid and Gahtani 2015) and are inferred to be deposited in an environment varying from intertidal to nearshore with an active interaction of marine and fluvial systems (Nagraj and Vijay 2014).

Studies of conventional cores from another deep well SR_B in shallow offshore in the southern part of KG Basin (also referred to as Pennar Basin) (Figure 16.2) have also suggested the existence of Late Permian sediments based on palynofloral assemblage. These Late Permian sediments have also been described to be texturally immature, poorly sorted feldspar-rich sandstones that are products of deposition of braided channels (Yadagiri 2017). In another well KB in offshore Pennar Basin (Figure 16.2), Triassic to Permian sediments have been encountered that are constituted mainly by feldspathic micaceous sandstones, carbonaceous shales with mafic igneous rock fragments. Sediments overlying the basement in well K4 near KB (Figure 16.2) have been classified as pre-Albian sediments that constitute chiefly arkosic texturally immature kaolinitic sandstones of a nonmarine depositional environment. Their similarity with Late Permian sediments of well SR_B may imply a similar genetic and age connotation.

Further south of the Palar basin and in the northern part of the Cauvery Basin, a shale section directly overlying pre-Cambrian basement encountered in the well P_B (Figure 16.2) had been dated to be of Lower Permian age based on palynostratigraphic studies. These sediments are believed to be equivalent to the Ongur Formation of the Palar basin and also bear signatures of a marine incursion during Asselian-Sakamarian (Early Permian) time (Basavaraju 1995; Prasad and Phor 1998). Similarly, Early Permian sediments were also observed in well KLI-B based on palyno-floral analysis (Prasad and Phor 2009). In another well PM_H eastward of P_B, 150 m sediments have been encountered that have been biostratigraphically classified as older than Oxfordian and envisaged to be deposited in marginal marine to fluvial environment implying pre-rift Gondwana sediments. Production testing of a horizon immediately above this zone yielded water with salinity similar to freshwater (3410 ppm) further reiterating its continental origin.

In well, SV_A in the northern part of the Cauvery Basin (Figure 16.2), undated dolerites/volcanics associated with slickensides were encountered below Lower Cretaceous formations.

In Sri Lanka, in Metibokka and Pussella areas of the Kurunegala district of west-central part, varve-like rhythmites have been observed with palynological characters suggesting an Early Triassic to Permian age (Dahanayake et al. 1989). Also in strata exposed near Tabbowa Tank, Tabbowa Basin, of western Sri Lanka (Figure 16.2), Glossopteris flora has been observed signifying the existence of Late Permian sediments in the area (Edirisooriya et al. 2018). Erratic boulders and planted surfaces have also been observed in parts of Sri Lanka (Figure 16.2) that indicate an Early Permian glaciofluvial environment (Katupotha 2013), similar to the Talchir boulder bed in India.

16.2.2 Rift Signatures: Faults, Lineaments, and Tectonic Elements

The outcrop and subcrop signatures of these pre-Cretaceous Permian to Middle Jurassic (Gondwana) sediments when plotted on a map define an approximate N–S alignment (Figure 16.6c). This linearity is more apparent when Sri Lanka is restored to its original tectonic position in pre-Cretaceous (Brandt et al. 2014) on a GIS platform. Since the Gondwana sediments in peninsular India are found to be bounded by rift systems, it is probable that these evidences of Permian to middle Jurassic sedimentation are bounded by a N–S-oriented rift system. The N–S-aligned Ongur sediments in the Palar basin may stand as one of the surface evidences of the proposed rift system. However, no other evidence of N–S fault systems on the surface is found to exist in the area.

16.2.2.1 Surface Expression of Faults

In an effort to delineate surface manifestations of subsurface faults, two separate GIS-based approaches independent of each other have been attempted. In one approach, morphostructural analysis based on drainage has been carried out assuming that neotectonic deformations would lead to reactivations of deep-seated structures formed due to an earlier phase of tectonism. This results in the formation of subtle structures in shallow sediments associated with predominantly vertical deformation (Shurr 1982; Schumm et al. 2002). Since drainage is the most sensitive of all geomorphic elements (Schumm et al. 2002), these deformations are imprinted on the regional drainage pattern as localized drainage anomalies like drainage offsets, compressed meanders, rectangular drainages, and rectilinear drainages (Figure 16.4b). In the present study, drainages have been extracted from 1 : 50 000 scale toposheets for the entire KG-Palar-Cauvery Basin (Figure 16.4a) area and, subsequently, analyzed based on principles discussed in Mazumder et al. (2013a, 2016), minor faults interpreted from drainage anomalies are delineated and joined to generate regional morpho-structural fault signatures from which N–S faults have been spatially filtered and mapped (Figure 16.4c) to represent probable fault signatures of the rift. For validation of these faults, detailed fieldwork has been carried out in parts of the Palar Basin and Cauvery Basin in Tamil Nadu as well as in the Krishna-Godavari Basin in Andhra Pradesh in which neotectonic evidences of faulting like incisions in lower-order drainages, development of unpaired terraces, and abrupt bends in drainages (Figure 16.4d) are used to corroborate the fault interpretations.

The second attempt is to identify N–S regional scale lineaments based on tonal variations in LANDSAT 8 OLI image data sets covering the entire area. Here regional lineaments are commonly used as aliases for surface expressions interpreted as surface expressions of geologic weak zones as well as faults that govern basin development (Masoud and Koike 2011). To make the analysis unbiased, lineament sets interpreted by other workers in the region that are associated with earthquake epicenters (Saravanavel and Ramasamy 2003; Reddy and Koteswara 2004; Kumaran et al. 2012) have been used to supplement the present interpretations and provide a tectonic connotation to the surface datasets (Figure 16.5a). Additionally, LISS 4 images of 5.8 m resolution depicting basement outcrops adjacent to that of Ongur Formation in the Palar Basin also indicated N–S orientations suggestive of effects of an NS tectonic trend (Figure 16.5b). In the case of Sri Lanka, it had undergone about 25°–28° of differential anticlockwise rotation w.r.t. India (as measured in a GIS environment). Based on this value, lineaments mapped by different workers (Vitanage 1985; Kumara and Dharmagunawardhane 2014) in Sri Lanka, on the clockwise rotation of 25°–28°, would be aligned N–S, which had been extracted and mapped (Figure 16.5c).

16.2.2.2 Subsurface Expression of Faults

The morphostructural faults and regional lineaments in the south-eastern margin of India and parts of Sri Lanka serve as the surface indicators of N–S faults that are associated with the proposed rift. These N–S faults are assumed to be deep-seated structural features affecting the pre-Cambrian basement of the area and, hence, imaged by geophysical methods.

16.2.2.2.1 Seismic Data

The morphology of the basement top surface in parts of both the Cauvery and the KG Basins has been derived using 3D reflection seismic data wherever available that can image down to 5.5–6 km. A slope map prepared using this surface and classified using slope amount indicates areas of high slope (55°–75°) to be suggestive of faults (Figure 16.6a, b). Most of these basement faults are found to be oriented NE–SW as a consequence of the Early Cretaceous rift episode affecting the entire east coast of India. This rift episode may have obliterated/reoriented any earlier structural fabric or signatures of any earlier tectonic deformation

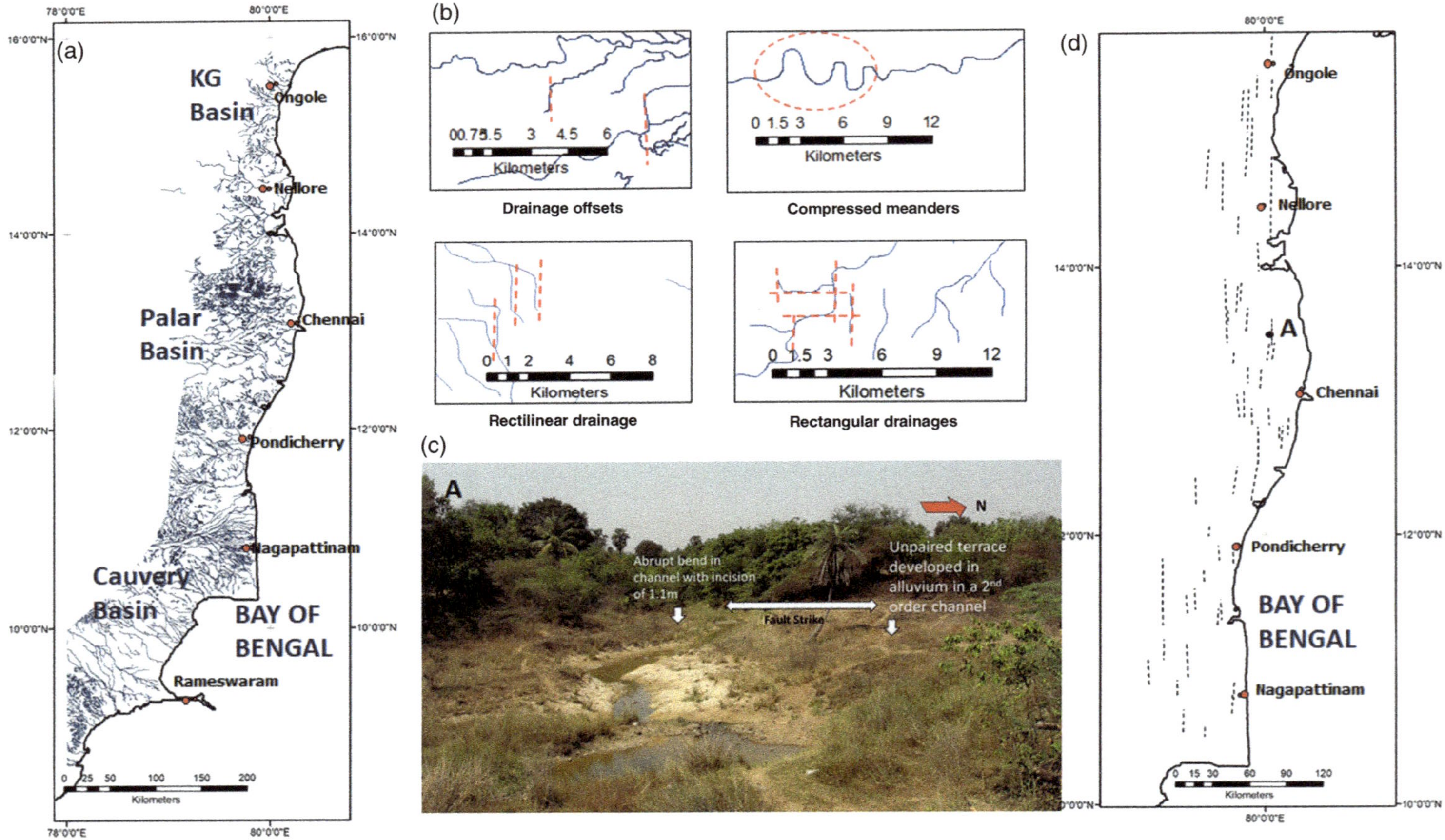

Figure 16.4 (a) Drainages extracted from 1 : 50 000 toposheets for the area covering Cauvery, Palar, Pennar, and southern KG basins. (b) Fault signatures delineated from localized drainage anomalies like compressed meanders, rectilinear drainages, rectangular drainages, and drainage offsets identified from the extracted drainages. (c) Regional NS morphostructural faults delineated from joining smaller/minor morphostructural fault signatures based on their trend and continuity. (d) Field verification of the interpreted faults at point A.

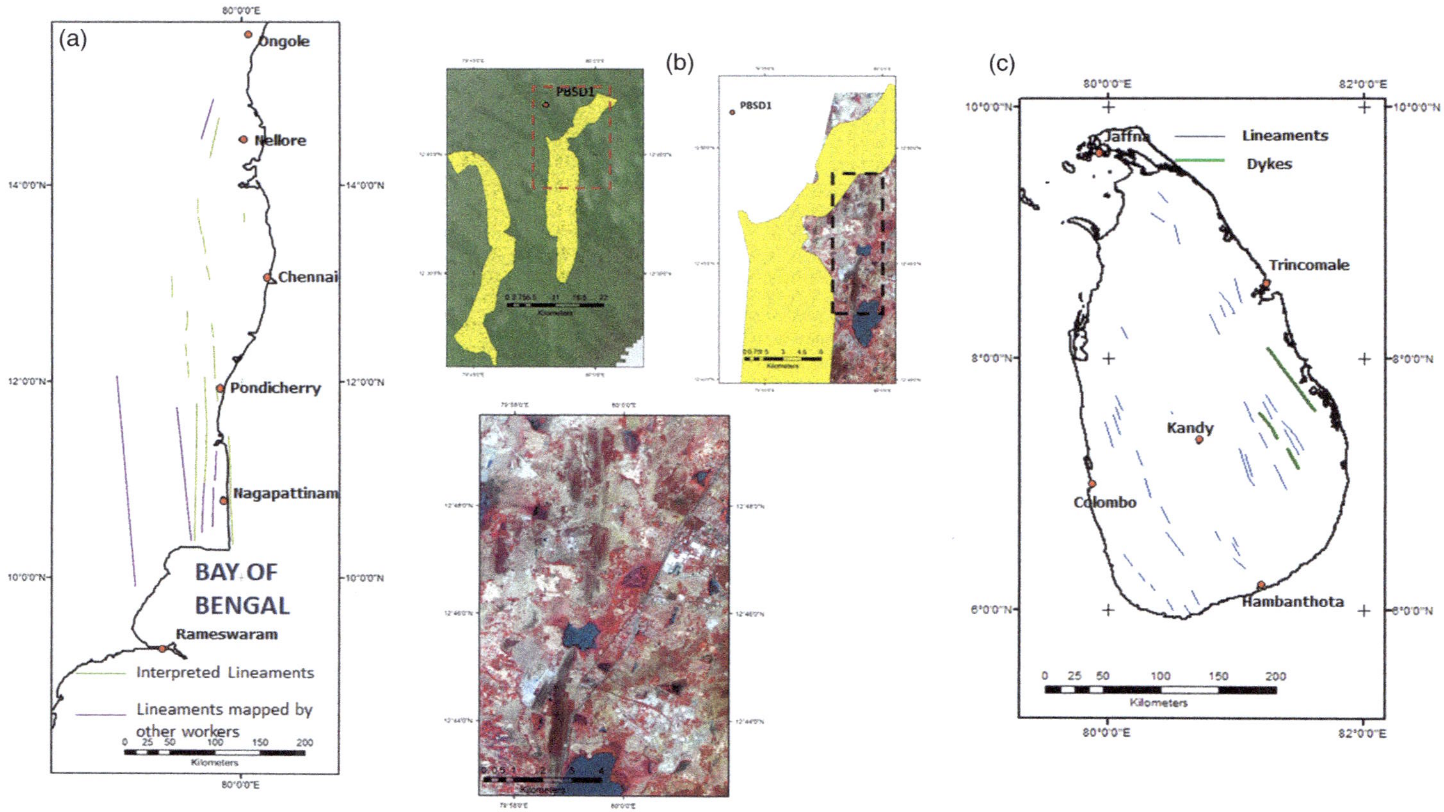

Figure 16.5 (a) N–S regional scale lineaments interpreted from LANDSAT 8 OLI data integrated with N–S lineaments associated with earthquake epicenters interpreted by other workers. Sources: Saravanavel Ramasamy (2003); Reddy and Koteswara (2004); Kumaran et al. (2012). (b) LISS 4 FCC (composite of bands 4, 3 and 2) of 5.8 m resolution NS oriented basement outcrops adjacent to outcrops of Ongur Formation in Palar Basin. (c) Lineaments interpreted in Sri Lanka (Vitanage 1985; Kumara and Dharmagunawardhane 2014) that on a clockwise rotation of 25°–28° would be aligned N–S.

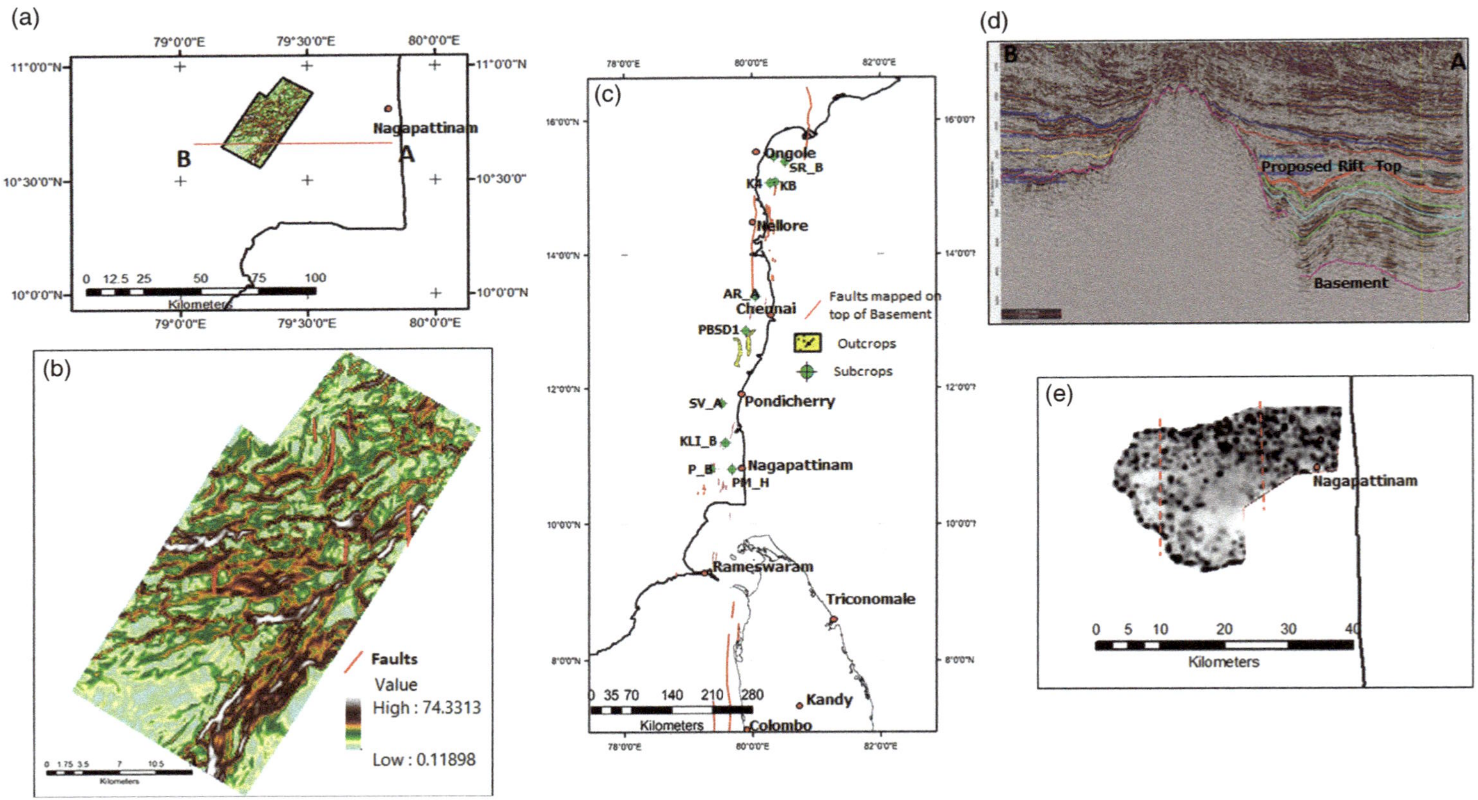

Figure 16.6 (a) Index showing the geographical footprint of the area of slope analysis on basement top and section AB (Figure 16.6d). (b) Slope map prepared on basement top near Nagapattinam in the Cauvery Basin based on which a number of N–S faults have been delineated. (c) Remnant N–S faults delineated on basement top in areas of Mannar, Cauvery, Palar, Pennar, and the southern part of the KG basin. (d) Seismic section along line AB distinguishes two sequences based on reflector geometry: the lower part shows parallel reflectors on the basement top that may signify sediments of the proposed rift, whereas further above, the reflectors are found to abut against a surface. The horizon separating them may be proposed rift top. Variance map prepared on rift top showing traces of N–S-oriented faults. (e) Seismic section SL-01 (after PRDS 2013) showing signatures of pre-rift (Gondwana) sediments above the basement top in the Mannar Basin.

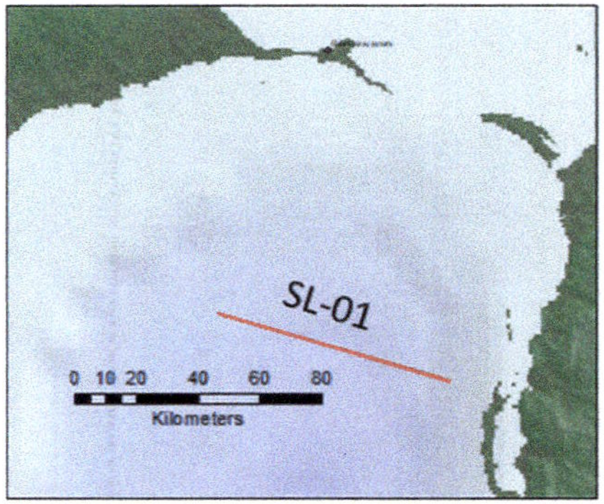

(f)

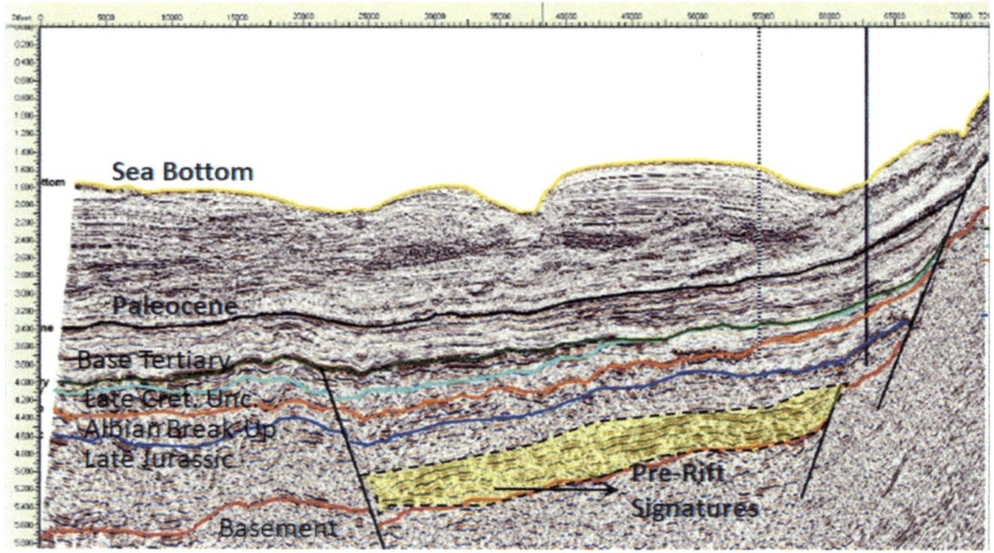

Figure 16.6 (Continued)

affecting the basement. However, remnant signatures of N–S faults can still be delineated based on the slope map as well as work done by other interpreters (Vairavan 1993; Baraik et al. 2011; PRDS 2013) and such N–S signatures have been demarcated along the entire study area (Figure 16.6c).

Seismic sections near theNagapattinam area, when interpreted, show parallel reflectors on basement top, whereas further above, the reflectors are found to abut against a surface (Figure 16.6d). This distinguishes them into two sequences where the earlier reflectors may suggest a sequence associated with the proposed Gondwana rift and the latter indicate sediments related to the Early Cretaceous rifting (Singha et al. 2017). Based on these interpretations, two areas of such Gondwana rift-related sediments with a maximum thickness of 1.5 km have been inferred in the graben areas of the Early Cretaceous rift system in the Cauvery Basin. A seismic variance map prepared on the proposed rift top in one of these areas also indicates evidences of N–S faulting (Figure 16.6e). Seismic section SL-01 in the Mannar Basin (PRDS 2013) also indicates signatures of at least 400 m of pre-rift sediments based on parallel reflectors above the basement top (Figure 16.6f). These sediments may consist of coals or organic-rich lacustrine mudrocks (Edirisooriya et al. 2018) of probable Late Paleozoic age (Baillie et al. 2004) with Gondwana affinities and, in turn, form the basement for the Mannar rift sediments and volcanics. Furthermore, interpretation of seismic characteristics in the western part of the north Palar Basin indicates magmatic intrusions associated with SW propagation of the KG Basin along an N–S trending high axes near the shoreline, thus implying pre-existing N–S faults in the area (Baraik et al. 2011). Signatures of parallel reflectors have also been identified in the southern part of the KG Basin near well SR_B, suggesting the continuation of the proposed Gondwana rift below the Early Cretaceous rift sediments (Shandilya et al. 2017). Evidences of N–S faults and their sympathetic fractures have also been observed in the FMI data of well SR_B.

16.2.2.2.2 Gravity Data

A horizontal gradient map of the Bouguer gravity field (Kumar et al. 2009) indicates a prominent N–S trend of high gravity gradient along the south-eastern Indian margin. This N–S trend is found to correlate with the alignment of the morpho-structural faults, regional lineaments as well as the faults mapped on basement top using seismic data (Figure 16.7a). This may imply the probable alignment of a major tectonic feature along this gravity trend. Additionally, a GIS-based overlay of the linear gravity gradient shows a remarkable correlation with outcrops of Ongur Formation. The subcrops representing Gondwana sediments also depict an association with the high gravity gradient trend. This probably signifies a rift-related deposition of these sediments. Subparallel to this trend in the Cauvery offshore, a prominent N–S trending linear gravity low is observed in the free-air gravity anomaly map (Figure 16.7b) (Murty et al. 2002).

16.2.2.2.3 Magnetic Data

The magnetic anomaly map compiled in the Cauvery Offshore area of the Bay of Bengal at 20 nT interval shows a major N–S lineament (Figure 16.7a) associated with a short wavelength medium amplitude anomaly of ~−300 nT. This is seen again further south with a reduced amplitude (−200 nT) (Subrahmanyam et al. 1993).

In areas where an active petroleum system exists in the basin, hydrocarbons from reservoir rocks may remigrate and escape to the surface through faults that cut across them in

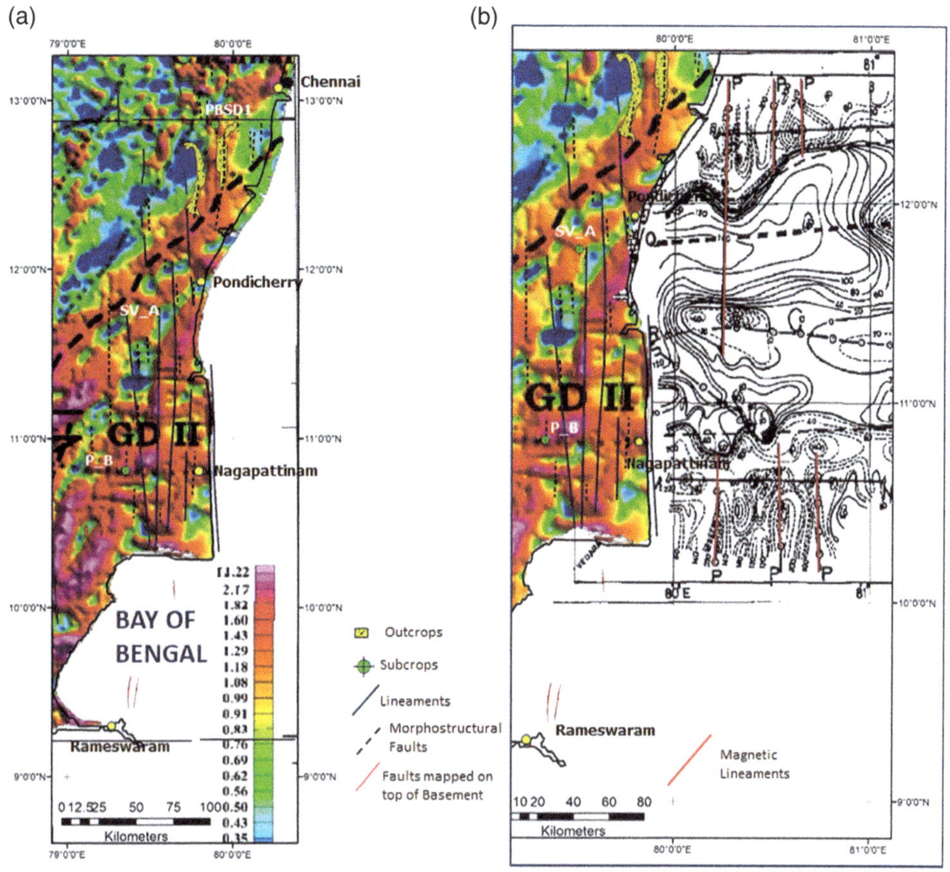

Figure 16.7 (a) Prominent N–S-oriented high gravity gradient (Kumar et al. 2009) found to correlate with morpho-structural faults, regional lineaments and faults mapped on basement top using seismic data. A strong correlation is observed with outcrops of Ongur Formation. Magnetic lineations based on magnetic anomaly map (Murty et al. 2002) also found parallel to the above. (b) Conspicuous N–S trending linear gravity low (Kumar et al. 2009) in free-air gravity anomaly map observed in Cauvery Offshore that is parallel to structural elements mapped in onland part and magnetic lineaments mapped in offshore (Subrahmanyam et al. 1993).

case of neotectonic reactivations of these faults. Some of the remigrated hydrocarbons may remain untrapped, whereas the light gaseous hydrocarbons are adsorbed by the soil matrix around the faults that act as a conduit. These areas around the faults are found to show an adsorbed gas maxima. In the Cauvery and Palar areas, such a high concentration of adsorbed gas measured in areas/blocks of ~500–600 km^2 is associated with the N–S faults interpreted from seismic or geomorphic data (Adsorbed Gas Anomaly Atlas 2017) validating the interpretations (Figure 16.8).

The above evidence of Gondwana sedimentation in outcrops and subcrops and faults that may be associated with the rift that accommodates those sediments, in a GIS-based overlay, suggests an N–S trending linear body. This N–S-oriented rift is aligned along the southeastern margin of India spanning the eastern part of KG, Palar, and Cauvery Basins extending on to offshore and in parts of Sri Lanka (Figure 16.2).

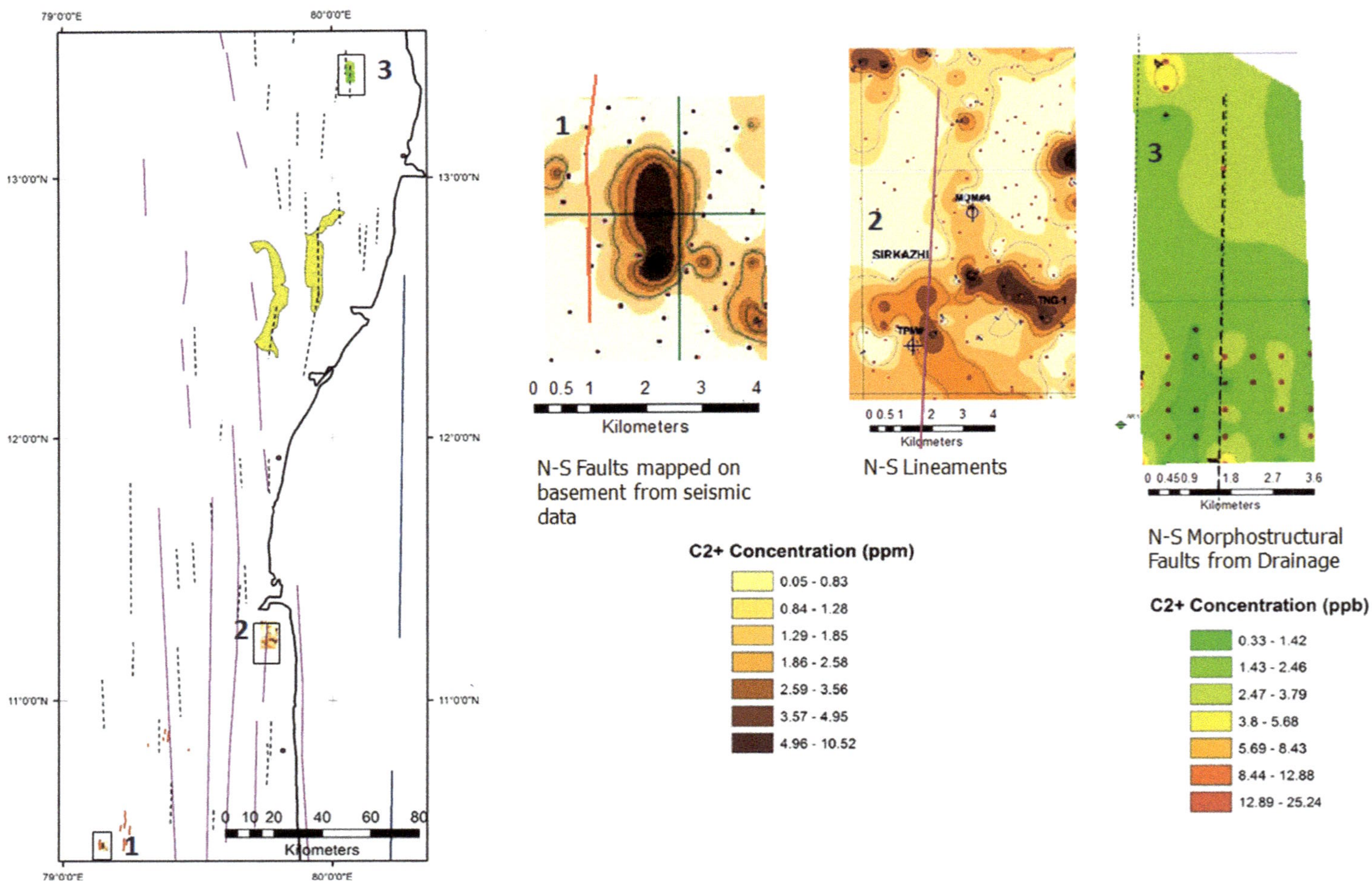

Figure 16.8 Anomalies with a high concentration of adsorbed gases found to be associated with N–S-oriented interpreted faults.

16.2.3 Correlation with Antarctica

Since the oldest sediments associated with the premised rift are Early Permian, rifting initiated presumably at that age. Also considering that the rift is aligned along the south-eastern margin of the Indian Plate, there may be a probability that some indications of rifting may be observed in parts of East Antarctica that formed a contiguous landmass with these areas. As per available plate tectonic templates (Mikhalsky et al. 2017), Enderby Land and Lutzow-Holm Bay part of Eastern Antarctica (Figure 16.9a) was associated with the areas that may have suffered the rift phase in the Indian Plate. A GIS-based plate

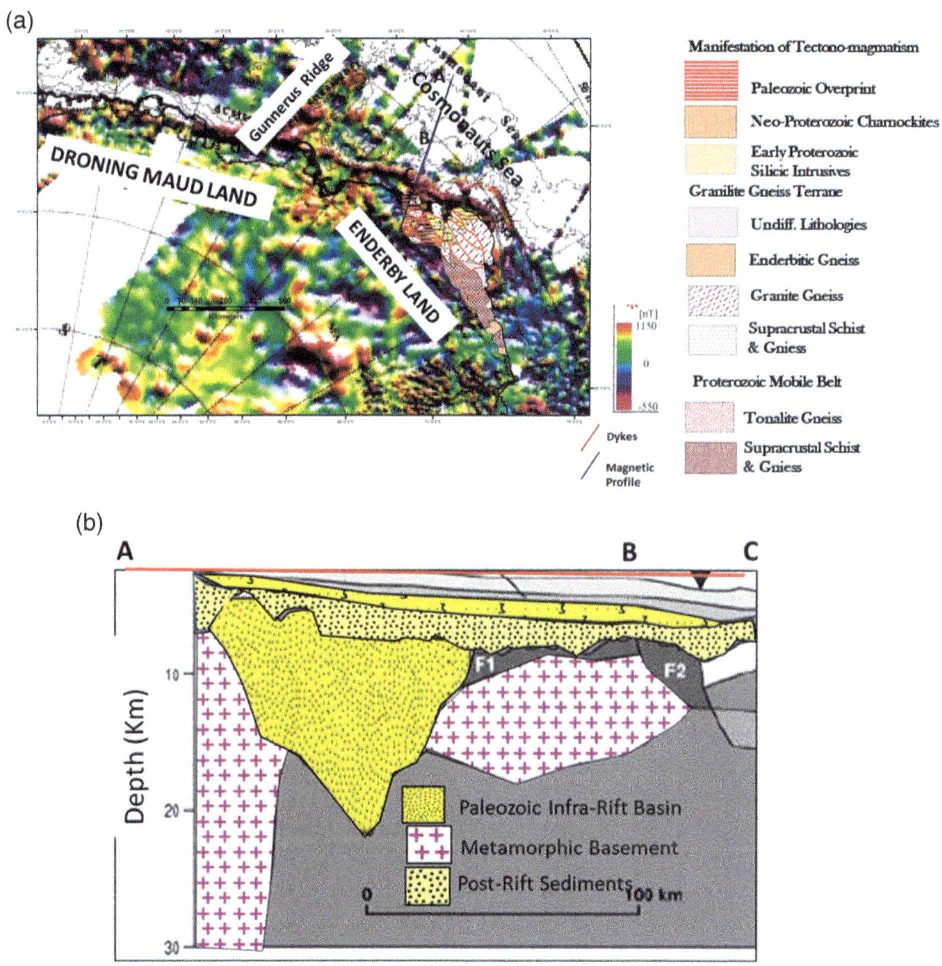

Figure 16.9 (a) Color shaded-relief map of the magnetic anomalies for East Antarctica (Golynsky et al. 2002) overlain on the geological map of Enderby Land (Mohanty 2015) showing a number of NW-oriented dykes. (b) Magnetic profile along line AC shows the development of normally magnetized units of Archaean-Proterozoic metamorphic basement rocks enclosing a thick, broad wedge of reversely magnetized rocks believed to be of a Paleozoic infra-rift basin. *Source:* Adapted from Stagg et al. (2004). (c) Early to Middle Permian showing a reverse magnetism (Gradstein and Ogg 2004) similar to the inferred Paleozoic infra-rift basin in magnetic profile AC.

(c)

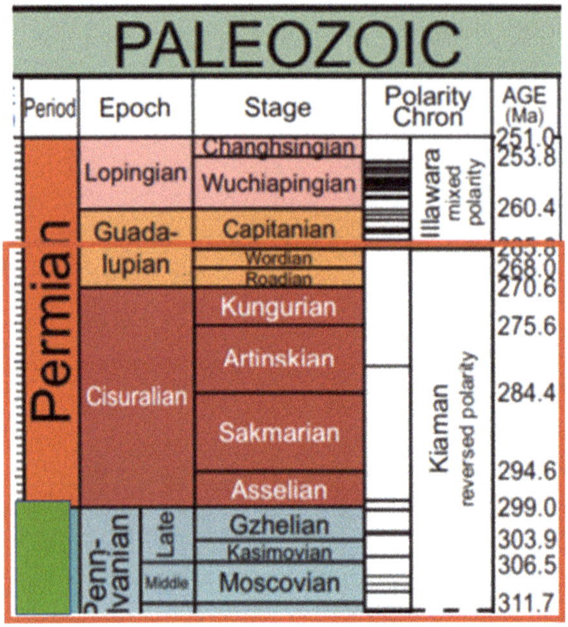

Figure 16.9 (Continued)

tectonic fit has been attempted involving rotation of themes representing rifting in the Indian part as well as those in Antarctica based on template positions (Figure 16.10).

16.2.3.1 Outcrops and Faults

In East Antarctica, the only Permo-Triassic sedimentary succession is preserved in the Lambert Graben within the Prince Charles Mountains region, which represents the southern extension of the presently NW–SE Mahanadi rift in India (Slater et al. 2015). In Enderby Land, there are exposures of granulites of the Napier and Rayner Complexes, which have been dated to be of the Archean age [51]. The Napier complex occupying most of the Enderby Land is constituted of gneissic complex and granulite facies rocks, intruded by a large number of mafic dike swarms (Mohanty 2015) that have a broad NW–SE orientation (Figure 16.9a). These dykes when rotated to make a plate tectonic fit are found to be parallel to the interpreted structural elements of the rift in the Indian part and appear to be in the same trend as the dykes mapped in Sri Lanka (Figure 16.10). As such, the fault sets along which the dykes had intruded may serve as pre-existing weak planes, which may have been reactivated during the rift phase.

16.2.3.2 Magnetic Profile Data

Magnetic field data helps in recognizing the distribution of major rifts along the continental margin and the interior of East Antarctica. In the absence/paucity of seismic and gravity surveys, magnetic data contributes the only evidence for inferring crustal structures and tectonics (Golynsky et al. 2002). Magnetic profile data along a line in the western part of Enderby Land

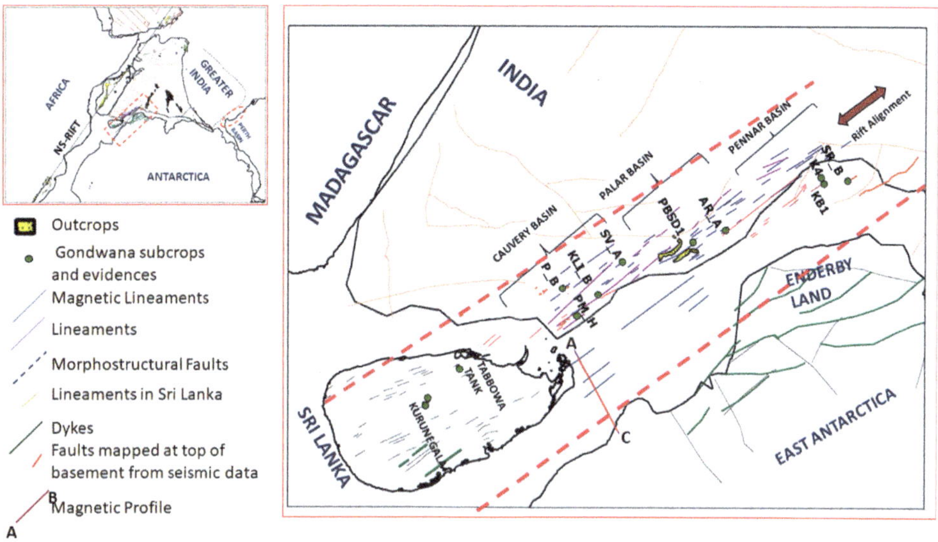

Figure 16.10 (a) GIS-based Permian plate tectonic fit involving rotation of themes representing India, Antarctica, and Sri Lanka to fit their template positions (Smith et al. 1981). Structural signatures of the proposed rift in India, Sri Lanka, and Antarctica (?) as well as offshore are found to show a parallel alignment. Magnetic profile AC near Enderbyland (Slater et al. 2015) rotated back to its Permian template position suggests the Paleozoic rift may strike/trend parallel to the proposed rift alignment. The outcrop and subcrops representing Gondwana sediments in both India (both onland and offshore) and Sri Lanka are also found to be aligned along with the proposed rift disposition. Intrusives in SV_A and the magnetic lineaments in Cauvery offshore represent volcanic/magmatic signatures (Subrahmanyam et al. 1993) related to rifting. Based on the rift signatures, the possible width of the rift is proposed and marked in red dotted lines. The inset plate tectonic fit shows the area mapped in red where the rift may exist. Also, NS rifts associated with the Indian plate (Mazumder et al. 2017); Prasad and Pundir 2017) have been marked in black dotted lines.

suggests normally magnetized two units of Archaean-Proterozoic metamorphic basement rocks, such as schists or gneisses enclosing a thick, broad wedge of reversely magnetized rocks. These enclosed rocks may be a Paleozoic infra-rift basin (Figure 16.9b), equivalents of which may occur on the conjugate East Indian coast (Stagg et al. 2004).

16.3 Discussions

A GIS-based overlay of the evidences of Gondwana sedimentations with the signatures of N–S faulting indicate that in most of the cases, there appears to be a close association between these two (Figure 16.2). Most of the outcrops or subcrops directly overlie or are adjacent to the interpreted traces of N–S faults. Conventional cores of well AR_A and SV_A, both of which overlie N–S faults, are reported to show slickensides indicating fault activity. This, in turn, implies a genetic relationship with the tectonics and sedimentation of the proposed rift. The oldest sediments that have been encountered in the rift are Talchir equivalent Early Permian sediments suggesting the origin of the rift during that time.

Further subcrops of Late Permian, Triassic to Middle Jurassic age have been encountered along the rift trend suggesting that the rift was active from Early Permian to at least Middle Jurassic times subsequent to which it may have been aborted.

The early sediments in this proposed rift, analogous to most Gondwana rifts, are glacio-fluvial in origin. This changed to a braided fluvial environment that characterizes a synrift deposition and a probable axial drainage system (Gawthorpe et al. 1997; Santos et al. 2014). This was followed by an estuarine or brackish lake environment with occasional tidal influence and nearshore indications (Basavaraju et al. 2016) implying widening of the rift basin or a marine incursion probably from the south. Subtle unconformities in Late Bajocian and between middle and late Jurassic encountered in well AR_A indicate uplifts (Raju et al. 2018), which may have resulted in abortion of the rift.

Post abortion of this rift, the India-Antarctica separation associated with a more dominant phase of rifting caused the eastern coastline to deform into a series of horst and grabens that developed oblique (~30°–40°) to the proposed rift trend (Figure 16.2). This may erode and remove these rift sediments from the uplifted areas or horsts, whereas they preserve in the grabens or basins. Also, this phase of tectonism resulted in a new fault pattern in which the earlier N–S faults plausibly reoriented or remained in small enclaves as remnant structures that have been regionally over-printed by the NE–SW fault pattern.

The Early Cretaceous India-Antarctica separation may also have caused part of the N–S rift to be preserved in offshore Enderbyland and Lützow-Holm Bay beneath Early-Cretaceous rift sediments. The magnetic profile (Figure 16.9b) in offshore Enderby Land indicates a Paleozoic rift basin between two Pre-Cambrian rift shoulders (Stagg et al. 2004). Since the supposedly rifted Paleozoic lithologic units in the profile are reversely magnetized they may fit in the Early Permian age as per the magnetostratigraphic time scale (Gradstein and Ogg 2004). GIS-based rotation of the magnetic profile line to fit its plate tectonic template suggests the units are aligned sympathetic to the structural evidences of the premised rift trend north of Sri Lanka (Figure 16.10).

Most rifts are associated with magmatism resulting from crustal stretching and thinning with zones of ponded magma centered below the rift (Maccaferri et al. 2014). The direction of minimum horizontal stress in such a tectonic setting is perpendicular to the rift and results in dykes and fissures parallel to the rift (Maccaferri et al. 2014; Wadge et al. 2016). In well SV_A (Figure 16.2), the dolerite intrusives encountered while drilling on the isotopic investigation are found to be petrogenetically older and different from volcanics/intrusives associated with younger events of India-Madagascar or India-Antarctica separation (Rathore et al. 2017). This may hint at this SV_A intrusive facies being related to the proposed rift volcanism. Additionally, the N–S magnetic lineaments in Cauvery Offshore (Figure 16.7a) suggest intrusions of high magnetization contrast and high susceptibility at 6–8 km depth, implying volcanic activity oriented along the rift axis (Subrahmanyam et al. 1993). These may serve as signatures of rift-related magmatism/volcanism associated with the proposed N–S rift.

This N–S rift will be parallel in orientation to the Gondwana rift in the Bengal Basin (Purnea-Rajmahal-Galsi) (Prasad and Pundir 2017) in which subcrop Gondwana sediments occur in N–S trending elongated grabens and are characterized with a similar sedimentation pattern at least in the early stages of development (Banerjee et al. 2013). Another envisaged N–S rift system spanning from the Arabian plate in the north and extending to the southern part of Africa that passes through the Indus Basin of Pakistan, offshore the Kutch-Saurashtra Basin of western India and Madagascar (Mazumder et al. 2017). This rift is also characterized by a similar sedimentation pattern can be considered to be coeval to this proposed rift. Note that the NNW–SSE-oriented transcurrent fault in the offshore Cauvery Basin along which Antarctica and Sri Lanka in Early Cretaceous moved southward (Lal et al. 2009) has a similar orientation to the rift trend and may be a result of reactivation of rift-related faults. Alternate theories nevertheless exist, according to which a Gondwana graben has been proposed to be aligned along the course of the Cauvery River (Prasad and Phor 2009), i.e. oriented broadly E–W. However, considering the disposition of outcrop and subcrop evidences of Gondwana only in the Cauvery and Palar Basins to represent its width dimension, it appears to be much more than even the Ethiopian-Kenyan rift (East African Rift) (Guseva 2013) and, hence, may not be plausible. In that case, there is a much higher probability of the premised rift to be oriented in an N–S direction that forms its long axis.

Considering the exploration point of view of these rift sediments, evidences of hydrocarbon have been observed during drilling of the interval assigned as Late Permian in well SR_B. Core samples in well KB were found to show hydrocarbon signatures. Similarly, in the Cauvery Basin, in well PM_H, sediments that have been classified as older than Oxfordian are also found to show hydrocarbon signatures. Geochemically, Bajocian-Bathonian sediments of well AR_A are found to show source rock characteristics. All these imply that a petroleum system may be expected to exist in the Gondwana sediments of the premised rift. An analogy could also be drawn with the Gondwana sediments of the KG-PG Basin, in which the Early Permian Kommugudem Formation forms the primary source rock and Mandapeta Sandstone of the Permo-Triassic age forms the reservoir rock of the commercial hydrocarbon producing Permo-Triassic Kommugudem-Mandapeta-Red Bed Petroleum System. A similar petroleum system also exists in the Perth Basin of Western Australia (Figure 16.10) that might be considered nearest or adjacent to the Greater Indian plate during Permo-Triassic. In this basin, Permian to Early Triassic rift-sag deposits of fluvial to marine siliciclastics, minor carbonates, and coals are associated with the primary petroleum system of the basin (Geoscience Australia 2019). Overall, Permian strata, rich in coals that are dominated by type II and III kerogens, are found to represent important natural gas source rocks in various sectors of the pan-Gondwana rift corridors (Edirisooriya et al. 2018).

16.4 Conclusions

The above analysis puts forward a model of an N–S oriented Lower Gondwana intracontinental rift basin or the ECGR oriented along the eastern boundaries of the KG, Pennar, Palar, and Cauvery basins that may have existed from Early Permian to Middle Jurassic. These rift sediments are mostly preserved in grabens formed during Lower Cretaceous rifting and may have a thickness of ~1.5 km. These sediments may serve as an additional target for hydrocarbon exploration in these basins other than the Early Cretaceous-Late Jurassic main rift phase sediments.

Similar GIS-based analysis and tectono-sedimentary reconstructions may be attempted as an analytical approach to redefining newer tectonic domains that may serve as prospective areas of hydrocarbon exploration in both space and time. The Indian coastal terrains (e.g. Misra et al. 2014) and sedimentary basins (e.g. Dasgupta and Mukherjee 2017) remain quite important to undertake such works.

Acknowledgments

The authors are grateful to ONGC and its management for the usage of its data, for reviewing the manuscript, and for the permission to publish it. Soumyajit Mukherjee (IIT Bombay) invited, handled, and reviewed this article. Merryl Le Roux, Frank Otmar Weinreich, Joss Everett, Andrew Harrison, and the proofreading team (Wiley Blackwell) are thanked for their assistance. This chapter has been encapsulated in Mukherjee et al. (2023).

The views expressed in this manuscript are solely those of the authors and not those of the organization they represent.

References

Aswal, H.S., Singh, K., and Mehrotra, N.C. (2001). Mesozoic-Cenozoic dinoflagellate cyst biostratigraphy of Krishna-Godavari basin. *ONGC Bulletin* 38 (2): 91–100.

Baillie, P., Barber, P.M., Deighton, I. et al. (2004). Petroleum systems of the deepwater MAnnar Basin, offshore Sri Lanka, Indonesian Petroleum Association. Proceedings, Deepwater and Frontier Exploration in Asia & Australasia Symposium, Jakarta (December 2004), DFE04-PO-015.

Banerjee, B., Marak, A., and Biswas, A.K. (2013). Subcrop Gondwana of Bengal Basin and their reservoir characteristics. *Journal Geological Society of India* 81: 741–754.

Baraik, A.K., Saxena, V.M., Sati, G.C. et al. (2011). Integrated analysis of available geo-scientific data to evaluate the hydrocarbon potential of Palar and Pennar basin. BRG, KDMIPE, ONGC Internal Report.

Basavaraju, M.H. (1995). Lower Gondwana palynofossils in subsurface sediments from well Pundi-2, Cauvery Basin, India. Regional Geoscience Laboratory, Cauvery Basin, Chennai, ONGC Internal Report.

Basavaraju, M.H., Jaiprakash, B.C., Chidambaram, L., and Ayyadurai, M. (2016). Biostratigraphy and depositional environments of subsurface sediments in well Arani-a, Palar basin, Tamil Nadu. *Journal Geological Society of India* 88: 407–420.

Bhattacharya, B., Bandyopadhyay, S., Mahapatra, S., and Banerjee, S. (2012). Record of tide-wave influence on the coalbearing Permian Barakar Formation, Raniganj Basin, India. *Sedimentary Geololgy* 267–268: 25–35.

Bhattacharya, H.N. and Bhattacharya, B. (2015). Lithofacies architecture and paleogeography of late Paleozoic glaciomarine Talchir Formation, Raniganj Basin, India. *Journal of Palaeogeography* 4: 269–283.

Brandt, S., Raith, M.M., Schenka, V. et al. (2014). Crustal evolution of the Southern Granulite Terrane, south India: new geochronological and geochemical data for felsic orthogneisses and granites. *Precambrian Research* 246: 91–122.

Casshyap, S. M. (1979). Pattern of sedimentation in Gondwana Basins. *Proceedings of Vth International Gondwana Symposium*, Wellington, New Zealand, (11–16 February), pp. 525–551.

Casshyap, S.M. and Tewari, R.C. (1988). Depositional models and tectonic evolution of Gondwana basins of peninsular India. *Palaeobotanist* 36: 59–66.

Chakraborty, C. and Ghosh, S. (2005). Pull-apart origin of the Satpura Gondwana basin, cntral India. *Journal of Earth System Science* 114 (3): 259–273.

Chand, S., Radhakrishna, M., and Subrahmanyam, C. (2001). India-East Antarctica conjugate margins: rift-shear tectonic setting inferred from gravity and bathymetry data. *Earth and Planetary Science Letters* 185: 225–236.

Chari, M.V., Sahu Narasimha, J.N., Banerjee, B. et al. (1995). Evolution of the Cauvery basin, India from subsidence modelling. *Marine and Petroleum Geology* 12 (6): 667–675.

Dahanayake, K., Jayasena, H.A.H., Singh, B.K. et al. (1989). A permo-triassic(?) plant microfossil assemblage from Sri Lanka. *Review of Palaeobotany and Palynology* 58: 197–203.

Dasgupta, S. and Mukherjee, S. (2017). Brittle shear tectonics in a narrow continental rift: asymmetric non-volcanic Barmer basin (Rajasthan, India). *The Journal of Geology* 125: 561–591.

Davison, I. and Underhill, J.R. (2013). Chapter 2: Tectonics and sedimentation in extensional rifts: implications for petroleum systems, tectonics and sedimentation: implications for petroleum systems. In: *Tectonics and Sedimentation in Extensional Rifts: Implications for Petroleum Systems, Tectonics and Sedimentation: Implications for Petroleum Systems*, AAPG Memoir 100 (ed. D. Gao), 33. Tulsa: AAPG.

De Lamotte, D.F., Fourdan, B., Leleu, S. et al. (2015). Style of rifting and the stages of Pangea breakup. *Tectonics* 34 (5): 1009–1029.

Desa M.A., Ismaiel M., Suresh Y., Krishna K. S., 2018, Oblique strike-slip motion off the Southeastern Continental Margin of India: implication for the separation of Sri Lanka from India, *Journal of Asian Earth Sciences*, 10.1016/j.jseaes.2018.01.015

Edirisooriya, G., Dharmagunawardhane, H.A., and Mcloughlin, S. (2018). The first record of the Permian Glossopteris flora from Sri Lanka: implications for hydrocarbon source rocks in the Mannar Basin. *Geological Magazine* 155 (4): 907–920.

Gaina C., Müller R D., Brown B., Ishihara T., Ivanov S., 2007, Breakup and early seafloor spreading between India and Antarctica, *Geophysical Journal International*, Volume 170, Issue 1, pp. 151–169, https://doi.org/10.1111/j.1365-246X.2007.03450.x

Gawthorpe, R.L., Sharp, I., Underhill, J.R., and Gupta, S. (1997). Linked sequence stratigraphic and structural evolution of propagating normal faults. *Geology* 25 (9): 795–798.

Geochemistry Group (2017). Adsorbed gas anomaly atlas of Indian sedimentary basins. KDMIPE, ONGC Internal Report.

Geoscience Australia (2019). Perth basin. www.ga.gov.au (accessed 4th January, 2019).

Gibbons, A. D., Whittaker J. M., and Muller R. D., 2013, The breakup of East Gondwana: assimilating constraints from Cretaceous ocean basins around India into a best-fit tectonic model, *Journal of Geophysical Research – Solid Earth*, 118, 808–822, doi:https://doi.org/10.1002/jgrb.50079.

Golynsky, A.V., Alyavdin, S.V., Masolov, V.N. et al. (2002). The composite magnetic anomaly map of the East Antarctic. *Tectonophysics* 347: 109–120.

Gradstein, F.M. and Ogg, J.G. (2004). Geologic time scale 2004 – why, how, and where next! *Lethaia* 37 (2): 175–181. https://doi.org/10.1080/00241160410006483.

Guseva, E.N. (2013). Comparative morphometric analysis of the rifts of Venus (Atla, Betaphoebe) and Earth (Kenyan-Ethiopian Rift). *44th Lunar and Planetary Science Conference*, The Woodlands, TX (18–22 March), 2258.pdf.

Katupotha, J. (2013). Planated surfaces and erratic boulders of Sri Lanka, Geological Society of Sri Lanka, at: IFS, Kandi, Sri Lanka. *Proceedings to 29th Technical Sessions*, Kandy, Sri Lanka (22 February 2013) 10.13140/RG.2.1.3088.6886.

Kularathna, E.K.C.W., Pitawala, H.M.T.G.A., Senaratne, A., and Weerasinghe, D. (2015). Gondwana breakup and development of sedimentary basins in Sri Lanka. *Proceedings of the 31st Technical Session of Geological Society of Sri Lanka*, Peradeniya, Sri Lanka (27 February 2013), GSSL-2015-06.

Kumar, N., Singh, A.P., and Singh, B. (2009). Structural fabric of the southern Indian shield as defined by gravity trends. *Journal of Asian Earth Sciences* 34: 577–585.

Kumara, S.M.P.G.S. and Dharmagunawardhane, H.A. (2014). A geostructural model for the Nelumwewa thermal spring: North central province, Sri Lanka. *Journal of Geological Society of Sri Lanka* 16: 19–27.

Kumaran, C.J., Saravanavel, J., Palanivel, K., et al. (2012). Uniqueness of ArcGIS 3D analyst module in exploring, analyzing and interpreting the basement structures in Cauvery Deltaic Region, Tamil Nadu, India. *13th Esri India User Conference,* New Delhi, India (5–6 December 2012). Noida: ESRI.

Lal, N.K., Siawal, A., and Kaul, A.K. (2009). Evolution of East Coast of India – a plate tectonic reconstruction. *Journal of Geological Society of India* 73: 249–260.

Langhi, L. and Borel G. D. (2004). Influence of the Neotethys rifting on the development of the Dampier Subbasin (North West Shelf of Australia), highlighted by subsidence modelling. *Tectonophysics* 397 (1): 93–111.

Maccaferri, F., Rivalta, E., Keir, D., and Acocella, V. (2014). Off-rift volcanism in rift zones determined by crustal unloading. *Nature Geoscience* 7: 297–300.

Masoud, A.A. and Koike, K. (2011). Auto-detection and integration of tectonically significant lineaments from SRTM DEM and remotely-sensed geophysical data. *ISPRS Journal of Photogrammetry and Remote Sensing* 66: 818–832.

Mazumder, S., Adhikari, K., Mitra, D.S. et al. (2016). A neotectonic based geomorphic analysis using remote sensing data to delineate potential areas of hydrocarbon exploration: Cachar area, Assam. *Journal of Geological Society of India* 88 (1): 87–97.

Mazumder, S., Dave, H.D., and Mitra, D.S. (2013a). A neotectonic based approach supplementing conventional seismic exploration in Bengal Basin with special reference to hydrocarbon migration and accumulation. *ONGC Bulletin* 48 (4): 100–111.

Mazumder, S., Tep, B., Pangtey, K.K.S. et al. (2017). Probable existence of a Gondwana transcontinental rift system in western India: implications in hydrocarbon exploration in Kutch and Saurashtra offshore: a GIS-based approach. *Journal of Earth System Sciences* 126: 81. Indian Academy of Sciences, https://doi.org/10.1007/s12040-017-0866-3.

Mazumder, S., Tep, B., Pangtey, K.K.S., and Mitra, D.S. (2013b). A morphotectonic based approach to derive tectonic framework and possible areas of hydrocarbon accumulation in onshore Palar Basin. *ONGC Bulletin* 48 (5): 112–123.

Mikhalsky, E., Krylov, D., Rodionov, N. et al. (2017). Refined geological history of the polyphase plutono-metamorphic complex in the Thala Hills area (Enderby Land, East Antarctica) from zircon SHRIMP dating and implications for Neoproterozoic amalgamation of Gondwanaland.

In: *Crustal Evolution of India and Antarctica: The Supercontinent Connection, Geological Society* (ed. N.C. Pant and S. Dasgupta), 457. London: Special Publications https://doi.org/10.1144/SP457.2.

Misra, A.A., Bhattacharya, G., Mukherjee, S., and Bose, N. (2014). Near N–S paleo-extension in the western Deccan region in India: does it link strike-slip tectonics with India-Seychelles rifting? *International Journal of Earth Sciences.* 103: 1645–1680.

Misra, A.A. and Mukherjee, S. (2015). *Tectonic Inheritance in Continental Rifts and Passive Margins*. Springerbriefs in Earth Sciences ISBN 978-3-319-20576-2.

Mohanty, S. (2015). Precambrian continent assembly and dispersal events of south Indian and East Antarctic shields. *International Geology Review* 57 (16): 1992–2027. https://doi.org/10.1080/00206814.2015.1048751.

Mukherjee, S., Dasgupta, S., Majumdar, C. et al. (2023). Introduction to: "petroleum industry updates: geoscientific and economic issues". In: *Petroleum Industry Updates: Geoscientific and Economic Issues* (ed. S. Mukherjee, S. Dasgupta, C. Majumdar, et al.). John Wiley & Sons, Hoboken, ISBN 978-1-119-68003-1.

Murty, G.P.S., Subrahmanyam, A.S., Murthy, K.S.R., and Sarma, K.V.L.N.S. (2002). Evidence of fault reactivation off Pondicherry coast from marine geophysical data. *Current Science* 83 (12): 1446–1449.

Nagendra, R. and Reddy, A.N. (2017). Major geologic events of the Cauvery Basin, India and their correlation with global signatures – a review. *Journal of Palaeogeography* 6 (1): 69–83.

Nagraj, M. and Vijay, M.M. (2014). Conventional core report CC3 and CC5 of well Arani-1. Regional Geoscience Laboratory, Cauvery Basin, Chennai, ONGC Internal Report.

Padhy, P. K., Das, S. K., Kumar, A. (2015). Krishna-godavari continental rift basin: shale gas and oil play systems. *10th Biennial International Conference & Exposition*, SPG Kochi (23–25 November 2013).Dehradun, India: Society of Petroleum Geophysicists, p. 327.

Prakash, S.R., Ramasamy, S., and Varghese, N.M. (2018). Provenance of the Gondwana sediments, Palar Basin, Southern India. *Arabian Journal of Geosciences* 11: 163. https://doi.org/10.1007 s12517-018-3494-6.

Prasad, B. and Phor, L. (1998). Palynostratigraphy of pre-cretaceous subsurface sediments of Cauvery Basin, Palynology Laboratory, Geology Division. KDM Institute of Petroleum Exploration, ONGC Internal Report.

Prasad, B. and Phor, L. (2009). Palynostratigrapity of the subsurface Gondwana and Post-Gondwana Mesozoics of the Cauvery Basin, India. *Journal of the Palaeontological Society of India* 54 (1): 4I–71I.

Prasad, B. and Pundir, B.S. (2017). Gondwana biostratigraphy of the Purnea Basin (Eastern Bihar, India), and its correlation with Rajmahal and Bengal Gondwana Basins. *Journal of Geological Society of India* 90 (4): 385–510.

PRDS (2013). Blocks offered in Sri Lanka's Second Licensing Round, Petroleum Resources Development Secretariat, Technical Bid Round Summary. http://www.prds-srilanka.com.

Premarathne, U., Suzuki, N., and Ratnayake, N.P. (2015). Modelling of hydrocarbon generation, migration and accumulation in the northern part of the Mannar Basin, offshore Sri Lanka. *Journal of Geological Society of Sri Lanka* 17, J.W. Herath Felicitation Volume: 139–158.

Raju, D. S. N., Bhatia M. L., Ali Q. A. et al. (1994). Sea-level changes, anoxic conditiond and their relation to hydrocarbon occurrences in KG and Cauvery basins. KDMIPE report no. GR1.03. R309, ONGC, Dehradun.

Raju, D.S.N., Reddy, A.N., Gilbert, H. et al. (2018). Age and paleoenvironmental evolution of the syn-rift fill sediments in East Coast basins of India. *ONGC Bulletin* 52 (2): 127–148.

Rangaraju, M.K., Agrawal, A., and Prabhakar, K.N. (1993). Tectono-stratigraphy, structural styles, evolutionary model and hydrocarbon habitat, Cauvery and Palar basins. In: *Proceeding. of second seminar on Petroliferous basins of India*, vol. 1 (ed. S.K. Biswas et al.), 371–388. Dehra Dun, India: Indian Petroleum Publishers.

Rathore, S.S. Gupta, P., Uniyal, G.C., et al. 2017, Genetic characterization and petrogenesis of basalts, east Coast basins. KDMIPE, Geology Group, ONGC Internal Report.

Ratnayake, A.S., Kularathne, C.W., and Sampeil Y. (2018). Assessment of hydrocarbon generation potential and thermal maturity of the offshore Mannar Basin, Sri Lanka. *Journal of Petroleum Exploration and Production Technology* 8: 641–654.

Reddy, P.R. and Koteswara, C. (2004). Seismicity in and around Ongole, Andhra Pradesh – an appraisal. *Journal of Indian Geophysical Union* 8 (2): 143–146.

Santos, M.G.M., Almeida, R.P., Godinho, L.P.S. et al. (2014). Distinct styles of fluvial deposition in a Cambrian rift basin. *Sedimentology* 61 (4): 881–891. ISSN 0037-0746.

Saravanavel, J., and Ramasamy, S.M. (2003). Probable seismo tectonic lineaments in South India. Conference: Annual Convection of ISRS and National Conference on Remote Sensing, CESS at: Trivandrum (9–12 December 2003).

Schumm, S.A., Dumont, J.F., and Holbrook, J.M. (2002). *Active Tectonics and Alluvial Rivers*. United Kingdom: Cambridge University Press.

Shandilya, A., Modak, R.S., Manish, K.P., and Vishnu Kumar, C.P. (2017). KG-PG basin: a synrift conundrum. *ONGC Bulletin* 52 (2): 103–116.

Shurr, G.W. (1982). Geological significance of lineaments interpreted from LANDSAT images near the Northern Black Hills. *Proceedings of Fourth International Williston Basin Symposium*, Regina, Saskatchewan (5–7 October 1982). Tulsa: AAPG, pp.313–320.

Singha, S., Prakash, R., Sahoo, A. et al. (2017). Syn-rift exploration in onland part of Cauvery Basin. *ONGC Bulletin* 52 (2): 39–60.

Slater, B.J., McLoughlin, S., and Hilton, J. (2015). A high-latitude Gondwanan lagerstatte: the Permian permineralised peat biota of the Prince Charles Mountains, Antarctica. *Gondwana Research* 27: 1446–1473.

Smith, A.G., Hurley, A.M., and Briden, J.C. (1981). *Phanerozoic Paleocontinental World Maps*. Cambridge University Press.

Stagg, H.M.J., Colwel, J.B., Direen, N.G. et al. (2004). Geology of the continental margin of Enderby and Mac. Robertson Lands, East Antarctica: Insights from a regional data set. *Marine Geophysical Researches* 25: 183–219.

Subrahmanyam, A.S., Lakshminarayana, S., Chandrasekhar, D.V. et al. (1993). Offshore structural trends from magnetic data over Cauvery Basin, East Coast of India. *Journal of Geological Society of India* 46: 269–273.

Subrahmanyam, C. and Chand, S. (2006). Evolution of the passive continental margins of India – a geophysical appraisal. *Gondwana Research* 10 (1–2): 167–178.

Tewari, R.C. (1998). Channel sandstone bodies in fluvial Permian–Triassic Gondwana succession of peninsular, India. *Journal of Geological Society of India* 51: 747–754.

Tripathi, A. and Vijaya (1997). Biostratigraphic age-determination of Sriperumbudur formation in borehole PBSD-1, Palar Basin, Tamil Nadu: a reassessment. *Journal of the Palaeontological Society of India* 42: 101–115.

Vairavan, V. (1993). Tectonic history and hydrocarbon prospects of Palar and Pennar basins, India. In: *Proceeding of second seminar on Petroliferous basins of India*, vol. 1 (ed. S.K. Biswas et al.), 391–396. Dehra Dun, India: Indian Petroleum Publishers.

Vitanage, P.W. (1985). Tectonics and mineralization in Sri Lanka. *Bulletin of Geological Society of Finland* 57 (Part 1–2): 157–168.

Wadge G., Biggs J., Lloyd R., Kendall J.-M., 2016, Historical volcanism and the state of stress in the East African Rift System, *Frontiers in Earth Sciences*, https://doi.org/10.3389/feart.2016.00086

Yadagiri, K. (2017). Reservoir characterization of synrift cores from KGS092NASRI-2, Shallow offshore, Ongole coast, Krishna-Godavari Basin. AP, Regional Geoscience Laboratory, KG-PG Basin, ONGC Internal Report.

Zaid, S.M. and Gahtani, F.A. (2015). Provenance, diagenesis, tectonic setting, and geochemistry of Hawkesbury Sandstone (Middle Triassic), southern Sydney Basin, Australia. *Turkish Journal of Earth Sciences* https://doi.org/10.3906/yer-1407-5, pp 72–98.

17

The Upper Assam Basin, Its Evolution, and Modification: A Review

Devojit Bezbaruah, Tapos Kumar Goswami, and Ranjan Kumar Sarmah*

Department of Applied Geology, Dibrugarh University, Dibrugarh, Assam, India

**Author for correspondence: taposgoswami@gmail.com*

Abstract

The Upper Assam Basin is divided into two parts: the Assam shelf and the Belt of Schuppen. During the Late Cretaceous, the Upper Assam Basin was initiated as a passive margin basin in the southern fringe of the continental shelf of the Assam-Meghalaya craton. During Paleocene-Eocene, with the WNW-ward advancement of the Burmese plate, an active margin was developed close to the Burmese plate beneath which the Indian plate subducted subsequently. Sedimentation commenced on this active margin. Toward the west, sedimentation also continued in the passive Upper Assam Basin coevally. As the Indian plate subducted, the sediments of the active margin basin came juxtaposed with sediments of the passive margin basin during the Oligocene Period. The Himalayan orogeny tilted the basin toward the north during the Miocene Period. Existing geological data propounds that from the Miocene onward, the Upper Assam Basin became an intermontane basin due to the amalgamation of the Himalayan foredeep and the Naga foredeep.

Keywords *Upper Assam Basin; Disang Thrust; Naga Thrust; Main Boundary Thrust; Himalayan Frontal Thrust*

17.1 Introduction

In a collisional tectonic setting, the progressive plate convergence is associated with deep marine, shallow marine, and sub-areal alluvial sediments (Sinclair 1997). This passage from marine to continental in the peripheral foreland basin is driven by progressive thrust propagation toward the foreland (Allen and Homewood 1986; Chakraborty et al. 2019). The Indo-Burmese collision thus constituted the foreland basin geometry preceded by passive margin sedimentation in the Assam and Assam-Arakan Basin. In this article, we

Handbook of Petroleum Geoscience: Exploration, Characterization, and Exploitation of Hydrocarbon Reservoirs, First Edition. Edited by Soumyajit Mukherjee, Swagato Dasgupta, Chandan Majumdar, Subhandip Mandal and Troyee Dasgupta.
© 2023 John Wiley & Sons Ltd. Published 2023 by John Wiley & Sons Ltd.

discuss briefly the evolution of the Upper Assam Basin where sediments record the involvement of shelf, slope, and deep marine facies.

The Upper Assam Basin is a foreland basin bordered in the north by the Himalayan orogen, in the east by the Mishmi block, and in the south by the Naga Schppen belt. The discovery of oil at Digboi in Upper Assam placed the state as an oil territory in 1889. The detection of oil located in a concealed structure by seismic survey in Naharkatiya-1 by Oil India Limited in 1953 is considered another milestone in the oil exploration in the region. The fact that the oil-bearing areas are extensive in the region is proved by subsequent discoveries of oil fields in the Surma valley by the Burma Oil Company (Krishna Rao and Rajkumar 1984). In Upper Assam Basin, the oil fields are divided into two significant groups: oil fields in the thrust sheets (Digboi and Kharsang) and oil fields in the foreland zone (Moran, Naharkatiya, Jorajan, Rudrsagar, Lakwa, and Galeki) (Handique and Borgohain 1984). Because of the discovery of the oil fields and the prospects for future hydrocarbon exploration and extraction, the Upper Assam Basin claims a unique status in the petroleum geology and tectonics of basin evolution in the region (Wandrey 2004).

The Upper Assam Basin is bounded by the three orogenic belts – the Himalayan belt in the north, the Mishmi belt in the northeast, and the Naga Schuppen belt in the south and southeast. The upper Assam Basin or Assam shelf foreland basin part is a flood-plain area of the river Brahmaputra (Figure 17.1). Sediments in the basin are 3–10 km thick. In the entire northeast India, thick pile of sediments ranging in age from Permo-Triassic Gondwana to Recent decode the changes in the depositional environment from marine to deltaic to fluviatile. This change in the facies architecture in sedimentary sequences conform to changing episodes of basin modifications. For example, within the sequence, a thick pile of recent

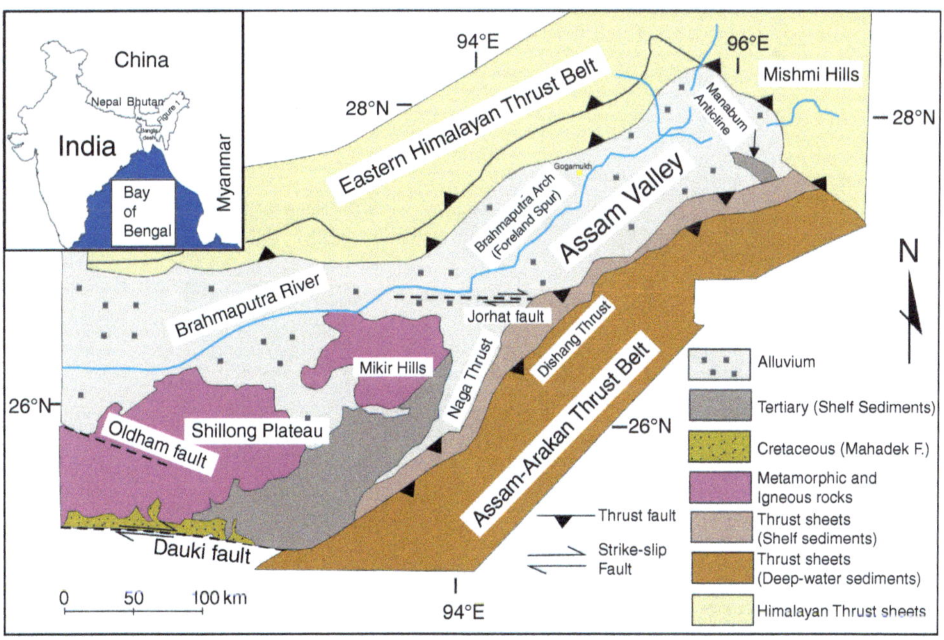

Figure 17.1 Regional map of the area modified from Berger et al. (1983) and Kent and Dasgupta (2004). The geographic location of the Upper Assam Basin (Assam valley) with respect to the surrounding countries is shown. The location of the area is further shown within a rectangle in the map of India (inset).

sediments of fluvial origin up to 2400 m has been recorded near Gogamukh (Nath and Krishna Rao 1981). But the present elevation of the Gogamukh area is ~99 m above the mean sea level. In most of the areas in Upper Assam, the recent fluvial sediments are >900 m thick. But the elevation of the area does not exceed 120 m above MSL. This indicates that for accommodating this thick pile of fluvial sediments, the basin is constantly sinking. The sedimentation history in the Upper Assam Basin and modification of the basin on the backdrop of collisional tectonics through time is a complex process. We address some of the key features of the formation, sedimentation, and modification of the Upper Assam Basin.

17.2 Basin Evolution

During the Early Cretaceous, the northern extent of the Indian plate is a landmass now covered by a part of the crystallines of the Lesser and the Higher Himalayas. Toward the northeast part and toward the south-southeastern edge of the Indian plate, there was the extended shoreline of the Bay of Bengal (Naik et al. 2004). Thus, the Shillong plateau, Mikir plateau, and the basement of the Upper Assam Basins together formed a single landmass (Figure 17.2 – shown as crystallines and metasediments). During the Upper Cretaceous, shallow marine sediments were deposited in the northern part of the basin (Figure 17.2). Toward the Naga hill side (south-southeastern part in Figure 17.2), the deeper part of the sea was present and deep marine sedimentation took place. Therefore, the oldest sediments toward the Indian basement craton or toward the northern part of the Assam-Arakan basin are Upper Cretaceous. These sediments are reported from three different localities of the area and accordingly exhibit three different facies. In the western part of the basin,

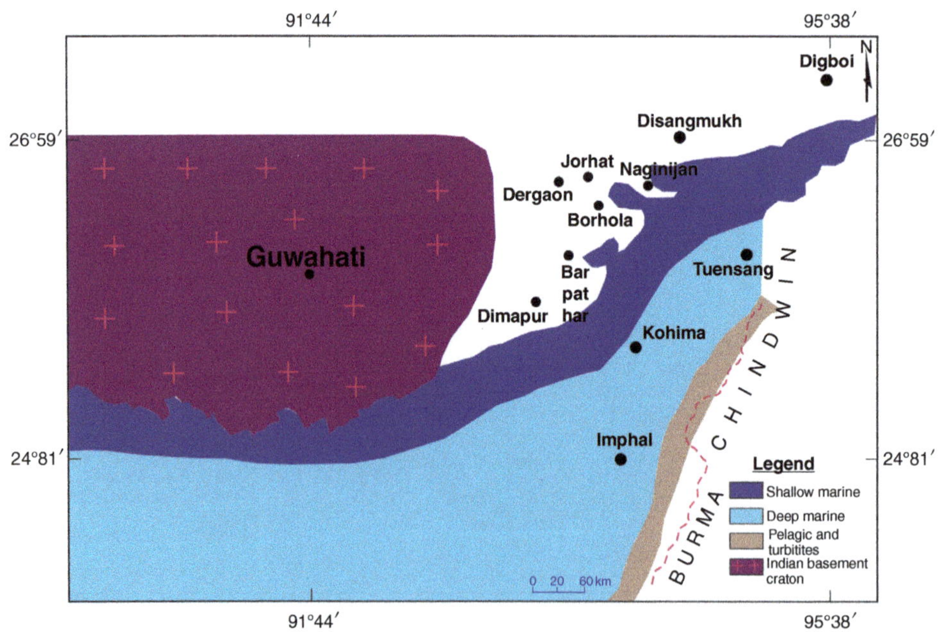

Figure 17.2 Generalized stratigraphic map – Upper Cretaceous. Assam and Assam-Arakan-fold belt. *Source:* Adapted from Naik et al. (2004).

sediments indicating the basin margin facies are located at Dergaon, Upper Assam plains (Figure 17.2). The flysch facies sediments are mapped in the eastern part of Nagaland and Manipur (Kohima and Imphal, Figure 17.2). Finally, the sediments connoting the pelagic facies are associated with the ophiolites of Nagaland and Manipur (Figure 17.2). Toward the north of the Indian plate, the sedimentation in the Tethys basin continued up to Eocene. The closure of the Tethys basin is marked by a transition from a lacustrine to a fluvial environment (Dewey et al. 1988). This change in sedimentation indicates that the Tethyan basin closed by the end of the Eocene and this, in turn, developed a continental sedimentary basin. Further, this indicates that by the end of Eocene, the intercontinental collision between the Indian and Asian plates had initiated. The river system (resulted in the process) drained the entire uplifted northern part and similar river systems also developed in the Tethyan basin (Mandal and Dasgupta 2013).

In the southern part, the sediments contributed from the Indian plate continued to deposit in the then sea between the Indian plate and the Burmese plate (Figure 17.2). The sedimentation continued from the Upper Cretaceous to the Eocene time and formed the sequence of Mahadek Formation of Upper Cretaceous and Jaintia Group of Eocene of Shillong and Mikir hills (Das Gupta and Biswas 2000) (Table 17.1). Contemporaneous

Table 17.1 The Tertiary Succession of Assam (Mathur and Evans 1964).

Age (approx.)	Group	Formation and Local Facies		
		Geosynclinal sediments		Shelf sediments
		Surma valley	Upper Assam and Naga Hills	
Recent and Pleistocene		Alluvium and high level terraces	Alluvium and high level terraces	Alluvium and high level terraces
		UNCONFORMITY		
Pliocene	Dihing	Not Subdivided	Not Subdivided	Dhekiajuli
		UNCONFORMITY		
MIO-PLIOCENE	Dupi Tila	Upper Dupi Tila Lower Dupi Tila	Namsang	Namsang
		UNCONFORMITY		
Miocene	Tipam	Girujan Clay Tipam Sandstone	Girujan Clay Tipam Sandstone	Girujan Clay Tipam Sandstone`
	Surma		Not Subdivided	Not Subdivided
		UNCONFORMITY		
Oligocene	Barail	Renji Jenam Laisang	Tikak Parbat Bargolai Naogaon }	Not Subdivided
Eocene	Disang			Jaintia { Kopili Sylhet Limestone Theria

The rank terms (Series, Stages etc.) are modified after Chakraborty and Bakshi (1972) and Bhandari et al. (1973).

Eocene sedimentation adjacent to the continental shelf of the Burmese plate developed the flysoid Disang Group of rocks and continued compression toward the eastern part of the Indian plate caused shallowing of the basin with the deposition of sub-flysoid Barails in Oligocene (Nandy 2001). The post-Indo-Burmese collision resulted the Disang Thrust (DT) in the buckled Indian plate, due to which the upliftment of Paleocene-Eocene Disang sediments took place (Figure 17.3). In other words, the part of the Assam-Arakan basin close to the Burmese plate got involved in the Naga orogeny. Disang sediments of the Eocene age are a tectonic mélange toward the southeastern parts of the Naga Hills and it is associated with the Naga ophiolites (Acharyya et al. 1991). This association indicates that the origin of the Disang sediments was close to or these sediments may be derived from the Burmese plate under which the subduction of the Indian plate was taking place during the time.

During the closure of the existing basin between the Indian and Burmese plates, sedimentation continued in the frontal basin of the DT. We construe that sediment supply to this basin was due to the uplifted Himalayan terrain in the north and from the uplifted Disang sediments, ophiolites, and the Naga metamorphics, which formed the shoulders of the basin. Therefore, extensive deltaic conditions prevailed in the Upper Assam Basin during the Oligocene Period (Figure 17.4a) (Ranga Rao 1983). On the other hand, toward the

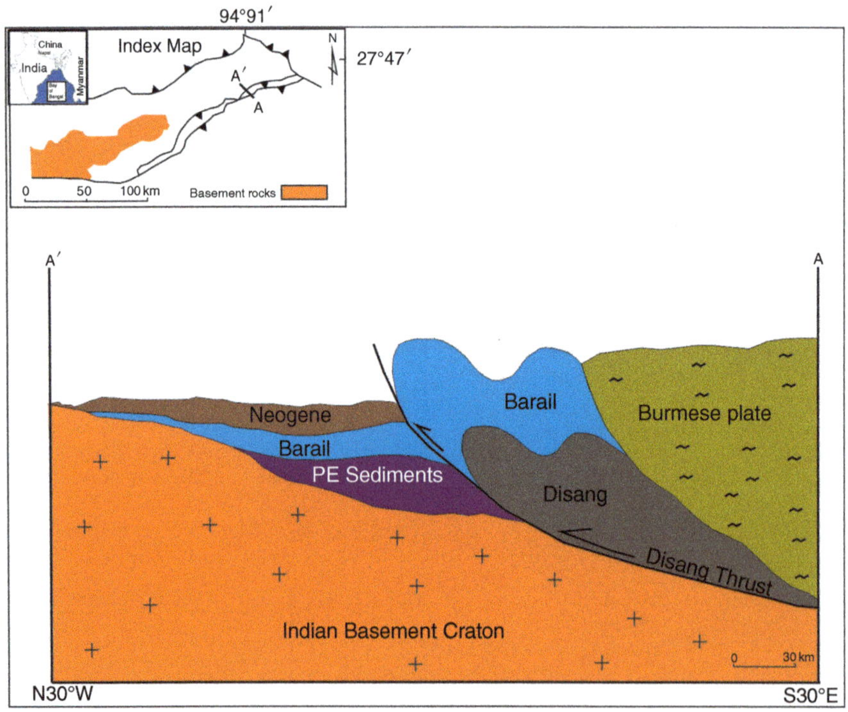

Figure 17.3 Schematic diagram of cross sectional view of thrusting of the Paleocene-Eocene sediments close to the Burmese plate (PE Sediments: Paleocene–Eocene sediments) *Source:* Adapted from Mathur and Evans (1964); Raju (1968); Dasgupta and Biswas (2000); Mandal and Dasgupta (2013).

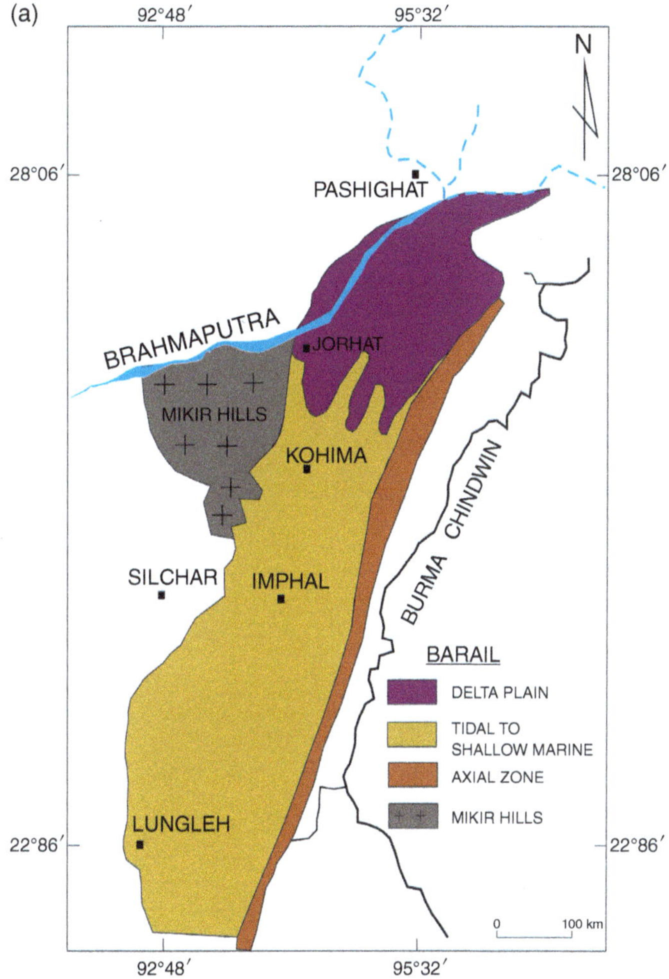

Figure 17.4 (a) Paleogeographic map during Oligocene time. (b) Paleogeographic map during Early Miocene time. The formation of tidal to shallow marine Surma Group of rocks indicates the regressive nature of the sea during the early Miocene. *Source:* Adapted from Ranga Rao (1983).

southwestern part of the basin, marine condition prevailed (Ranga Rao 1983). The presence of the Upper Cretaceous, Eocene and Oligocene sediments in the Upper Assam Basin is recorded in the drilled wells. Considering the distribution of the sediments from the Upper Assam Basin up to the Bengal Basin, it has been observed that there was regression of the sea from the northeast side to the southwest side, and this phase of regression from Oligocene (Barail) to the early part of the Miocene is coeval with the formation of the Surma Group (marine) (Figure 17.4b) (Ranga Rao 1983; Uddin 2007). The extensive development of the Surma Group (tidal to shallow marine) in the Surma Valley on the (southwest) side further indicates the regressive nature of the sea during the time (Figure 17.4b). This regression might have occurred because of the uplift of the Oligocene sediment due to the closing

(b)

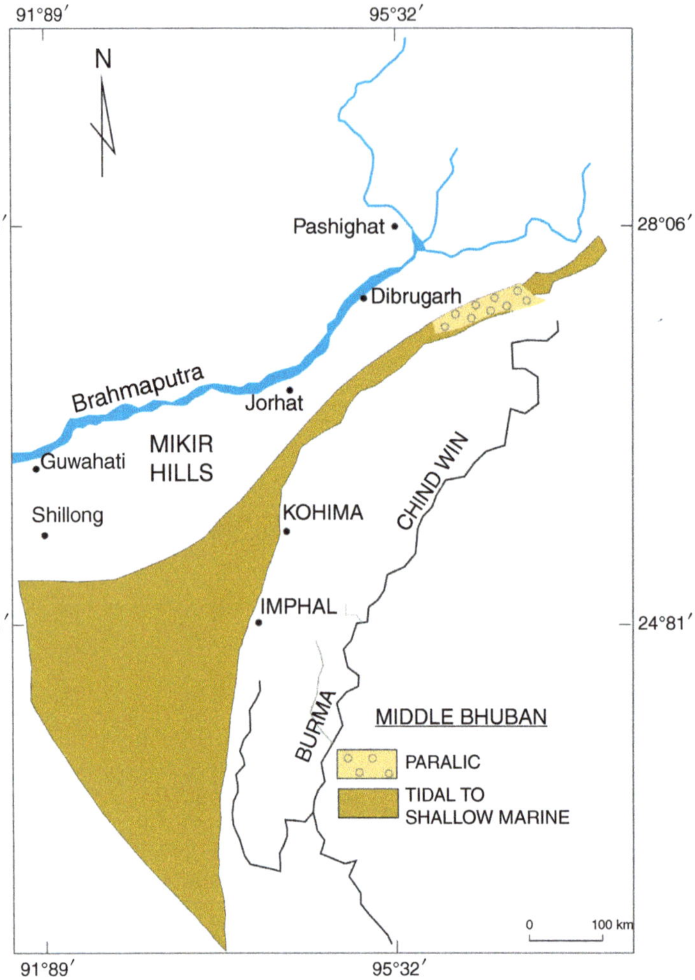

Figure 17.4 (Continued)

of the northeastern part of the basin as the Indian plate continued to subduct below the Burmese plate during this period. In the frontal part of the DT, the Barail sediments were uplifted by DT and splayed in response to the WSW-directed compression due to the Indo-Burmese subduction and collision. On the other hand, in the southwestern part of the basin, marine conditions prevailed during the same time where the Surma sediments were deposited. This process of regression continued and the complete closure of the marine basin from the northeastern part took place by the end of the Oligocene to the Early Miocene (Bhattacharjee 1991). The southwestern part of the "sea inlet" continued to receive sediments during this time. In the Naga fold and Thrust belt, the continued compression developed the first foredeep of the Naga orogeny, in the frontal part of the thrust and folded Barail sequence, and the frontal Neogene molasse basin is thus developed. The phase of initial sagging of the basin coincided with the emplacement of the DT. Thus, the eastern margin of this foredeep is marked by DT. The DT extends over 480 km along NE–SW (Srinivasan 2007). Simultaneously, the development of Main Boundary Thrust (MBT) in the Himalayan side also took place (Mandal and Dasgupta 2013). The initial displacement along the MBT occurred at ~11 Ma at a time when there is a substantial increase in the rate of sediment accumulation, the sandstone-shale ratio, and the formation of amalgamated sandstone bodies (Valdiya 2010, Yin et al. 2010). The emplacement of MBT also led to the development of a foredeep in the northern part of the basin. The formation of two foredeeps – in the north (Siwalik basin) and in the south (frontal Neogene basin) – respectively indicates sagging both in the northern and southern parts of the basin and the eventual formation of a "high" in between these basins (Das Gupta and Biswas 2000). The "high" between these two foredeeps is known as the "Brahmaputra arch" trending E-W and extends for ~300 km in the Upper Assam Basin (Das Gupta and Biswas 2000). In the Naga foredeep, the Neogene sediments were deposited during the Miocene Period and were coeval with the deposition of Siwalik sediments in the Himalayan foredeep (Figure 17.5). During this period, the Upper Assam Basin was formed by merging these two foredeeps. Integration of the two foredeeps is a major modification of the Upper Assam sedimentary basin history during its evolutionary history. Therefore, basin modification during the Miocene time is crucial. It would be worth noting that during post-Pleistocene, foredeeps are formed in the regions of frontal Naga Thrust (NT) and frontal HFT separated by the Brahmaputra arch. In the Upper Assam Basin, during Mio-Pliocene, the fluvial to lacustrine sediments were deposited (Ranga Rao 1983) (Figure 17.6). Due to the continuous growth of the south vergent thrusts in the

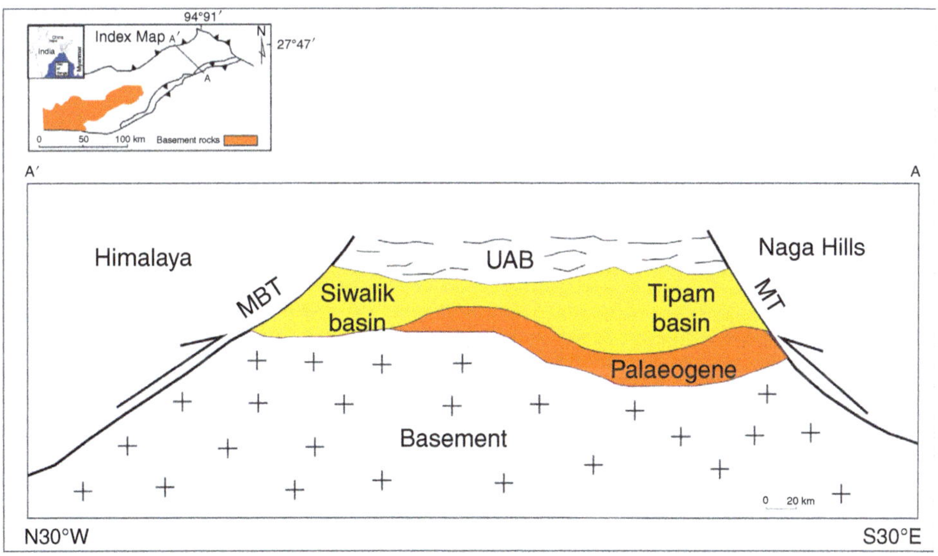

Figure 17.5 Schematic cross section showing the formation of Mio-Pliocene Siwalik and Tipam basin (MBT: Main Boundary Thrust, MT: Margherita Thrust, UAB: Upper Assam Basin). Section line A–A' is shown in the index map.

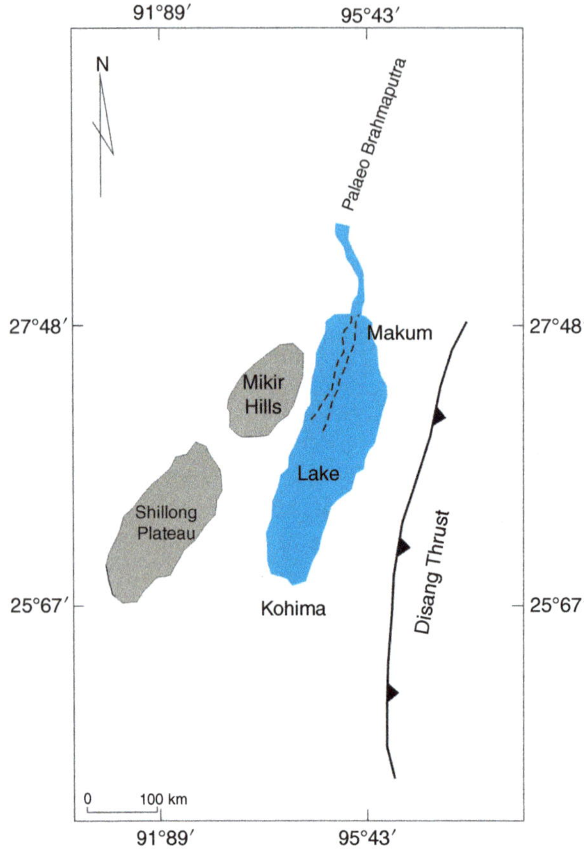

Figure 17.6 Schematic diagram showing paleogeography during Mio-Pliocene. Locations of Shillong Plateau, Mikir Hills, Paleo-Brahmaputra, and Disang Thrusts are shown.

Himalayas in the north and northwest vergent thrust in the Naga Hills in the southwest, the sagging of the basin and the sediment-fill continued. During the Late Pliocene, the Siwalik sediments were thrust over the sediments of the Upper Assam along the Himalayan Frontal Thrust (HFT) (Nakata 1972). This thrusting developed the sag in the south of the HFT. In the southwestern part, whereas, overthrusting of the sediments along the NT led to the sagging of the Upper Assam Basin toward the northwestern part. It is interpreted that the emplacement of the Tipam and Dihing Group of rocks over the Brahmaputra alluvium by the NT might have taken place during Pleistocene/Post-Pliocene time (Das Gupta and Biswas 2000). On the basis of field documentation by the authors and existing data, it may be well expounded that the NT cannot form a major decollement surface for the imbricate structures to develop in the Upper Assam Basin, unlike reported by Uddin et al. (2007). As can be seen in Figure 17.7, initially, the DT imbricated from a major decollement surface. The emplacement of DT is followed by the subsequent tectonic activity due to continued Indo-Burmese post-collisional compression. As a result, subsequent thrusts of

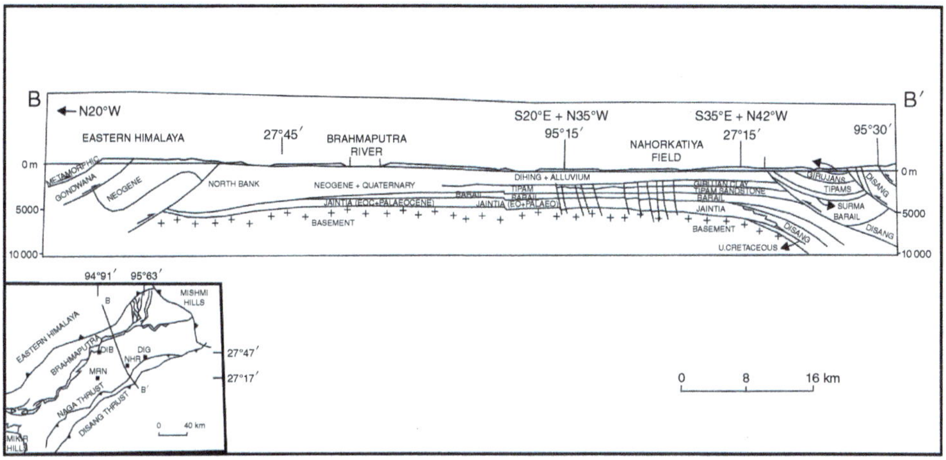

Figure 17.7 Generalized section across the Assam Valley showing possible structural and stratigraphic configuration. *Source:* Adapted from Das Gupta (1976); Das Gupta and Biswas (2000). Inset: the index map: DIG, Digboi; DIB, Dibrugarh; NHR, Naharkatiya; MRN, Moran.

the Belt of Schuppen from SE to NW came into existence, the youngest being the NT emplaced at post-Pleistocene time (Figure 17.7). Due to continued overthrusting from north and SE, the whole area gradually sunk further, which led to the deposition of thick Quaternary sediments in the Upper Assam Basin. This process of sediment fill in the basin is still continuing.

There is hardly any record of earthquake epicentral location within the Upper Assam Basin. The area shows very low seismicity in comparison to the dense epicentral plots in the northern, eastern, and southern parts of the basin and this phenomenon is described as the Assam Gap (Khattri et al. 1983; Nandy 2001; Angelier and Baruah, 2009). This low seismicity may be attributed to the fact that subsequent to the overthrusting of the tertiary sequences both along the NT and the HFT over the Brahmaputra alluvium, there is remarkably less accumulation of stress taking place in the area (Figure 17.8).

From Eocene to Oligocene, the sediment thickness toward the Naga Hills was more and it tapers toward the north or Himalayan side. However, during Miocene to recent the sediment pile thickened toward the Himalayan side and thinned toward the southeastern side or the Naga Hill side (Nath and Krishna Rao 1981; Das Gupta and Biswas 2000) (Figure 17.9). This indicates that during the post-Oligocene time, the rate of subsidence of the Himalayan foredeep was more than the sedimentation in the foredeep formed due to the Naga orogeny.

During the second phase of the Naga orogeny, the Barail sediments were folded (Bhattacharjee 1991), i.e. a part of the Upper Assam Basin got involved in the WNW-directed compression responsible for the evolution of the Naga orogeny (Figure 17.2). The marginal part of the Barail basin in the Upper Assam Shelf and its adjoining belt of Schuppen has been tilted and exposed to subareal erosion as a consequence of a major uplift of the Indo-Burma Ranges (IBR) that occurred around the Oligocene-Miocene boundary (Najman et al. 2019). This upheaval event might have caused the post-Oligocene unconformity at the southern edge of the basin, while in the deeper part of the basin, correlative

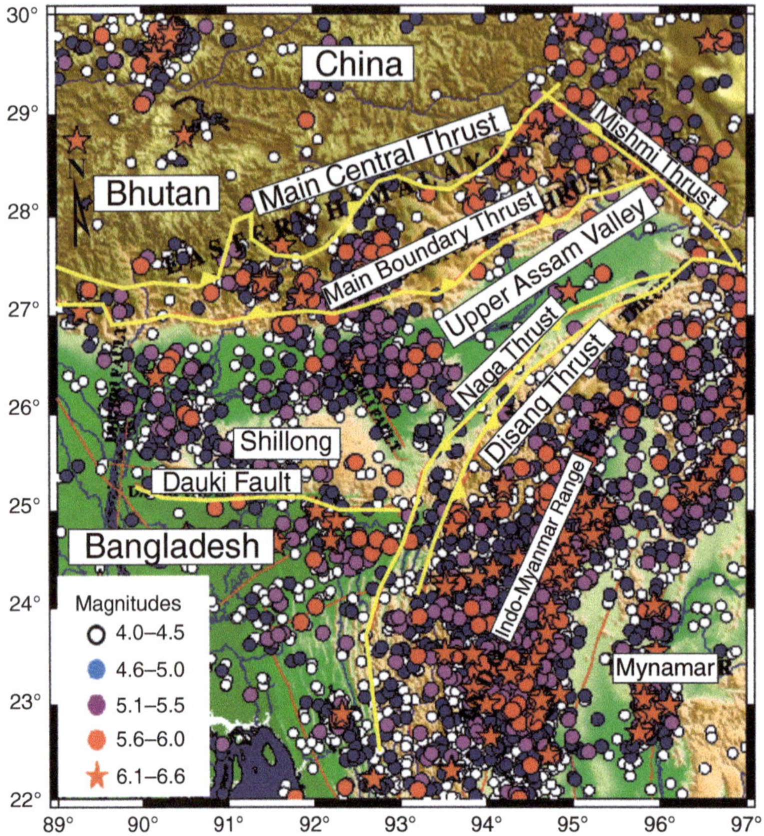

Figure 17.8 The epicentral plot (the ISC and USGS database) displaying variation in magnitude in the northeast India for the time range of 1901–2018. The Upper Assam valley is shown within the rectangle.

conformity is present. It can be interpreted that by the end of Oligocene, (sub) areal unconformity is formed at the south-eastern edge of the basin (Naga Hills side), while in the north-eastern part, correlative conformity occurs. Through this interpretation, the Tipam Group of rocks of Upper Assam became homotaxial with the Surma and Tipam Groups of Surma Valley (Pandey 1982).

17.3 Paleo-Brahmaputra and Dispersal of Sediments

Tectonic activities during Mio-Pliocene were responsible for the formation of the N–S folding in the Surma valley and Chin hill region (Nandy 1986). This phase of tectonism is probably responsible for the change in NE–SW course of the Brahmaputra River and formed the present catchments of the Barak River in the south (Figure 17.10). This view is supported by the fact that both heavy mineral occurrences and fossil assemblages of Late

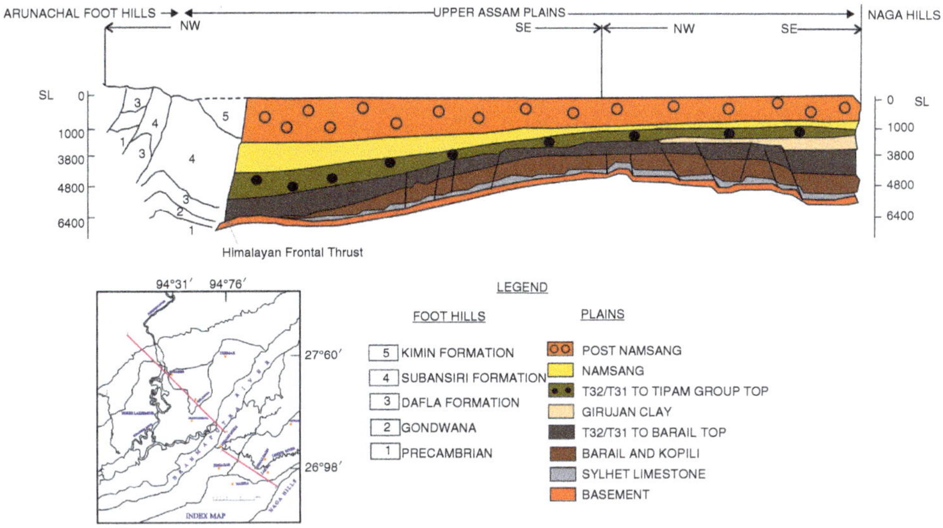

Figure 17.9 Geological cross section along AB and CD (*Source:* After Nath and Krishna Rao 1981; Kent and Dasgupta 2004) (Inset: section lines).

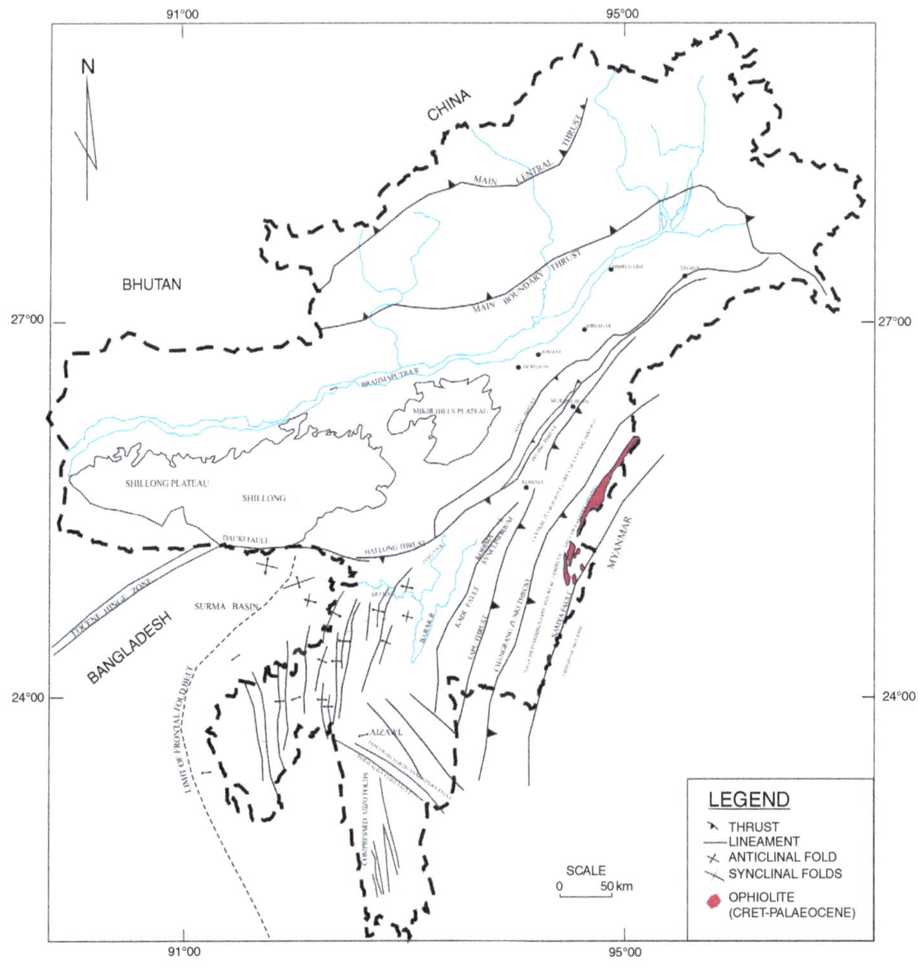

Figure 17.10 Tectonic map of Assam-Arakan Basin. *Source:* Adapted from Das Gupta and Biswas (2000). The Surma Basin and Barak River in southern Assam are shown.

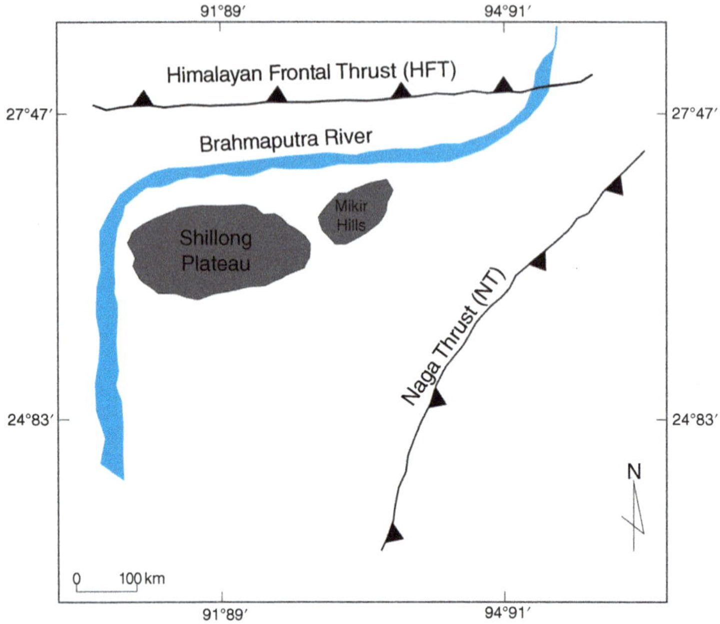

Figure 17.11 Schematic diagram showing paleogeography during Pleistocene to recent. Himalayan Frontal Thrust (HFT) and Naga Thrust (NT) are shown.

Tertiary deposits of Surma Valley and Tripura differ from those of Nagaland and the Upper Assam valley (Dutta 1978).

During the Mio-Pliocene time, Namsang and Dihing Group of rocks were deposited unconformably over the Tipam Group in shallow lacustrine and fluviatile basins. This lacustrine basin extends to the northeast of Kohima (Figure 17.6). Therefore, during Pliocene, the paleo-Brahmaputra River and its tributaries probably emptied into lacustrine basins formed during that period.

The present configuration of the Brahmaputra River took place only from the Pleistocene onward (Brahmaputra Board 1986). The present-day Brahmaputra River is evidently younger than the Chappor Formation (Pleistocene) through which it flows. This change in the direction of flow from south–west to westerly in the course of Brahmaputra River was probably due to the development of sagging in front of the HFT during Pleistocene, which created lows and increase of obstruction in the SW side due to folding in the Surma valley and in the Chin region (Figure 17.11). Both NT and the HFT developed in the same time, i.e. post-Pleistocene.

Thus, from the occurrence and distribution of the sedimentary units, it can be inferred that the paleo-Brahmaputra River and its tributaries originated possibly at the end-Eocene (Figure 17.12). This is based on the fact that Yarlung Tsangpo, prior to its capture by the Brahmaputra, was connected to Irrawaddi and responsible for sedimentation in Central Myanmar during Miocene (Zhang et al. 2019 and references therein). This indicates that origin of paleo-Brahmaputra must be prior to Miocene and we are interpreting this timing as Eocene. During the end of Eocene, the Brahmaputra River and its

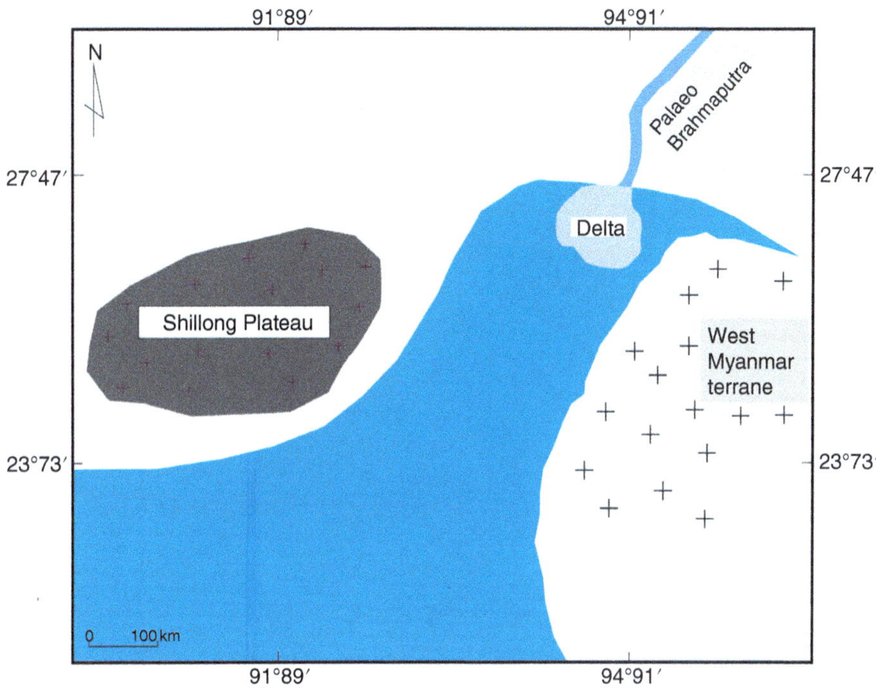

Figure 17.12 Schematic diagram showing paleogeography during Eocene to Oligocene. Paleo-Brahmaputra River is shown.

tributaries discharge into the sea that covered the area between Shillong Plateau and West Burma terrane. As a result, deposition of widespread deltaic deposits took place in the Upper Assam Basin.

During Miocene, the Brahmaputra River was flowing toward the SW direction (Figure 17.13). As a result, the deposition of the fluvial Tipam Group of rocks took place in the Upper Assam Basin. The Brahmaputra drained into the sea which subsequently regressed up to the eastern part of the Bengal basin (Figure 17.13).

The presence of fluvial and deltaic sandstones and conglomerates in the eastern fold belt and in the Sylhet trough was reported by Holtrop and Kaizer (1970). Further, the continuous increase of Miocene sediment in the eastern Bengal indicates that large river systems like paleo-Brahmaputra, paleo Dibang, paleo Lohit, paleo Subansisri and other river system drained the uplifted Eocene and Oligocene sequences of the eastern Himalaya and the IBR, which acted as a provenance (Figures 17.1 and 17.10). Sediments were thus carried to the Bengal fan and as a result, a major delta complex formed (Akhtar et al. 1998; Gani and Alam 1999; Alam et al. 2003). The shift in the course of the Brahmaputra River from Eocene through Pleistocene to recent is shown in Figures 17.11–17.13. A recent study of the paleo-drainage evolution of the Brahmaputra River from the provenance studies of the various sedimentary sections of the northeast India indicates a major shift in the course of the river from prior to 7–2.0 Ma (Chirouze et al. 2013; Bracciali et al. 2015; Coutand et al. 2016; Govin et al. 2018).

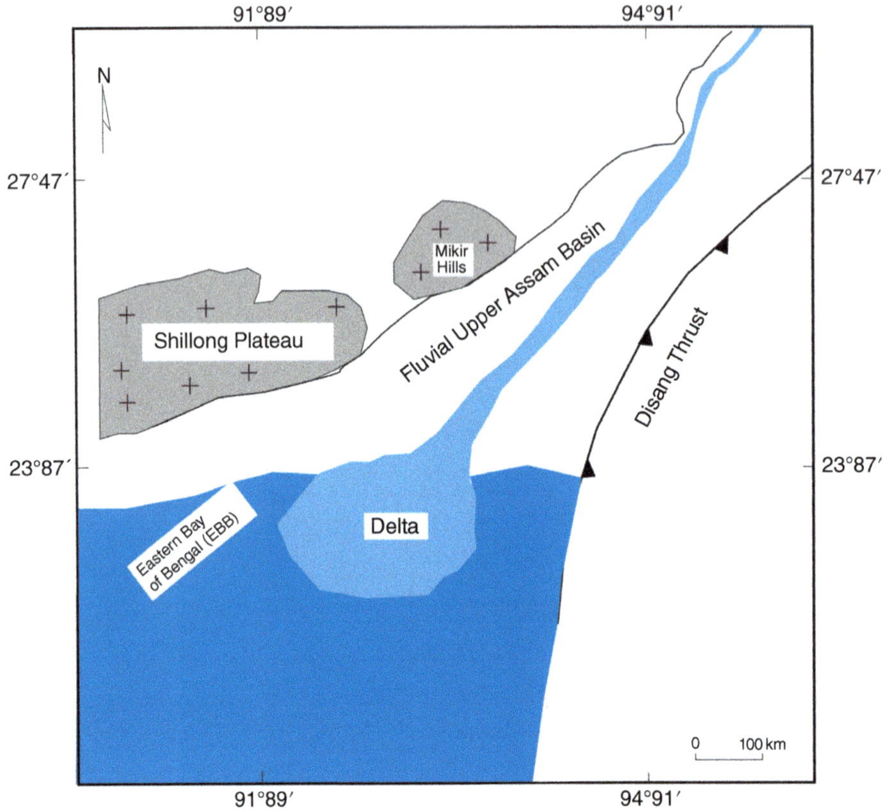

Figure 17.13 Schematic diagram paleogeography during Miocene. Shillong Plateau, Disang Thrust, Eastern Bay of Bengal, and Upper Assam Basin are shown.

17.4 Hydrocarbon Potential

Oil exploration in India commenced with the discovery of the Digboi oil field (1889) in Upper Assam. During the post-Independence period, systematic exploration carried out by ONGC and OIL led to the discovery of >100 oil and gas fields in the Upper Assam Shelf and in the adjoining Naga-Patkai fold-thrust belt (Figure 17.14). However, the areas to the north of the Brahmaputra River up to the foothills of the Eastern Himalayas have remained poorly explored. In the fold-thrust belt of Schuppen, Kharsang, Kumchai, and Dirok oil fields having oil accumulations in the Upper Miocene-Pliocene reservoirs and the Borholla-Champang oil field with fractured basement rocks as reservoirs have been discovered. Oil accumulations in these oil fields are combination/strati-structural traps. The subtle litho-stratigraphic traps are not yet properly explored and assessed. The exploration efforts so far made have reached a middle stage of exploration maturity. Even after such intense exploration activities made till date, the Upper Assam Shelf has a high potential for "yet-to-find" hydrocarbons. The Tinsukia-Sadiya area, which partly falls in the Sadiya low in the Mishmi Hills front, the Dhansiri valley and the belt of Schuppen possess high hydrocarbon

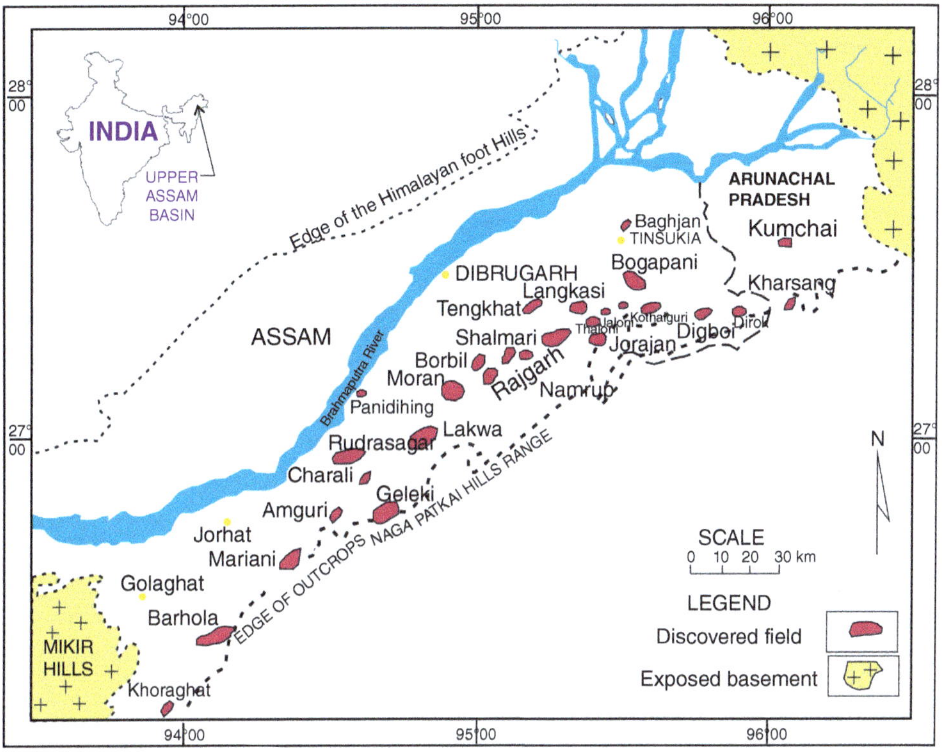

Figure 17.14 Map shows the distribution pattern of oil fields in Upper Assam. *Source:* After Handique et al. (1989); Dasgupta and Biswas (2000); DGH (2011).

prospects for further intensive exploration. The prognosticated resource base of the Upper Assam Shelf and the belt of Schuppen is ~3180 MMt, of which about 27% has been converted to in-place geological reserves. Recent studies (Pahari et al. 2008; Sahoo and Gogoi 2011; Brahma et al. 2013; Avadhani and Deo 2017) indicate that the favorable targets for future exploration for hydrocarbons lie in areas close to the NT and sub-thrust structures in the Naga fold-thrust belt and in strati-structural play in the Tura-Sylhet sequence and Bokabil sequence in Khoraghat to Mekrang of the Dhansiri Valley.

17.5 Conclusions

In the Upper Assam Basin, a thick pile of sediments from Late Cretaceous to Recent is present. The sedimentary basin was formed during the Late Cretaceous Period. The sedimentation in the basin was influenced by the episodes of Naga and Himalayan orogenesis. Till the end Oligocene, the process of sedimentation was influenced by the Naga orogeny. Consequently, up to the end of Oligocene, the basin modification was also primarily influenced by the Naga orogeny with little influence from the Himalayan orogeny. Due to the continued northward movement of the Indian plate, post-collisional (India-Asia) southward propagation of the major Himalayan thrusts created the formation of the Himalayan fold-thrust belt in the northern part of the Upper Assam Basin. The emplacement of MBT during the Early Miocene developed sag in the northern edge of the Upper Assam Basin. From the Miocene onward, the sedimentary basin is modified by the Himalayan orogeny and the Naga orogeny. The development of HFT and NT resulted in the formation of the Quaternary basin of Upper Assam by merging of Post Siwalik Himalayan foredeep and Post-Pleistocene Naga foredeep. This process is still continuing, which is indicated by the presence of a thick pile of Quaternary sediments in the Upper Assam Basin. Therefore, the Upper Assam Basin is an amalgamation of the Himalayan foredeep and the Naga foredeep.

Acknowledgments

The authors thank the Head (Department of Applied Geology, Dibrugarh University) for permission and encouragement. The authors are thankful to the Editors of the Petroleum Geoscience book. Soumyajit Mukherjee (IIT Bombay) made two rounds of review. Merryl Le Roux, Frank Otmar Weinreich, Joss Everett, Andrew Harrison, and the proofreading team (Wiley Blackwell) are thanked for their assistance. CPDA grant (IIT Bombay) supported SM. This chapter has been encapsulated in Mukherjee et al. (2023).

References

Acharyya, S.K., Ray, K.K., and Sengupta, S. (1991). The Naga hills and Andaman ophiolite belt, their setting, nature and collisional emplacement history. *Phys. Chem. Earth* 18 (Pt. 1): 293–315.

Akhtar, M.H., Bhuiyan, A.H., Hussain, M., and Imam, M.B. (1998). Turbitide sequence located in Bangladesh, oil and gas. *Tectonophysics* 21: 109–111.

Alam, M.M., Alam Mustafa, M., Curray, J.R. et al. (2003). An overview of the sedimentary geology in the Bengal basin in relation to the regional tectonic framework and basin fill history. *Sediment. Geol.* 155 (3–4): 179–208.

Allen, P.A. and Homewood, P. (1986). *Foreland Basins: Special Publication No. 8 of the International Association of Sedimentologists*. John Wiley & Sons.

Angelier, J. and Baruah, S. (2009). Seismotectonics in Northeast India: Astress analysis of focal mechanism solutions of earthquakes and its kinematic implications. *Geophys. J. Int.* 178: 303–326.

Avadhani, V.L.N. and Deo, P.P. (2017). *Exploration Challenges in Tripura Fold Belt*. Assam and Assam Arakan Bsin: SPG India http://www.researchgate.net/pub./333642812.

Berger, P., Cerny, A., Hook, P. et al. (1983). *Well Evaluation Conference, India, December 1983*, 263pp. Singapore: Schlumberger Technical Services.

Bhandari, L.L., Fuloria, R.C., and Sasrty, V.V. (1973). Stratigraphy of Assam valley, India. *AAPG Bull.* 57 (4): 642–652.

Bhattacharjee, C.C. (1991). The ophiolites of Northeast India- a subduction zone ophiolite complex of the Indi-Burman orogenic belt. *Tectonophysics* 191: 213–222.

Bracciali, L., Najman, Y., Parrish, R.R. et al. (2015). The Brahmaputra tale of tectonics and erosion: early miocene river capture in the Eastern Himalaya. *Earth Planet. Sci. Lett.* 415: 25–37. https://doi.org/10.1016/j.epsl.2015.01.022.

Brahma, J., Sircar, A., and Karmakar, G.P. (2013). Hydrocarbon prospectivity in central part of Tripura, India, using an integrated approach. *J. Geog. and Geol.* 5/3: 3113–3116.

Brahmaputra Board (1986). 3.1 Master plan of Brahmaputra basin, part 1. Annual Report, Government of India, 112pp.

Chakraborty, A. and Bakshi, S. (1972). Stratigraphy of the cretaceous--tertiary sedimentary sequence, southwest of Shillong plateau. *Q.J. Geol. Min. Metall. Soc. India* 44: 109–127.

Chakraborty, P.P., Tandon, S.K., and Saha, S. (2019). Development of phanerozoic sedimentary basins of India. *J. Asian. Earth Sci.* https://doi.org/10.1016/j.jseaes.2019.103991.

Chirouze, F., Huyghe, P., van der Beek, P. et al. (2013). Tectonics, exhumation, and drainage evolution of the eastern Himalaya since 13 Ma from detrital geochemistry and thermochronology Kameng river section, Arunachal Pradesh. *Geol. Soc. Am. Bull.* 125: 523–538. https://doi.org/10.1130/B30697.1.

Coutand, I., Barrier, L., Govin, G. et al. (2016). Late miocene-pleistocene evolution of India-Eurasia convergence partitioning between the Bhutan Himalaya and the Shillong plateau new evidences from foreland basin deposits along the Dungsam Chu section, Eastern Bhutan. *Tectonics* 35: 2963–2994. https://doi.org/10.1002/2016TC004258.

Das Gupta, A.B. (1976). Petroleum occurrences in the Brahmaputra valley and the petroleum prospects in the North Eastern Himalayas. Himalayan geology seminar, New Delhi, G.S.I. misc. pub.

Das Gupta, A.B. and Biswas, A.K. (2000). *Geology of Assam*, 167p. Bangalore: Geology Society of India.

Dewey, J.F., Shakleton, R.M., Chengfa, C., and Sun, Y. (1988). Tectonic evolution of the Tibetan plateau. *Proc. Math. Phys. Eng. Sci.* 327: 379–413.

DGH (2011). Directorate General of HydrocarbonsBasin Information: Assam-Arakan Basin, 2: 1–12.

Dutta, S.K. (1978). Palaeogeography of North East India a subduction zone ophiolite complex. *J. P.G., Dib. Univ.* 1: 1–7.

Gani, A.R. and Alam, M.M. (1999). Trench slope controlled deep sea clastics in the exposed lower surma group in the south eastern fold belt of Bengal basin, Bangladesh. *Sediment. Geol.* 127: 221–236.

Govin, G., Najman, Y., Copely, A. et al. (2018). Timing and mechanism of the rise of the Shillong plateau in the Himalayan foreland. *Geology* 46: 279–282.

Handique, G.K. and Borgohain, J.K. (1984). *Sub Surface structure and its Role in Hydrocarbon Entrapment in Parts of Upper Assam Oil Field Region, Sem. Recent Advances on the study of Cenozoic Geology in N.E.R. India*, 66–88. Dibrugarh University.

Handique, G.K., Sethi, A.K., and Sarma, S.C. (1989). Review of tertiary stratigraphy of parts of upper Assam valley. *Geol. Surv. Ind. Sp. Pub.* 23: 23–36.

Heidbach, O., Reinecker, J., Tingay, M., Muller, B., Sperner, B., Fuchs, K., 2007. Plate boundary forces are not enough: second- and third-order stress patterns highlighted in the world stress map database. *Tectonics*, 26, TC6014 doi: https://doi.org/10.1029/2007TC002133.

Holtrop, J.F. and Kaizer, J. (1970). *Some Aspects of the Stratigraphy and Correlation of the Surma Basin Wells, East Pakisttan, Stratigraphic Correlation Between Sedimentary Basins of ESCAP Region*, Min. Res. Dev. Ser. No.36, vol. 2nd vol., 143–154. New York: U.N.

Kent, W.N. and Dasgupta, U. (2004). Structural evolution in response to fold and thrust belt tectonics in Northern Assam a key to hydrocarbon exploration in the Jaipur anticline area. *Mar. Pet. Geol.* 21: 785–803.

Khattri, K.N., Wyss, M., Gaur, V.K. et al. (1983). Local seismic activity in the Assam gap, NE India. *Seis. Soc. Amer. Bull.* 73: 459–469.

Mandal, K. and Dasgupta, R. (2013). Upper Assam basin and its basinal depositional history. *10th Biennial International Conference & Exposition*, Koshi, 1–7. Geo India

Mathur, L.P. and Evans, P.(1964). Oil in India. *22nd Session International Geological Congress Proceedings* 7–52, New Delhi.

Mukherjee, S., Dasgupta, S., Majumdar, C. et al. (2023). Introduction to: Handbook of Petroleum Geoscience: Exploration, Characterization, and Exploitation of Hydrocarbon Reservoirs. In: *Handbook of Petroleum Geoscience: Exploration, Characterization, and Exploitation of Hydrocarbon Reservoirs*, Petroleum Industry Updates: Geoscientific and Economic Issues (ed. S. Mukherjee, S. Dasgupta, C. Majumdar, et al.). John Wiley & Sons, Hoboken, ISBN 978-1-119-68003-1.

Naik, G.C., Kumar, R., Soren, M.N. (2004). Tectonic setting and petroleum system significance of pre-tertiary sediments, Assam Basin, Northeast India. *5th Conference & Exposition on Petroleum Geophysics*, Hyderbad, India, pp. 365–371. Society of Petroleum Geophysicists.

Najman, Y., Sobel, E., Millar, I. et al. (2019). The uplift of indo-Burman ranges, Myanmar. *Geophys. Res. Abstr.* 21: 2029–2059.

Nakata, T. (1972). Geomorphic history and crustal movements of the foot hills of the Himalayas. *Tohoku Univ. Inst. of Geog.* 39–77.

Nandy, D.R., 1986. Geology and tectonics of Arakan Yoma –a reappraisal. GEOSEA \1 Proceedings Vol. II, Geol. Soc. Malaysia, Bulletin, 137–148.

Nandy, D.R. (2001). Geodynamics of North East India and the adjoining region. *J. Geol. Soc. India* 209.

Nath, R. and Krihna Rao, V.V (1981). Seismic stratigraphy in Dhakuakhana-Gogamukh area of North Lakhimpur district. Assam. workshop on geological interpretation of the geophysical data, institute of petroleum exploration, Dehradun, India.

Pahari, S., Singh, H., Prasad, I.V.S.V., and Sing, R.R. (2008). *Petroleum Systems of Upper Assam Shlf*, 14–21. Dehradun: Geohorizons.

Pandey, J. (1982). Chronostratigraphic correlation of Neogene sedimentaries of Western Indian shelf, Himalayas and upper Assam. in workshop on cenozoic stratigraphy and palynology in India. *Pal. Soc.India spl.* 1: 95–129.

Raju, A.T.R. (1968). Geological evolution of Assam and Cambay tertiary basins of India. *Am. Assoc. Pet. Geol. Bull.* 52 (12): 2422–2437.

Rangarao, A. (1983). Geology and hydrocarbon potential of Assam-Arakan basin and its adjacent region. In: *Petroliferous Basins of India: Petroleum Asia* (ed. L.L. Bhandari), 127–158. Dehradun: Himachal Times Group.

Rao, V.V.K. and Rajkumar, J. (1984). Exploration for hydrocarbon in North East India by O.N.G.C. *Proc. Vol. Sem. Recent Advances on the study of Cenozoic Geology in N.E.R. India.* Dibrugarh University, p. 1–37.

Sahoo, M. and Gogoi, K. D. 2011 Structural styles and its implicationson North Assam shelf, upper Assam basin. Extended Abstract Presented at GEO-India, Greater Noida, New Delhi, India (12–14 January 2011), 1–16.

Sinclair, H.D. (1997). Flysch to molasse transition in peripheral foreland basins: the role of the passive margin versus slab breakoff. *Geology* 25 (12): 1123–1126.

Srinivasan, V. (2007). Regional structural configuration and seismic pattern of belt of schuppen in Northeast India. *Geol. Soc.* 70: 801–814.

Uddin, A., Kumar, P., and Sharma, J.N. (2007). Early orogenic history of the eastern Himalayas: compositional studies of palaeogene sanstones from Assam north East India. *Int. Geol. Rev.* 49: 798–810.

Valdiya, K.S. (2010). *The Making of India, Geodynamic Evolution*, vol. 426. Switzerland: Springer International Publishing AG.

Wandrey C.J. (2004), Sylhet-Kopili/Barail-Tipam Composite Total Petroleum System, Assam Geological Province, India, USGS Open File Report, 2208-D.

Yin, A., Dubey, C.S., Webb, A.A.G. et al. (2010). Geologic correlation of the Himalayan orogen and Indian craton: part 2. Structural geology, geochronology, and tectonic evolution of the eastern Himalaya. *Geol. Soc. Am. Bull.* 122: 360–395.

Zhang, P., Najman, Y., Mei, L. et al. (2019). Palaeodrainage evolution of the large rivers of East Asia, and Himalayan-Tibet tectonics. *Earth-Science Reviews* 192: 601–630. ISSN 0012-8252.

18

Basement Tectonics in the Assam Shelf and Its Implications in Hydrocarbon Exploration – A Remote-Sensing and GIS-Based Perspective

Subhobroto Mazumder, Blecy Tep, D.S. Mitra, and K.K.S. Pangtey*

Remote Sensing and Geomatics, Geology Group, KDMIPE, ONGC, Dehradun, Uttarakhand, India

**Author for correspondence: subhom007@gmail.com*

Abstract

The Assam Shelf in the NE part of India is geomorphically manifested as the Brahmaputra Valley and forms the foreland of two major thrust belts: the Main Boundary Thrust in the north and the Naga-Disang Thrust system in the south. The Precambrian basement underlying the Assam Shelf had suffered multiple episodes of tectonism/deformation with outcrops of it exposed in the Shillong-Mikir massifs. The present study attempts to integrate various geological and geophysical aspects of the basement of the Assam Shelf in a GIS framework to derive a geological model of it and based on that, identify favorable areas of hydrocarbon exploration in the basement. In the Dhansiri Valley or the South Assam Shelf, clockwise rotation of the adjacent Mikir Massif block led to the development of a number of E–W zones of dextral strike-slip faulting. Additionally, to accommodate this clockwise rotation of the Mikir Block, a number of NW–SE structural trends were created. Both these trends are highly fractured and zones of their intersection form regions of maximum fracture porosity in the basement. Structural highs associated with such fractured areas form promising zones of basement exploration. The South Assam Shelf is further separated from the North Assam Shelf by the Jorhat Fault, which acts as a tectonic divide between these two regions. The North Assam Shelf is supposedly deformed by a transtensional wrench system defined by the Jorhat Fault in the south and an E–W strike-slip fault in the north near Tinsukia, which itself had been further distorted/modified by compression due to the Naga Thrust in the east. Neotectonic reactivations related to this wrench system resulted in the formation of folds in the shallower sediments, whereas the (sub)-vertical strike-slip faults of the NW–SE to E–W wrench system that are parallel to the S_{Hmax} direction act as conduits to secondary hydrocarbon remigration. Hydrocarbons remigrating from the breached deeper and larger primary accumulation structures reaccumulate in the shallower smaller folds and form reservoirs at multiple levels. Associations of morphostructurally interpreted faults and geomorphic highs act as a guide to identifying these potential areas of hydrocarbon exploration.

Keywords *Assam Shelf; basement tectonics; hydrocarbon exploration; remote sensing*

Handbook of Petroleum Geoscience: Exploration, Characterization, and Exploitation of Hydrocarbon Reservoirs, First Edition. Edited by Soumyajit Mukherjee, Swagato Dasgupta, Chandan Majumdar, Subhandip Mandal and Troyee Dasgupta.
© 2023 John Wiley & Sons Ltd. Published 2023 by John Wiley & Sons Ltd.

Abbreviations Used

NAS	North Assam Shelf
UAN	Upper Assam North
SAS	South Assam Shelf
UAS	Upper Assam Shelf
MBT	Main Boundary Thrust
MFT	Main Frontal Thrust
OLI	Operational Land Imager
ETM+	Enhanced Thematic Mapper +
BSZ	Barapani Shear Zone
BGC	Basement Gneissic Complex
LMF	Lower Metapelitic Formation
UQF	Upper Quartzitic Formation
ONGC	Oil and Natural Gas Corporation Limited
ASTER	Advanced Spaceborne Thermal Emission and Reflection Radiometer
DEM	Digital Elevation Models
FMI	Formation Microimager
DIF	Drilling-Induced Fracture
GYF	Golden Yellow Fluorescence
TWT	Two-Way Time

18.1 Introduction

The Assam Shelf in the NE part of India forms the northeastern prolongation of the Indian Peninsular Shield bound on its edges by the Main Boundary Thrust (MBT) in the north, the Mishmi Thrust in the east, the Naga-HaflongThrust in the southeast and the Shillong Massif at its south (Figure 18.1a). Geomorphically, the shelf-region also forms the Brahmaputra Valley with the river aligned along its strike and the valley shoulders being defined by the Eastern Himalayan Fold belt in the north, the Mishmi Hills in the east and the Schuppen belt comprising the Naga Patkai Hills in the south (Figure 18.1a). Tectonically, it also forms the foreland of two major thrusts, MBT spanning the entire northern extent of the valley and the Naga-Disang Thrust system that borders its eastern-southeastern boundary. The compression due to thrusting from both ends of the shelf deformed the basement underlying the shelf area in a broad arch known as the Brahmaputra Arch (Sahoo and Gogoi 2011). The basement on either side of the arch/ridge is found to slope away from the ridge (Figure 18.11b), i.e. toward the Himalayan foothills in the north and Naga Hills in the south forming two separate basinal provinces. This shelf region in the south bank is further divided into two parts by an E–W trending basement-controlled Jorhat Fault (Figure 18.1a). The area north of it called the North Assam Shelf (NAS) or Upper Assam North (UAN) occupies the northern Brahmaputra valley, whereas part of the shelf south of it within the Dhansiri Valley is the South Assam Shelf (SAS) or Upper Assam South (UAS) enveloped between the Naga Thrust and the Mikir Massif. The basement in both cases dips southeast with the slope of the basement underlying the SAS found to be much steeper than that of the NAS (Pangtey et al. 2016). The sedimentary sequence in both the areas ranges in age from Paleocene to Recent with inliers of Gondwana sediments and Early Cretaceous Traps that rest

(a)

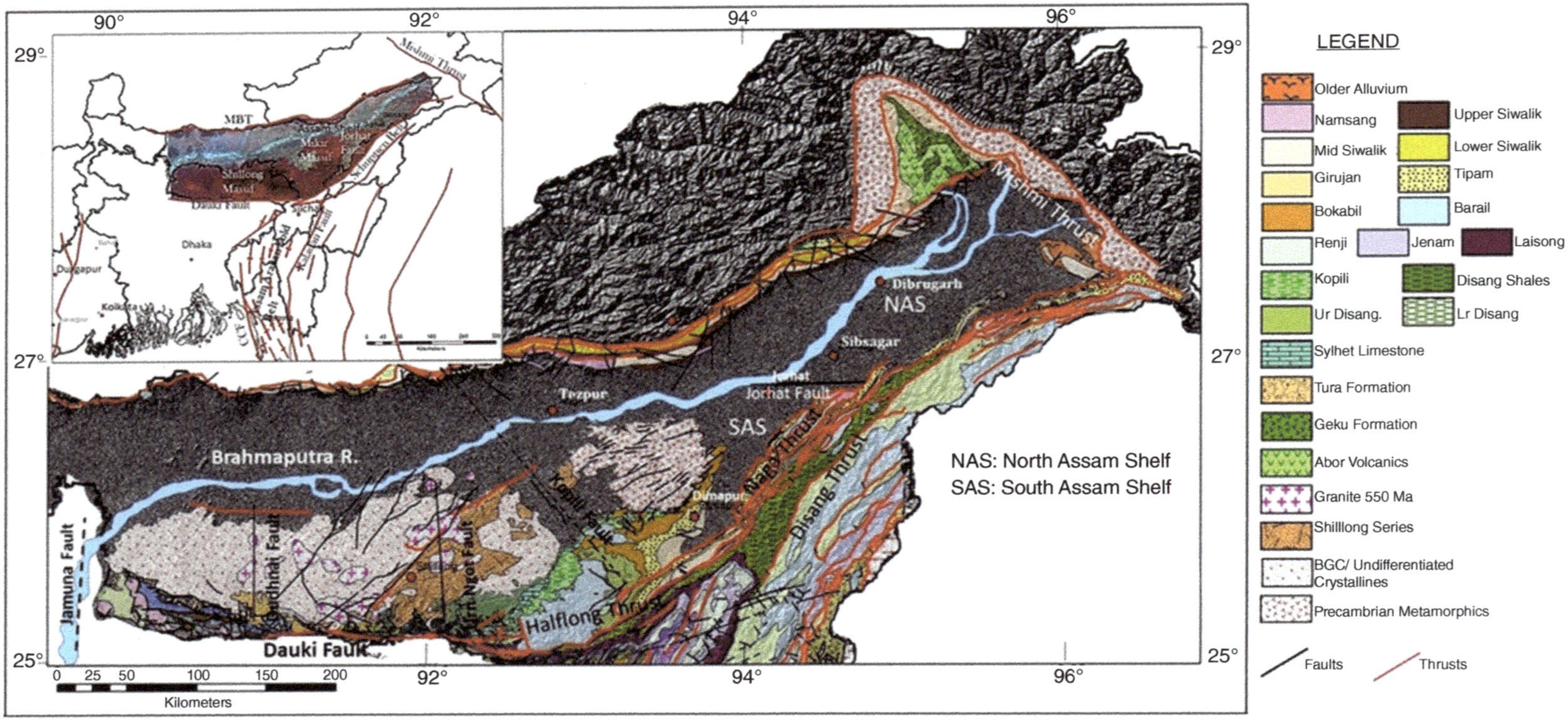

Figure 18.1 (a) Geological and tectonic layout of the Assam Shelf overlain on a mosaicked hillshade derived from ASTER DEM of 30 m resolution. The inset figure shows LANDSAT ETM + image composite of bands 4, 3, and 2 of the Assam Shelf depicting the Brahmaputra Valley and the disposition of different tectonic elements in the north-eastern part of India. (b) Brahmaputra Arch/Ridge formed due to compression from NW and SE divided the Assam Shelf into two apparently disconnected sub-basins north and south of the Ridge.

(b)

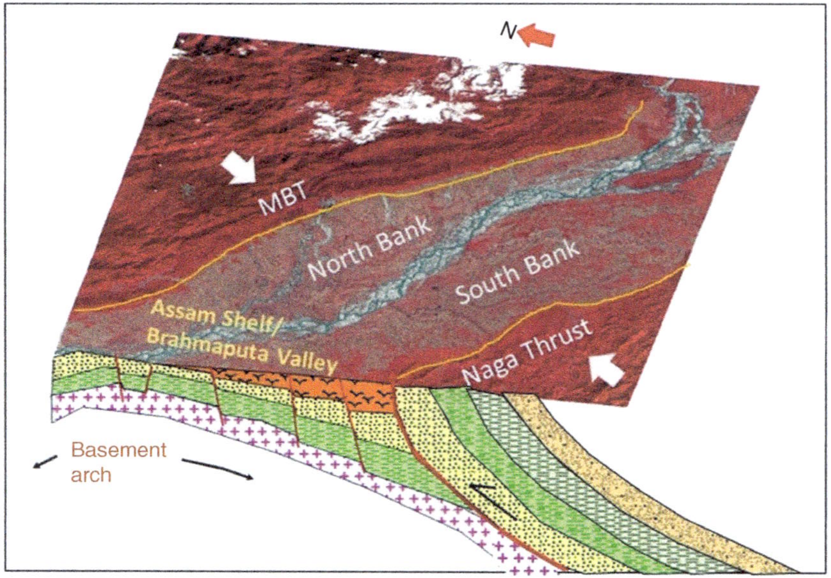

Figure 18.1 (Continued)

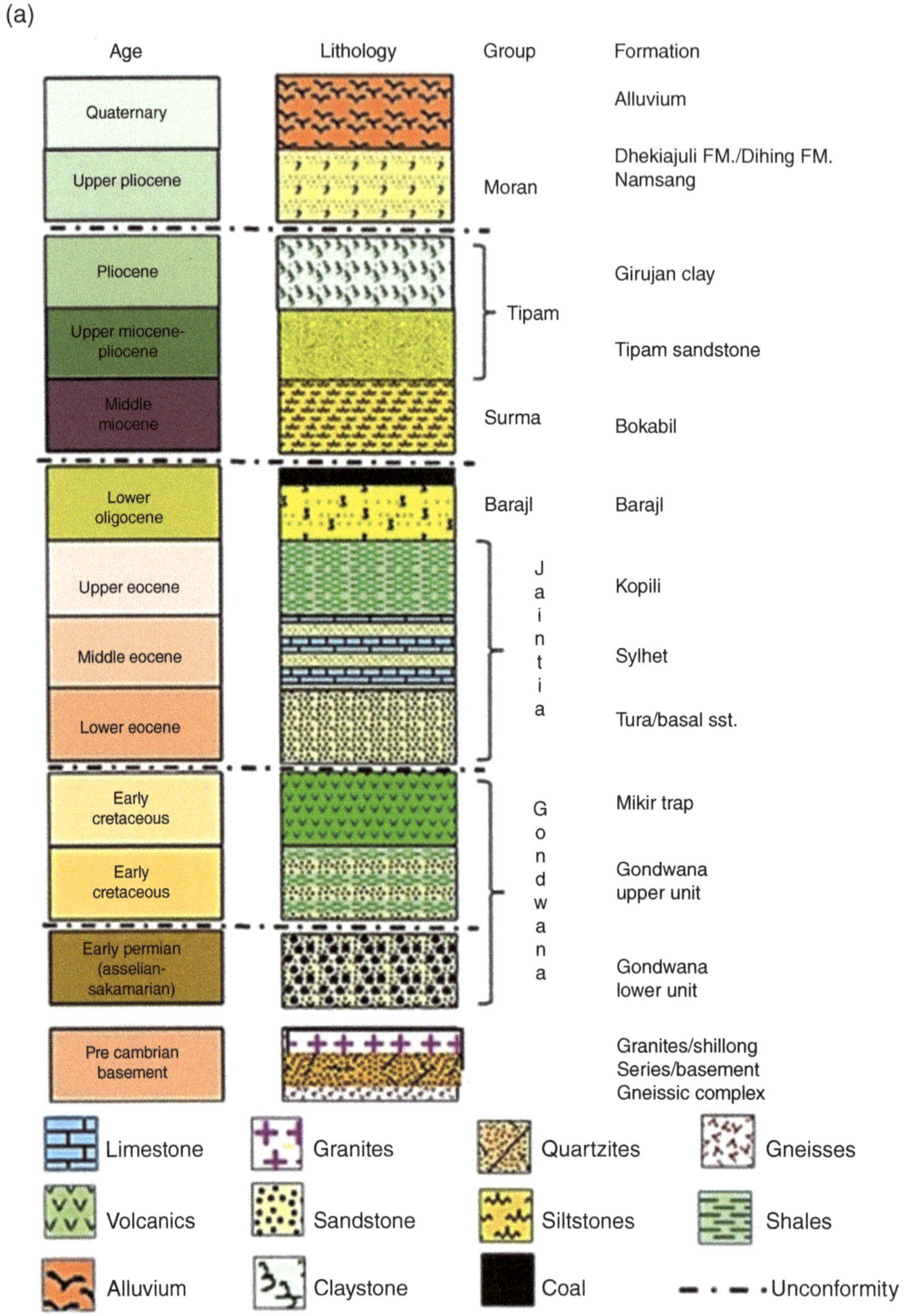

Figure 18.2 (a) Generalized stratigraphy of the Assam Shelf. *Source:* After Mathur et al. (2001), Roy Moulik et al. (2009) and various ONGC unpublished reports. (b) Map of hydrocarbon fields of the Assam Shelf, mostly found concentrated in the south bank of the Brahmaputra River. Among these, the Borholla-Champang field (encircled in black) produces mostly from a fractured basement (outcrop legends same as Figure 18.1a).

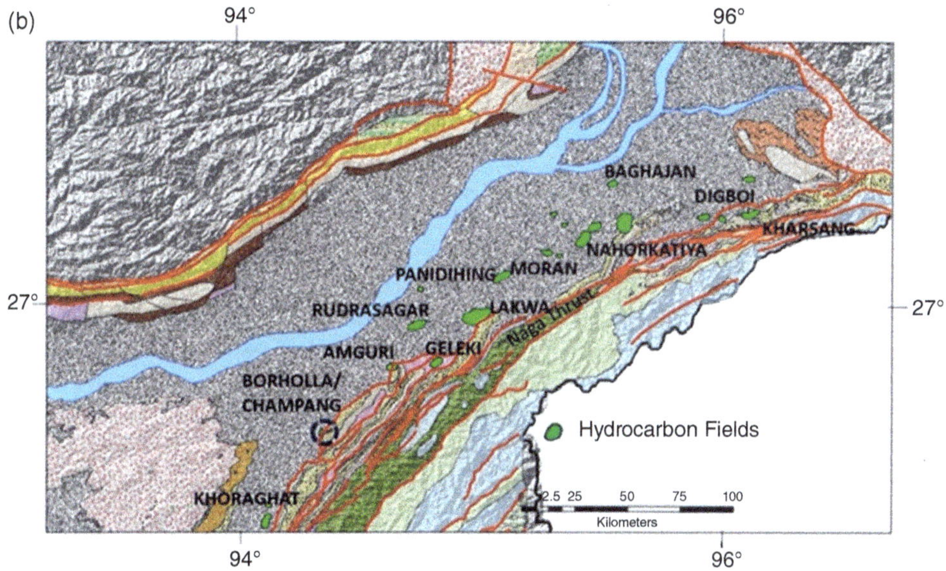

Figure 18.2 (Continued)

unconformably over a Precambrian granitic or gneissic basement (Figure 18.2a). The south bank shelfal basin also forms a rich petroliferous province with most of the oil and gas fields being situated on the southeastern slope of the Brahmaputra arch. Borholla-Champang field in the SAS hosts significant reservoirs in its Fractured Precambrian basement. Basement-derived oil has also been discovered in the Khoraghat field (Mortha et al. 2015) and Dayalpur area in SAS (Figure 18.2b).

Though a number of workers had attempted structural/tectonic analysis of basement rocks in the Assam Shelf, most of these studies are independent outcrop/remote sensing-based surface interpretations (Kakati 2013; Goswami 2014) or seismic/gravity-magnetic based subsurface analysis (Kumar et al. 2012). This chapter attempts an integrated and regional approach to study the different episodes of basement tectonics affecting both the NAS and SAS. The results of the integrated analysis have subsequently been used to develop a regional GIS-based geological model about the impact of basement tectonics in the formation of probable areas of hydrocarbon interest similar to those above.

18.2 Basement and Basement Tectonics in Assam Shelf

In the Assam Shelf, the Precambrian basement in the basinal part crops out in the Mikir and Shillong Massif areas west of the Assam Shelf. These Precambrian outcrops in Shillong Massif are representative of the craton underlying the Assam Shelf is considered similar to that of the Indian Peninsula but is tectonically disconnected from it by the Garo-Rajmahal Gap/depression that hosts the Bengal and the Purnea basins (Tiwari and Jassal 2001; Hossain et al. 2019; Shahriar et al. 2020).

Both the Shillong and Mikir Massifs initially formed a single uplifted cratonic block. This had later been separated by a NW–SE, ~300 km long- and 50 km-wide Kopili Fault. The southern part of the massif is marked by an E–W Dauki Fault Zone that separates it from a ~17 km-thick Tertiary section in the Bengal basin to its south (Uddin and Lundberg 2004; Hossain et al. 2020b). The Dauki Fault is presumably a Precambrian lineament reactivated by the compressive tectonic forces generated by the Himalayan and Indo-Burman subduction systems. It has accommodated ~25% of the Indian plate-related crustal shortening. The E–W-oriented Brahmaputra fault probably marks the northern boundary of the Shillong Massif (Islam and Shinjo 2012, Hossain et al. 2019; Hossain et al. 2020a).

Satellite images (LANDSAT OLI AND ETM + multispectral data of 30 m resolution from https://earthexplorer.usgs.gov/) depict both Shillong and Mikir massifs traversed by several fracture sets. Of these fracture sets, N–S-oriented fractures cut across the entire width of the basement outcrop of Shillong Massif with some of them even cutting across its southern bounding limit of the Dauki Fault Zone (Hossain et al. 2020b). The N–S-oriented dextral Urn Ngot Fault dissects the Shillong Massif with abundant magmatic intrusions aligned along this fault. Another N–S trending Dudhnoi Fault in the western part of Shillong Massif, crosses the entire width of the massif and offsets the Dauki Fault dextrally. Carbonatite intrusions along these faults indicate that they are deep-seated. The Jamuna fault apparent from the straight N–S course of the river Brahmaputra on the western margin of the Shillong massif marks the western extent of the Shillong massif (Figure 18.1a) (Das 1995; Hossain et al. 2020b). Satellite imageries also depict a prominent NE–SW trending basement-controlled Barapani Shear Zone (BSZ) with sinistral slip. (Devi and Sarma 2010). The effect of shearing is found to continue into the Mikir Hills but appears to die out southward at latitude 25°25′ N where shearing is untraceable in imagery. (Das et al. 1995, for details of all these faults, please see Hossain et al. (2019)).

Stratigraphically, the oldest rocks of the Shillong-Mikir massifs belong to Basement Gneissic Complex (BGC) constituted metapelites, pyroxene granulites, quartzo-feldspathic gneiss, migmatites, and augen-gneiss metamorphosed to amphibolite to granulite facies. The rocks that overlay the BGC comprise the Shillong Group and are constituted of a Lower Metapelitic Formation (LMF) and Upper Quartzitic Formation (UQF) separated by an inter-formational conglomerate. Both these groups are intruded by a number of granite bodies known in different parts of the massif as Kyrdem granite, Nongpoh granite, and Mylliem granite in Neo-Proterozoic. The general trend of the granite bodies is E–W or ENE–WSW and, thereby, follows the tectonic axis of the Shillong Massif (Wallang 2009). However, as per Prasad and Srivastava (1987), ENE–WSW and NE–SW fractures serve as the principal conduits for these igneous intrusions. These Precambrian formations are overlain by rocks of Tertiary age in the eastern and northeastern parts of the Shillong Massif, whereas in the southern part, they are partly overlain by Mesozoic rocks (Figure 18.1a) (Devi et al. 2009).

The Precambrian rocks of the Shillong-Mikir massif deformed several times. In the Mikir Massif, three generations of folds are discernable in the BGC and two generations in the Proterozoic Shillong Group (Prasad and Srivastava 1987). The First Phase of Folding (F1) are tight isoclinal folds with the axis trending NNE–SSW and are restricted to the BGC. F2 is tight subvertical with an axis oriented E–W to ENE–WSW and are present both in BGC and the Shillong Group. F3 are broad open warping with a fold axis trending SE and axial planes nearly vertical. These again are found both in the BGC and the Shillong Group (Prasad and Srivastava 1987). All these Precambrian trendshave been structurally inherited and play a significant role in guiding later structures and building up interesting areas especially for basement hydrocarbon exploration.

Post-development of the Precambrian basement fabric, the initial basin formation in the Assam Shelf can be considered to have started in the Permo-Triassic with the development of a number of Gondwana horst and grabens parallel to the eastern boundary of that part of the plate (Mishra et al. 2016) which define the overall basement structure (Figure 18.3a) in the basin (Kumar et al. 2012). In the Dhansiri Valley or the SAS sediments encountered in sub-crops in a number of ONGC wells on palynological analysis shows variation in age from an Early Permian (equivalent to Talchir Formation of Gondwanas) to Early Cretaceous (equivalent to Rajmahal Formation) age. Re-worked Early Permian palynomorphs suggesting sub-tropical/temperate climate and humid freshwater environment have been observed in Upper Bhuban sediments near Kohima (Singh et al. 2014) suggesting derivation of Gondwana outcrops in shelfal part.

Seismic and electrolog data (Figure 18.3b) suggest that these Gondwana sediments were deposited in narrow rifted grabens associated with a general trend of E–W to NE–SW. In Matimekhana-Deohalarea near Tinsukia operated by Oil India Limited, analysis of seismic data indicated Pre-Tertiary sequences around 700–800 m expected to be equivalent of Early Cretaceous age that had been deposited in a narrow N–S trending graben (Figure 18.3b) (Dasgupta et al. 2011). See Misra and Mukherjee (2018) for seismic data interpretation techniques. Thick volcanic traps have been observed in the grabens in Dhansiri Valley that were found to be stratigraphically equivalent to Late Jurassic-Early Cretaceous Sylhet Traps exposed in the Shillong Massif. Undated pyroclastic rocks interbedded with volcaniclastic sediments overlying the gneissic basement were encountered in some wells in the NAS (Figure 18.3b) and can also be considered to be stratigraphically similar (Raja et al. 2012) to that in the Dhansiri Valley. This graben formation and volcanism were probably associated with the breakup of theIndian Plate from Antarctica and Australia (Das et al. 2004) prior to its northward drift.

From Eocene onward, the continual northward drift of the Indian Plate subsequent to its breakup resulted in the initiation of subduction of the largely oceanic Greater India Basin of the Indian Plate (Van Hinsbergen et al. 2012) below the Tibetan Plate in the north as well as beneath the BurmesePlate in the east (Yang et al. 2020). This drift continued with environmental conditions in Assam Shelf varying from shelfal to near shore due to movement in basement faults till the Indian Platecollided terminally with the Tibetan block initiating the Himalayan orogeny or the Hard Collision of Indiawith a thicker and contiguous Indian continental lithosphere (Van Hinsbergen et al. 2012). These continents initially collided along the western syntaxis of the Indian plate around the beginning of Oligocene and then progressively sutured eastward (Science China Press 2017). By mid Oligocene, collision with the Burma lithospheric plate also began with a northwestward compression of the Assam Shelf. This led to epierogenic uplifts, moderate inversion of normal faults and structural dissection of the Assam Shelf along major basement tectonic trends.

Miocene marked the start of significant uplift of the Himalayas with the MBT that marks the northern extent of the Assam Shelf being formed in Middle to Late Miocene (Meigs et al. 1995). Also around Middle Miocene, entire oceanic crust of the Indian plate at the south

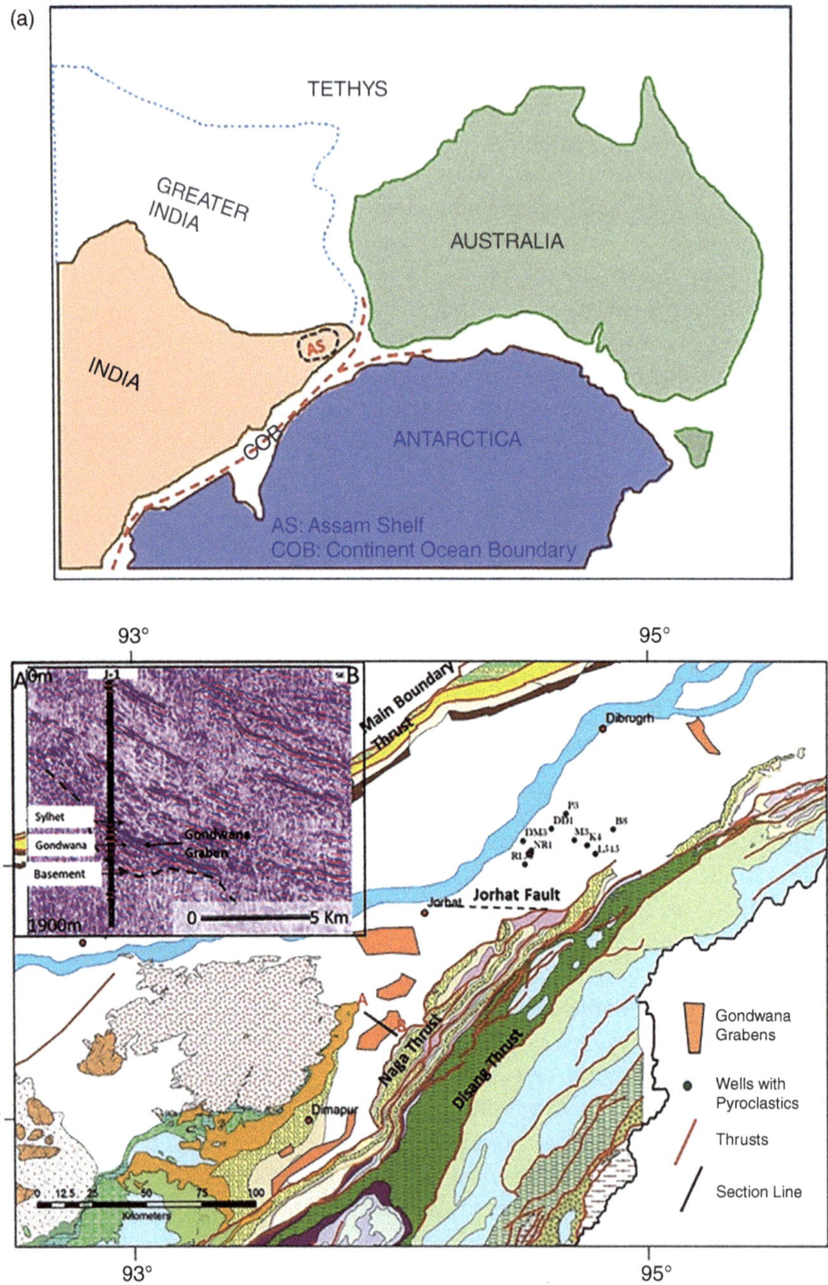

Figure 18.3 (a) Separation of the Indian Plate from Antarctica during Early Cretaceous led to the development of the NE–SW basin forming horsts and grabens in the Assam Shelf. *Source:* After Alam et al. (2003)/with permission of Elsevier. (b) The E–W, NE–SW to N–S narrow rifted Gondwana grabens formed in the south bank of the Brahmaputra River in the Assam Shelf. *Source:* After Das et al. (2004) and Dasgupta et al. (2011). Thick volcanic traps in the SAS and pre-Tertiary pyroclastic rocks in the NAS equivalent to the Sylhet Traps have also been encountered (outcrop legends same as in Figure 18.1a). Inset figure shows a seismic section along line AB and electrolog data depicting the development of a Gondwana graben in the basement of the SAS.

eastern margin of the Assam Shelf had subducted under West Burma Plate thus initiating a hard and extremely oblique, collision between these two plates (Figure 18.4). This led to the prominent formation of the Brahmaputra Arch that defines a buried basement ridge with its axis correlating with the course of the Brahmaputra River (Figure 18.1b). Theformation of this ridge results in two apparently disconnected basins on either side of the ridge, filled mostly with Tertiary sediments with a comparable sedimentation history and fairly well correlatable stratigraphic sequences (Roychoudhury and Mukherjea 2011) covered with a thick layer of alluvium. The continued push of the Indian Plate on the Burmese Platealso led to the development of several sub-horizontal thrust sequences with a westward transport direction (Figure 18.4b). The most prominent among them is the Disang Thrust which forms the sole or decollement thrust. The north-westward propagation of tectonic forces acting on the sediments in the immediate foreland of the Disang Thrust caused the formation of numerous imbricate thrust faults with upliftment and successive overriding of the older sediments (Ghosh and Hazra 2011; Gosh et al. 2014). The Naga Thrust probably formed the frontal thrust of the thrust sheets and displays prominent salient and re-entrants

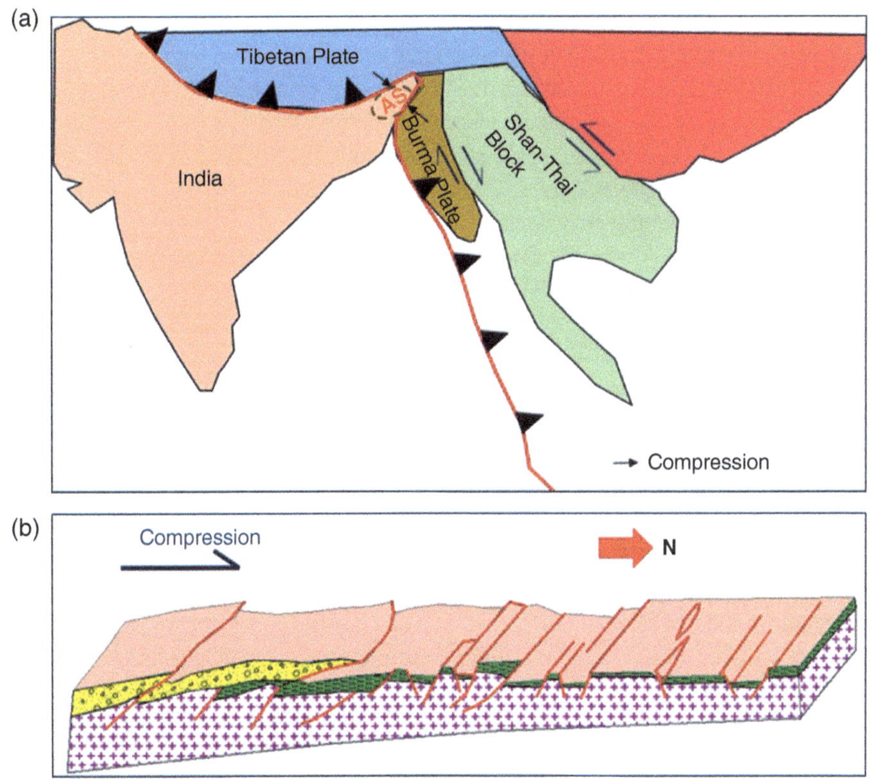

Figure 18.4 (a) Plate position at the end of Tertiary shows simultaneous compression from the NW and SE due to Himalayan orogeny and subduction under the Burmese Plate, respectively leading to the formation of the Brahmaputra Ridge/Arch. *Source:* After Alam et al. (2003)/with permission of Elsevier. (b) Schematic block layout of the Assam Shelf south of the Brahmaputra Ridge in which the thrust sheets are found to move over a rifted foreland basin with a horst and graben fabric.

(convex and concave notches of the thrust sheet) on its trajectory that are believed to be formed due to the differential velocity of the thrust sheet with the re-entrants being associated with structural lows. The Naga and Disang Thrust merge in the southwest forming HalflongThrust which further merges with the Dauki Fault westward. The significant displacement along the Dauki Fault is premised to have started only after the Miocene. The Shillong-Mikir block emerged as an outcrop of the Precambrian basement sometimes in the Late Pliocene probably as a result of compressional deformation due to the Indo-Eurasia collision (Govin et al. 2018). The Mishmi Thrust bounding the northern extremity of the Assam shelf was also active during the late Tertiary resulting in a southwestward compression of the basin. Due to upliftment of the Mishmi granitic complex along the Mishmi Thrust, the Brahmaputra basement ridge plunged toward the northeast, creating a prominent depression in front of the Mishmi Hills.

A second and more intense deformation began in the Plio-Pleistocene. In this stage, normal faults bounding foreland structures of the thrust sequences were reactivated and inverted by compression associated with the thrust faults (see Misra and Mukherjee (2015); Dutta and Mukherjee (2019) for global review on such reactivations) propagating from the southeast, north and northwest. The NW–SE compressional movement generated a transverse system of basement faults oriented NW–SE that dissected the earlier strike-wise developed NE–SW longitudinal fault system. The loading of the advancing thrust sheets also led to a major reactivation along the E–W Jorhat Fault with a major strike-slip component (Lahiri and Sinha 2014). Movement across the Jorhat Fault detached the Mikir forebulge or the convex flexural deflection developed in front of the Mikir Massif, thus separating NAS and SAS (Pahari et al. 2008). Following this event, the NAS and SAS behaved differently across the Jorhat Fault (Kent and Dasgupta 2004) acting as two separate sub basins.

Subsequently or simultaneously the Shillong-Mikir block was separated by the Kopili Fault. The fault is characterized by a right-lateral displacement and a steep, near-vertical fault plane (Singh et al. 2017) that developed the Kopili Valley/Graben in between the two massifs with the Mikir Massif moving further NE compared to the Shillong Massif. Seismic profiles establish it to be wider toward the southeast than toward its northwest end. The Kopili Fault also differentiates the Assam Valley into a Shillong block and an Assam block (Mikir Block) on either side of it with both the blocks rotating clockwise relative to Indian Plate (Barman 2014; Jade et al. 2017). GPS data sets indicate that the Shillong block rotates clockwise at 1.15° Myr^{-1} whereas the Assam block/Brahmaputra Block has a rotation rate of ≈1.13° Myr^{-1} (Vernant et al. 2014). The Shillong block along with theDauki Fault accommodates 25% strain resulting from the Indian Platemovement (Islam and Shinjo 2012) and hence is tilted slightly northward. This results in the compression of the Brahmaputra Valley and the shifting of the Brahmaputra River northward by at least 5 km at present (Das 1995).

These different tectonic phases in their respective geological periods have their individual sympathetic structural signatures that constitute part of the fabric of the basin. This control its evolution, subsequent tectonic trends, sedimentation pattern and also the distribution of hydrocarbon systems in the Assam Shelf. Deformations associated with the different episodes of tectonism may result in favorable basement exploration parameters viz., development of areas of maximum fracturing, creation of conduits for migration, generation of basement highs and reactivated shallower structural highs (Mazumder et al. 2019). As such, analysis of basement tectonics may enable to delineate conducive areas where basement or areas affected by basement tectonics may act as a potential hydrocarbon reservoirs. In the subsequent sections, different data sets like remote sensing and geomorphology, geological fieldworks, seismic data, image-logs, gravity and magnetics have been used to demarcate such favorable areas of the basement for exploration in Assam Shelf.

18.3 Principles and Methodology

Since the neotectonic Assam Shelf has undergone multiple episodes of tectonism, it can be premised that basement related structural features are reactivated and are manifested on the surface as outcrop or geomorphic features. Additionally, the study is based on the assumption that impressions/evidences of a particular phase of basement tectonics will be identical in pattern but can vary in scale subject to the methods/tools employed to analyze them. Such signature patterns of a particular phase of tectonics observed in outcrops may be repeated on different scales in subsurface seismic or electro-log data or in remote sensing images. As per the above presumption, structural patterns in the basement are extracted from different geophysical and geological data sets. These are subsequently compared with field analogues and integrated to prepare a conceptual model for basement exploration.

A two-phase methodology has been followed during the course of this study (Figure 18.5).

First, the basinal areas of the NAS and SAS as well as the basement outcrop areas of Shillong and Mikir Massifs are analyzed to delineate structures exposed or are manifested on the geomorphic surface.

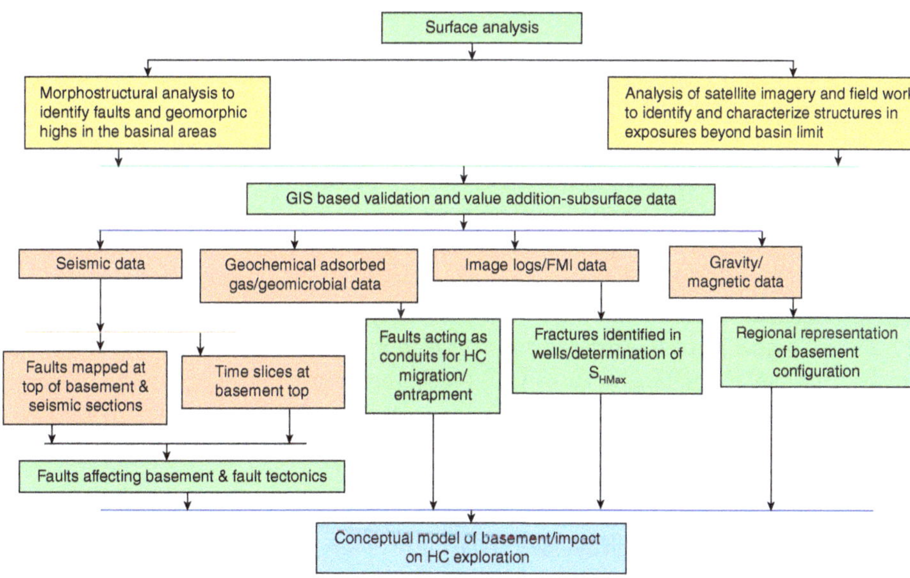

Figure 18.5 Flowchart showing workflow for the study.

In the basinal area of the Brahmaputra Valley, a neotectonics-based (morpho) structural analysis is carried out based on Mazumder et al. (2011, 2013a, b, 2016) to detect surface manifestations of the deep-seated/basement-related structures. These basement-related structures include both faults and structural highs. Both these elements reactivate repeatedly in the neotectonic regime and are manifested on the surface as anomalies in geomorphic elements e.g., topography, soil and especially drainage, which form the most sensitive parameters (Schumm et al. 2002). Such neotectonic manifestations like fluvial anomalies and seismicity are well studied and documented in the Assam Shelf (Gowd et al. 1998; Sarma and Sharma 2018 and others).

In this study, structure controlled drainage geometries (e.g. rectangular drainages, drainage offsets, compressed meanders) (Figure 18.6a), abrupt changes in surface slope (Figure 18.6c) derived from digital elevation models (30 m resolution ASTER DEMs downloaded from the EarthExplorer, USGS) and lineaments extracted from tonal variations from 30 m resolution Landsat 8 OLI images (EarthExplorer, USGS) (Figure 18.6d) are used to delineate fault segments.

All these data sets can be considered as surface representations of a small part of subsurface fault or building blocks of a regional fault that can be joined based on their apparent continuity and similarity in alignment.

Correspondingly, geomorphic highs signify surface manifestations of probable structural highs in basement that are identified from features such as radial and peripheral drainages (Figure 18.6b). Associations with features like barbed drainages, rectangular drainage patterns, a lower drainage density and low moisture contents are used to identify geomorphic highs with a higher degree of confidence based of representing basement structural highs. These morpho-structural interpretations are subsequently validated by extensive field checks where the mapped structures are checked for neotectonic faulting imprints and evidences like unpaired terracing (Figure 18.6e), multiple terrace levels, abnormal incisions and sudden appearance of braiding.

In areas of Shillong and Mikir massif where the basement is exposed, field work and image/DEM analysis have been used to identify structural features in the outcropping basement rocks. Different band composites of LANDSAT 8 OLI image sets (EarthExplorer, USGS) have been pan-sharpened, contrast stretched and edge enhanced to identify different regional scale structural features. Instances of folding in basement lithology from band tracings (Figure 18.6f), abrupt termination of basement outcrops to delineate faults and prominent lineament patterns have been deciphered from different image data sets. As a part of fieldwork in these outcrops, attitude of joint sets and fractures developed in basement lithology were measured, characterized and used to identify trends of their causative regional faults and for modeling of subsurface basement features.

These surface interpretations (Figure 18.7) are further extrapolated and validated using ancillary geological and geophysical information to delineate their subsurface continuity/disposition of the basement related structural features and develop an integrated basement model.

For an initial level of correlation, the faults mapped using morphostructural analysis were overlain with faults that had been mapped at the top of different horizons using seismic data (Figure 18.8a). A positive correlation not only implies the subsurface continuity of the surface fault down to the correlated level but also indicates repeated activation of the

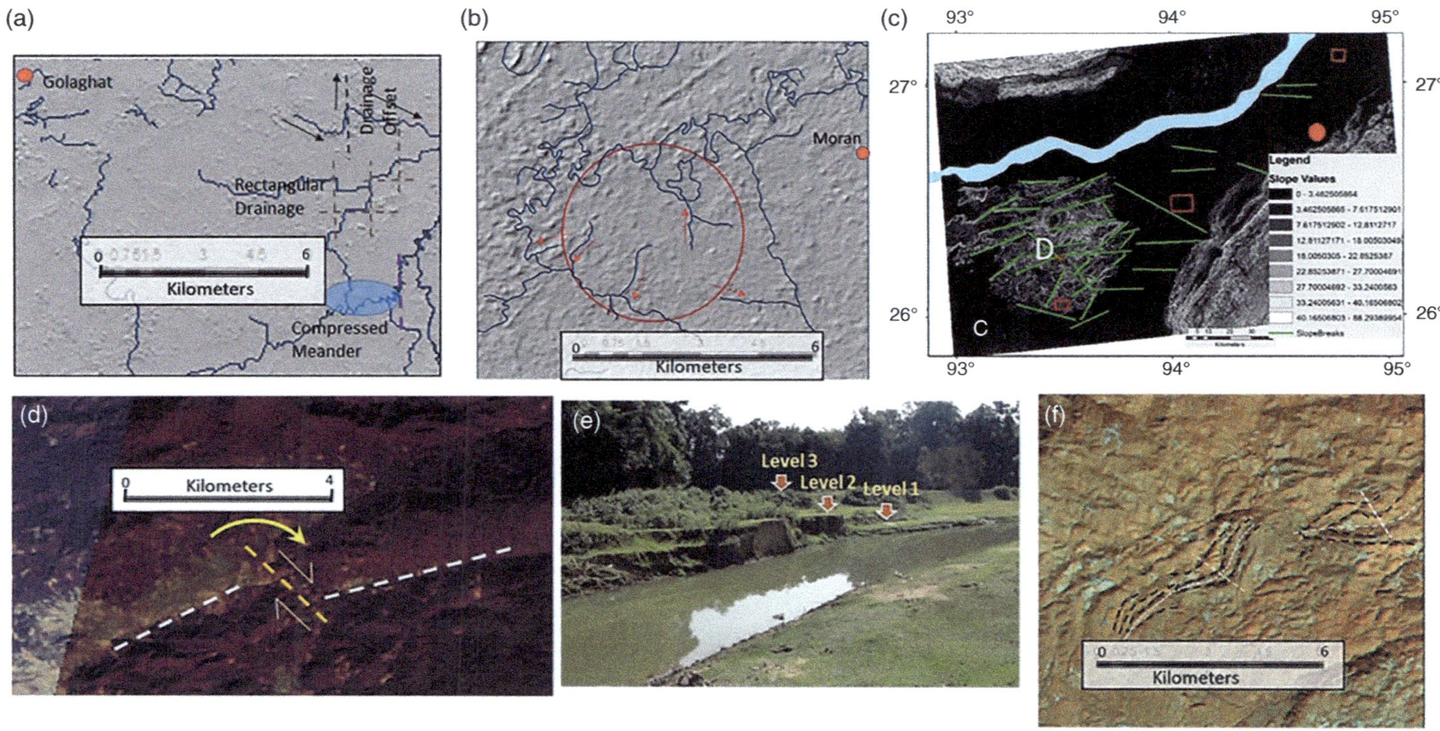

Figure 18.6 (a) Morphostructural faults near Golaghat, SAS identified from structurally controlled drainage geometries like drainage offsets, rectangular drainages, and compressed meanders. (b) Geomorphic highs implying underlying structural highs identified from semi-radial drainages near Moran in NAS. (c) Lines of abrupt slope breaks identified from the slope map derived from ASTER 30 m resolution ASTER Digital Elevation Models where lighter shades indicate a higher slope and darker shades stand for a gentler slope from the outcropping Mikir Massif area. (d) Lineaments identified from LANDSAT 8 OLI composites of bands 7, 4, and 1 in the Mikir Massif show a clockwise rotation of trends. (e) At least 3 levels of unpaired terracing observed along a channel near Geleki suggestive of at least 3 phases of neotectonic activations. (f) Band tracing in LANDSAT 8 OLI image composite of bands 4, 5, and 7 in Mikir Massif shows the development of simple and superposed folding. The black dotted lines indicate the fold geometry, whereas the white dotted lines indicate axial traces.

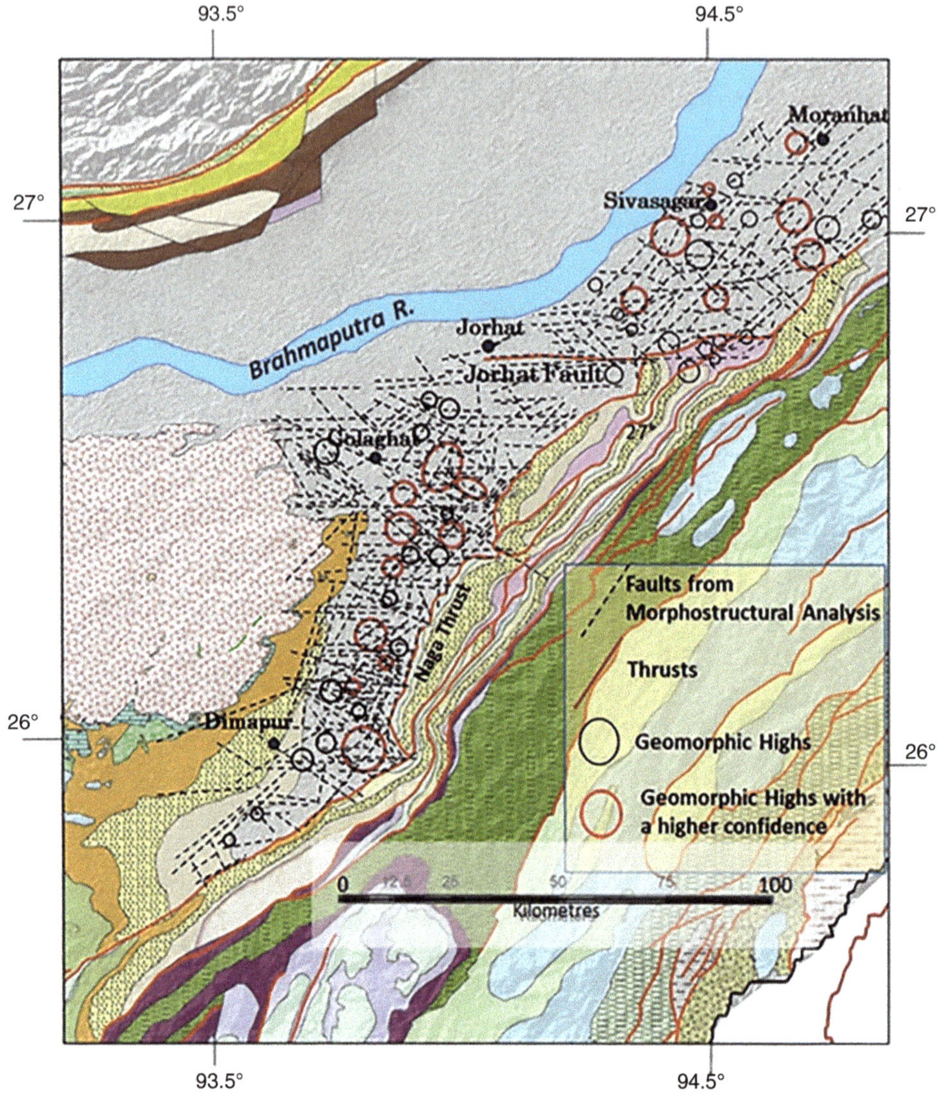

Figure 18.7 Faults and Geomorphic Highs that had been mapped using morphostructural analysis in the North and SAS (outcrop legends same as Figure 18.1a).

subsurface faults mapped using seismic data and its subsequent manifestation on the surface. Morphostructurally interpreted geomorphic highs when geospatially superposed with structural highs that have been mapped at basement (Figure 18.8b) or any other deep-seated horizon display a good amount of correlation. Such a correlation implies that surface geomorphic highs can act as indicators of subsurface structural highs.

Formation Microimager (FMI) logs recorded in deep wells principally to study fracture patterns. In the present study, FMI has been used to analyze the fracture patterns in the basement encountered in the various deep wells of Assam Shelf in which it has been

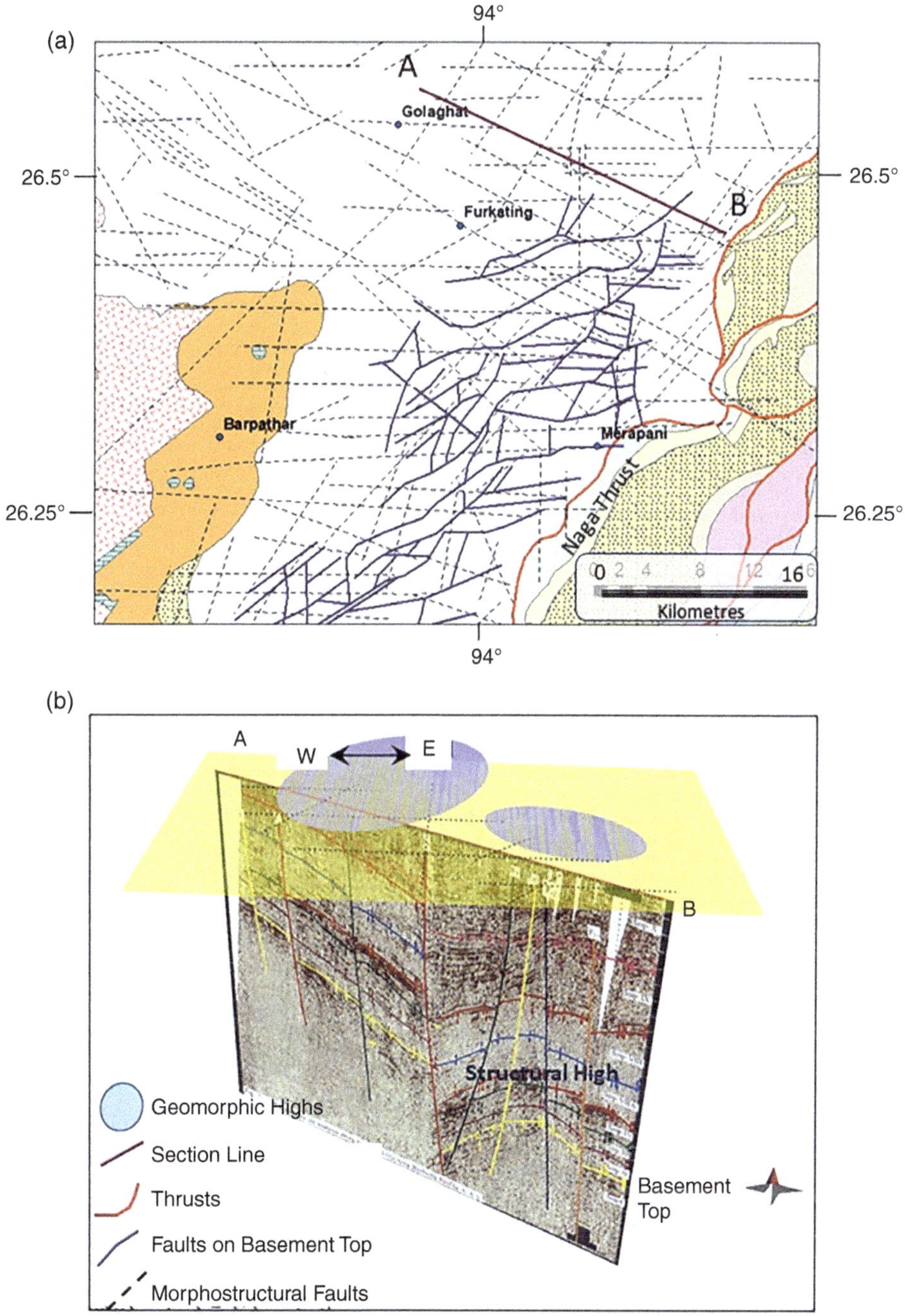

Figure 18.8 Correlation of faults mapped at the top of the basement with morphostructural faults. (Outcrop legends same as that in Figure 18.1a and b). Structural highs at the basement top level in seismic sections found to correlate with geomorphic highs at the surface. The line of section AB is marked in the map in red. *Source:* Adapted from Mazumder et al. (2018).

recorded. These fracture patterns interpreted from FMI logs are further correlated with the surface fracture patterns from outcrops as well as that deciphered from morphostructural analysis. These correlations help in the characterization of the different fracture sets and correlate them with sympathetic faults to subsequently derive episodes of the tectonism suffered by the area.

The FMI/Borehole Image data sets in this study have also been used for tectonic stress analysis. Borehole breakouts and drilling-induced fractures (DIFs) observed in the image logs (Figure 18.9) serve as important indicators of horizontal tectonic stress orientation (Alizadeh et al. 2015; Dasgupta et al. 2019) locally. Such logs have been recorded up to about 3200 m depth reaching Basement in SAS and a maximum of 4400 m in NAS in Basement or upto shallower depths in Tipams and Barails. Borehole breakouts are enlargements and elongation of a borehole in the direction parallel to the minimum (least) horizontal stress (S_{hmin}) (Prensky 1992). DIFs are fractures (Figure 18.9) observed in image logs as subvertical narrow zones of low reflectivity that are separated by 180° and are aligned subparallel to the borehole axis (Tingay et al. 2008) with their orientation indicating maximum horizontal stress (S_{Hmax}). However, this analysis provides that state of stress in a localized area around the well. For a further regional analysis of tectonic stress, passive seismic studies or fault plane solutions of earthquakes in this area have been used. Since most parts of the Assam shelf area had been considered to be a seismic gap due to the non-occurrence of major earthquakes over a long time (~100 years), focal plane solutions from the adjacent fold and thrust belt have been used to determine the direction. From the focal plane solutions (Angelier and Baruah 2009; Thingbaijam et al. 2008) as well as the World Stress Map (Heidbach et al. 2016) that provides a geospatial compilation of present-day stress indicators, an overall average of NW-oriented Maximum Compressive Stress (S_{Hmax}) had been deciphered and delineated in both the NAS and SAS.

The direction of the S_{Hmax} has been used to determine the orientation of the basement faults that would act as conduits for vertical remigration of hydrocarbons vis-à-vis alignments of faults that would act as entrapments. Faults that are (sub)-parallel to the maximum principal stress are assumed to be conduits for migration, whereas those at high-angle serve as entrapments (Zhang et al. 2011). Such remigration along the faults is manifested and validated by a number of parameters such as shallow-level hydrocarbon shows observed in wells during drilling, and hydrocarbon (Figure 18.9b) on the surface. Additionally, geomicrobial and adsorbed gas anomalous concentrations in soils have been proved to be reliable indicators of oil and gas in the subsurface. The alignment of high adsorbed gas or geomicrobial concentration along a morphostructural fault suggests that the fault acts as a conduit for vertical remigration (Figure 18.9c). Also oil-to-oil correlation studies, where the different oils from different horizons are correlated based on some biomarkers, can serve as instruments to identify faults acting as conduits. As a part of the last parameter, molecular-level characterization of organofacies and maturation level of oils produced from wells drilled at different horizons along the remigration faults in the Lakwa field of Sivsagar District, Assam were tested. Based on the results, the oils encountered in the Barails were found to be similar in parameters mentioned above to that of Tipam Sandstone, which were subsequently found to be similar to that of the Girujan Clay Formation (Goswami 1998, 2001 and 2005). This implies that the same hydrocarbon from the Barail reservoirs had remigrated upward.

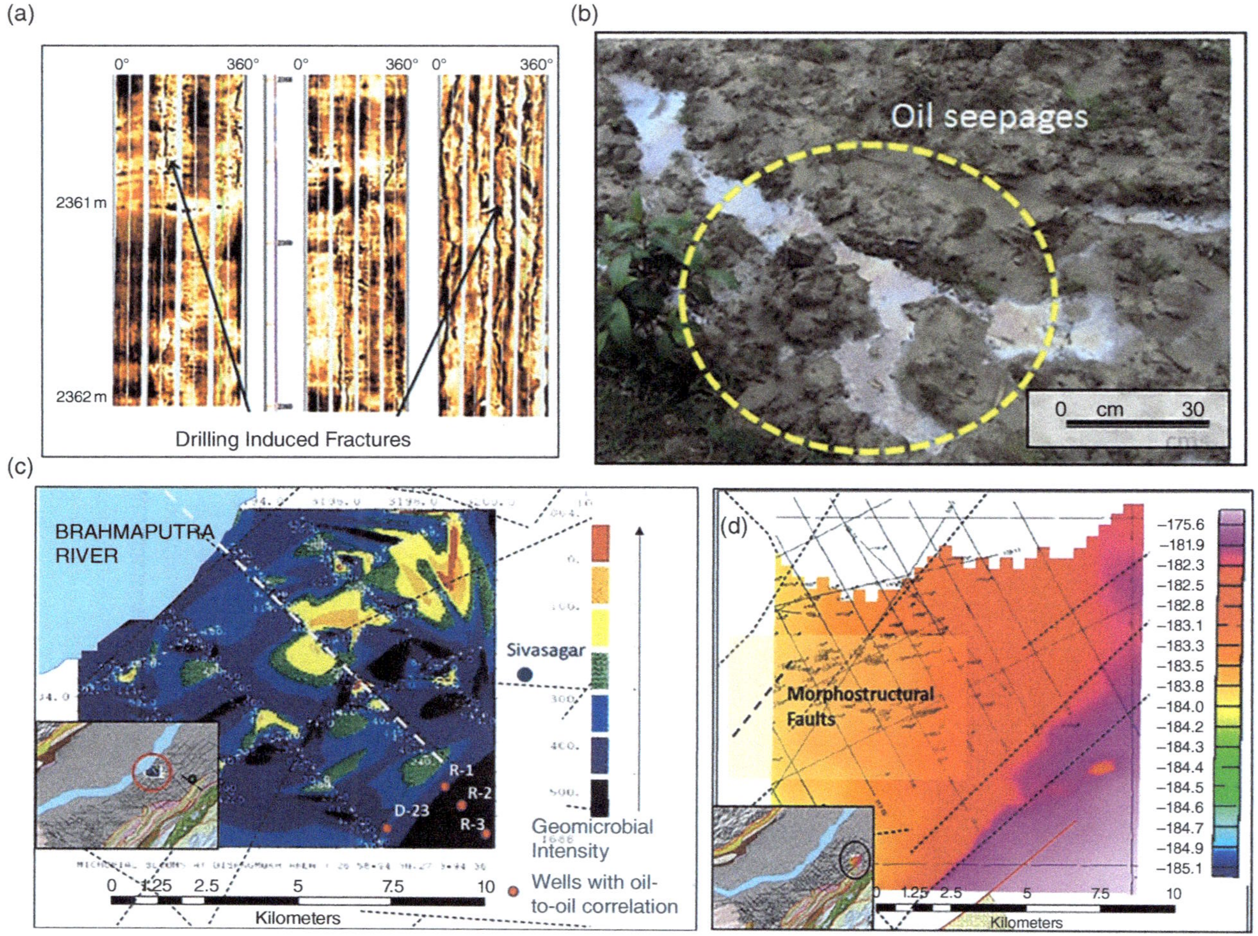

Figure 18.9 (a) Drilling-induced fractures in image logs of Geleki Field showing S_{Hmax}-oriented NW. *Source:* After Tep et al. (2015). (b) Oil seepages observed along the NW–SE fault in NAS with its location marked by a green dot in the inset figure of (c). (c) Geomicrobial anomalies aligned along the NW–SE fault marked in dotted white line. Additionally, wells with oil-to-oil correlation between deeper and shallower horizons are also observed to be oriented along the fault suggesting that the fault acts as a conduit for vertical migration. *Source:* Nathaniel et al. (1997). Location of the map is marked in the insert figure highlighted by a red circle. (d) Correlation of the surface morphostructural faults with Bouguer anomaly of Lakwa field of the NAS. The NE–SW trending faults are found to show a fair amount of correlation with the Bouguer Gravity variations implying that these faults are manifestations of deep-seated features. Inset map shows the position of the Bouguer Gravity map.

Gravity and magnetic data help build a regional representation of basement configuration. Deep-seated crustal-level structural elements affecting basement and associated basement uplifts/subsidences are found to be manifested as gravity or magnetic signatures. As such, the signatures from gravity and magnetic maps have been used as tectonic guidelines for the surface interpretations. These gravity-magnetic signatures have also been used for correlation with surface signatures (Figure 18.9d) in order to determine whether the interpreted surface features are penetrative down to the basement or below levels and are, thus, being controlled by the regional basement fabric of the area.

18.4 Results and Discussions

18.4.1 E–W Trends in South Assam Shelf

Results of the band tracings depict that the F2 folding with an E–W axial traces form the most dominant trend. This implies the axial plane foliation corresponding to F2 folding forms the most dominant foliation in the basement of the area along which the later structures may be oriented (Figure 18.10). Such F2 foldings are also found to be dominant in the Shillong Massif especially in the Shillong Series and are associated with E–W-trending fractures parallel to the axial trends (Figure 18.10b). Fieldwork carried out in the accessible parts of Mikir Massif to characterize the fractures indicated that most of the E–W-trending fractures are vertical, whereas the other two trends are found to show some amount of dip (Figure 18.10c).

In the Dhansiri Valley part (SAS) of the Assam Shelf, seismic time slices studied to draw an analogy with the exposed basement depict a folded basement fabric with a dominant imprint of E–W-oriented fold axis implying an F2 phase of folding (Figure 18.11). The digitized time slice of the Borholla area at 1900 ms when overlain with morphostructural faults mapped at the surface indicates the E–W faults coincide with the F2 axis forming the predominant foliation of the area. Additionally, some degree of strike-slip movement is also apparent along these E–W faults indicating a probable accommodative clockwise rotational movement in this area (Figure 18.11b). See Mukherjee and Khonsari (2017) for rotational fault kinematics. Deep wells B-22 and B-26 drilled by ONGC and placed in line with these E–W-oriented faults are found to show predominantly E–W-oriented fractures/joints, along with subsidiary NW–SE faults (Figure 18.11b). Conventional cores recovered from the basement in wells B-11 and B-12 that occur in the extrapolated trend of the E–W faults (position documented in Figure 18.11b) show dominantly vertical fractures (Figure 18.11) along which maximum golden yellow fluorescence (GYF) indicative of hydrocarbon occurrence is concentrated. Since outcrop field studies in Mikir Massif reveal that most of the E–W oriented fractures are vertical, in a converse analogy, it could be conjectured that the vertical fractures that are observed in the recovered cores were probably E–W-oriented. Deep well F-3 (position documented in Figure 18.11b) is also placed along an E–W fault and the F2 fold axis shows evidences of strike slip along vertical fracture planes that are again predicted to be E–W (Figure 18.11c). As such, these E–W-oriented faults have a major strike slip/shear component and play a major role in increasing basement fracture porosity and localizing hydrocarbon at least in the SAS area.

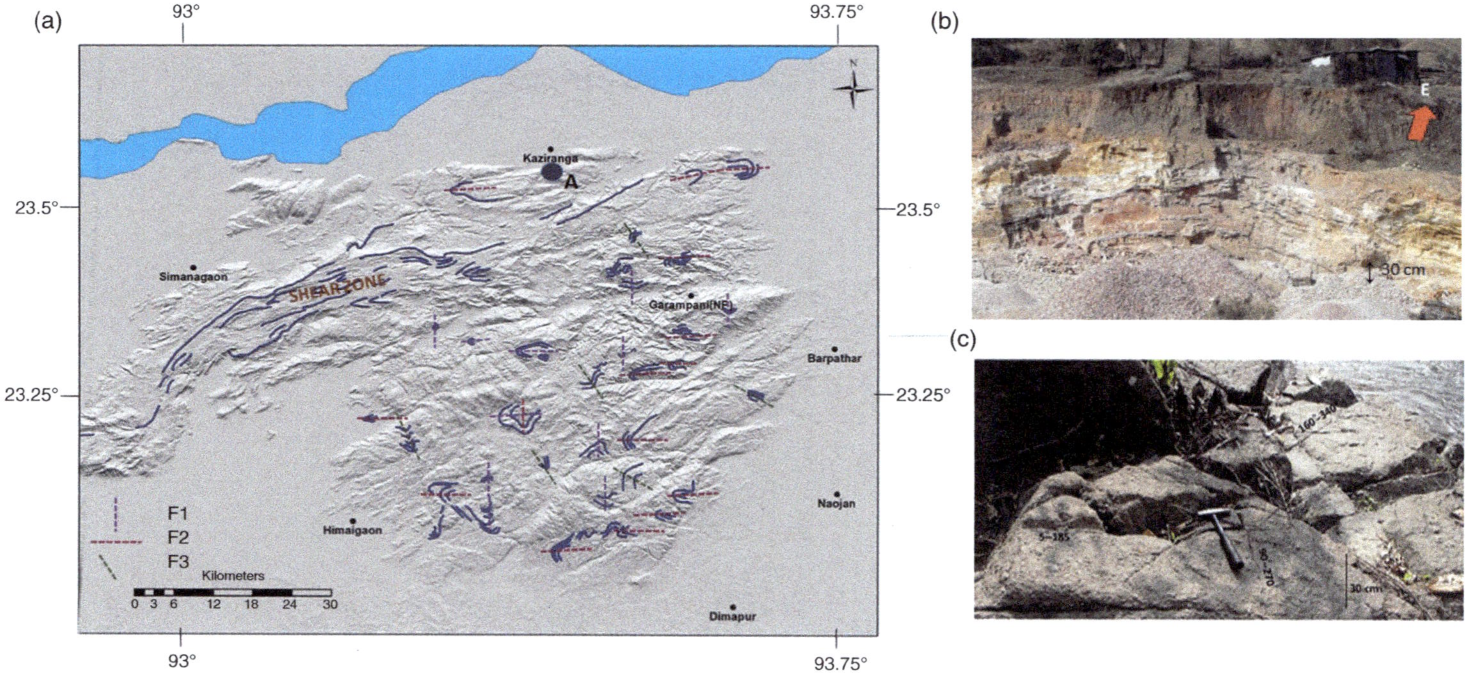

Figure 18.10 (a) Band tracing in the ASTER DEM mosaic of 30 m resolution representing Mikir Massif. Results indicate different generations of folding in the Mikir Massif area. At least three generations of folding are identified of which the F2 trend is the most dominant. Superimposition structures of folds F1 and F2 are also observed. (b) F2 folding observed in quartzite of Shillong Series on the Shilllong–Cherrapunjee Road showing well-developed axial plane foliation and jointing in E–W orientation. (c) Joint sets developed in basement outcrop in the river bank near Kohora in Mikir Massif-oriented E–W, N–S, and NW–SE.

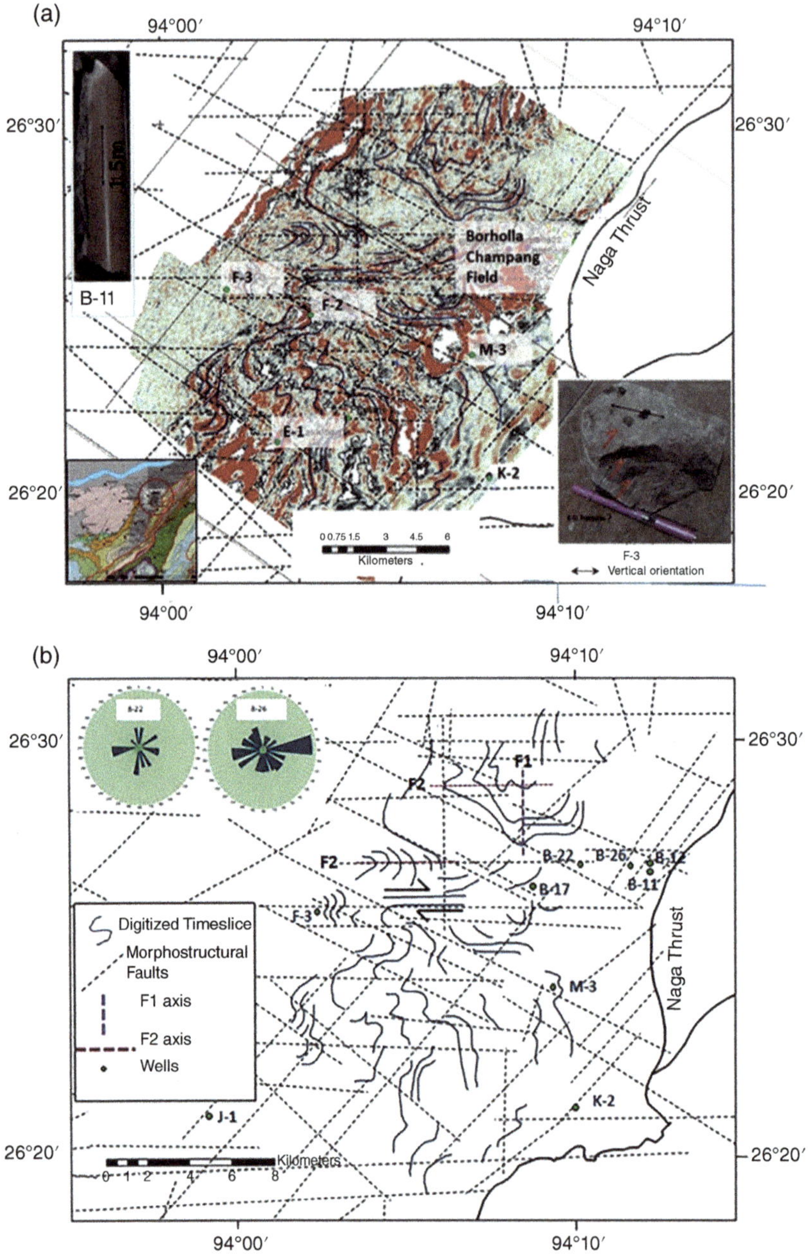

Figure 18.11 (a) Timeslice of Borholla area in Dhansiri Valley (SAS) at 1900 ms correlated with morphostructurally interpreted faults in the area. (b) Digitized Timeslice of Borholla area at 1900 ms shows the development of a dominant F2 generation of folding with an E–W fold axis. Correlation with morphostructural faults indicates the E–W faults correlating with the F2 axis act as shear planes along which strike-slip behavior is observed. Superposition of F1 and F2 axis is also observed. Inset figures in Figure 18.11 show the map position of the time slice and also depicts conventional cores recovered from well B-11 and F-3 (positions depicted in Figure 18.11b) showing the development of vertical fractures. In B-11, GYF indicating presence of hydrocarbon has been observed along the vertical fractures. In well F3, evidences of strike slip/shearing have been observed along the vertical fractures. Inset figures in (b) depict stereograms of B-22 and B-26 oriented along these E–W-oriented faults showing a dominance of E–W-oriented fractures. *Source:* Akbar et al. (2015).

18.4.2 Rotation of Structural Trends and Bomdila Lineament

Subsequent to the rise of the conjoined Shillong Mikir massifs, it was split by the Kopili Fault in Late Pliocene or later. The Assam block on the east of the Kopili Fault is found to rotate at a slightly faster rate than the Shillong block west of it resulting in a clockwise rotation of the Mikir Block compared to the Shillong Block (Vernant et al. 2014). A plot of epicenters along the Kopili Graben shows that seismic activities concentrated more toward its narrow northwestern part than the wider southeastern areas implying the northwestern part to be more seismically neotectonic (Duarah and Phukan 2011). Also along the Kopili Graben, a series of NNW–SSE gravity lows and highs (Singh and Barua 2001) have been interpreted suggesting smaller secondary horsts and grabens bounded by faults underlying the Kopili Graben area. These horsts and grabens are found to show a dextral offset against a centrally located ENE–WSW fault zone (Figure 18.12). This is also observed in the vertical component of magnetic data and aeromagnetic datasets (Singh and Barua 2001). The movement along the Kopili Fault is found to be associated with a clockwise rotation of the Mikir Block compared to the Shillong Block. The major lineament and tectonic elements when correlated with each other are found to show a change in trend from NE–SW in the Shillong Massif to almost E–W in the Mikir Massif (Figures 18.6d and 18.13). This change in trend (Ganju and Ram 1987) can be attributed to the clockwise rotation of the Mikir Block (Figure 18.14) during the post Pliocene.

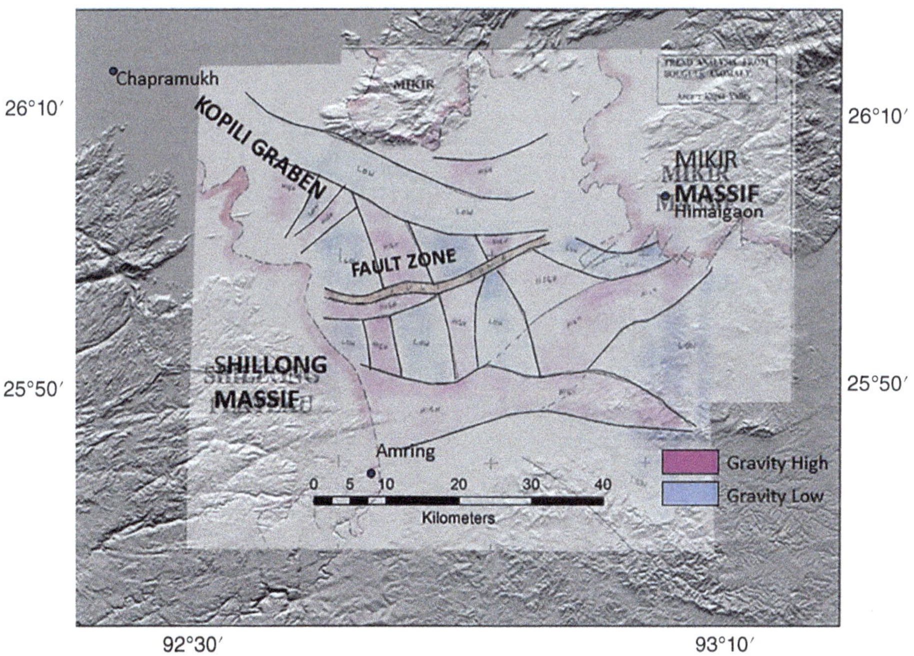

Figure 18.12 Gravity field studies (Singh and Barua 2001) indicate a series of NNW–SSE gravity lows and highs implying smaller secondary horsts and grabens bounded by faults underlying the Kopili Graben area. These small and narrow horsts and grabens are eventually offset by an ENE–WSW-oriented fault zone continuing from Shillong Massif to Mikir Massif.

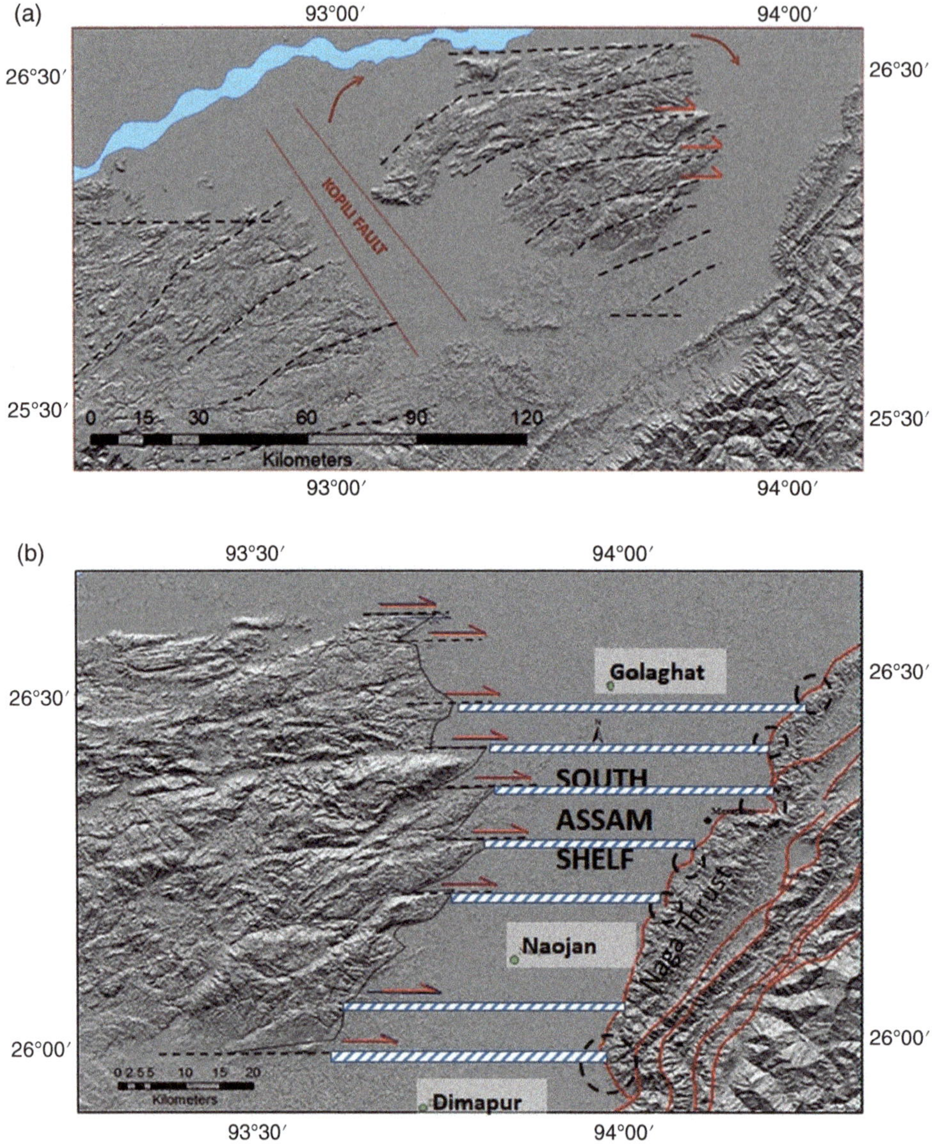

Figure 18.13 (a) Probable clockwise rotation of the Mikir Massif and Dhansiri Valley part of SAS with respect to the Shillong Massif that has led to the clockwise rotation of the NE–SW tectonic elements (marked in dotted lines) in the Shillong Massif to almost E–W in the Mikir Massif. This rotation leads to a dextral translational movement of the faults in the Mikir Massif. (b) This movement is evident in the eastern edges of the Mikir massif showing a serrated border with one block appearing to slip past the other as a result of this clockwise rotation. The causative strike-slip faults probably continue in the basement of the basinal part of the SAS occurring as bands or zones of E–W fracturing (marked as striped zones) continuing from their serration in the Mikir Massif to the Naga Thrust, which marks the frontal boundary of the Thrust Zone. Corresponding areas of Naga Thrust that lie across these serrations show the development of re-entrants (marked in black dashed circles) due to the effect of these E–W zones that are abundantly reactivated and highly fractured.

Figure 18.14 Block diagram attempting to explain the clockwise rotation of the structural trends from NE–SW to E–W.

This clockwise rotation is laterally converted/translated to a dextral strike-slip translational motion that is evident at the eastern edge of the Mikir Massif, where different blocks appear to slip past each other in a strike-slip manner (Figure 18.13b) as observed in the seismic time slice described earlier. The causative strike-slip faults continue in the basement of the basinal part of the Dhansiri Valley/SAS resulting in fracturing of the basement along its strike and also result in the corresponding development of a re-entrant in the Naga Thrust. These E–W faults probably occur as bands or zones of fractures continuing from their serration in the eastern margin of Mikir Massif to their corresponding re-entrant in the Naga Thrust (Figure 18.13b), the entire zone being abundantly reactivated and hence highly fractured.

Remote-sensing image analysis of the Assam Shelf shows that the BSZ continues from the Shillong Massif where it has a NE–SW trend to the Mikir Massif where it changes its trend and rotates clockwise. This trend of the BSZ probably continues in the subsurface of the Dhansiri Valley (SAS) as suggested by the dominance of E–W trend at its corresponding position in the eastern margin of Mikir Massif. Further eastward, it correlates with the subsurface trace of the Jorhat Fault creating a large re-entrant in the Naga Thrust (Figure 18.15).

To accommodate this dextral or clockwise rotation of the Mikir Massif and SAS, a number of NW–SE regional strike-slip faults might have been formed in the shelf area. One such major fault is the Bomdila Lineament, which forms a very regional tectonic element extending from Arunachal Pradesh across the Naga Thrust. This lineament is more or less parallel to the Kopili Lineament and is found to show offsets across the MBT as well as in the Siwaliks, near Bomdila, Arunachal Pradesh. The lineament further creates drainage offset in the Brahmaputra River, of the order of ~22 km along with other geomorphic evidences in the Dhansiri Valley (for details, see Sarma and Sharma 2018; Mazumder et al. 2018) and further

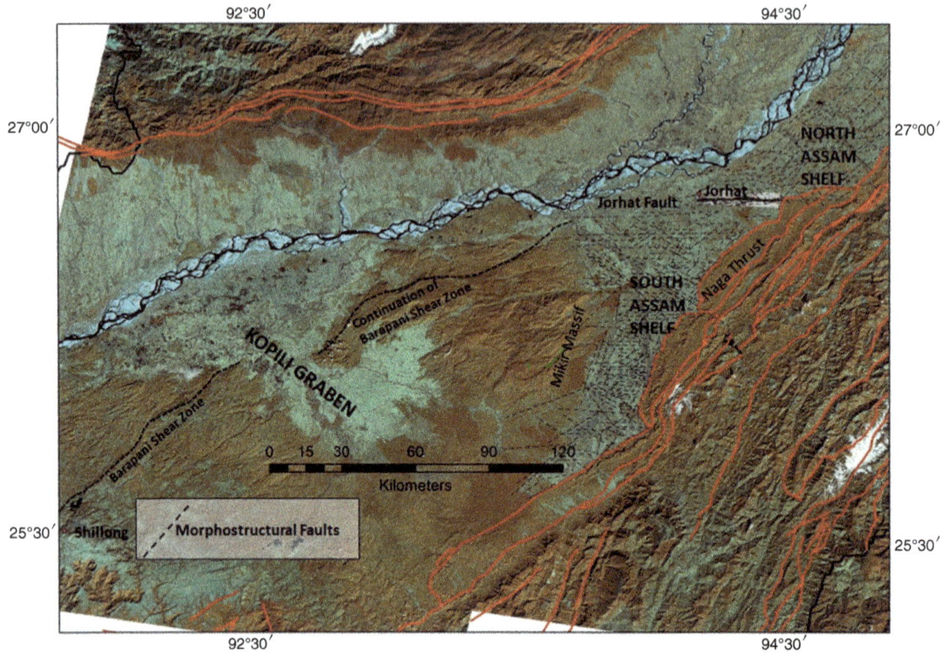

Figure 18.15 LANDSAT 8 OLI image composite of bands 4, 5, and 7 shows the BSZ continuing from Shillong Massif along a NE trend that swings clockwise in the Mikir Massif and further continues E–W in the subsurface of the SAS. This correlates with the subsurface trace of the Jorhat Fault (mapped at Sylhet-level limstone) and creates a re-entrant in the Naga Thrust.

southward as it intersects the Naga Thrust as a re-entrant (Figure 18.16). A number of interpreted morphostructural faults are found to be aligned along this lineament suggesting it forms a broad zone of faulting and associated fracturing. Its surface evidences are observed mostly as geomorphic manifestations in the form of cliffs (Figure 18.16b) or as incisions in drainages aligned along the Bomdila Lineament. In the subsurface, the Bomdila Lineament is found to have a significant signature on the basement top, forming a depression of at least 6 km width and ~500 m seconds in time units of two-way time (TWT), roughly equivalent to 600–650 m. (Figure 18.16c). Faults interpreted in the seismic section are found to be concentrated along the depression forming geometries like negative flower structures suggesting distinct effects of transtensional strike-slip deformation. Plots of earthquake epicenters that are aligned along the Bomdila Lineament indicate foci as deep as 33 km reaching lower crustal levels (Figure 18.16). This implies that the Bomdila Lineament is a deep-seated feature that had been reactivated multiple times and might have caused extensive fracturing in the basement and is presently still active. This is corroborated from dipmeter log data used in general to identify fractures in deep wells like B-15 drilled in the vicinity of the Bomdila Lineament, depicting that basement is fractured with most of the fracture azimuths oriented in the NW–SE direction parallel to the Bomdila Lineament (Prasad and Srivastava 1987). Sympathetic NW–SE trends that occur further south of the Bomdila lineament resulting from the rotation of the Mikir Block can be assumed to be equally fractured. These faults

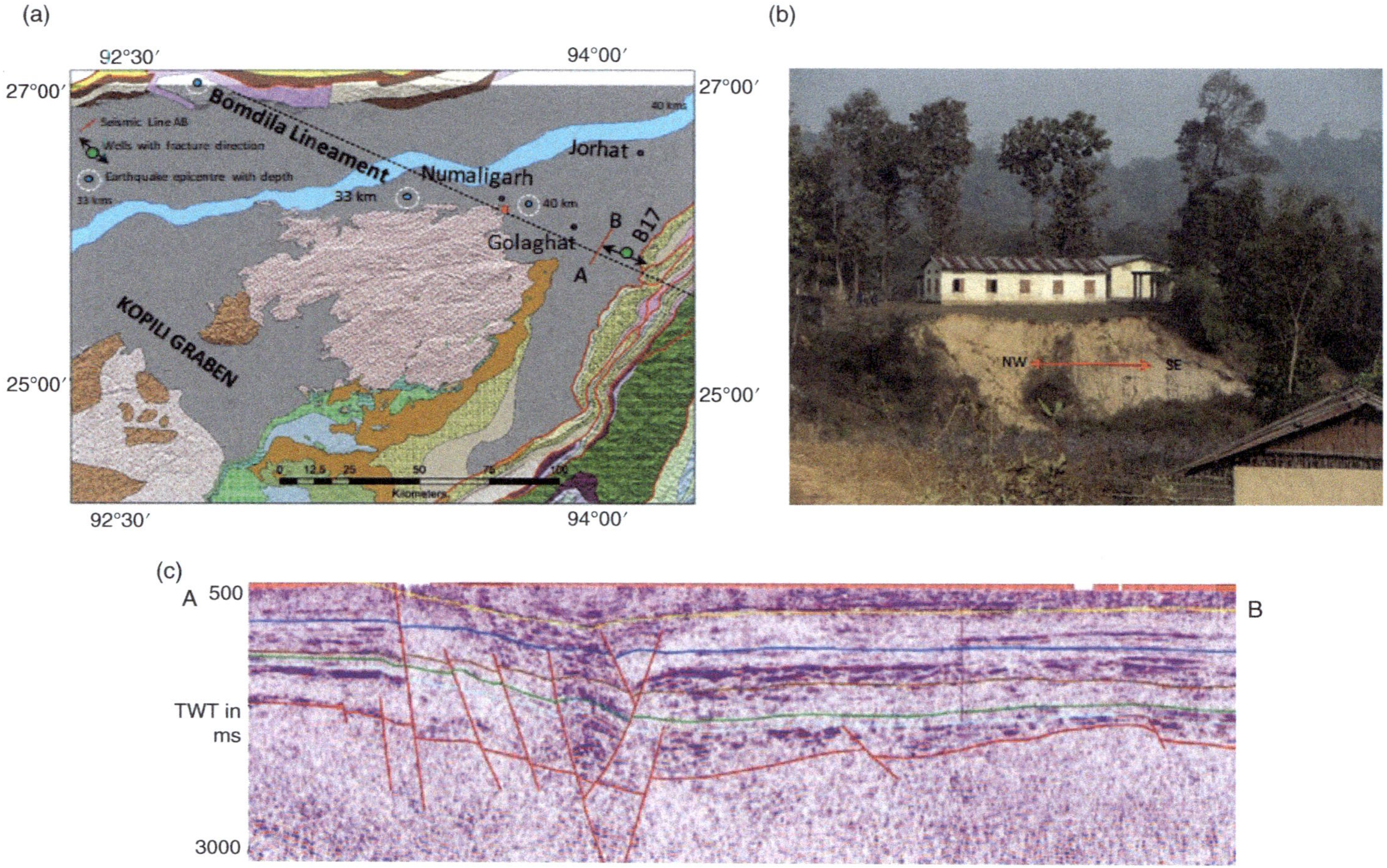

Figure 18.16 (a) Map of the Bomdila Lineament showing an offset of about 22 km in the River Brahmaputra. Effects of the lineament are evident in features like earthquake epicenters with foci as deep as 40 km and fractures in wells. (b) About 6 m high cliff-oriented NW–SE developed in alluvium near Numaligarh due to the effect of the Bomdila Lineament. The location marked as a red point in Figure 18.16. (c) The seismic section along the line AB across the lineament suggesting a depression in the basement of about 6 km and development of faults (negative flower) in the zone indicative of strike-slip kinematics.

will also follow the surface morphostructural trends and will be similarly fractured as Bomdila Lineament but probably be rotated more clockwise than it.

18.4.3 Implications in Hydrocarbon Exploration from the Basement in SAS

The World Stress Map (Heidbach et al. 2016) indicates that the S_{Hmax} in the study area is probably oriented NW to E–W (Figure 18.17a). This indicates that the fractures that are oriented in trends varying between these two orientations plausibly act as conduits of vertical migration of hydrocarbon. Correlation of the morphostructural faults with results of geomicrobial surveys (indicating hydrocarbon signatures) suggests that the geomicrobial anomalies are either aligned along or oriented at low angles to the NW–SE faults (Figure 18.17b and c). This further implies that the NW faults probably act as conduits for upward migration of hydrocarbon in that area.

The E–W trending strike-slip faults traversing the Dhansiri Valley from the Mikir Massif to Naga Thrust form linear zones of basement fracturing (Figure 18.13b). Similarly, the NW–SE faults that are sympathetic to the Bomdila Lineament are also instrumental in creating fractured basement zones along its trajectory. Hence, areas of intersection between these two trends along with the earlier created NE–SW trend will create areas of the basement that are supposed to be intensely fractured. These areas will have very high basement porosity and also may be more probable to be associated with hydrocarbon accumulations. Well D1 drilled in such an E–W zone at the junction of these three trends shows evidences of the fractured basement at about 3000 m during drilling and an interplay of NW–SE, NE–SW, and ENE–WSW to E–W faults and fractures in FMI logs and results of seismic attribute analysis (Figure 18.17d) (Kataruka et al. 2017). Hydrocarbon shows were observed during drilling of its basement, but testing results of basement interval showed a commercially unviable flow of feeble gas (based on unpublished well completion reports). However, one of the salient features of the basement acting as a good reservoir, other than it being intensely fractured, is that it should be associated with a basement high. The geomorphic highs which had been categorized to be of higher confidence to represent subsurface structural highs may indicate such basement highs underlying them. Also, a density map of the surface morphostructural faults may stand as an alias to represent the fracture density of the basement. As such areas of overlay of the high-density zones of morphostructurally interpreted faults with geomorphic highs of higher confidence occurring in an E–W zone of faulting (connecting serrations in Mikir Massif to corresponding re-entrants in Naga Thrust) (Figure 18.17e) may imply the most promising areas of basement exploration in the SAS.

18.4.4 Basement Deformation in the North Assam Shelf

The basement in the NAS or the southern bank of the Brahmaputra Valley can be believed to be less deformed than that of the SAS/Dhansiri Valley. The simultaneous uplift of the conjoined Shillong-Mikir massifs on the western boundary and tectonic loading of the thrust sheets on the eastern boundary in the Late Miocene–Pliocene result in a much steeper dip of the basement in the SAS than in the NAS. A comparison between the directional analysis of the morphostructural faults of the NAS and SAS shows a distinct variation in the tectonic trends implying these two areas had suffered separate styles of tectonism (Figure 18.18). The basins forming NE–SW faults are much more dominant in the NAS

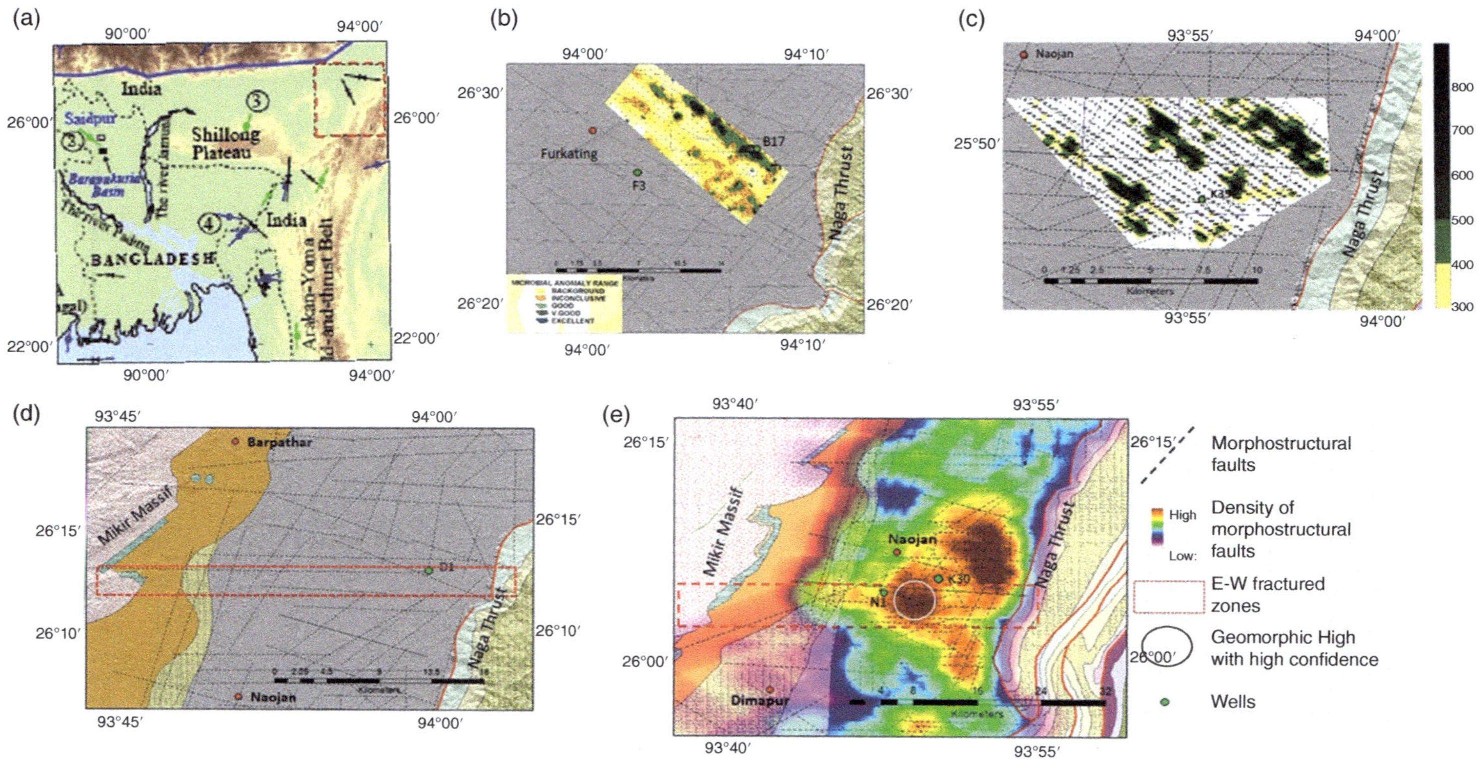

Figure 18.17 (a) Part of World Stress Map (Heidbach et al. 2016) based on present-day stress indicators, indicates that the SHmax in the SAS area is probably oriented NW to E–W (b) and (c) Geomicrobial anomalies assumed to be reliable indicators of oil and gas in the subsurface are found to be oriented NW–SE in the Dhansiri Valley (SAS) in both the Borholla area in the north and Khorghat-Nambar area in the south suggesting that the NW–SE faults are probably the principal conduits for vertical migration of hydrocarbons. (d) Well D1 drilled in the E–W-fractured zone at the intersection of the NW–SE fault shows a fractured basement with oil shows. (e) An overlay of high-density morphostructural faults and geomorphic high with high confidence occurring in an E–W zone of faulting may imply the most promising areas of basement exploration in the SAS. *Source:* Heidbach et al. 2016/World Stress Map/CC BY 4.0.

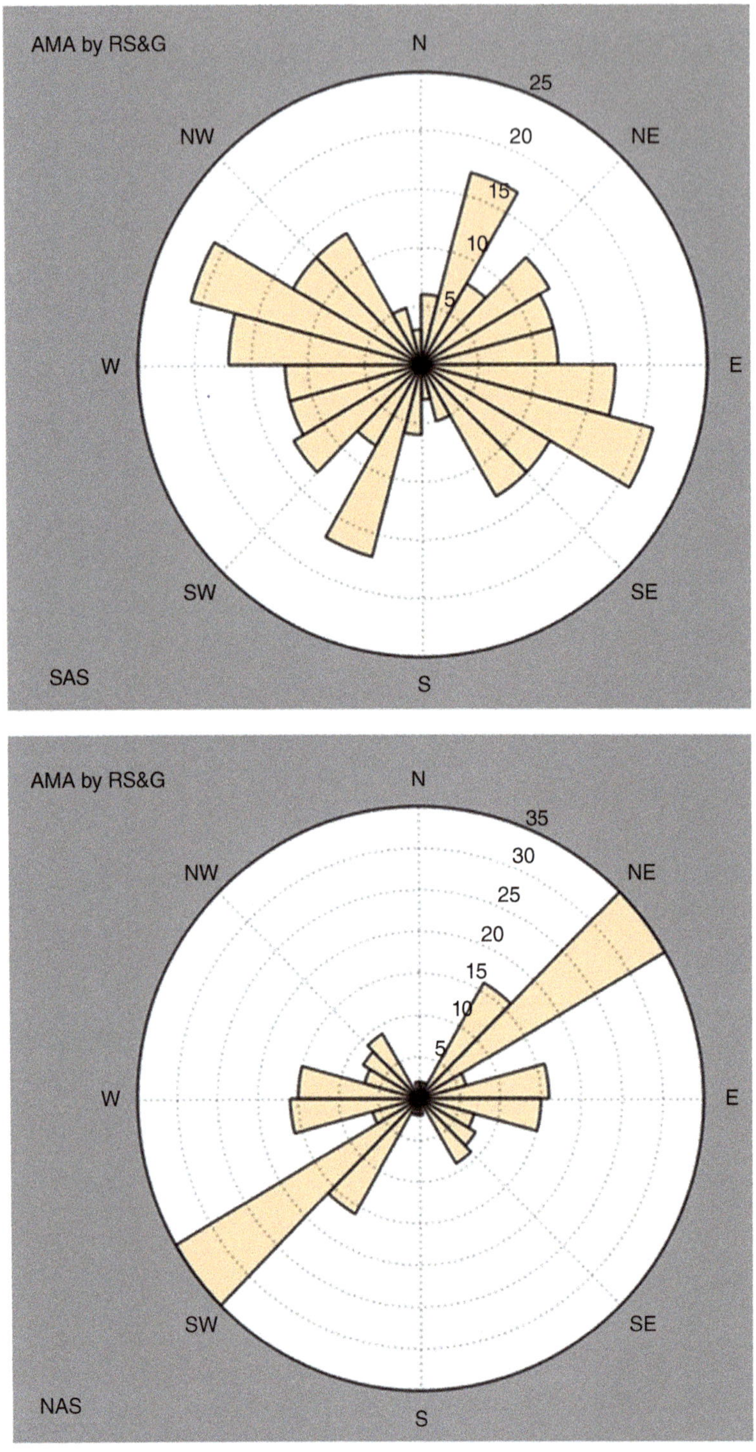

Figure 18.18 Rose diagram showing directional analysis of SAS and NAS structural trends. The NE–SW basin-forming trend in the NAS is found to be more dominant and more clockwise rotated with respect to the SAS.

and are found to be rotated more clockwise with respect to the SAS. The Jorhat Fault that separates these two areas probably acts as a tectonic barrier.

The morphostructural faults that had been mapped in NAS in the south bank of the Brahmaputra when correlated with faults mapped on Girujan top in the North Bank, it is observed that the faults mapped from seismic data in the north bank of the Bramhaputra can be directly correlated with its morphostructural counterpart in the south bank. These correlated faults are broadly E–W-oriented and can be traced on both sides of the Brahmaputra and hence may be assumed to be a part of a single fault (Figure 18.19). Since these E–W faults strike parallel to the Jorhat Fault, it can also be believed to be characterized by a high component of dextral strike slip in it.

In case there are two E–W-oriented bounding dextral strike-slip faults overlapping each other, it is possible that it will lead to a top-to-the-right shear with the development of a transtensional component between the bounding strike-slip faults. This might lead to the development of en-echelon NE–SW faults with associated development of E–W-trending strike-slip faults parallel to the bounding faults and another NW–SE system of strike-slip faults perpendicular to the extensional direction (Figure 18.19b). The fault pattern in the study area as derived by morphostructural analysis appears very much similar to the above hypothetical and modeled fault pattern (Figure 18.19c). Hence, it might be possible that this pattern is a result of the above stress or faulting mechanism acting upon the NAS or the entire northern Brahmaputra Valley. A similar E–W fault pattern is also observed on top of the Sylhet Limestone northward on the north bank of Brahmaputra River (Kumar and Borgohain 2005) suggesting a further northward-bounding strike-slip fault (Figure 18.19d). In a larger regional context, two major bounding strike slips creating a wrench mechanism can be postulated from the mosaic of LANDSAT 8 OLI images of the area based on the sigmoid geometry of the Brahmaputra River. The one on the south coincides with the Jorhat Fault zone, whereas the fault on the north occurs probably northward of Tinsukia (Figure 18.19e). The signatures of these two faults are also apparent in residual gravity data of the region (Figure 18.19f). These two strike-slip faults might be considered to be the bounding faults of the system and are instrumental in creating the fault geometry of which the aforesaid pattern may be a part. This may also be the cause of the clockwise rotation of the NE–SW faults. This model also premises that these E–W faults transect the Brahmaputra ridge and connect the two sub-basins on either side of the ridge. This present wrench model however has an additional thrust component on its sides due to the effect of Naga Thrust in the southeast and the MBT on its northwest. This might lead to a greater compression in the areas of overlap between the NE–SW-oriented faults leading to the development of NE–SW-oriented structures between them compartmentalized by the NW–SE and E–W faults, some of which are found to be hydrocarbon-bearing (Figure 18.19g and h).

18.4.5 Implications of Wrench Model on Petroleum System of NAS

Since the strike-slip faults postulated here as a part of the wrench model and are rooted in basement, these faults can connect stratigraphically deeper source rocks with reservoirs in shallow parts of the sedimentary section to form petroleum pools. Wrench-related anticlines can form traps favorable for petroleum accumulation. Such wrenching effects also lead

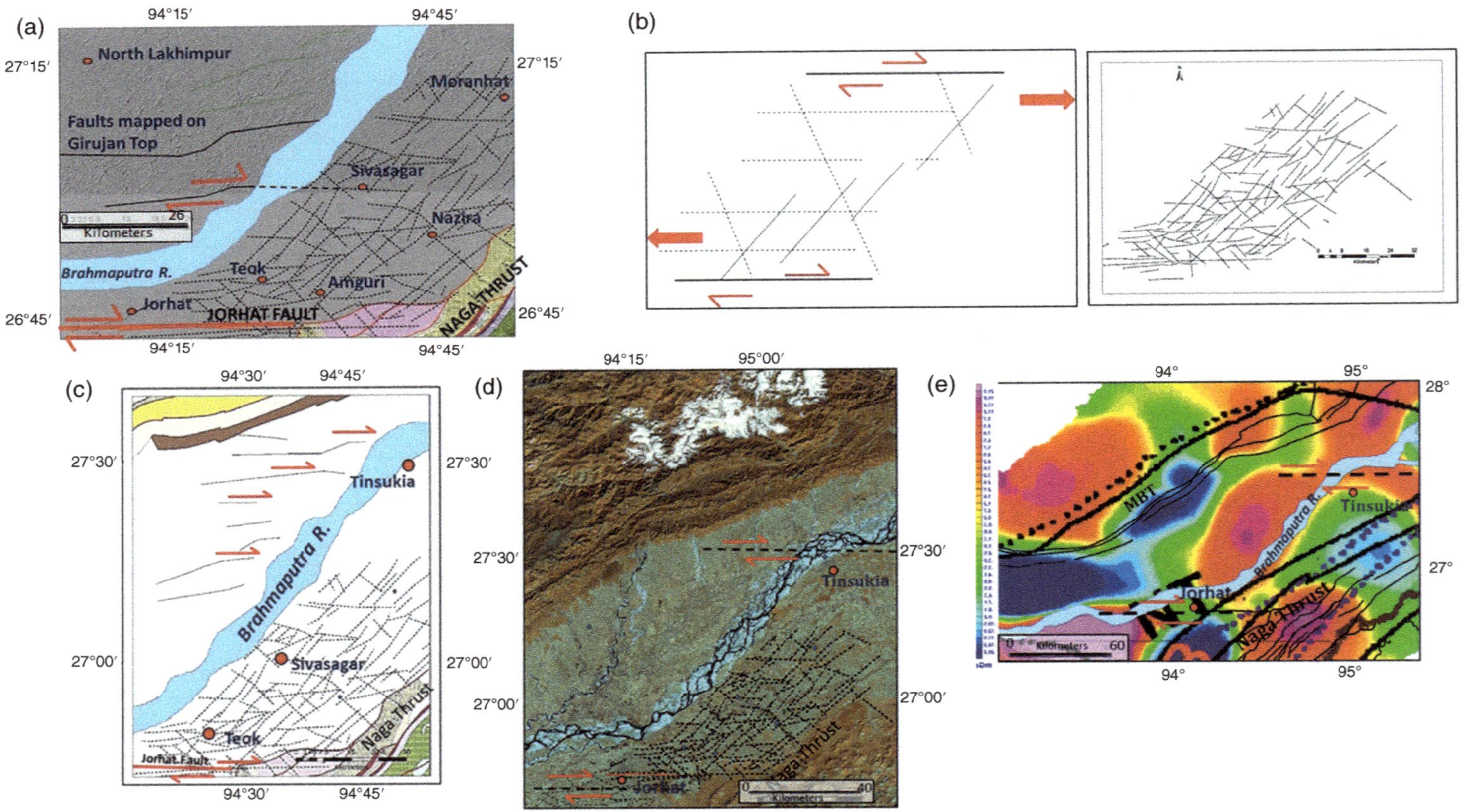

Figure 18.19 (a) Faults mapped from seismic data on top of the Girujan Clay horizon on the north bank of the Brahmaputra River correlating with the E–W morphostructural faults on the south bank implying that they are part of a single fault. (b) Model of faults that would have formed due to wrenching of two E–W-oriented bounding dextral strike-slip faults overlapping each other. (c) Morphostructural fault network in the NAS similar to the modeled faults. (d) E–W faults mapped on the Sylhet Limestone Top in the north bank of the Brahmaputra River (Kumar and Borgohain, 2005) suggesting a further northward bounding strike-slip fault. (e) LANDSAT 8 OLI image composite of bands 4, 5, and 7 postulating two major bounding E–W strike-slip faults creating a wrench mechanism based on the sigmoidal geometry of the Brahmaputra River. The southern bounding fault coincides with the Jorhat fault zone, whereas the northern fault occurs near Tinsukia. (f) Signatures of the E–W bounding faults apparent in the residual gravity data of the Assam Shelf. (g) The effect of the Naga Thrust on the wrench model may result in a greater compression in the areas of overlap between the NE–SW oriented fault leading to the development of NE–SW oriented structures between these faults dissected by the NW–SE and E–W faults. (h) Analogous structures found to be developed in the North Assam Shelf.

(f)

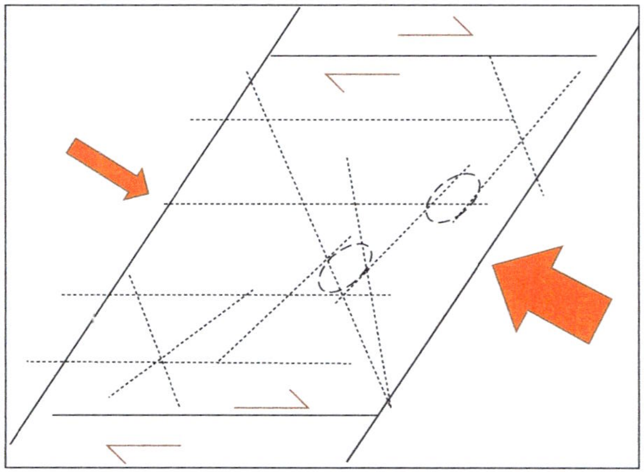

(g)

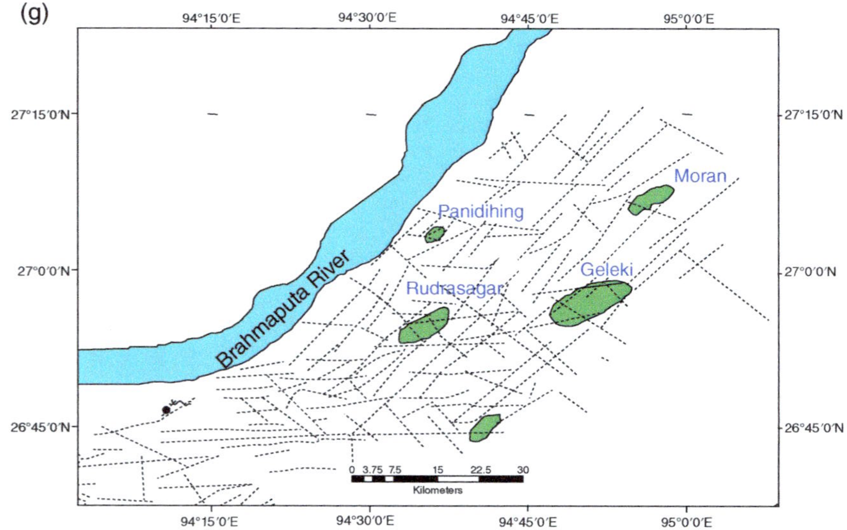

Figure 18.19 (Continued)

to pre-existing structures or deeper/older areas of hydrocarbon accumulation getting reoriented and eventually destroyed, whereas new folded structures and faults are formed at a shallower depth. Seismic profiles across the major faults of wrench zones indicate the formation of flower structures with fan-like, steeply dipping faults that show negligible to almost zero separations of horizons across it and converge downward into a single and subvertical fault (https://www.files.ethz.ch/structuralgeology/JPB/files/English/5wrench.pdf, 2017) (Figure 18.20b).

An analogy of the Assam Shelf can be drawn with the Bohai sea basin in South China Sea that forms one of the most petroliferous basins in China with 41% hydrocarbons occurring in wrench-related structures under the control of the Tanlu strike-slip fault zone (Guosheng et al. 2014). The late Neogene-Quaternary age folds associated with wrench offsets formed favorable traps along these wrench zones, whereas the inherited structure of the strike-slip fault along with its main faults acted as the principal routes for hydrocarbon remigration (Figure 18.20c) and redistribution from the underlying source rock or larger primary accumulation structures to smaller Neogene-Quaternary age folds (Guosheng et al. 2014). These remigrations are a result of seismic pumping due to neotectonic deformations that also controlled the accumulation of oil and gas in these secondary structures (Guosheng et al. 2011).

In the NAS, similar to the Bohai sea basin, active tectonics related to the wrench system might have formed shallow reactivated structures that serve as locales of hydrocarbon entrapment (Figure 18.20d and e). Faults that are at low angles to the S_{HMax} direction derived from FMI data sets and World Stress Maps act as conduits or pathways of hydrocarbon migration. The prevailing strike-slip tectonic regime helps enhance the vertical flow through the association of subvertical faults and extensional fractures formed due to wrenching (Sibson and Scott 1998). Continued neotectonics caused the breaching and subsequent destruction of deeper and older accumulations and resulted in the formation of smaller and shallower folded structures. Remigration of hydrocarbon from the breached primary accumulation may thus constitute multiple-level hydrocarbon reservoirs in the wrench-related structural highs/closures.

Geomorphic highs that are associated with morphostructural faults identified as migration conduits can be considered potential areas of interest for hydrocarbon exploration in the NAS (Figure 18.19e). These imply the existence of smaller secondary wrench-related structural highs at multiple levels in the subsurface, some of which may be beyond seismic resolution in shallower horizons like the upper part of Tipam Sandstone. The faults in these cases may act as the conduits of remigration that create these secondary areas of accumulation.

18.5 Conclusions

Basement tectonics in the Assam Shelf plays a significant role in defining areas of hydrocarbon exploration. In the SAS, a clockwise rotation of the Mikir Block developed E–W zones of dextral strike-slip and N–W-aligned strike-slip faults. Both these trends are highly fractured and areas of their intersections along with the older basin forming NE–SW trend create regions of maximum fracture porosity in the basement. Structural highs associated

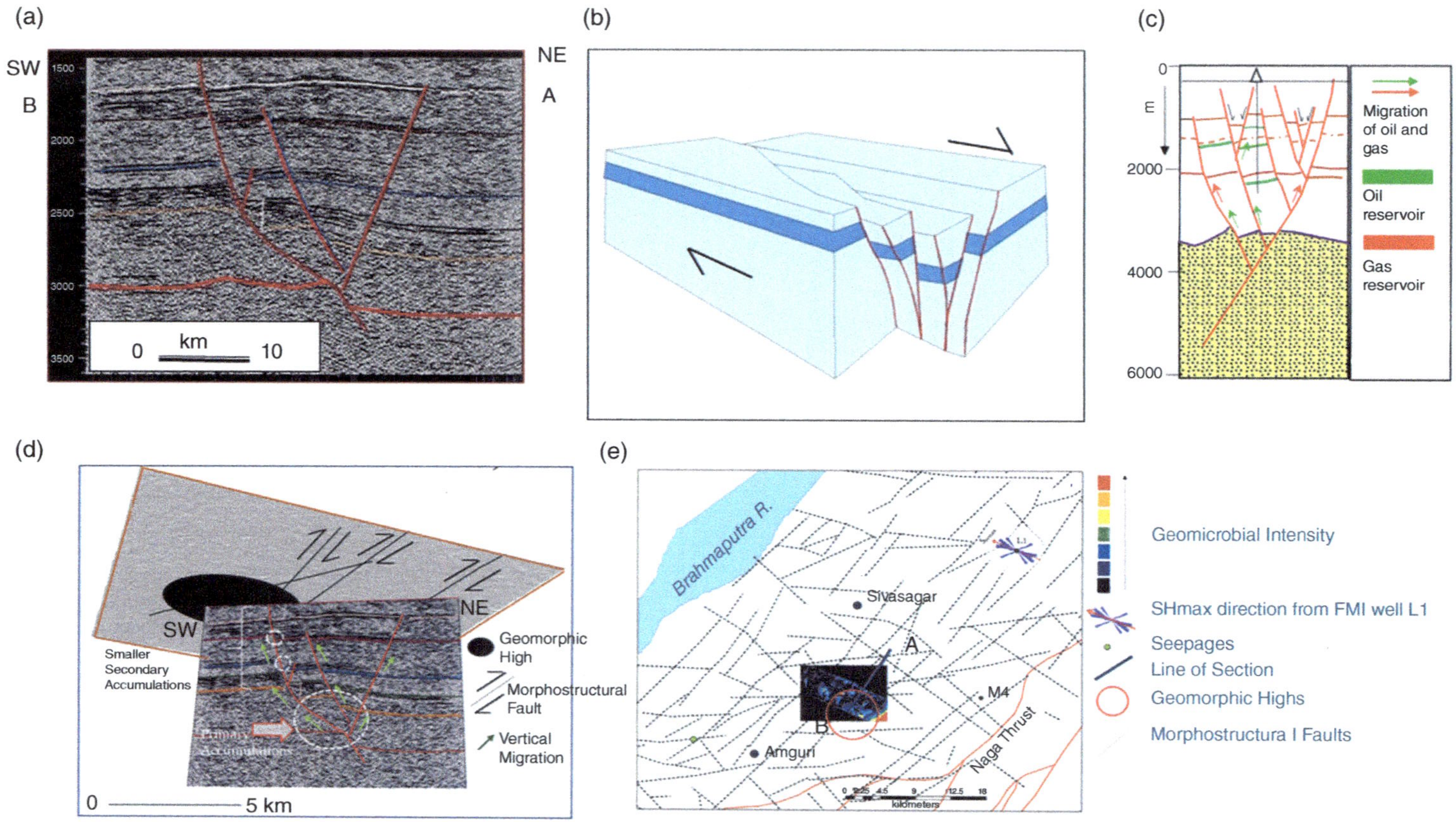

Figure 18.20 (a) Seismic section along line AB (Figure 18.19e) in NAS showing the development of a flower structure with fan-like, steeply dipping faults converging downward into a single and subvertical fault rooted in basement. (b) A typical model of a negative flower structure in a wrench fault zone analogous to that postulated in the NAS. *Source:* After https://www.files.ethz.ch/structuralgeology/JPB/files/English/5wrench.pdf, 2017. (c) Late Neogene to Quaternary age folds associated with wrench offsets in the Bohai Sea Basin, China form favorable traps at multiple levels in the wrench zones with the inherited structure of the strike-slip fault acting as principal routes for hydrocarbon remigration. *Source:* Adapted from Guosheng et al. (2014). (d) Deeper and larger primary hydrocarbon accumulations get breached and destroyed due to repeated reactivations of the wrench-generated faults. The smaller folds formed during wrenching serve as newer but smaller and shallower accumulations. The fault structure acts as the conduit for vertical migration resulting in multiple-level reservoirs. Geomorphic highs help in delineating such shallow folded structures some of which may be beyond seismic resolution, whereas the morphostructural faults help in delineating the subsurface faults. (e) Layout of the seismic trace overlain on the morphostructural faults and geomorphic highs. Directional analysis of DIFs from image logs of wells helps in delineating faults that are probable to act as conduits, whereas geomicrobial maps and seepages help in validating them.

with such fractured areas form promising zones of basement exploration. The Jorhat Fault that acts as a tectonic divide separates the NAS from the SAS. In the NAS, the Jorhat Fault and an E–W strike-slip fault near Tinsukia form a transtensional wrench system that has been further modified by compression due to the Naga Thrust. Neotectonic reactivations associated with the wrench system resulted in the formation of folds associated with wrench offsets in the shallower sediments, whereas the (sub)-vertical strike-slip faults that are parallel to the S_{Hmax} direction (NW–SE to E–W) act as conduits to hydrocarbon remigration. Hydrocarbons remigrating from the breached deeper and larger primary accumulation structures reaccumulate in the shallower smaller folds forming reservoirs at multiple levels. The morphostructurally interpreted faults and geomorphic highs help have an idea about these potential areas of hydrocarbon exploration based on basement tectonics.

Acknowledgments

The authors express their gratitude to Dr. Soumyajit Mukherjee (IIT Bombay) for his invitation to submit this manuscript in the edited book and for reviewing this article. Merryl Le Roux, Frank Otmar Weinreich, Joss Everett, Andrew Harrison, and the proofreading team (Wiley Blackwell) are thanked for their assistance. This chapter has been encapsulated in Mukherjee et al. (2023).

The views expressed in the chapter are solely of the authors themselves and not necessarily of the organization they represent.

References

Akbar M., Nurmi R., Standen E. et al. 2015. Finding fractures in deep and tight reservoir, Schlumberger. https://fdocuments.in/document/rel-pub-mewer14-3.html (accessed 3 August 2020).

Alam, M., Alam, M.M., Curray, J.R. et al. (2003). An overview of the sedimentary geology of the Bengal Basin in relation to the regional tectonic framework and basin-fill history. *Sedimentary Geology* 155: 179–208.

Alizadeh, M., Movahed, Z., Junin, R. et al. (2015). Finding the drilling induced fractures and borehole breakout directions using image logs. *Journal of Advanced Research in Applied Mechanics* 10 (1): 9–29.

Angelier, J. and Baruah, S. (2009). Seismotectonics in Northeast India: a stress analysis of focal mechanism solutions of earthquakes and its kinematic implications. *Geophysical Journal International* 178 (1): 303–326.

Barman, P. (2014). Patial and temporal variation of atmospheric precipitable water vapor and active surface deformation studies in Northeast India using GPS. Ph.D thesis. Submitted in Department of Physics, Tezpur University. https://shodhganga.inflibnet.ac.in/handle/10603/99800 (accessed 10 August 2020).

Baruah, S., Baruah, S., Kalita, A., and Kayal, J.R. (2011). Groundmotion parameters in the Shillong–Mikir plateau, Northeastern India. *Geomatics, Natural Hazardsand Risk* 2 (4): 349–363. https://doi.org/10.1080/19475705.2011.568070.

Das, J. (1995). Tectonics of the Shillong plateau and adjoining regions in northeast India using field and remotely sensed data. Ph.D. thesis. Department of Earth Sciences, IIT Roorkee.

Das, K.C., Sarma, V.S.B. and Ayyadurai, M. (2004). Gondwana sediments: a promising hydrocarbon exploration target in Assam shelf. *5th Conference & Exposition on Petroleum Geophysics*, SPG Hyderabad-2004, India (4–6 December). Dehradun: Society of Petroleum Geophysicists, pp. 468–472.

Das, J.D., Saraf, A.K., and Jain, A.K. (1995). Fault tectonics of the Shillong plateau and adjoining regions, North-East India using remote sensing data. *International Journal of Remote Sensing* 16 (9): 1633–1646.

Dasgupta, T., Dasgupta, S., and Mukherjee, S. (2019). Image log interpretation of geomechanical issues. In: *Teaching Methodologies in Structural Geology and Tectonics* (ed. S. Mukherjee), 237–251. Singapore: Springer. ISBN 978-981-13-2781-0.

Dasgupta, K., Samin, S. and Bharali, B.R. (2011). An analysis of pre-tertiary plays in Matimekhana-Deohal area – a case study from OIL's operational area in Northeast India, Search and Discovery Article 10303. *AAPG International Conference and Exhibition*. Calgary, Alberta, Canada (12–15 September 2010). Tulsa: AAPG.

Devi, N.R. and Sarma, K.P. (2010). Strain analysis and stratigraphic status of Nongkhya sumer and mawmaram conglomerates of Shillong basin, Meghalaya, India. *Journal of Earth System Sciences* 119 (2): 161–174.

Devi, N.R., Kakati, P., Borah, P. et al. (2009). *Earth System Sciences: Felicitation Volumes in Honour of Professor V. K. Verma*, vol. I (ed. A. Kumar, R.S. Kuswaha and B. Thakur), 324–341. New Delhi: Convept Publishing Company.

Duarah, B.P. and Phukan, S. (2011). Understanding the tectonic behaviour of the Shillong plateau, India using remote sensing data. *Journal of the Geological Society of India* 77 (2): 105–101.

Dutta, D. and Mukherjee, S. (2019). Opposite shear senses: geneses, global occurrences, numerical simulations and a case study from the Indian Western Himalaya. *Journal of Structural Geology* 126: 357–392.

Ganju, J.L. and Ram, J. 1987. Fracture Pattern Analysis of Shillong-Mikir Massif Based on Remotely Sensed Data. Geology division, KDMIPE, ONGC Internal Report.

Ghosh, T. and Hazra, S. (2011). Estimating basal friction and fluid pressure from the geometry and spacing of frontal faults in Naga fold thrust belt, India. *Marine and Petroleum Geology* 28 (7): 1376–1380.

Ghosh, T., Basu, S., and Hazra, S. (2014). Geological mapping of the Schuppen belt of North-East India using geospatial technology. *Journal of Asian Earth Sciences* https://doi.org/10.1016/j.jseaes.2013.09.004.

Goswami, B.G. (1998). *Oil/Gas Source Correlation Studies of Oils and Gases Discovered within the Girujan Clay Formation of Lakwa-Lakhmani Field*. Upper Assam: ONGC Internal Publication.

Goswami, B.G. (2001). *Geochemival Characterization and Source Investigation of Oils Discovered in Khoraghat, Panidihing, Safrai and Nambar structures of Upper Assam Basin*. ONGC Internal Publication.

Goswami, B.G. (2005). *Oil to Oil and Oil to Source Correlation in Barail of Upper Assam*. ONGC Internal Publication.

Goswami, T. (2014). Deformation of the basement gneissic complex of the Assam Meghalaya craton in and around Sumer-Nayabunglow area, Meghalaya, India. *International Journal of Science* 3 (2): 149–151.

Govin, G., Najman, Y., Copley, A. et al. (2018). Timing and mechanism of the rise of the Shillong plateau in the Himalayan foreland. *Geology* 46 (3): 279–282. https://doi.org/10.1130/G39864.1.

Gowd, T.N., Srirama Rao, S.V., and Chary, K.B. (1998). Seismotectonics of Northeastern India. *Current Science* 74 (1): 75–80.

Guosheng, X., Xu, Z., Gong, D. et al. (2014). Relationship between tan-Lu fault and hydrocarbon accumulation in Liaozhong sag, Bohai Bay, Eastern China. *Journal of Earth Science* 25 (2): 324–336.

Guosheng, X., Ma, R., Gong, D. et al. (2011). Features of the fault system and its relationship with migration and accumulation of hydrocarbon in Liaodong Bay. *Petroleum Science* 8 (3): 251–263.

Heidbach, O., Rajabi, M., Reiter, K. and Ziegler, M. (2016). World stress map 2016. GFZ Data Services. https://doi.org/10.5880/WSM.2016.002.

Hossain, M.S., Khan, M.S.H., Abdullah, R., and Chowdhury, K.R. (2020a). Tectonic development of the Bengal Basin in relation to fold-thrust belt to the East and to the North. In: *Structural Geometry of Mobile Belts of the Indian Subcontinent* (ed. T.K. Biswal, S.K. Ray and B. Grasemann), 91–109. Cham, Switzerland: Springer Nature Switzerland AG https://doi.org/10.1007/978-3-030-40593-9_4.

Hossain, M.S., Khan, M.S.H., Chowdhury, K.R., and Abdullah, R. (2019). Synthesis of the tectonic and structural elements of the Bengal Basin and its surroundings. In: *Tectonics and Structural Geology: Indian Context* (ed. S. Mukherjee), 135–218. Cham, Switzerland: Springer International Publishing AG https://doi.org/10.1007/978-3-319-99341-6_6.

Hossain, M.S., Xiao, W., Khan, M.S.H. et al. (2020b). Geodynamic model and tectono-structural framework of the Bengal Basin and its surroundings. *Journal of Maps* 16 (2): 445–458. https://doi.org/10.1080/17445647.2020.1770136.

Hossain, M.S. (2017). Strike-slip and oblique-slip tectonics. http://www.files.ethz.ch/structuralgeology/JPB/files/English/5wrench.pdf (accessed 12 August 2020).

Islam, M.S. and Shinjo, R. (2012). The Dauki fault at the Shillong plateau-Bengal Basin boundary in North-Eastern India: 2D finite element modelling. *Journal of Earth Science* 23: 854–863.

Jade, S., Shrungeshwara, T.S., Kumar, K. et al. (2017). India plate angular velocity and contemporary deformation rates from continuous GPS measurements from 1996 to 2015m. *Scientific Reports* 7: 11439. https://doi.org/10.1038/s41598-017-11697-w.

Kakati, P. (2013). Microstructural behaviour of lower metapelitic formation of the proterozoic Shillong basin Meghalaya India. Ph.D thesis. Department of Geological Science, Guwahati University.

Kataruka, P., Pamdey, A., Rao, S.C. et al. (2017). Basement Prospectivity of Suphatam-Telihal Area in South Assam Shelf. Centre of Delivery-Basement Exploration, ONGC Internal Report, Mumbai.

Kent, W.N. and Dasgupta, U. (2004). Structural evolution in response to fold and thrust belt tectonics in Northern Assam. A key to hydrocarbon exploration in the Jaipur anticline area. *Marine and Petroleum Geology* 21 (7): 785–803.

Kumar, M. and Borgohain, R. (2005). Palynofacies analysis and depositional environment in Bihupuria well a, North Bank of Brahmaputra, upper Assam Basin. *Geological Society of India* 65: 70–82.

Kumar, T.S., Bharali, B.R. and Verma, A.K. 2012. Basement configuration and structural style in OIL's operational areas of upper Assam, Search and Discoveiy Article 50739. *GEO-India*. Greater Noida, New Delhi, India (12–14 January 2011). Dehradun: Association of Petroleum Geologists.

Lahiri, S.K. and Sinha, R. (2014). Morphotectonic evolution of the Majuli Island in Brahmaputra valley of Assam, India inferred from geomorphic and geophysical analysis. *Geomorphology* 227: 101–11112.

Mathur, N., Raju, S.V., and Kulkarni, T.G. (2001). Improved identification of payzones through integration of geochemical and log data: a case study from upper Assambasin, India. *AAPG Bulletin* 85 (2): 309–323.

Mazumder, S., Mitra DS, A.K., Mahapatra, S., and Pangtey, K.K.S. (2016). A neotectonic based geomorphic analysis using remote sensing data to delineate potential areas of hydrocarbon exploration: Cachar area, Assam. *Journal Geological Society of India* 88: 87–97.

Mazumder, S., Dave, H.D., Samal, J.K., and Mitra, D.S. (2011). Delineation of shallow structures as exploratory input in Ganga Basin based on Morphotectonic studies. *ONGC Bulletin* 46: 190–198.

Mazumder, S., Pangtey, K.K.S., Mitra, D.S. (2013a) Delineation of a possible subsurface ridge in onshore Palar Basin based on morphotectonic studies and its implications. Paper ID: 001, In: *Proceedings of 10th Biennial International Conference & Exposition*, SPG, Kochi (23–25 November 2013). Dehradun: Society of Petroleum Geophysicists.

Mazumder, S., Tep, B., Pangtey, K.K.S., and Mitra, D.S. (2013b). A Morphotectonic based approach to derive tectonic framework and possible areas of hydrocarbon accumulation in onshore Palar Basin. *ONGC Bulletin* 48: 113–123.

Mazumder, S., Tep, B., Pangtey, K.K.S., and Mitra, D.S. (2019). Basement tectonics and shear zones in Cauvery Basin (India): implications in hydrocarbon exploration. In: *Tectonics and Structural Geology: Indian Context* (ed. S. Mukherjee), 279–311. Springer Geology, https://doi.org/10.1007/978-3-319-99341-6_9.

Mazumder, S., Tep, B. and Pangtey, K.K.S. 2018. Basement tectonics in South Assam shelf to decipher promising zones for hydrocarbon exploration: a GIS based approach. Paper ID: AU215, *GeoIndia 2018, the 4th South Asian Geosciences Conference*, India Exposition Mart, Greater Noida, NCR, India (6–8 September 2018). Dehradun: Association of Petroleum Geologists.

Meigs, A.J., Burbank, D.W., and Beck, R.A. (1995). Middle-late Miocene (>10 Ma) formation of the main boundary thrust in the western Himalaya. *Geology* 23 (5): 423–426.

Mishra, S.K., Phukon, J., Bhoktiari, P., and Rahaman, A. (2016). Integrated structural analysis of fractured basement reservoir in the South Assam shelf India. *Proc of Indian National Science Academy* 82 (3): 923–933. https://doi.org/10.16943/ptinsa/2016/48493.

Misra, A.A. and Mukherjee, S. (2015). *Tectonic Inheritance in Continental Rifts and Passive Margins*. Spam: Springer briefs in Earth Sciences. ISBN 978-3-319-20576-2.

Misra, A.A. and Mukherjee, S. (2018). *Atlas of Structural Geological Interpretation from Seismic Images*. Wiley Blackwell. ISBN 978–1–119-15832-5.

Mukherjee, S., Dasgupta, S., Majumdar, C. et al. (2023). Introduction to: Handbook of Petroleum Geoscience: Exploration, Characterization, and Exploitation of Hydrocarbon Reservoirs. In: *Handbook of Petroleum Geoscience: Exploration, Characterization, and Exploitation of Hydrocarbon Reservoirs*, Petroleum Industry Updates: Geoscientific and Economic Issues (ed. S. Mukherjee, S. Dasgupta, C. Majumdar, et al.). John Wiley & Sons, Hoboken, ISBN 978-1-119-68003-1.

Mortha, P.K., Borah, R. and Renjith, K.R. (2015). An integrated approach to basement exploration in Dhansiri Valley, South Assam Shelf, A & AA basin. *11th Conference & Exposition on Petroleum Geophysics*, SPG Jaipur-2015, India (4–6 December). Dehradun: Society of Petroleum Geophysicists.

Mukherjee, S. and Jha, V.C. (2011). Role of lithology and rock structure in drainage development in the Kaliani River basin, Assam, India. *Ethiopian Journal of Environmental Studies and Management* 4 (2): 61–68.

Mukherjee, S. and Khonsari, M.M. (2017). Brittle rotational faults and the associated shear heating. *Marine and Petroleum Geology* 88: 551–554.

Murty, K.N. (1983). Geology and hydrocarbon resources of Assam shelf-recent advances and present status, petroliferous basins of India. *Petroleum Asia Journal* I: 1–14.

Nandy, D.R. and Dasgupta, S. (1991). Seismotectonic domains of northeastern India and adjacent areas. *Physics and Chemistry of the Earth* 18 (Part 1): 371–384.

Nathaniel D.E., Goswami P.K., Pegu S.K. and Hazarika P.K. (1997). Geomicrobial Prospecting for Hydrocarbons in Disangmukh Area. Upper Assam, Institute of Biotechnology and Geotectonic Studies, ONGC, Jorhat, Unpublished Internal Report.

Pahari, S., Singh, H., Prasad, I.V.S.V., and Singh, R.R. (2008). Petroleum systems of upper Assam shelf, India. *Geohorizons* Dec 2008: 14.

Pangtey, K.K.S., Sahoo, M. and Kumar, S. (2016). Morphotectonic analysis of South Assam Shelf for Identification of Prospective Areas in Basement and Geomechanical and Fracture Modelling of Basement in South Assam Shelf. KDMIPE, ONGC, Unpublished Report.

Prasad D.N. and Srivastava A.K. (1987). Pattern and Genesis of Fractures in the Basement of Assam Foreland: From a Study in Mikir Hill Basement. KDMIPE, ONGC Internal Report.

Prensky, S. (1992). Borehole breakouts and in-situ rock stress – a review. *The Log Analyst* 33 (3): 304–312.

Raja, A.S., Deka, P. and Das, A. (2012). Identification of Pyroclastic Rocks in the Subsurface of North Assam Shelf. Regional Geological Laboratories, Sivsagar, ONGC Internal Report.

Roy Moulik S.K., Singh H.J., Rawat R.K.S. et al. (2009). Sand distribution pattern and depositional Model of kopili formation (Eocene) with special reference to sequence stratigraphic framework from North Assam Shelf, Assam Arakan Basin, India, Search and Discovery Article 50196. *AAPG Annual Convention*, Denver, Colorado (7–10 June 2009).

Roychoudhury, S.C. and Mukherjea, A. (2011). Diversities of petroleum habitat in the West Bengal-Bangladesh Assam and Assam-Arakan megabasin complex and future frontiers. *ONGC Bulletin* 46 (3): 12.

Sahoo, M. and Gogoi, K.D. (2011). Structural and sedimentary evolution of Upper Assam Basin, India and Implications on Hydrocarbon Prospectivity, Search and Discovery Article 10315 (2011). AAPG Datapages Inc. *GEO-India*, Greater Noida, New Delhi, India (12–14 January 2011)

Sarma, J.N. and Sharma, S. (2018). Neotectonic activity of the Bomdila fault in Northeastern India from geomorphological evidences using remote sensing and GIS. *Journal of Earth System Science* 127: 113. https://doi.org/10.1007/s12040-018-1008-2.

Schumm, S.A., Dumont, J.F., and Holbrook, J.M. (2002). *Active Tectonics and Alluvial Rivers*. United Kingdom: Cambridge University Press.

Science China Press (2017). The initial collision between Indian and Asian continental. https://phys.org/news/2017-04-collision-indian-asian-continental.html (accessed 15 August 2020).

Shahriar, M.S., Ameen, S.M.M., Hossain, M.S. et al. (2020). Revealing the basement in Barapukuria: a geochemical study of a Gondwana coal basin basement from Northwest Bangladesh. *Journal of the Geological Society of India* 95 (6): 571–586. https://doi.org/10.1007/s12594-020-1484-2.

Sharma, S. and Sarma, J.N. (2012). Drainage analysis in a part of the Brahmaputra Valley in Sivasagar District, Assam, India, to detect the role of Neotectonic activity. *Journal of Indian Society of Remote Sensing* 41 (4): https://doi.org/10.1007/s12524-013-0262-7.

Sibson, R.H. and Scott, J. (1998). Stress/fault controls on the containment and release of overpressed fluids: examples from gold-quartz vein systems in Juneau Alaska; Victoria, Australia and Otago, New Zealand. *Ore Geology Reviews 1998* 13: 293–306.

Singh, A.P., Rao, N.P., Kumar, M.R. et al. (2017). Role of the Kopili fault in deformation tectonics of the indo-Burmese arc inferred from the rupture process of the 3 January 2016 Mw 6.7 Imphal earthquake. *Bulletin of the Seismological Society of America* 107: 1041–1047.

Singh, P. and Barua, B.N. (2001). Computation of Depth to Basement Using Potential Field along Existing Seismic Profiles of Kopili Valley. Geophysics Division, KDMIPE, ONGC Internal Report.

Singh, Y.R., Singh, B.P., Murthy, S. et al. (2014). Reworked early permian palynomorphs and tertiary palynomorphs from the upper Bhuban formation (Miocene), Nagaland, India. *Himalayan Geology* 37 (1) 2016: 35–41.

Tep B., Mazumder S. and Pangtey K.K.S. (2015) Study of the effect of neotectonism in the re-migration of hydrocarbon in North Assam Shelf, ID 248. *11th Biennial International Conference and Exposition*, Society of Petroleum Geophysicists, SPG Jaipur, India (4–6 December). Dehradun: Society of Petroleum Geophysicists.

Thingbaijam, K.K.S., Nath, S.K., Yadav, A. et al. (2008). Recent seismicity in Northeast India and its adjoining region. *Journal of Seismology* 12: 107–123.

Tingay, M., Reinecker, J. and Müller, B. (2008). Borehole breakout and drilling-induced fracture analysis from image logs, World Stress Map, Guidelines: Image Logs.

Tiwari, S. and Jassal, G.S. (2001). Origin and evolution of the Garo-Rajmahal gap. *Journal of the Geological Society of India* 57 (5): 389–403.

Uddin, A. and Lundberg, N. (2004). Miocene sedimentation and subsidence during continent–continent collision, Bengal Basin, Bangladesh. *Sedimentary Geology* 164 (1–2): 131–146.

Van Hinsbergen, D.J.J., Lippert, P.C., Dupont-Nivet, G. et al. (2012). Greater India Basin hypothesis and a two-stage cenozoic collision between India and Asia. *Proceedings of the National Academy of Sciences of the United States if America*. 109 (20): 7659–7664.

Vernant, P., Bilham, R., Szeliga, W. et al. (2014). Clockwise rotation of the Brahmaputra Valley relative to India: tectonic convergence in the eastern Himalaya, Naga Hills, and Shillong plateau. *Journal of Geophysical Research: Solid Earth* 119 (8): https://doi.org/10.1002/2014JB011196.

Wallang, V.J.J. (2009). Geo-environment and metabolism of Shillong city ecosystem. Ph.D. thesis. submitted to Centre for Environmental Studies, North-Eastern Hill University.

Yang, L., Xiao, W., Rahman, M.J.J. et al. (2020). Indo-burma passive amalgamation along kaladan fault: insights from provenance of Chittagong-Tripura fold belt (Bangladesh). *Geological Society of America Bulletin* 132 (9–10): 1953–1968. https://doi.org/10.1130/B35429.1.

Zhang, L., Luo, X., Vasseur, G. et al. (2011). Evaluation of geological factors in characterizing fault connectivity during hydrocarbon migration: application to Bohai Bay basin. *Marine and Petroleum Geology* 28 (9): 1634–1647.

19

Taphonomy, Petrophysics, and the Relationship of Dense Shell-Accumulation with Reservoir Quality

Rodrigo Scalise Horodyski[1], *Guilherme Furlan Chinelatto*[2], *Hugo Schmidt-Neto*[1], *Eduardo Guareschi*[1], *and Sudipta Dasgupta*[3,*]

[1] *Geology Graduate Program, Universidade do Vale do Rio Dos Sinos (UNISINOS), São Leopoldo, Brazil*
[2] *Geology Institute, Universidade Estadual de Campinas (UNICAMP), São Paulo, Brazil*
[3] *Department of Earth Sciences, Indian Institute of Technology Bombay (IIT Bombay), Mumbai, Maharashtra, India*

**Author for correspondence: sudipta.dasgupta@gmail.com*

Abstract

Dense accumulations of skeletal remains represent the complex depositional histories within the short- or long-time intervals ("taphofacies") and provide with highly valuable pieces of information in the paleoenvironmental and stratigraphic analyses. These accumulations have a variety of taphonomic characteristics, e.g. geometry, biofabric, and different taphonomic signatures. In addition, the analysis of strata containing the dense shell remains presents clues leading to the petrophysical characteristics (porosity and permeability) and, thus, appears strikingly utilitarian to the reservoir geoscientists. The purpose of this chapter is to furnish a synthesis of the main taphonomic and petrophysical aspects of the fossiliferous accumulations, i.e. the shell beds, to revisit and refine our understanding of their genesis and reservoir potential. We try here to explain the concepts of taphonomically active zone, taphofacies, taphonomic signatures, and intrinsic taphonomic damage vis-à-vis the biofabric, geometry, and packaging degrees of the shell accumulations. These taphonomic characteristics must essentially be integrated with the reservoir petrophysics of shell beds. We try to demonstrate the relationship between taphonomy and petrophysics by exemplifying the Barremian-Aptian pre-salt shell beds of Brazil, e.g. the Morro do Chaves Formation (in the Sergipe-Alagoas basin) and the Coqueiros and the Itapema formations (in the Campos and Santos basins, respectively).

Keywords *Taphonomyshell beds; coquina; reservoir sedimentology; cold-water carbonate deposits; taphofacies*

19.1 Introduction

The exploration of the petroleum resources is dependent on the petrographic and petrophysical studies of subsurface rocks and several other advanced geoscientific analyses at different scales including the analog outcrop studies. In this way, geoscientists aim to

Handbook of Petroleum Geoscience: Exploration, Characterization, and Exploitation of Hydrocarbon Reservoirs, First Edition. Edited by Soumyajit Mukherjee, Swagato Dasgupta, Chandan Majumdar, Subhandip Mandal and Troyee Dasgupta.
© 2023 John Wiley & Sons Ltd. Published 2023 by John Wiley & Sons Ltd.

discover and characterize the suitable and exploitable reservoir lithofacies and how they are distributed in space and time in an acreage area and/or a basin. Therefore, evaluating the nature of depositional settings, the distribution and association of facies, and their petrophysical characteristics are fundamental to determining the quality of a reservoir architectural element. As a result of these efforts, the basins with potential reservoirs have hitherto been discovered, characterized, announced, and exploited. One such example is the discovered carbonate reservoirs belonging to the pre-salt horizons of the offshore Campos and Santos basins of southeast Brazil (e.g. Abrahão and Warme 1990; Guardado et al. 2000; Riccomini et al. 2012; Thompson et al. 2015) – the two most prolific-producing basins of Brazil (Chinelatto 2020). These carbonate rocks are composed partly of coquinas. The coquinas are sedimentary rocks composed of both the whole and/or fragmented shells as the framework grains arranged in a matrix of secondary and/or primary carbonate or siliciclastic grains and cement (Kidwell (1991a); Nebelsick et al. 2011). For the petroleum industry, its interest in the study of shell beds or coquinas encompasses the improvement in their genetic understanding, involving the petrographic characterization, vis-à-vis the spatial distribution in the basins (Chinelatto 2020, and the references therein).

The petrographic features of the depositional setting, e.g. the variety of carbonate grains and variations in the diagenesis, influence the complex system of porosity and, thereby, the permeability of rocks resulting in the petrophysical heterogeneity and anisotropy (Lucia 1999; Lønøy 2006). In this sense, a large number of researchers and authors (e.g. Corbett et al. 2016, 2017; Luna et al. 2016; Herlinger et al. 2017; Belila et al. 2018; Zielinski et al. 2018) have adopted a petrophysical approach in the studies of coquinas in evaluating the distribution of the pores and pore throats, pore sizes, and pore connectivity. However, the relationship between the sedimentary and the diagenetic processes involved in the formation of coquinas, porosity, and permeability in varying intensities in the fossiliferous rocks is still poorly understood. Hence, the concept of taphonomy can be and has been applied in establishing a relationship between the different patterns of exoskeletal accumulations and the corresponding petrophysical characteristics.

Furthermore, this chapter article intends to furnish a brief dataset of how the taphonomy can be used as one of the powerful tools of reservoir characterization and what the main factors are in taphonomic analyses along with some examples of the taphonomic studies related to the oil and gas industry.

19.2 Taphonomically Active Zone

Taphonomic signatures are inherently related to the taphonomic damages caused to the bioclasts present in the taphonomically active zone (TAZ; Figure 19.1; Davies et al. (1989)). In marine environments, the TAZ occupies the space from the sea bottom to the immediate subsurface, where physical reworking caused by wave and/or currents, chemical reactions (e.g. dissolution process), and/or biological perturbations (i.e. combinations of bioturbations, bioerosion, and incrustation) can modify the organic detrital remains by inflicting or obscuring the injuries on the shells before the onset of fossilization by processes

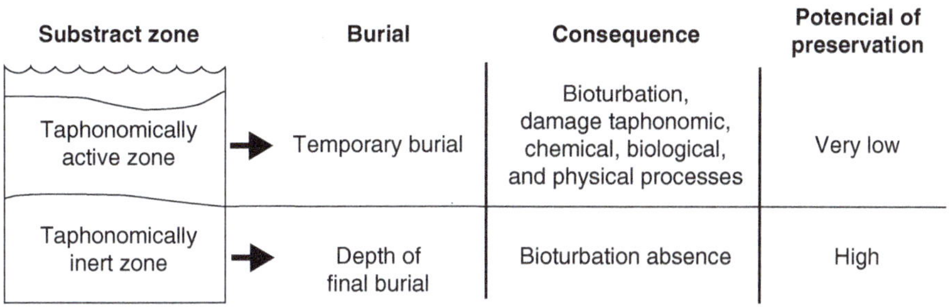

Figure 19.1 All features of the taphonomically active zone and inert zone.

known to be "fossil diagenesis" (Parsons-Hubbard 2005). The bioclasts become secured and separated from these phenomena only when they are incorporated in the taphonomically inert zone (TIZ), generally because of sustained and/or elevated sedimentary rates.

The quantitative and qualitative characteristics of fossiliferous deposits are directly related to the residence time of bioclasts in the TAZ and are controlled by the fluctuations in the sedimentation rate, in the oxygenation condition (Eh), and of the kinetic energy condition of the corresponding depositional environment. In this sense, under the low-sedimentation-rate condition, there is an increase in the hard skeletal input (Kidwell 1986); therefore, a higher reworking of bioclasts is expected in these deposits of mostly disarticulated, fragmented, and bioeroded bioclasts (Speyer and Brett 1986, 1988). Alternatively, at an elevated sedimentary rate, bioclasts are buried fast and the shell concentration tends to be dissolved and reprecipitated as the matrix (Kidwell 1986). In addition to that, the chances of bioclasts entering the TIZ also increase. The petrographic manifestation of these phenomena validates the use of taphonomic signatures in paleoecological and paleoenvironmental studies. As described later, the taphonomic signatures show a close relationship with the paleoenvironmental conditions and the stratigraphic systems (Holland 1995, 2000, and the references therein).

19.3 Taphofacies

The concept of taphofacies was originally presented by Speyer and Brett (1986) in their high-resolution taphonomic study of the trilobite deposits from the Devonian Hamilton Group (New York, USA). In that study, the authors sought to understand the genesis of fossil assemblages and demonstrated the advantages of using taphonomic signatures in solving the questions related to the sedimentary processes of the depositional setting. They concluded that the characteristics of fossiliferous deposits are the result of variations in the (i) degree of energy condition of the depositional environment, (ii) sedimentation rate, and (iii) oxygenation condition of the setting.

Later, Speyer and Brett (1988) presented another article on this topic. The pair of those contributions formed the background, which motivated later workers toward new studies. It was possible because the authors used methodology as basic as the use of taphonomic signatures for different marine taxa and associated them with the sedimentary facies concept while applying the methodology on the mid-Paleozoic epeiric-sea deposits. Thus, the final product came out as a new paleontological concept known as "taphofacies."

In this way, if the preservation modes of fossil assemblages are in response to the dynamics of the sedimentary settings, the assumption holds good that the taphonomic signatures are like the "snapshots" of the conditions, which the bioclasts were subjected to. Therefore, the taphofacies consist of a "fossil assemblage with distinct taphonomic signatures preserved in a delimitated sedimentary succession" that evidences the genesis of the fossiliferous deposits (Speyer and Brett 1986). As described later, the taphonomic signatures can be related to the physical, chemical, and/or biological perturbations, therefore resulting in the disarticulation, fragmentation, corrosion, and/or reorientation of the bioclasts. For this reason, taphofacies analysis enables the researchers in interpreting the paleoenvironmental settings and the paleoecological relations in addition to its contributions in the paleogeographic, stratigraphic, and evolutionary studies (Einsele et al. 1991; Speyer and Brett 1991; Brett and Walker 2002).

Ultimately, taphofacies also elucidates the sequence-stratigraphic contexts like the progradational, aggradational, and retrogradational sedimentary regimes (Brett 1995, 1998). Thus, the variation of the accommodation space (controlled by glacioeustasy and tectonic activity) can be recognized in the different types of fossil assemblages (e.g. Brett 1995; Simões and Torello 2003; Schmidt-Neto et al. 2018; Horodyski et al. 2019). In this sense, any fossil record is singular or unique and the recognition of its specific taphofacies can expectedly lead to the delineation of each depositional system in the sequence.

19.4 Taphonomic Signatures

Marine organisms (at least the majority of them) tend to have hard parts as pieces of their endoskeletons (e.g. fishes and echinoderms) or exoskeletons (e.g. crustaceans and mollusks). While these parts are preserved (to be as the fossil record and as modern organic record), a set of information about the processes, which modified them, can be registered. These recorded alterations of the skeletal remains are known as the taphonomic signatures that can reflect the environmental conditions and/or intrinsic nature of the hard parts. The hard parts are termed the "bioclasts" by the taphonomists while describing the organic remains. The main factors somehow making changes to the bioclasts can be listed as: (i) physical oxidation; (ii) hydrolysis; (iii) UV radiation; (iv) fouling by encrusting organisms; (v) microbial maceration (in aerobic and anaerobic conditions); (vi) dissolution and diagenetic alterations of mineralogical structures (e.g. Speyer and Brett 1986; Kidwell et al. 1986; Davies et al. 1989; Fürsich and Oschmann 1993; Behrensmeyer et al. 2000). The set of signatures can also be subdivided to be physical, chemical, and biological damages as described later in further detail. It is possible to identify the occurrence of only one of these damages as well as their simultaneous occurrences in the same bioclast.

Therefore, we can recognize that the taphonomic signatures together constitute an important tool for paleoecological studies by providing information on the depositional environments, living niche, intertaxa and intrataxon ecological relationships, ontogenic processes (e.g. Kondo 1997; Simões and Torello 2003; Schmidt-Neto et al. 2014, 2018a, 2018b, and interpretations of the fossiliferous intervals in the system-tract context of sequence stratigraphy (e.g. Holland 1995, 2000; Fürsich and Pandey 2003; Mandic et al. 2004; Parras and Casadío 2006; Zabini et al. 2012; Scarponi et al. 2012; Aucoin et al. 2016). The versatile utilities of taphonomic signatures along with the sedimentary facies analyses helped Speyer and Brett (1986, 1988) introduce the concept of "taphofacies," which is widely used in the paleoenvironmental interpretations based on taphonomic signatures of the fossil record.

In a higher resolution, the taphonomic signatures can help in making a diverse set of interpretations as exemplified next:

Physical damages are caused because of fragmentation, abrasion, and different degrees of disarticulation of the bioclasts. Hence, the physical damages of bioclasts point toward the energy condition of the depositional environment, sedimentation rate, frequency of events, degree of reworking of bioclasts, time-average of deposition, and residence time in the TAZ, and how agents overprint their specific signature on bioclasts. For example, the presence of articulated multielement skeletons, e.g. of echinoderms, is due to the fast sedimentation and burial of organisms, thereby making obruption deposits (e.g. Schäfer 1972; Brett 1995), which correspond to a singular event. Such an obruption event reduces the reworking or residence time of a dead organism in the TAZ to zero resulting in spectacular preservation of the most delicate skeletons (Table 19.1). Conversely, in fossiliferous tempestite deposits, a high frequency of fragmented and disarticulated bioclasts can be observed due to the high kinetic energy of the depositional event and the high reworking potential of the bioclasts due to the action of storm waves (e.g. Aigner 1985; Fürsich and Oschmann 1993; Simões and Torello 2003; Schmidt-Neto et al. 2018a).

Table 19.1 Taphonomic signatures and the processes related to the origin of the fossil record.

Damage	Signature	Cause	Inference
Absent			
	Articulated skeletons	Fast burial; obruption deposit event	Very short residence time in the taphonomic active zone (TAZ)
	High preservation degree of details	Lacking physical, chemical, or biological damages	Miss recovering; fast burial in very fine-grained sediments along with short residence time in the TAZ
Physical			
	Fragmentation	Mechanical shock	Reworking
	Shatter	Mechanical pressure	Crushed by high sedimentary episodic deposition; crushed by trampling
	Rounding of the margin	Abrasion	Subaerial exposure
	Rounding of bioclast	Abrasion	Continuous reworking by flows

(*Continued*)

Table 19.1 (Continued)

Damage	Signature	Cause	Inference
	Disarticulation	Mechanical working	Reworking by scavengers, winds, flows, or seismic activity
	Orientation		
	Polymodal	Randomly flows	Storm events
	Bimodal	Bidirectional flows	Wave currents
	Unimodal	Unidirectional flows	Unidirectional currents: river flows; longshore current; hyperpycnal currents; turbidity
Chemical			
	Less of color	Acidification of the water	Increasing of C rate; long residence time in the TAZ
	Pitting	Acidification of the water	Increasing of C rate; long residence time in the TAZ
	Corrosion	Acidification of the water	Increasing of C rate; long residence time in the TAZ
	Dissolution	Acidification of the water	Increasing of C rate; long residence time in the TAZ
	Etching	Biological secretion	Anchoring by fouling organisms; long residence time in the TAZ
Biological			
	Drill hole	Predation	Ecological relationships
	Scar or broken	Predation or predatory behavior	Ecological relationships
	Boring	Anchoring	Long residence time in the TAZ
		on the outside of the shell	Colony developed in living shells
		on the inside of the shell	Colony developed in dead shells
	Encrustation	Anchoring	Long residence time in the TAZ
		on the outside of the shell	Colony developed in living shells
		on the inside of the shell	Colony developed in dead shells
	Any kind of lumps	Parasitism	Ecological relationships

Chemical damages essentially indicate the chemical environmental changes, e.g. salinity fluctuations, oxy oscillation (fluctuations between anoxic and highly oxygenated conditions), and changes in the isotopic carbon cycles. As demonstrated by Ritter et al. (2013), shells suffer higher dissolution in the estuarine setting than in the marine sedimentary environments due to the lower aquatic pH ranges in the former. This happens as a consequence of decaying organic matter in the estuaries and embayments that release a lot of

C into the environment. Signs of dissolution, corrosion, pitting or loss of color are the common taphonomic signatures resulting from chemical damages. However, Brett et al. (2012) explained how the iron-rich content in the sediments can produce the pyrite-coated fossils, making a "protective cover" for the bioclasts instead of damaging them. Some biological damages can be classified as chemical damages. For example, barnacles anchor onto the hard substrates using their corrosive secretion, thereby producing chemical scars on the surface of bioclasts.

Biological damages are related to the ecological interactions and demonstrate the predator–prey, commensalistic, and parasitic relationships. An example of this can be illustrated in the deposits related to taphonomic feedback phenomena (i.e. the interaction of a living community with dead skeletal remains; Kidwell (1991b)). In this case, the higher concentration of shells (as a result of high biologic production or as sorting deposits; see Kidwell (1991b) for more details) change the substrate characteristics from unconsolidated (also known as soft ground or loose ground) to hardground, thereby allowing the attachment of sclerobionts (e.g. sponges, corals, and bryozoans), pedunculate (e.g. crinoids), or byssate (e.g. bivalves) organisms. These communities will let their behavioral traces on the shells. Another good example is related to the interspecific interactions such as between the hermit crabs and sclerobiont communities that coinhabit on and inside the empty gastropod shells. According to Walker (1988), shells inhibited by hermit crabs tend to show preferential colonization by bryozoans and serpulids near the open shell. Also, Schmidt-Neto et al. (2018b), in their study of the ecological significance of bioerosion in the Lower Permian shell beds from Paraná Basin, indicated how the bioerosion can help in interpreting the paleotemperature of the ambient water. Borings also reveal a complex history behind the diagenesis of fossils as follows:

1) The borings can occur in both living and dead shells. When the borings are observed only on the outside of a shell, the bioerosion likely occurred during the lifetime of the shelled animal. If the borings are observed on both inside and outside of the shell, it signifies that the animal was already dead when its shell was bioeroded.
2) If bioerosions were made into the dead shells, it means that the shells remained at the sediment–water interface (SWI) for a sufficient duration of time favorable for the bioeroders like sponges to bore (to know more about the time of development of sponges in shells, see Schmidt-Neto et al. (2018b) and references therein).
3) Finally, if the fossil record is composed of the molds of shells and the casts of the bioerosions, it implies that the borings were filled and lithified before the dissolution of the shells.

The set of taphonomic signatures from different kinds of damages also reveals important pieces of information to geoscientists. For example, even with the low sedimentation rate in the offshore and open shelf zones (i.e. basinward at and beyond the depth of storm wave base), the well-preserved fossil assemblages can be expected because of (i) low probability of physical damage in this zone; and (ii) the higher chances of pyritization or ferric incrustations of the bioclasts at the relatively stagnated ocean floor (e.g. Brett 1995; Brett et al. 2012). On the other hand, a mixture of physically and biologically damaged shells and associated well-preserved bioclasts is a typical taphonomic signature of the fossiliferous lag deposits (e.g. Kidwell 1991a; Banerjee and Kidwell 1991; Parras and Casadío 2006).

Transgressive ravinement, particularly, the maximum flooding-generated deposition on top of a transgressive erosion surface during the transgressive system tract (TST), causes a high rate of skeletal deposition with well preservation of hard parts. In the offshore or on the shelf during this time, the bioclasts are the most bioeroded ones due to the lowest rates of sedimentation while the clastic-sediment supply is being trapped landward, primarily in the estuaries. Conversely, shelly deposits of the regressive phases, particularly during the falling-stage-system tract (FSST), tend to be composed of the poorly preserved bioclasts (fragmented and loosely packed) due to the higher kinetic energy condition and higher frequency of erosion events. Therefore, recognition of these taphonomic features in the fossil assemblages allows the geoscientists to apprehend and evaluate the relation of such signatures with the specific system tracts in a sequence stratigraphic framework.

19.5 Intrinsic Taphonomic Damages

The intrinsic features of skeletons can interfere with (at least in part) how the physical, chemical, and biological damages will affect the integrity of a bioclast and how intense this damage will be. Among the main characteristics, we highlight the mineralogical composition, the shell type (i.e. univalve, bivalve, multielement, or arborescent), and the skeletal anatomy responsible for joining the parts in bivalves and multielement shells.

Mineralogy imparts a strong influence on the preservation potential of the bioclasts. For example, bivalve shells exhibit the high probabilities of preservation because their shells, in a majority of taxa, are composed of crossing layers of aragonite and calcite. This bi-mineralogical network provides a higher resistance to the mechanical wear and tear (Cox 1967) in addition to imparting a higher resistance from dissolution than the exoskeletons of other taxa having different mineralogical makeups. Mollusk shells always have the risk of breaking due to their hardness. The opposite character can be observed in the shells of lingulid brachiopods having chitinous organic phosphatic shells – the kind of shells that are more flexible, nearly unbreakable, although their valves can wrinkle or tear apart (Kocsis et al. 2012). However, the mineralogical composition of the lingulid shells results in rapid degradation of the organic matrix and complete disintegration within two to three weeks. A third example can be observed in echinoderms. Their skeleton being sufficiently irrigated with blood is resistant while the animal is living. Nonetheless, after death, an echinoderm skeleton becomes fragile due to its multielement nature and the mineralogical composition of high-Mg calcite with $MgCO_3$ contents (Ebert 2007; Kroh and Nebelsick 2010).

Other intrinsic factors, as already mentioned earlier, are the kind of skeleton and the structures that join its pieces together. These features reflect the probability of skeletons being preserved articulated or disarticulated. Moreover, there is an inverse relationship between the probability of articulate preservation and the time of residence at the SWI. This understanding allows the paleontologists and taphonomists infer the exposition time of the bioclasts at the TAZ (e.g. Schäfer 1972; Allen 1990; Brett 1990; Ghilardi and Simões (2000)). Due to the presence of a strong ligament in the lunule and the one-toothed hinge, bivalves can keep articulated for a relatively long time, which may vary from days to months (e.g. Ghilardi and Simões 2000). The closed articulated shells can also be transported at a distance of more than 20 km (Cadée 2002). The chances of echinoderms with multielement shells to be preserved articulated are bleak and shells can be diminished within a temporal window varying from a few hours to days (e.g. Schäfer 1972; Brett 1990, 1995). Marine arthropods like trilobites and crustaceans were/are more resistant to disarticulation than the echinoderms, but less resistant than mollusks. Vertebrate endoskeletons are dependent on the strength of the tendons to maintain the bones joined and the skeletal disarticulation is strongly influenced by the scavengers and bacterial activities (e.g. Behrensmeyer et al. 2000).

Finally, the morphological attributes also play as the important factors related to the intrinsic taphonomic damages. Although the thicker bioclasts should make a higher resistance to physical damages and chemical alterations, the influence of shell thickness against the biological damages should be considered and further studied. Bioclasts, which remain at the SWI, are likely to be colonized by fouling organisms. However, many taxa – from microalgae to macro-invertebrates – avoid shells having extremely smooth or armed (by robust ribs) exterior. Such morphological features make it difficult for the anchoring epibionts to colonize and, therefore, diminish the biological damages to bioclasts.

19.6 Biofabric, Geometry, and Packing of Shell-Accumulation

The descriptive nomenclature for dense skeletal concentrations fundamentally includes four categories (Figure 19.2): (*sensu* Kidwell et al. (1986))

1) Taxonomic composition;
2) Bioclastic fabric (especially, close-packing);
3) Geometry; and
4) Internal structure of the deposit.

Internal structures vary from simple to complex. A simple skeletal concentration can usually be internally homogeneous or monotypic in nature among its attribute features (Kidwell et al. 1986). Complex concentrations exhibit complicated patterns of variation in one or more features, e.g. alternating horizons of articulated, disarticulated, and reoriented hard parts, in any combination.

Biofabric refers to attributes, e.g. skeletal orientation, close-packing, sorting by size and shape, and 3-dimensional arrangement of skeletal elements in the matrix (Figure 19.3; Kidwell et al. (1986)). On the bedding surfaces, orientation is usually described using rose diagrams. For the cross sectional view, Kidwell suggests: (i) concordant (parallel to subparallel alignment of the long axes of hard parts with respect to the bedding surface); (ii) perpendicular; and (iii) oblique. The term "fabric" includes imbrication, edgewise (vertically positioned), stacking, and nesting.

Monotypic and polytypic are the skeletal concentrations that are respectively characterized by single and multiple types of skeletons (Figure 19.2; Kidwell et al. (1986)). Kidwell extended Fagersrom's (1964) original definition of monotypy to apply it to any taxonomic

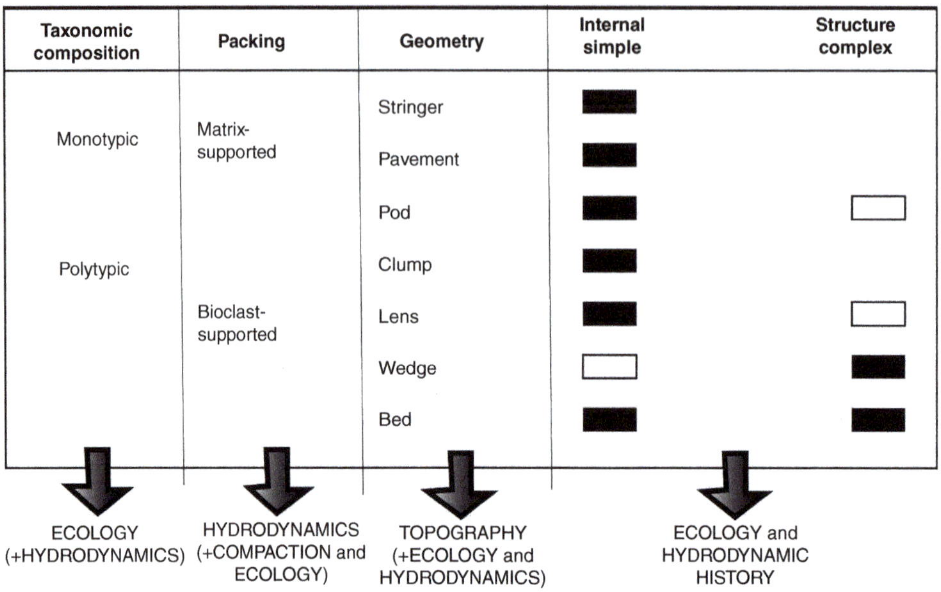

Figure 19.2 Procedure for the description of skeletal concentrations, with standardized terms for major features and their genetic significance. Not all combinations of features are likely. For example, stringers and pavements are almost always simple in internal structure (shaded box), whereas, among the thicker beds, both simple and complex internal structures are common. *Source:* Adapted from Kidwell (1986).

level. The lower the taxonomic level, the greater is the ecological or hydrodynamic significance of monotypy.

The geometry of the deposit depends on a number of biological and physical factors (Figure 19.3; Kidwell and Holland (1991)), which include:

1) Topography of the depositional surface;
2) Mode of life of the bioclast producer;
3) Duration in the TAZ; and
4) Physical process of hard parts' concentration.

Following are the main definitions of terms shown in Figure 19.3.

- **Stringer:** A 2-dimensional, elongated, and localized deposit, which is not laterally continuous and corresponds to specific physical and biological dynamics. The thickness does not exceed one shell or skeletal element.
- **Clump:** A small-scale deposit with poorly defined margins.
- **Pods:** The small-scale irregular concentration with well-defined edges.
- **Lens:** A small or large lens-shaped deposit having a relatively regular geometry along with its tapered lateral terminations (laterally pinching out and/or bevelled by erosion). Lens type of deposits include the ray-pit fills, channels and scour fills, isolated shoals, and biothermal structures.

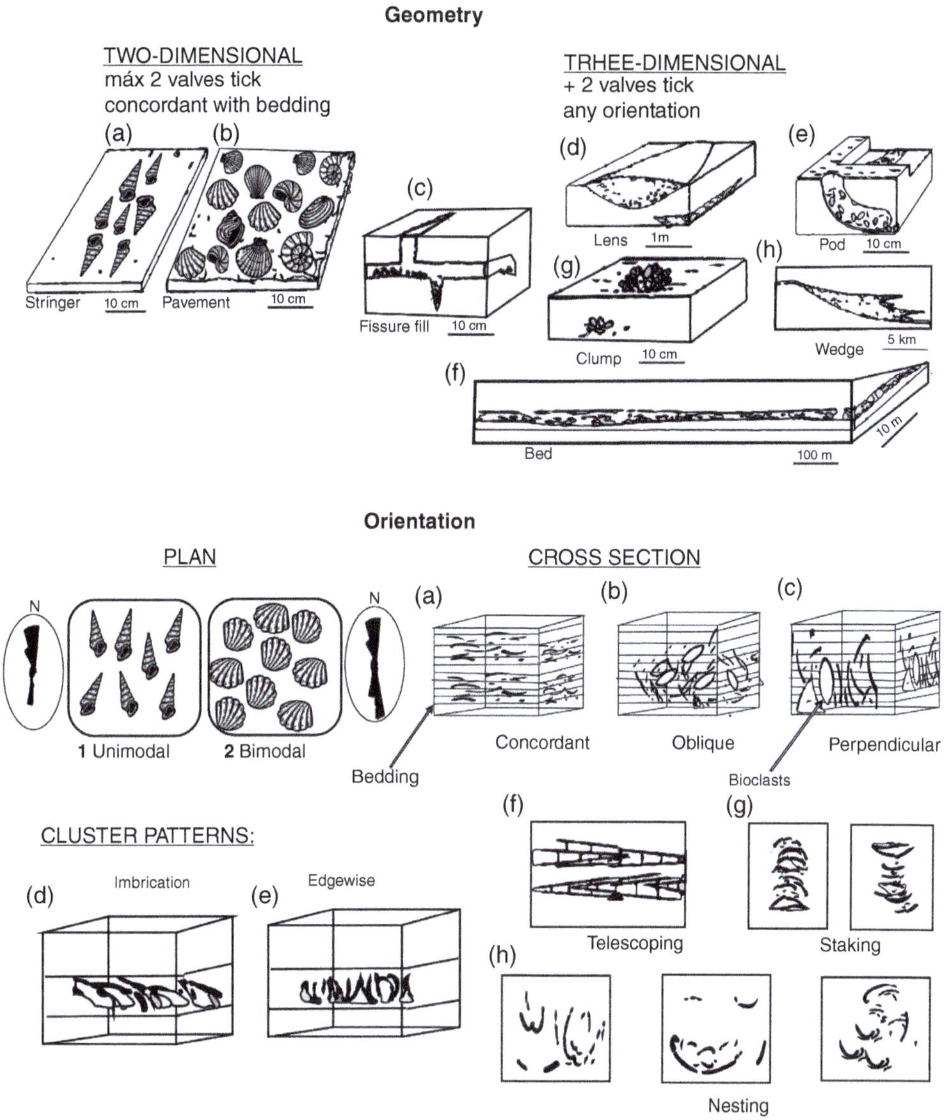

Figure 19.3 Terminology for hard-part orientation on bedding planes and in cross section of the bed. *Source:* Adapted from Kidwell et al. (1986).

- **Wedges:** Accumulations taper laterally only in one direction, whereas the beds are continuous, tabular, or sheet-like.
- **Beds:** Thickness variation depends upon the topography of their upper and lower boundaries and is distinguished from horizons of physically discrete lenses or pods.
- **Densely and loosely packed** (Figures 19.4–19.5): The term "densely packed" describes the deposits that are bioclast-supported, i.e. the array of coarse bioclasts clearly delivers the mechanical support needed for the bed, along with the finer matrix particles and/or

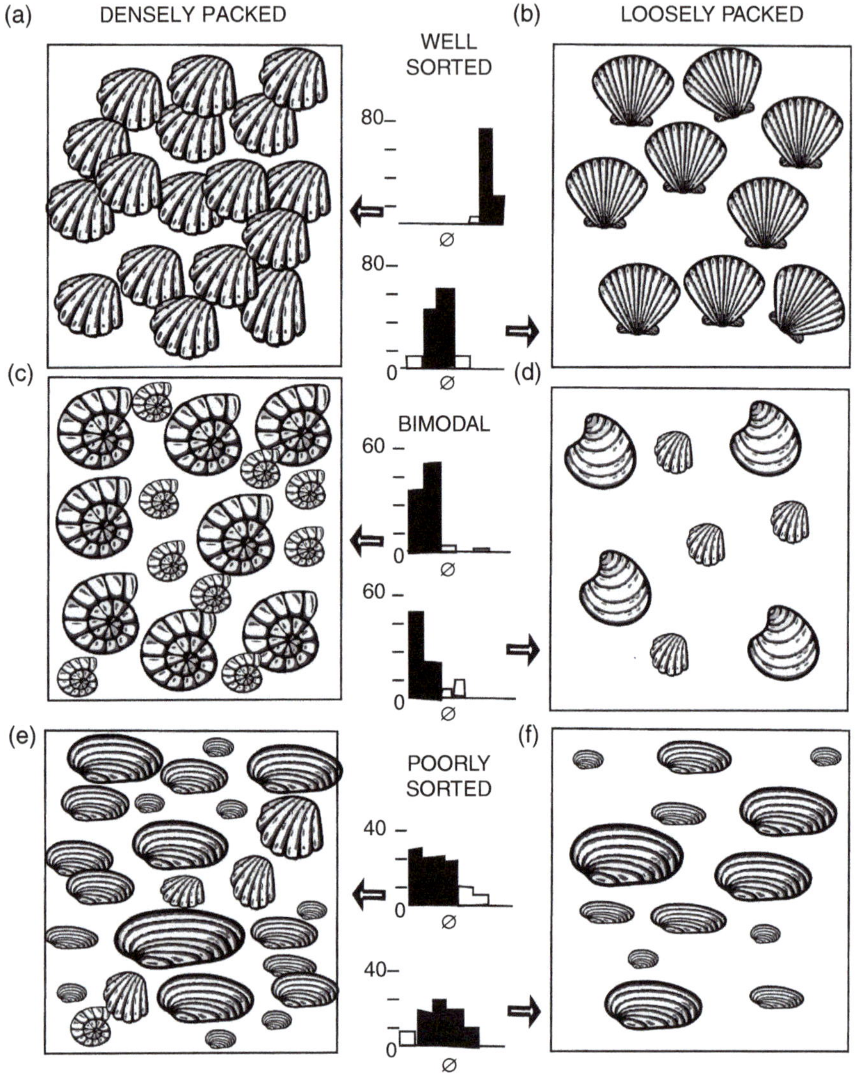

Figure 19.4 Terminology for the geometry of skeletal concentrations. *Source:* Adapted from Kidwell et al. (1986).

cement filling the interstices. Bioclast–bioclast contacts are common. The term "loosely packed" describes the deposits that are matrix-supported, but with coarse bioclasts that are closely associated. Most of these bioclasts are within one diameter.

- **Dispersed** (Figure 19.5): The term "dispersed" describes the deposits that are matrix-supported with sparsely distributed bioclasts. More than the one-diameter distance or the body length separates most bioclasts from each other.
- Internal structures vary from simple to complex. Simple skeletal concentrations are internally homogeneous or have a monotonic trend in features. Complex concentrations exhibit the complicated patterns of variation in one or more features, like the alternating horizons of articulated and disarticulated hard parts and reoriented hard parts (Kidwell et al. 1986).

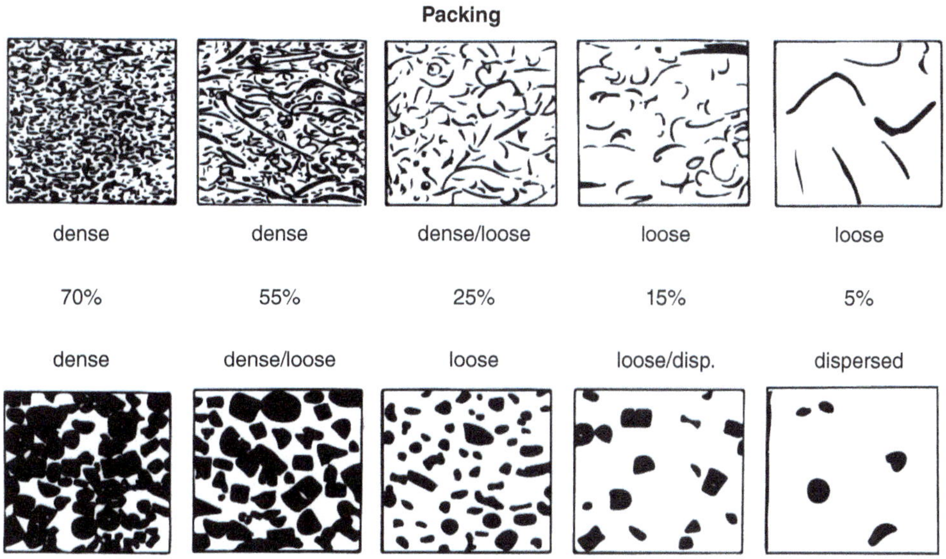

Figure 19.5 Packing of a volume rock. Bioclast of different shapes may constitute the same percent volume of rock (line drawings of Schäfer (1969)), but exhibit significantly different degrees of close-packing (our terms). *Source:* Adapted from Kidwell et al. (1986).

19.7 Taphonomy and Petrophysics

The sedimentary geology is essential for the exploration and development of petroleum fields. Understanding the depositional environment, the facies distribution, and their petrophysical characteristics is essential to determine the quality of a reservoir. The role of geologists in this regard becomes essential, which, through the use of diverse techniques, can obtain a detailed study of a field, where it is possible to predict the facies distribution and to identify the target reservoir lithosomes present in an exploration acreage. As pointed earlier, in this chapter, the use of taphonomy and stratigraphy provides significant information on the origin of condensed shell beds. The accumulation history can therefore be constructed and a sedimentological facies distribution can be predicted, hence, proving taphonomy to be an important tool for the reservoir sedimentologists in the paleoenvironment reconstruction.

The description of condensed shell reservoirs is unusual in the literature. However, the series of discoveries of the pre-salt carbonate reservoirs along the offshore of Brazilian eastern margin as well as in its western African offshore counterpart has instilled enthusiasm among the reservoir geologists and petrophysicists for the understanding of how these deposits developed, their spatiotemporal occurrences, and their petrophysical properties. Shell-bed reservoirs have been explored along the Brazilian and the west African offshore margins (e.g. the Campos and Congo basins) since 1980s and 1990s and are considered to be the leading lithofacies in hydrocarbon production (Tigre et al. 1983; Guardado et al. 1989; Dale et al. 1992). These shell beds have been described as the bivalve-rich grainstones, rudstones, and packstones interbedded with shales and sandstones deposited under the lacustrine settings (Abrahão and Warme 1990; Carvalho et al. 2000; Harris 2000; Thompson et al. 2015; Ceraldi and Green 2016; Saller et al. 2016; Kattah 2017; Muniz and Bosence 2018). Recently, newly published studies have tried to characterize these shell-bed reservoirs according to their corresponding taphonomic characteristics and taphofacies association (Muniz 2013; Muniz and Bosence 2017, 2018; Mizuno et al. 2018; Oliveira et al. 2019; Chinelatto et al. 2020a, 2020b). Analog instances of similar deposits have also been used to improve the understanding of reservoir architectures, heterogeneities and anisotropies, and spatial petrophysical trends (Kinoshita 2010; Tavares et al. 2015; Corbett et al. 2016, 2017; Chinelatto et al. 2018a, 2018b; Rigueti et al. 2020). In this section, we will present a series of studies on the use of taphonomy in the prediction of depositional facies, depositional interpretation, and its relationship with porosity and permeability.

19.8 Relationship of Taphonomy with Petrophysics: Examples from Barremian-Aptian Shell Beds of Brazilian Pre-Salt Reservoirs

19.8.1 Morro do Chaves Formation (Sergipe-Alagoas Basin, Brazil)

The shell beds (or coquina beds) of the Morro do Chaves Formation in the Sergipe-Alagoas basin are considered to be the geological analog for the pre-salt reservoirs of the Santos and Campos basins in Brazil (Abrahão and Warme 1990). The Sergipe-Alagoas passive-margin basin is located toward the north-northeast of the Campos hydrocarbon province between the latitude parallels 9° and 11°S. Its origin is associated with the breakup of Western Gondwana between the late Jurassic and early Cretaceous (Feijó 1994; Campos Neto et al. 2007). The stratigraphic succession of this basin has been subdivided into four megasequences, viz. (i) pre-rift, (ii) syn-rift, (iii) gulf or restricted sea, and (iv) open sea phases. The Morro do Chaves Formation belongs to the Barremian to Aptian syn-rift phase.

The Morro do Chaves Formation is interpreted to be deposited under the shallow-water lacustrine condition. The deposits consist of condensed shell beds, mudstones, sandstones, and conglomerates (Figueiredo 1981; Kinoshita 2010; Thompson 2013; Thompson et al. 2015; Tavares et al. 2015; Chinelatto et al. 2018a; Rigueti et al. 2020). Some of these aforementioned works are focused on using the taphonomic criteria to interpret the depositional environment of condensed shell beds and to explain the evolution of deposits. Tavares et al. (2015) were the first to present the use of taphonomy in their facies classification with the fragmentation of shells and the composition of micritic or siliciclastic grains as the criteria. In their study, four coquina facies were described: (i) Cf: fragmented coquina with no micrite; (ii) Cfi: impure fragmented coquina with no micrite; (iii) Cm: non-fragmented coquina with micrite; and (iv) Cmi: impure non-fragmented coquina with micrite. Later, Chinelatto et al. (2018a) proposed a taphofacies model for the Morro do Chaves Formation by using additional taphonomic features placing the carbonate lithofacies into six taphofacies categories (i.e. Tf-1 to Tf-6). Recently, Rigueti et al. (2020) recognized 14 sedimentary facies through the use of taphonomic, sedimentological, and stratigraphic aspects. Table 19.2 shows a summarized comparison between the principal condensed shell facies described in these studies. In general, the model of facies distribution is related to the kinetic energy of the aquatic

Table 19.2 Principal condensed shell facies described in the Morro do Chaves Formation and their main taphonomic features.

Tavares et al. (2015)			Chinelatto et al. (2018)			Rigueti et al. (2020)		
Facies	Taphonomic characteristics	Interpretation	Facies	Taphonomic characteristics	Interpretation	Facies	Taphonomic characteristics	Interpretation
Cf	Fragmented and abraded shells (80%); densely packed;	Reworked shells by traction flows, in an intermediate subaqueous lacustrine domain.	T1	Fragmented, abraded and concordantly oriented shells; densely packed	Shallow water lake environments, above fair-weather wave base	cApp	Concordant, or oblique to bedding, rounded fragmented bivalve shells; densely packed	Nearshore, above fair-weather wave base
Cfi	40% of fragmented shells, less than 10% of micrite and more than 10% of siliciclastics	Reworked shells by traction flows, in an intermediate subaqueous lacustrine domain, with siliciclastic input.	T2	Fragmented, abraded and concordantly oriented shells, interbedded with thin laminae of mudstone; densely packed	Shallow water lake environments, above fair-weather wave base	cRdc	Concordant, oblique to bedding or randomly oriented, fragmented and whole bivalve shells; densely packed	Nearshore, above fair-weather wave base
Cm	More than 20% of non-fragmented shells, more than 10% of micrite and less than 10% of siliciclastics.	Deep subaqueous lacustrine domain.	T3	Fragmented and concordantly oriented shells; articulated valves, densely packed; open or closed, are rare.	Shallow water lake environments, above fair-weather wave base	cRdm	Randomly oriented, whole bivalve shells; densely packed	Nearshore, below fair-weather wave base and above storm wave base
Cmi	More than 20% of non-fragmented shells, more than 10% of micrite and more than 10% of siliciclastics.	Deep subaqueous lacustrine domain, with siliciclastic input.	T4	Fragmented and whole, randomly oriented shells; densely packed	Storm-dominated lower shoreface, below fair-weather wave	cRm	Moderately to poorly sorted, randomly oriented, fragmented and whole bivalve	Offshore, below storm wave base

(*Continued*)

Table 19.2 (Continued)

Tavares et al. (2015)			Chinelatto et al. (2018)			Rigueti et al. (2020)		
Facies	Taphonomic characteristics	Interpretation	Facies	Taphonomic characteristics	Interpretation	Facies	Taphonomic characteristics	Interpretation
					base and above storm wave base		shells; densely to loosely packed	
			T5	Low fragmentation, and randomly oriented shells; packing varies from dense to loose; closed valves are rare	Storm-dominated lower shoreface, below fair-weather wave base and above storm wave base	cRpm	Poorly sorted, randomly oriented, fragmented and whole bivalve shells; densely to loosely packed	Offshore, below storm wave base
			T6	Whole randomly oriented shells; loose packing	Storm-dominated lower shoreface, below fair-weather wave base and above storm wave base			

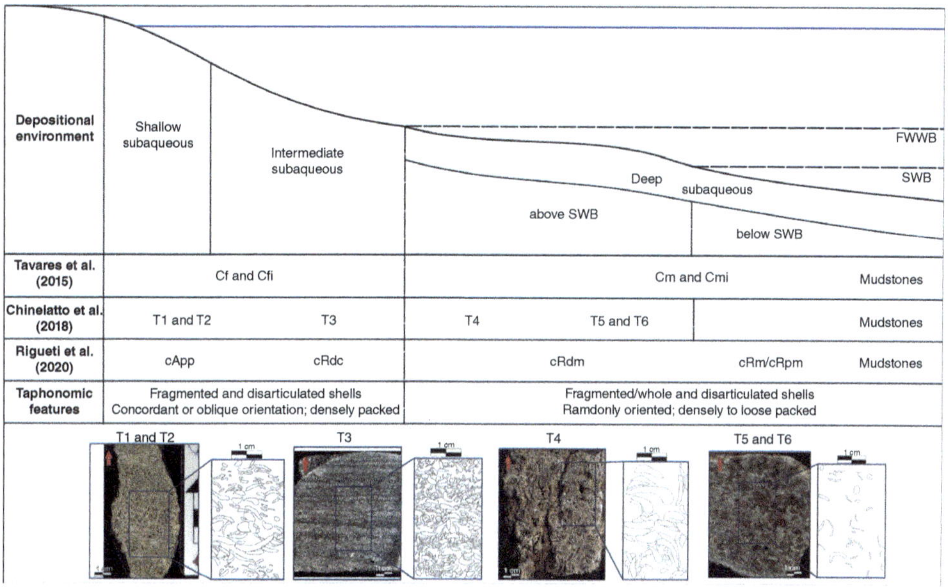

Figure 19.6 Depositional interpretation of condensed shell beds under fair-weather and storm conditions.

depositional environment involved, which influences directly the taphonomic characteristics. They illustrate the evolution of condensed shell beds as the shallow, deep, and storm deposits – corresponding to the varying textural and taphonomic characteristics. The geological model of these deposits is shown in Figure 19.6.

The shallow water facies are composed mainly of bivalve-rich grainstones and rudstones with disarticulated and fragmented shells, oriented concave-down and oblique valves, and a general absence of finer-grained sediments. In these deposits, high-energy flows favor the transport of bioclasts, break the shells, and strip away the finer grains. The taphonomic characteristics of described facies reinforce the idea of depositional environments being located above the fair-weather wave-base that exhibit highly fragmented, abraded, and disarticulated shells, oriented valves, well to moderate sorting, and absence of preserved valves in life position (Kidwell et al. 1986; Kidwell 1991a; Fürsich and Oschmann 1993; Fürsich 1995).

The storm deposits were described as rudstones, grainstones, packstones, and wackestones with moderately to poorly sorted grains, occurrences of micrite, moderate fragmentation, moderate abrasion, and chaotically arranged shells. The rare presence of reworked articulated valves (both the closed and butterfly ones) has also been reported. The occurrence of erosional surfaces at the bottom of tempestite shell beds is common. Storm events rework shells and promote the deposition of chaotically oriented and fragmented bioclasts. Depending upon the storm energy, it can winnow the finer grains favoring the accumulation of densely packed shell beds (Brett and Baird 1986; Fürsich and Oschmann 1986, 1993). Two main storm deposits occur in the described coquinas of Morro do Chaves Formation: the storm winnowed and poorly sorted storm deposits (Chinelatto et al. 2018a). The deep deposits below the storm-wave base are described mainly as mudstones with absence of bivalve shells.

The porosity of shallow facies (i.e. above the fair-weather wave base) ranges from 5 to 25% (Kinoshita 2010; Tavares et al. 2015; Belila et al. 2018; Zielinski et al. 2018). Generally, the porosity occurs as interparticle and intraparticle pores, moldics pores, and vugs. Porosity can be strongly influenced by diagenetic processes such as cementation and dissolution (Tavares et al. 2015; Chinelatto et al. 2018b; Zielinski et al. 2018). As pointed out by Corbett et al. (2016), a single bed may have porosity ranging between 9 and 19% and permeability between 4 mD and 1.4 D. Such a variation is related to the primary porosity preservation, creation of secondary porosity, and porosity obliteration. The condensed shell beds with well-sorted grains may preserve primary pores that enhance the connectivity with secondary pores (moldics and vugs), although the calcite cementation can reduce the porosity and permeability.

The storm deposits exhibit a high range of porosity and permeability similar to the shell beds deposited above the fair-weather wave base. The winnowed storm deposits exhibit the highest range of values of porosity and permeability (i.e. 5–10% and 100 mD) because of the preservation of primary pores due to winnowing and of secondary pores due to dissolution. The poorly sorted facies shows the lowest value ranges of porosity and permeability (<5% and <10 mD). Pores are associated with the diagenetic dissolution of grains resulting in moldic pores and vugs (Tavares et al. 2015; Belila et al. 2018; Zielinski et al. 2018; Chinelatto et al. 2018b). Figure 19.7 shows some thin sections of well-sorted and poorly sorted storm-deposit facies.

According to the genetic classification of porosity in carbonate rocks proposed by Ahr (2008), these results reveal that the porosity in all shell beds varies due to depositional and diagenetic controls over porosity. For the Morro do Chaves Formation, taphonomy helps in identifying the origin of deposits (shallow, storm-winnowed, and storm-poorly sorted) and the origin of porosity can be related to depositional and diagenetic processes (Chinelatto et al. 2020a). Due to the depositional process, well-sorted condensed shell beds (shallow and winnowed storm deposits) can preserve its primary pores. As mentioned earlier, the energy of currents can sweep and strip the fine grains off, thereby having sedimentation-related control over porosity. However, as the carbonate sediments are highly susceptible to diagenetic modifications between shallow (epidiagenesis) and burial diagenesis, both the cementation and compaction incur changes in the final porosity (Chinelatto et al. 2018b, 2020a; Zielinski et al. 2018). The poorly sorted storm deposits generally exhibit high content of micrite that obstructs the primary pores and reduces the porosity – even close to zero. In these deposits, diagenesis is the main precursor for any secondary porosity generation.

19.8.2 Coqueiros and Itapemaformations (Campos and Santos Basins, Brazil)

The Campos and Santos basins are located at the offshore Brazilian eastern and southern margins. These basins were formed during the breaking up of Western Gondwana between the late Jurassic and early Cretaceous (Moreira et al. 2007). During this separation process, the thinning of the continental crust occurred, which resulted in a system of faults and

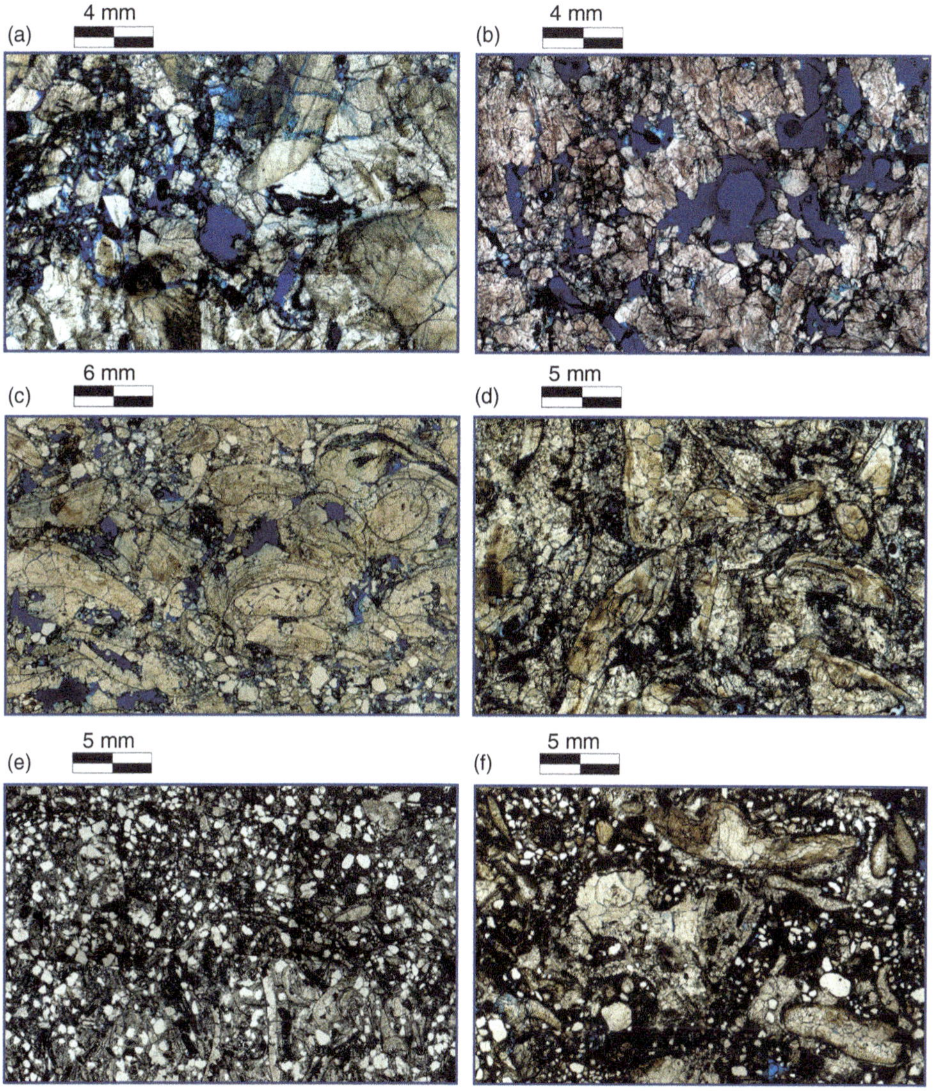

Figure 19.7 Thin section from the main pore type found in condensed shell beds of Morro do Chaves Formation. (A–B) rudstones with interparticle, moldic, and vugs. (C–D) grainstone-packstone with interparticle pores. (E–F) low interparticle pores due to occurrences of finer grains. *Source:* Chinelatto et al. (2018b).

fractures forming the tectono-physiographic structures like horsts and grabens and caused intense volcanism in the south and southeast regions (Cainelli and Mohriak 1999; Mohriak et al. 2008). This period is characterized tectonically as the syn-rift stage, which can be found in all basins on the Brazilian eastern coastal margins and offshore. In this period, the Coqueiros and Itapemaformations were deposited between the Barremian and Aptian ages.

The Coqueiros Formation is described to be lacustrine deposits consisting mainly of shales and carbonates (i.e. bioclastic rudstones and grainstones) that interfinger and intercalate with the proximal clastic sediments (Abrahão and Warme 1990; Carvalho et al. 2000, Moreira et al. 2007). These carbonates consist primarily of condensed bivalve bioclastic shell beds and secondarily of gastropod and ostracod bioclasts. Locally, pure calcite deposits (pure coquina) occur alongside shell beds containing abundant talc and stevensitic peloids (Dias et al. 1988). The condensed shell beds have been interpreted as being deposited atop the structural highs under shallow water and high-energy conditions (Abrahão 1987; Carvalho et al. 2000; Muniz 2013; Thompson et al. 2015; Muniz and Bosence 2018; Mizuno et al. 2018; Oliveira et al. 2019; Souza et al. 2020). Apart from the documentation facies description and textural variation of shell beds in the literature, only over the last one decade, some research has been published using taphonomic criteria for facies description and environmental interpretation (Muniz 2013; Muniz and Bosence 2017, 2018; Mizuno et al. 2018; Oliveira et al. 2019; Olivito and Souza 2020). The facies model depicting variation from the proximal to the distal lake margin (i.e. above fair-weather wave base and above storm-wave base, respectively) is significantly similar to the one shown in the Morro do Chaves Formation. Figure 19.8 exhibits the facies model as well as the taphonomic characteristics (Muniz 2013; Muniz and Bosence 2017, 2018).

The shallow taphofacies (i.e. above fair-weather wave base) comprises well-sorted grainstones and with rudstones in subordinate cementation. The taphofacies exhibits the

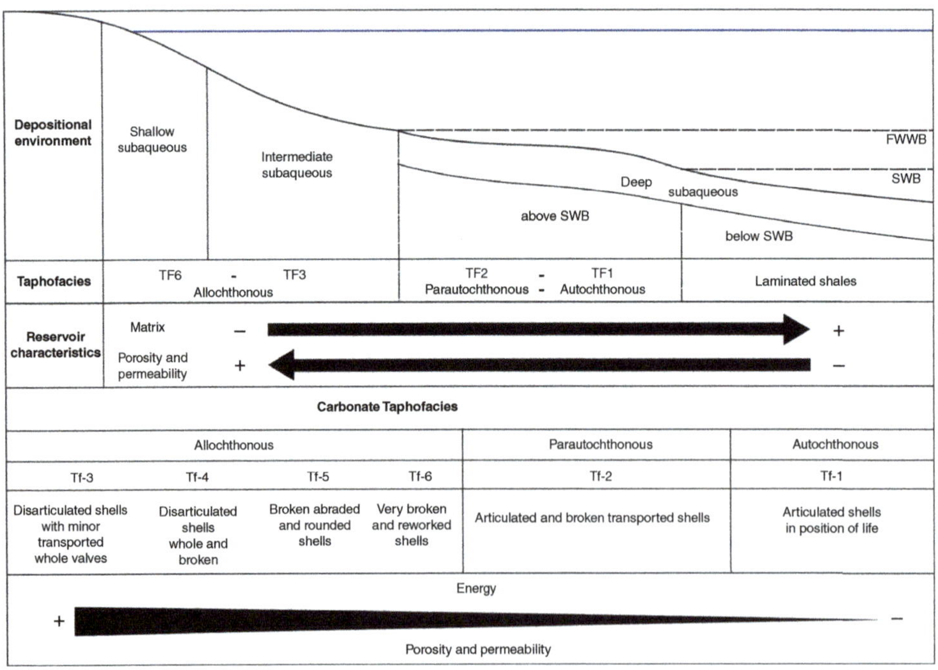

Figure 19.8 Schematic facies model for the Coqueiros Formation illustrating the effect of the hydraulic gradient and reservoir characteristics. *Source:* Modified from Muniz (2013); Muniz and Bosence (2017, 2018).

occurrence of disarticulated and reworked shells as the main taphonomic features with the degree of fragmentation varying in accordance with kinetic energy of the depositional subenvironments. The main pore types observed in these shallow facies are the interparticle, moldic, and vugs. Herlinger et al. (2017) pointed out that the samples with preserved interparticle porosity exhibit the best values of permeability – also being associated with partially preserved primary porosity and the low degrees of compaction and cementation. The moldic pores contribute to total porosity. The occurrences of vugs resulted in the enlargement of primary pores, thus increasing the permeability.

The deep taphofacies (i.e. sediments deposited between the fair-weather and storm wave bases) includes moderately to poorly sorted packstones, floatstones, and cemented rudstones. The taphofacies exhibits articulated and broken shells toward its proximal areas, whereas the most distal deposits exhibit articulated valves in their life position. The main pore type is moldic and intercrystalline (Muniz and Bosence 2017, 2018). The grainstones and rudstones with moldic and vugs exhibit good values of porosity but low values of permeability in comparison to the samples of preserved interparticle pores. Generally, this can also be associated with the cementation and compaction that partially or totally occludes the interparticle pores, thereby reducing the pore throat (Herlinger et al. 2017). Figure 19.9 exhibits the results obtained by Herlinger et al. (2017).

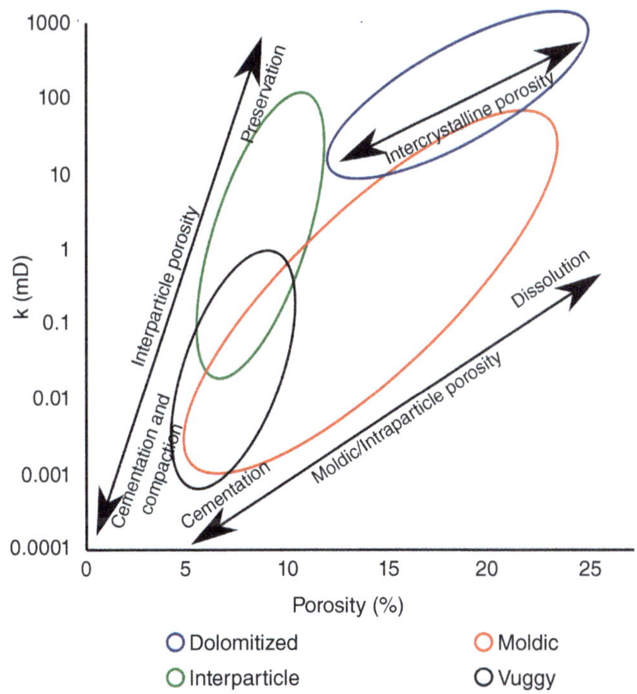

Figure 19.9 Main pore type in condensed shell beds of Coqueiros Formation (Campos Basin) and their relationship with porosity (%), permeability in millidarcys (mD), and diagenetic features. The green circle corresponds to the domain of shell beds with interparticle porosity, black with vuggy, red with moldics, and blue with dolomitized shell beds. *Source:* Adapted from Herlinger et al. (2017).

The Itapema Formation is equivalent to the Coqueiros Formation of Campos basin and it is described as the lacustrine deposits composed mainly of carbonates and shales (Moreira et al. 2007). A few works are available with a detailed description of how these deposits occur. Recently, Chinelatto et al. (2020b) proposed a taphofacies interpretation model for the condensed shell beds of Itapema Formation and related the taphofacies to their corresponding petrophysical characteristics. They described the condensed shell layers as the bivalve-rich grainstones and rudstones, with a high degree of fragmented and abraded valves and shells organized in three different ways: oriented, oblique, and distributed in a chaotic manner. Additionally, they also classified sorting according to their textural components, e.g. (i) well-sorted and (ii) unsorted biofabrics. The well-sorted biofabric has shells and their fragments as the main grain composition, whereas the unsorted biofabric exhibits a high amount of other carbonate particles (ooids, peloids, and ostracods) and/or siliciclastic grains. In this way, two main groups of taphofacies were described, viz. the well-sorted (Tf-1, Tf-3, and Tf-5) and unsorted (Tf-2, Tf-4, and Tf-6) taphofacies. Well-sorted taphofacies were interpreted as the foreshore and shoreface deposits and unsorted as the protected or restricted (i.e. lagoon and washover) and offshore deposits. Figure 19.10 provides the taphofacies interpretation of the Itapema Formation in the fair-weather and storm conditions. The pores were described according to Choquette and Pray (1970) and Ahr (2008) classification schemes. The values of porosity and permeability were obtained by means of gas injection, high-resolution computed tomography (H-CT), and nuclear magnetic resonance (NMR) imaging and characterization techniques.

The origin of porosity in all taphofacies varies from being depositional to diagenetic in nature with the great majority of it being "hybrid" pores. The depositional domain is observed in well-sorted taphofacies where the main pore type is interparticle as a result of the depositional processes, mainly waves and currents, that remove the finer grains. The unsorted taphofacies have finer grains that fill inside the primary pore space and reduce the depositional porosity to values even close to zero. Therefore, the unsorted taphofacies do not originate during the depositional domain. Diagenesis is incipient during the depositional domain with low dissolution and cementation. The H-CT analysis shows that pore volume in sorted taphofacies is low. The NMR imaging demonstrates that microporosity has a low contribution to total porosity. In general, samples from the depositional domain exhibit good pore connectivity that has a positive influence on the permeability values.

The hybrid domain exhibits samples of well-sorted and poorly sorted taphofacies with porosity produced by both the depositional and diagenetic processes. Diagenesis here is responsible for increasing or reducing the porosity through dissolution and cementation. The calcite cementation in the form of rims and/or druse is responsible for reducing the primary pore space, although late dissolution of shells and cement promotes the creation of the vugs and moldic pores. The H-CT images exhibit large pores correlated with the dissolution of shells and cement and the smaller pores with some occurrences of intraparticle and interparticle pores. The NMR results of samples show that the lithology underwent porosity enhancement by dissolution. The samples exhibit large quantities of macropores and little influence of microporosity. The samples affected by cementation exhibit a similar distribution of micro- and macropores.

The diagenetic domain has, as its main characteristic, enhanced or reduced volumes of porosity due to diagenetic modification. In this group, the main pore types are moldics and

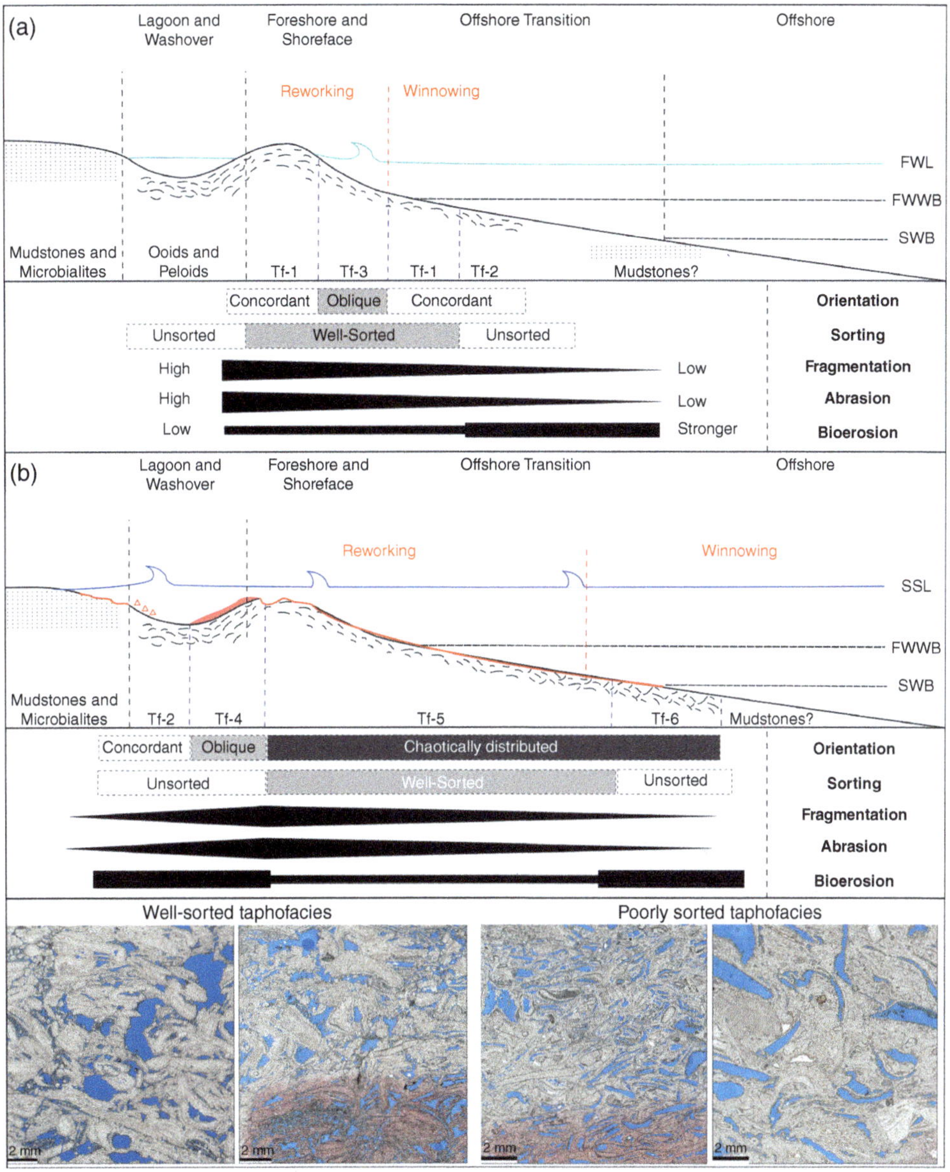

Figure 19.10 Depositional interpretation of taphofacies for shell beds of Itapema Formation (Santos Basin). (a) Fair-weather conditions; (b) storm conditions. *Source:* Chinelatto et al. (2020b), Elsevier.

vugs, whereas the interparticle pores are reduced by calcite cementation. Samples of poorly sorted taphofacies belong to the diagenetic domain. The H-CT images exhibit large pores being moldic and vuggy in nature. The NMR results reveal the micropores making up one-third and macropores two-thirds of the total porosity. The permeability value is usually lower in the diagenetic domain compared to the depositional and hybrid domains.

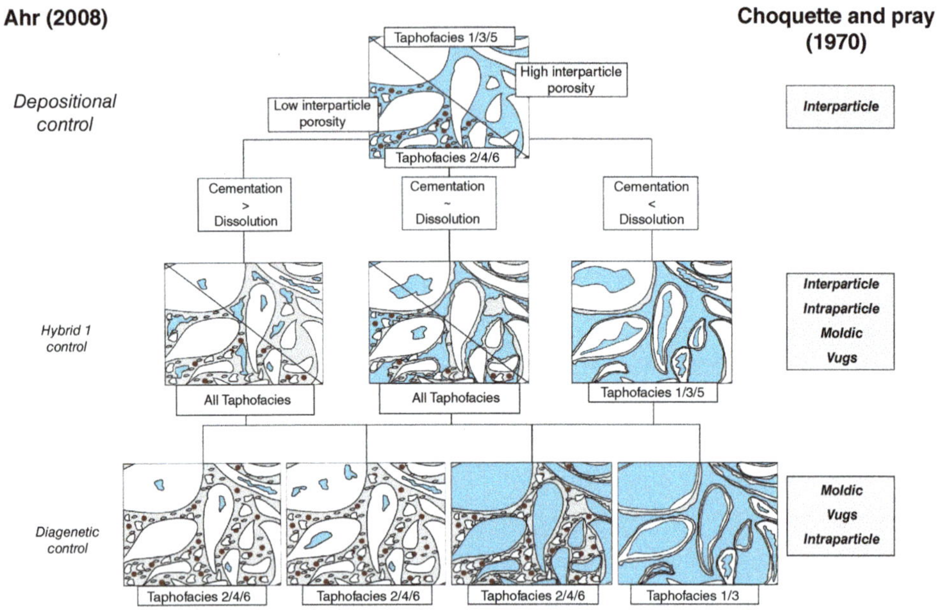

Figure 19.11 Control of porosity between depositional to diagenetic in condensed shell beds of Itapema Formation (Santos Basin). The figure shows the degree of calcite cementation (C) and dissolution (D) and the main pore types found in each situation. The blue areas correspond to the pore space, white the bioclasts and gray the cement. *Source:* Adapted from Chinelatto et al. (2020b).

Figure 19.11 shows the influence of calcite cementation and dissolution and the main pore types in all the described taphofacies. While comparing the described taphofacies with the NMR logs, there is a good observable correlation between the well-sorted and unsorted taphofacies with the permeability values. As shown in Figure 19.12, the lowest values of permeability are generally related to the unsorted taphofacies due to the poor connectivity of pores (isolated moldic and vugs), whereas the highest values are related to the well-sorted taphofacies that exhibit good pore connectivity.

19.9 Final Remarks

The sedimentary geology is essential to the exploration and development of petroleum fields. In this way, knowledge about the depositional settings, the facies distribution, as well as their petrophysical characteristics, is fundamental to estimating the quality of a reservoir. The role of geologists in this regard becomes crucial. The uses of diverse techniques can obtain data for a detailed study of a field, where it is possible to predict the facies distribution in the subsurface and to identify the target lithologies present in an exploratory area. With the advancement of paleontology, taphonomy has become an essential tool in the sedimentological and stratigraphic approaches. The genetic characteristics of fossiliferous dense deposits can be detected along with their paleoecological features, which are intrinsic to the ancient communities and populations. It can be evaluated if a community was

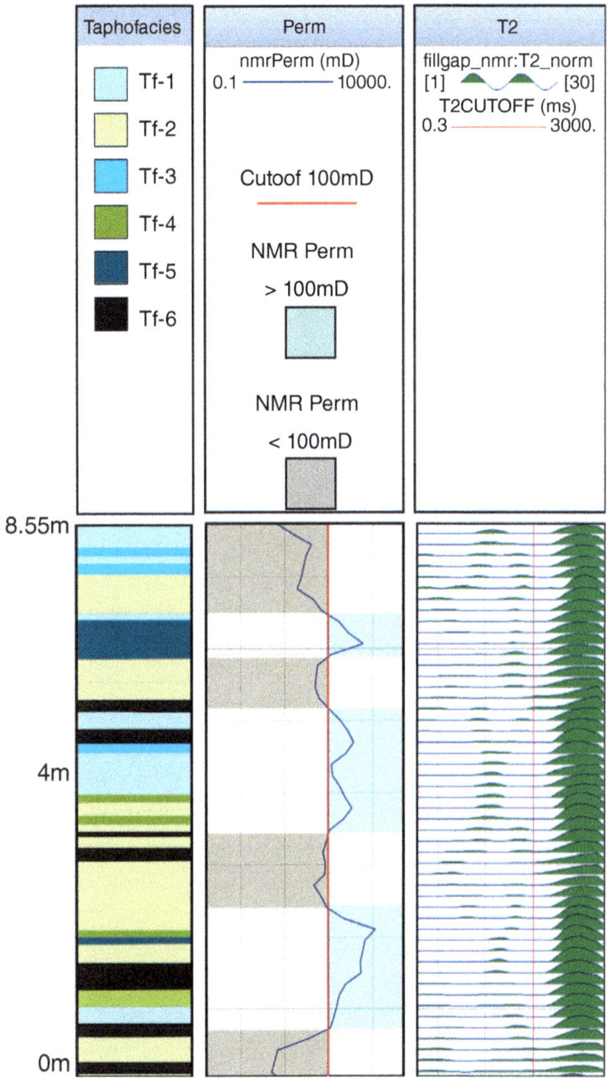

Figure 19.12 Correlation between well-sorted and unsorted taphofacies with permeability values obtained through nuclear magnetic resonance log. *Source:* Adapted from Chinelatto et al. (2020b).

syn-depositionally accumulated and/or was modified by the post-depositional events. Taphonomic analysis of shell beds plays an important role in the paleoenvironmental studies due to the fact that the densely packed shelly accumulations are facies related to the underlying beds, and thus can be considered distinct from the previous depositional events.

Additionally, densely packed fossil assemblages can be the long-term product, as the shell beds can be interpreted as depositional gaps or hiatuses from different stratigraphic orders (e.g. second order and third order in sequence stratigraphy). In this sense, the petrophysical characteristics like porosity and permeability of the rock matrix need to be analyzed together with the taphonomic signatures (e.g. geometry, packing, and sorting). However, the approach has so far been not explored to its full possible utilization by geoscientists, thereby highlighting its importance as a new and promising tool in petrophysical surveys.

Acknowledgments

Soumyajit Mukherjee (IIT Bombay) invited to submit this article. The authors thank the reviewers for their contributions that allowed us to improve the manuscript. RSH is grateful to the CAPES (Coordenação de Aperfeiçoamento de Pessoal de Nível Superior), Brazil, and to the MCTIC/CNPq n° 28/2018 (project # 420748/2018-0). This article will contribute to be the part of his research project "Heterogeneididadenaqualidade do registrofóssil e suaaplicaçãonageologia." GFC is grateful to the CAPES as well (project # 1579147) and to the CNPq (Conselho Nacional de Desenvolvimento Científico e Tecnológico), Brazil (project # 132487/2014-4). HSN and EG also are grateful to the CAPES for the PNPD grant and for the graduate grant (# 88887.372443/2019-00), respectively. SDG expresses his gratitude to IRCC, IIT Bombay, Mumbai. Merryl Le Roux, Frank Otmar Weinreich, Joss Everett, Andrew Harrison, and the proofreading team (Wiley Blackwell) are thanked for their assistance. This chapter has been encapsulated in Mukherjee et al. (2023).

References

Abrahão, D. (1987). *Lacustrine and Associated Deposits in a Rifted Continental-margin: The LagoaFeia Formation, Lower Cretaceous, Campos Basin*. Colorado School of Mines: Offshore Brazil https://books.google.com.br/books?id=PlRZuAAACAAJ.

Abrahão, D. and Warme, J., E. (1990). Lacustrine and associated deposits in a rifted continental margin – lower cretaceous LagoaFeia formation, Campos Basin, Offshore Brazil. In: *Lacustrine Basin Exploration: Case Studies and Modern Analogs*, vol. 50 (ed. B.J. Katz). American Association of Petroleum Geologists https://doi.org/10.1306/M50523C18.

Ahr, W.M. (2008). *Geology of Carbonate Reservoirs: The Identification, Description, and Characterization of Hydrocarbon Reservoirs in Carbonate Rocks*, 296. Hoboken: Wiley.

Aigner, T. (1985). *Storm Depositional Systems, Dynamic Stratigraphy in Modern and Ancient Shallow-marine Sequences. Lecture Note in Earth Science*, 174. Berlin: Springer-Verlag.

Allen, J.R.L. (1990). Transport – hidrodynamics: shells. In: *Palaeobiology a Synthesis* (ed. D.E.G. Briggs and C. Peter), 227–229. Blackwell Science Ltd.

Arthur E.B. (2015). Morphodynamics. (eds. Adolf Seilacher and Alan D. Gishlick), In *Integrative and Comparative Biology*, 55 (3): 543–544. https://doi.org/10.1093/icb/icv090.

Aucoin, C. D.; Brett, C.E.; Dattilo, B.F; Thomka, J.R. 2016. Sequence stratigraphic model for repeated "butter shale" Lagerstätten in the Ordovician (Katian) of the Cincinnati, Ohio region, USA. *Canadian Journal of Earth Sciences*, 53: 1–11. https://http://dx.doi.org/https://doi.org/10.1139/cjes-2015-0219

Banerjee, I. and Kidwell, S.M. (1991). Significance of molluscan shell beds in sequence stratigraphy: an example from the lower cretaceous Mannville Group of Canada. *Sedimentology* 38 (5): 913–934. https://doi.org/10.1111/j.1365-3091.1991.tb01879.x.

Behrensmeyer, A.K., Kidwell, S.M., and Gastaldo, R.A. (2000). Taphonomy and paleobiology. In: *Deep Time: Paleobiology's Perspective, Paleobiology Supplement 4*, vol. 26 (ed. D.H. Erwin and S.L. Wing), 103–147. https://doi.org/10.1017/S0094837300026907.

Belila, A.M.P., Kuroda, M.C., Souza, J.P.P. et al. (2018). Petrophysical characterization of coquinas from Morro do Chaves formation (Sergipe-Alagoas Basin) by X-ray computed tomography. *Revista do Instituto de Geociências – USP* 18 (3): 3–13.

Brett, C.E. (1990). Destructive taphonomic processes and skeletal durability. In: *Palaeobiology a Synthesis* (ed. D.E.G. Briggs and C. Peter), 223–226. Blackwell Science Ltd.

Brett, C.E. (1995). Sequence stratigraphic, biostratigraphy, and taphonomy in shallow marine environments. *Palaios* 10: 597–616. https://doi.org/10.2307/3515097.

Brett, C.E. (1998). Sequence stratigraphy, Paleoecology, and evolution: biotic clues and responses to sea-level fluctuations. *Palaios* 13: 241–262.

Brett, C., and Baird, G. (1986). Comparative taphonomy: a Kkey to paleoenvironmental interpretation based on fossil preservation. *PALAIOS*, 1(3), 207–227. doi:https://doi.org/10.2307/3514686.

Brett, C. and Walker, S. (2002). Predators and predation in paleozoic marine environments. *Paleontological Society Papers* 8. https://doi.org/10.1017/S1089332600001078: 93–118.

Brett, C.E., Zambito, J.J., Hunda, B.R., and Schindler, E. (2012). Mid-Paleozoic trilobite lagerstätten: models of diagentically enhanced obruption deposits. *Palaios* 27: 326–345. https://doi.org/10.2110/palo.2011.p11-040r.

Cadée, G.C. (2002). Floating articulated bivalves, Texel, North Sea. *Palaeogeography, Palaeoclimatology, Palaeoecology* 183 (2002): 355–359. https://doi.org/10.1016/S0031-0182(01)00487-4.

Cainelli, C. and Mohriak, W.U. (1999). Some remarks on the evolution of sedimentary basins along the Eastern Brazilian continental margin. *Episodes* 22: 206–216.

Campos Neto, O.P.A.C., Lima, W.S., and Cruz, F.E.G. (2007). Bacia de Sergipe-Alagoas. *Boletim de Geociências da Petrobras* 15: 405–415.

Carvalho, M.D., Praça, U.M., Silva-Telles, A.C. et al. (2000). Bioclastic carbonate lacustrine facies models in the campos basin (Lower Cretaceous), Brazil. In: *Lake Basins Through Space and Time*, vol. 46 (ed. E.H. Gierlowski-Kordesch and K.R. Kelts), 245–256. AAPG Studies in Geology.

Ceraldi, T.S. and Green, D. (2016). Evolution of the South Atlantic lacustrine deposits in response to early Cretaceous rifting, subsidence and lake hydrology. *Petroleum Geoscience of the West Africa Margin* 438: 77–98.

Chang, H.K., Kowsmann, R.O., Figueiredo, A.M.F. And Bender, A.A. (1992). Tectonics and stratigraphy of the East Brazil Rift system: an overview: *Tectonophysics*, v. 213, p. 97–138.

Chinelatto, G.F. (2020). Análise petrofísica em coquinhas abordando o conceito de tafofácies. In: *Tese de doutoramento*, 131. Campinas: UniversidadeEstadual de Campinas.

Chinelatto, G.F., Vidal, A.C., Kuroda, M.C., and Basilici, G. (2018a). A taphofacies model for coquina sedimentation in lakes, (lower Cretaceous, Morro do Chaves formation, NE Brazil). *Cretaceous Research* 85: 1–19.

Chinelatto, G.F., Kuroda, M.C., and Vidal, A.C. (2018b). Relação entre biofábrica e porosidade, coquinas da Formação Morro do Chaves (Barremiano/Aptiano), Bacia de Sergipe-Alagoas, NE-Brasil. *Geologia USP. Série Científica* 18: 57–72.

Chinelatto, G.F., Belila, A.M.P., Basso, M., and Vidal, A.C. (2020a). Relationship between biofabric and petrophysics in coquinas, insights on Brazilian Barremian-Aptian carbonates of the Santos and Sergipe-Alagoas Basins. *International Journal of Earth Sciences* (In press).

Chinelatto, G.F., Belila, A.M.P., Basso, M. et al. (2020b). A taphofacies interpretation of shell concentrations and their relationship with petrophysics: a case study of Barremian-Aptian coquinas in the Itapema Formation, Santos Basin-Brazil. *Marine and Petroleum Geology* 116.

Corbett, P.W.M., Estrella, R., Rodriguez, A.M. et al. (2016). Integration of Cretaceous Morro do Chaves rock properties (NE Brazil) with the Holocene Hamelin Coquina architecture (Shark Bay, Western Australia) to model effective permeability. *Petroleum Geoscience* 22 (2): 105–122.

Corbett, P.W.M., Wang, H., Câmara, R.N. et al. (2017). Using the porosity exponent (m) and pore-scale resistivity modelling to understand pore fabric types in coquinas (Barremian-Aptian) of the Morro do Chaves Formation, NE Brazil. *Marine and Petroleum Geology* 88: 628–647.

Cox, R. (1967). General features of Bivalvia. In: *Treatise on Invertebrate Paleontology. Pt N, V 1 (of 3), Molusca 6, Bivalvia* (ed. R.C. Moore). Lawrence: Geological Society of America and University of Kansas Press.

Dale, C.T., Lopes, J.R., and Abilio, S. (1992). Takula oil field and the greater Takula Area, Cabinda, Angola. In: *Giant Oil and Gas Fields of the Decade 1978-1988*, AAPG Memoir, 54 (ed. M.T. Halbouty), 197–215. Tulsa, Oklahoma USA: AAPG.

Davies, J.D., Powell, E.N., and Stanton, R.J. (1989). Taphonomic signature as a function of environmental process: shells and shell beds in a hurricane-influenced inlet on the Texas coast. *Palaeogeography, Palaeoclimatology, Palaeoecology* 72: 317–356. https://doi.org/10.1016/0031-0182(89)90150-8.

Dias, J.L., Oliveira, J.Q., and Vieira, J.C. (1988). Sedimentological and stratigraphic analysis of the LagoaFeia formation, rift phase of Campos Basin, offshore Brazil. *Revista Brasileira de Geociências* 18 (3): 252–260.

Ebert, T.A. (2007). Growth and survival of postsettlement sea urchins. In: *Edible Sea Urchins: Biology and Ecology*, Developments in Aquaculture and Fisheries Science, 2e (ed. J.M. Lawrence), 95–134. Amsterdam: Elsevier.

Einsele, G., Ricken, W., and Seilacher, A. (1991). Cycles and events in stratigraphy-basic concepts and terms. In: *Cycles and Events in Stratigraphy* (ed. G. Emsele, W. Ricken and A. Seilacher), l–19. Berlin: Springer.

Fagerstrom, A. (1964). Fossil communities in paleoecology: their recognition and significance. *Geol. Soc. Am. Bull.* 75: 1197–1216.

Feijó, F.J. (1994). Bacias de Sergipe e Alagoas. *Boletim de Geociências da Petrobras, Rio de Janeiro* 8 (1): 149–161.

Figueiredo, A.M.F. (1981). Depositional systems in the lower Cretaceous Morro do Chaves and CoqueiroSeco formations, and their relationship to petroleum accumulations, middle rift sequence, Sergipe-Alagoas Basin, Brazil. Ph.D. thesis. The University of Texas in Austin, 275p.

Fürsich, F.T. (1995). Shell concentrations. *Eclogae Geologicae Helvetiae* 88 (3): 643–655.

Fürsich, F.T. and Oschmann, W. (1986). Storm shell beds of Nanogyra virgule in the Upper Jurassic of France. *NeuesJahrbuch fur Mineralogie Geologie und Palaontologie. Abhandlungen* 172: 141–161.

Fürsich, F.T. and Oschmann, W. (1993). Shell beds as tools in basin analysis: the Jurassic of Kachchh, western India. *Journal of the Geological Society* 150 (1): 169–185. https://doi.org/10.1144/gsjgs.150.1.0169.

Fürsich, F.T.; Pandey, D.K. 2003. Sequence stratigraphic significance of sedimentary cycles and shell concentrations in the Upper Jurassic-Lower Cretaceous of Kachchh, wstern India., *Palaeogeogr., Palaeoclimatol., Palaeoecol.*, 193: 285–309. https://http://dx.doi.org/https://doi.org/10.1016/S0031-0182(03)00233-5

Ghilardi, R.P. and Simões, M.G. (2000). Bivalves Fósseis como indicadores da dinâmica sedimentar: um estudo de caso do Paleozóico Superior da Bacia do Paraná, Brasil. *Acta Geológica Leopoldensia* 23: 3–20.

Guardado, L.R., Gamboa, L.A.P., and Lucchesi, C.F. (1989). Petroleum geology of the Campos Basin Brazil, a model for a producing Atlantic type basin. In: *Divergent/Passive Margin Basins* (ed. J.D. Edwards and P.A. Santogrossi), 3–80. Amsterdam: Elsevier.

Guardado, L.R., Spadini, A.R., and Brandão, . And Mello, J.S.L., M.R. (2000). *Petroleum systems of South Atlantic margins*, vol. 73 (ed. M.R. Mello and B.J. Katz), 317–324. AAPG Memoir.

Harris, P.M. (2000). Toca carbonate, Congo Basin: response to an evolving rift lake. In: *Petroleum Systems of South Atlantic Margins*, vol. 73 (ed. M.R. Mello and B.J. Katz), 341–360. AAPG Memoir.

Herlinger, R.J., Zambonato, E.E., and De Ros, L.F. (2017). Influence of diagenesis on the quality of lower cretaceous pre-salt lacustrine carbonate reservoirs from northern Campos Basin, offshore Brazil. *J. Sediment. Res.* 87 (12): 1285–1313.

Holland, S.M. (1995). The stratigraphic distribution of fossils. *Paleobiology* 21: 92–109. https://doi.org/10.1017/S0094837300013099.

Holland, S.M. (2000). The quality of the fossil record: a sequence stratigraphic perspective. *Paleobiology* 26 (4): 148–168. https://doi.org/10.1017/S0094837300026919.

Horodyski, R.S., Brett, C.E., Sedorko, D. et al. (2019). Storm-related taphofacies and paleoenvironments of Malvinokaffric assemblages from the Lower/Middle Devonian in southwestern Gondwana. *Palaeogeography, Palaeoclimatology, Palaeoecology* 514: 706–722. https://doi.org/10.1016/j.palaeo.2018.10.001.

Kattah, S. (2017). Exploration opportunities in the pre-salt play, deepwater Campos Basin, Brazil. *The Sedimentary Record* 15 (1): 4–8.

Kidwell, S.M. (1986). Models for fossil concentrations: paleobiologic implications. *Paleobiology* 12 (1): 6–24.

Kidwell, S.M. and Holland, S.M. (1991). Field Description Of Coarse Bioclastic Fabric. *Palaios* 6: 426–434. Olszewski, T.D. (1999). Taking advantage of time-averaging. *Paleobiology*, 25:226–238.

Kidwell, S.M. (1991a). The stratigraphy of shell concentrations. In: *Taphonomy: Releasing the Data Locked in the Fossil Record* (ed. P.A. Allison and D.E.G. Briggs), 211–290. Plenum Press.

Kidwell, S.M. (1991b). Taphonomic feedback (live/dead interactions) in the genesis of bioclastic beds: keys to reconstructing sedimentary dynamics. In: *Taphonomy: Releasing the Data Locked in the Fossil Record* (ed. P.A. Allison and D.E.G. Briggs), 211–290. Plenum Press.

Kidwell, S.M., Fürsich, F.T., and Aigner, T. (1986). Conceptual framework for the analysis and classification of fossil concentrations. *Palaios* 1: 228–238. https://doi.org/10.2307/3514687.

Kinoshita, M.E. (2010). Modelagem sísmica-geométrica de fácies dos carbonatos lacustres da Formação Morro do Chaves, Bacia de Sergipe-Alagoas. *Boletim de Geociências da Petrobrás* 18 (2): 249–269.

Kocsis, L., Dulai, A., Bitner, M.A. et al. (2012). Geochemical compositions of neogene phosphatic brachiopods: implications for ancient environmental and marine conditions. *Palaeogeography, Palaeoclimatology, Palaeoecology* 326–328: 66–77. https://doi.org/10.1016/j.palaeo.2012.02.004.

Kondo, Y. (1997). Inferred bivalve response to rapid burial in a Pleistocene shallow-marine deposit from New Zealand. *Palaeogeography, Palaeoclimatology, Palaeoecology* 128: 87–100. https://doi.org/10.1016/S0031-0182(96)00039-9.

Kroh, A. and Nebelsick, J.H. (2010). Echinoderms and Oligo-Miocene carbonate systems: potential applications in sedimentology and environmental reconstruction. *Int. Assoc. Sedimentol. Spec. Publ.* 42: 201–228.

Lucia, F.J. (1999). *Carbonate Reservoir Characterization*, 226. Berlin: Springer-Verlag.

Luna, J.L., Perosi, F.A., Ribeiro, M.G.S. et al. (2016). Petrophysical rock typing of coquinas from the Morro do Chaves Formation, Sergipe-Alagoas Basin. *Brazilian Journal of Geophysics* 34 (4): 509–521.

Lønøy, A. (2006). Making sense of carbonate pore systems. *American Association of Petroleum Geologists* 90 (9): 1381–1405.

Mandic, O., Harzhauser, M., and Roetzel, R. (2004). Taphonomy and sequence stratigraphy of spectacular shell accumulations from the type stratum of the Central Paratethysstage Eggenburgian (Lower Miocene, NE Austria). *Cour. Forsch. Inst. Senckenberg* 246: 69–88.

Mizuno, T.A., Mizusaki, A.M.P., and Lykawka, R. (2018). Facies and paleoenvironments of the Coqueiros formation (Lower Cretaceous, Campos Basin): a high frequency stratigraphic model to support pre-salt "coquinas" reservoir development in the Brazilian continental margin. *Journal of South American Earth Sciences* 88: 107–117.

Mohriak, W., Nemčok, M., and Enciso, G. (2008). South Atlantic divergent margin evolution: rift-border uplift and salt tectonics in the basins of SE Brazil. *Geological Society, London, Special Publications* 294: 365–398.

Moreira, J.L.P., Madeira, C.V.M., Gil, J.A., and Machado, M.A.P. (2007). Bacia de Santos. *Boletim de Geociências da Petrobras* 15 (2): 531–549.

Mukherjee, S., Dasgupta, S., Majumdar, C. et al. (2023). Introduction to: Handbook of Petroleum Geoscience: Exploration, Characterization, and Exploitation of Hydrocarbon Reservoirs. In: *Handbook of Petroleum Geoscience: Exploration, Characterization, and Exploitation of Hydrocarbon Reservoirs*, Petroleum Industry Updates: Geoscientific and Economic Issues (ed. S. Mukherjee, S. Dasgupta, C. Majumdar, et al.). John Wiley & Sons, Hoboken, ISBN 978-1-119-68003-1.

Muniz, M.C. (2013). Tectono-Stratigraphic evolution of the Barremian-Aptian Continental Rift Carbonates in Southern Campos Basin, Brazil. Ph.D. thesis. Royal Holloway University of London, Earth Sciences Department, p 343.

Muniz, M.C. and Bosence, D.W.J. (2017). Reservoir layering controlled by the taphonomy and taphofacies of coquinas from the early Aptian Campos Basin, Brazil. *AAPG ICE 2017A*, London, UK.

Muniz, M.C. and Bosence, D.W.J. (2018). Lacustrine carbonate platforms: Facies, cycles, and tectonosedimentary models for the presaltLagoaFeia Group (Lower Cretaceous), Campos Basin, Brazil. *American Association of Petroleum Geologists Bulletin* 102 (12): 2569–2597.

Nebelsick, J.H., Bassi, D., and Rasser, M.W. (2011). Microtaphofacies: exploring the potential for taphonomic analysis in carbonates. In: *Taphonomy*, Aims and Scope Topics in Geobiology Book Series, vol. 32 (ed. P.A. Allison and D.J. Bottjer). Dordrecht: Springer https://doi.org/10.1007/978-90-481-8643-3_9.

Oliveira, V.C.B., Silva, C.M.A., Borghi, L.F., and Carvalho, I.S. (2019). Lacustrine coquinas and hybrid deposits from rift phase: Pre-Salt, lower Cretaceous, Campos Basin, Brazil. *Journal of South American Earth Sciences* 95: 102254.

Olivito, J. and Souza, F. (2020). Depositional model of early Cretaceous lacustrine carbonate reservoirs of the Coqueiros formation - Northern Campos Basin, southeastern Brazil. *Marine and Petroleum Geology* 111: 414–439.

Parras, A. (2005). Taphonomy and sequence stratigraphic significance of oyster-dominated concentrations from the San Julián Formation, Oligocene of Patagonia, Argentina. *Palaeogeography Palaeoclimatology Palaeoecology* 217: 47–66. https://doi.org/10.1016/j.palaeo.2004.11.015.

Parras, A. and Casadío, S. (2006). The Oyster Crassostrea? Hatcheri (Ortmann,1897), a physical ecosystem engineer from the upper Oligocene-Lower Miocene of Patagonia, southern Argentina. *Palaios* 21: 168–186. https://dx.doi.org/10.2110/palo.2005.p05-48.

Parsons-Hubbard, K. (2005). Molluscan taphofacies in recent carbonate reef/lagoon systems and theirapplication to sub-fossil samples from reef cores. *Palaios* 20 (2): 175–191.

Pietzsch, R., Oliveira, D.M., Tedeschi, L.R. et al. (2018). Palaeohydrology of the lower Cretaceous pre-salt lacustrine system, from rift to post-rift phase, Santos Basin, Brazil. *Palaeogeography, Palaeoclimatology, Palaeoecology* 507: 60–80.

Riccomini, C., Lucy, G.S. And Tassinari, C.C.G., 2012, Pré-sal: geologia de exploração: *Revista USP*, v. 95, p. 33–42.

Rigueti AL, Dal'Bó PF, Borghi L, Mendes M (2020) Bioclastic accumulation in a lake rift basin: the early Cretaceous coquinas of the Sergipe-Alagoas Basin, Brazil, *Journal of Sedimentary Research*, v90 n2 228–249

Ritter, M.N., Erthal, F., and Coimbra, J.C. (2013). Taphonomic signatures in molluscan fossil assemblages from the Holocene lagoon system un the northern part of the coastal plain, Rio Grande do Sul State, Brazil. *Quaternary International* 305: 5–14. https://doi.org/10.1016/j.quaint.2013.03.013.

Saller, A., Rushton, S., Buambua, L. et al. (2016). Pre-Salt stratigraphy and depositional systems in the Kwanza Basin, offshore Angola. *AAPG (Am. Assoc. Pet. Geol.) Bull* 100: 1135–1164. https://doi.org/10.1306/ 02111615216.

Simões, M.G. and Ghilardi, R.P. (2000). Protocolo Tafonômico/Paleoautoecológico com Ferramenta nas Análises Paleossinecológicas de Invertebrados: Exemplos de Alpicação em Concentrações Fossilíferas do Paleozóico da Bacia do Paraná, Brasil. *Pesquisas em Geociências, Porto Alegre* 27 (2): 3–s. ISSN 1807-9806. https://doi.org/10.22456/1807-9806.20186.

Simões, M.G. and Torello, F.F. (2003). Modelo de tafofácies para os moluscos bivalves do Grupo Passa Dois (Formações Serra Alta, Teresina e Corumbataí), Permiano Superior, Bacia do Paraná, Brasil. *Revista Brasileira de Geociências* 33: 371–380. https://doi.org/10.5327/rbg.v33i4.1083.

Souza, I.M., Sainz-Díaz, C.I., Viseras, C., and Pergher, S.B. (2020). Adsorption capacity evaluation of zeolites as carrier of isoniazid. *Microporous Mesoporous Materials* 292: 109733. https://doi.org/10.1016/j.micromeso.2019.109733.

Scarponi, D., Kaufman, D., Amorosi, A., and Kowalewski, M. (2012). Sequence stratigraphy and the resolution of the fossil record. *Geology* https://doi.org/10.1130/G33849.1.

Schäfer, W. (1972). *Ecology and Paleoecology of Marine Environments*, 450. Chicago, IL: University of Chicago Press.

Schmidt-Neto, H., Netto, R.G., and Tognoli, F.M.W. (2014). Análise tafonômica das concentrações fossilíferas da Formação Rio Bonito na região de Taió, Sul do Brasil. *Revista Brasileira de Paleontologia* 17 (2): 207–224. https://doi.org/10.4072/rbp.2014.2.08.

Schmidt-Neto, H., Netto, R.G., and Dasgupta, S. (2018a). Storm-related taphofacies in estuarine settings: an integrated analysis on the early Permian deposits of the Rio Bonito Formation (Paraná Basin, S Brazil). *Journal of South American Earth Sciences* 85: 263–277. https://doi.org/10.1016/j.jsames.2018.05.005.

Schmidt-Neto, H., Netto, R.G., and Villegas-Martín, J. (2018b). Bioerosion in early Permian shells from the Rio Bonito Formation of Brazil: paleoecological, paleobiological, and taphonomic implications. *Palaeogeography, Palaeoclimatology, Palaeoecology* 505: 256–264. https://doi.org/10.1016/j.palaeo.2018.06.003.

Speyer, S.E. and Brett, C.E. (1986). Trilobite taphonomy and middle Devonian taphofacies. *Palaios* 1: 312–327. https://doi.org/10.2307/3514694.

Speyer, S.E. and Brett, C.E. (1988). Taphofacies models for epeiric sea environments: middle paleozoic examples. *Palaeogeography, Palaeoclimatology, Palaeoecology* 63: 225–262. https://doi.org/10.1016/0031-0182(88)90098-3.

Speyer, S.E. and Brett, C.E. (1991). Taphofacies controls background and episodic processes in fossil assemblage preservation. In: *Taphonomy: Releasing the Data Locked in the Fossil Record*, Topics in Geobiology, vol. 9 (ed. P.A. Allison and D.E.G. Briggs), 501–545. New York: Plenum Press.

Parsons-Hubbard, K. (2005). Molluscan taphofacies in recent carbonate reef/lagoon systems and their application to sub-fossil samples from reef cores. *Palaios* 20 (2): 175–191.

Tavares, A.C., Borghi, L., Corbett, P. et al. (2015). Facies and depositional environments for the coquinas of the Morro do Chaves Formation, Sergipe-Alagoas Basin, defined by taphonomic and compositional criteria. *Brazilian Journal of Geology* 45 (3): 415–429.

Tedeschi, L.R., Jenkyns, H.C., Robinson, S.A. et al. (2017). New age constraints on Aptian evaporites and carbonates from the South Atlantic: implications for oceanic anoxic event 1a. *Geology* 45: 543–546.

Tigre, C.A., Schaller, H.A., Del Lucchese, C., and Possato, S.A. (1983). Pampo, linguado, and badejo fields: their discoveries, appraisals, and early production systems. *Offshore Technology Conference*. https://doi.org/10.4043/4627-MS.

Thompson, D.L. (2013). The stratigraphic architecture and depositional environments of non-marine carbonates from Barremian-Aptian Pre-Salt strata of the Brazilian continental margin. Ph.D. thesis. Monash University, Melbourne, Victoria, Australia, 277pp.

Thompson, D.L., Stilwell, J.D., and Hall, M. (2015). Lacustrine carbonate reservoirs from early Cretaceous rift lakes of western Gondwana: pre-salt coquinas of Brazil and West Africa. *Gondwana Research* 28 (1): 26–51.

Walker, S.E. (1988). Taphonomic significance of hermit crabs (ANOMURA: PAGURIDEA): epifaunal hermit crab – infaunal gastropod example. *Palaeogeography, Palaeoclimatology, Palaeoecology* 63: 45–71. https://doi.org/10.1016/0031-0182(88)90090-9.

Zabini, C., Holz, M., Bosetti, E.P. et al. (2012). Sequence stratigraphy and taphonomic signatures of marine invertebrates: a devonian (Pragian/Eifelian) example of the Paraná Basin, Brazil. *Journal of South American Earth Sciences* 33: 8–20. https://doi.org/10.1016/j.jsames.2011.08.001.

Zielinski, J.P.T., Vidal, A.C., Chinelatto, G.F. et al. (2018). Evaluation of pore system properties of coquinas from Morro do Chaves formation by means of x-ray microtomography. *Brazilian Journal of Geophysics* 36 (4).

20

Tectonic Evolution of Jaisalmer Basin (Rajasthan, India)

Rajesh Pandey

Frontier Basin, Oil and Natural Gas Corporation, Dehradun, Uttarakhand, India

**Author for correspondence: pandey_rajesh@ongc.co.in*

Abstract

The evolution of the NW margin of India began with the rifting of the Gondwana and Laurasia (about 180 Ma ago) continental landmass. The imprint of this tectonic episode can be understood using a comprehensive and integrated approach from the sedimentary basins of the NW India. This chapter presents the evolution of the NW India margin (Jaisalmer Basin) using subsurface seismic and well data. The study suggests multiphase evolution of the western Rajasthan shelf. The stages are (i) Late Proterozoic rifting, (ii) Permo-Triassic rifting, (iii) Early Jurassic rifting, (iv) Passive margin phase, and (v) Collision phase starting from ~55 Ma.

Keywords *Petroleum geosciences; gas emission; hydrocarbon field; tectonics; geodynamics*

20.1 Introduction

The western Rajasthan shelf encompasses three basins of different ages and tectonics, viz. the Bikaner-Nagaur Basin, the Jaisalmer Basin, and the Barmer Basin. The development of each basin was constrained by its stratigraphy and subsurface structural pattern (e.g. DasGupta 1975; Misra and Sharma 1986; Roy and Jakhar 2002; Boruah and Deb 2012; Raju et al. 2014; Sharma et al. 2016; Dasgupta and Mukherjee 2017, 2019; Biswas et al. 2022a; Kar et al. 2022). Three major tectonic events in different time period (one in Proterozoic and others in Mesozoic and Tertiary) in the NW margin of India produced/affected these basins (Gaina et al. 2013, 2015).

The Proterozoic Bikaner-Nagaur Basin was the outcome of Infra-Cambrian rifting over the northeastern edge of the incipient Gondwanaland involving the Indian plate (Leggett 2012). It evolved in three phases, phase-I from Infra-Cambrian to Cambrian, phase-II from Permo-Triassic to Jurassic, linked to Gondwana break-up, and phase-III of EarlyTertiary period. Phase-I is the main development phase of the Bikaner-Nagaur Basin (Casshyap 1979; Casshyap and Tewari 1988; Bhattacharya et al. 2012; Pangtey et al. 2013;

Handbook of Petroleum Geoscience: Exploration, Characterization, and Exploitation of Hydrocarbon Reservoirs, First Edition. Edited by Soumyajit Mukherjee, Swagato Dasgupta, Chandan Majumdar, Subhandip Mandal and Troyee Dasgupta.
© 2023 John Wiley & Sons Ltd. Published 2023 by John Wiley & Sons Ltd.

Bhattacharya and Bhattacharya 2015). There are large hiatus/erosional events at the interface of each phase (Singh and Tewari 2011).

Recenty Biswas et al. (2022b) provided a detail morphometric analysis from the Jaisalmer basin. The authors also reviewed stratigraphic succession for this basin. The Jaisalmer Basin also evolved in three main phases, namely the Permo-Triassic to Lower Jurassic rifting phase, the Middle Jurassic to Cretaceous passive margin phase and the Tertiary foreland phase (Biswas 1987; Siawal and Kaul 2006). The first two phases are linked with the rifting of East Africa from India-Madagascar and the last phase is linked with the collision of the Indian plate with the Kohistan-Ladakh arc and then with the Eurasian plate (Khan et al. 2013; Farshori 1972).

The Barmer Basin was developed within the continental part of NW India due to the Early Tertiary rifting of Seychelles from India (Kothari et al. 2015). Ideally, it has developed in multiple rifting phases during the Early Cretaceous and Late Cretaceous–Paleocene periods (Dasgupta and Mukherjee 2017) followed by a post-rift phase during the collision of the Indian plate with the Eurasian plate.

Since all the three basins are situated over the onland part of NW India and are linked with the evolution of the NW continental margin of India, it is observed that while some part of the NW margin was under extension, the other parts were under compression at a different geological time (Pandey et al. 2019). This article focuses on the tectonic evolution of the Jaisalmer Basin.

The Jaisalmer Basin lies in the Eastern part of the Great Indus Basin (Figure 20.1). It is a pericratonic basin, with the sedimentary succession ranging from Permo-Triassic to Recent, which is punctuated by several unconformities (Pandey et al. 2018, 2019, 2020). Major unconformities are Early Cambrian to Permian, Jurassic/Cretaceous boundary, Early/Late Cretaceous Boundary, Late Cretaceous to Early Paleocene, and Late Eocene to Pleistocene (Aswal et al. 2008). The number of sedimentological and biostratigraphic studies have been performed by eminent geoscientists of the world (Blanford 1876; Singh 1984; Oldham 1886; Aswal et al. 2008; Verma et al. 2012; Ahmad et al. 2013; Khan and Khan 2015; Ahmad et al. 2017; Pandey et al. 2018). The basin is subdivided into four tectonic units, namely Jaisalmer Mari High, Shahgarh, Kisangarh, and Miajlar sub-basins (Misra and Singh 1993; Singh 2000; Singh 2006) using regional gravity and seismic data. Structural trends within the basin are associated with major trends of the Indian subcontinent. Three trends, viz. NE–SW (Aravalli grain), ENE–WSW/E–W (Satpura grain), and NNW–SSE (Dharwarian grain) influenced the basin evolution and structural configuration from time to time (Dasgupta 1975; Roy and Jakhar 2002). These trends resulted in distinct fault patterns with multiple orthogonally/obliquely related faults straddling the area. The structural style of the Jaisalmer Basin can be traced on outcrops and in the subsurface by faults and the maximum effect of faulting is limited between Ramgarh and Kanoi Faults (Misra and Sharma 1986; Sharma et al. 2016). Major Faults in the region are Mansurian Fault and Ghotaro Fault in the west, and Kanoi Fault and Manpiya Ramgarh Fault near Jaisalmer (Figure 20.1). Surface faults demonstrate sinistral movement trending NW–SE (Pandey et al. 2019). Majorly linked to the strike-slip faulting, thermal doming and hotspot activity in the basin are associated with the separation of the Indian and Madagascan plates (Raval and Veeraswamy 2000; Roy 2003; Khan et al. 2013). The Himalayan collision (Oligocene to present-day) has also caused sinistral transpression in the west of the Lower Indus Basin, due to which fold-thrust

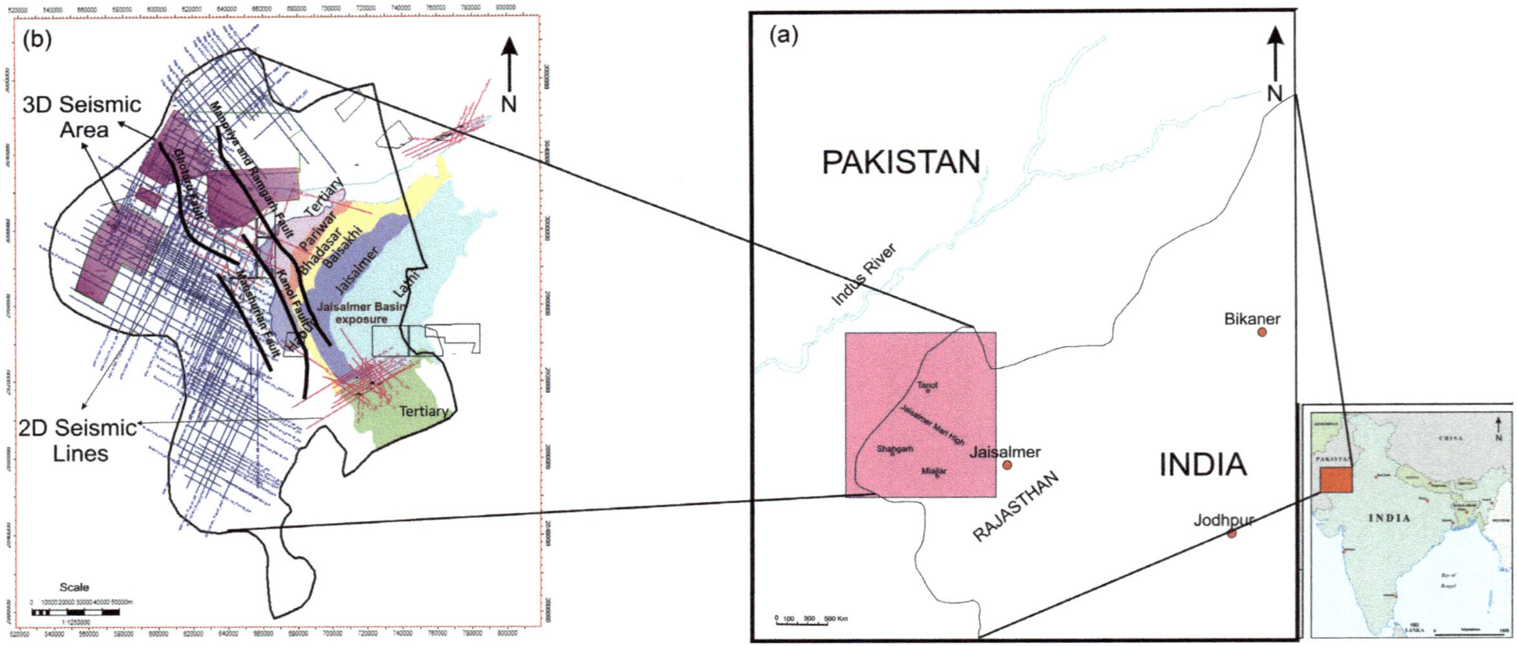

Figure 20.1 (a) Location map of the Jaisalmer Basin with geographical boundaries. (b) Map representing 2D and 3D seismic coverage along with surface exposure of the Jaisalmer Basin.

structures are overprinted by sinistral flower structures in the Jaisalmer basin (Farshori 1972; Pandey et al. 2019).

The chapter discusses the regional tectonostratigraphic status and salient aspects of the evolution of the Jaisalmer Basin. Moreover, some tectonic events associated with the area are also listed and discussed, which give rise to the Bikaner-Nagaur and Barmer Basins.

20.2 Methodology

The data used for the current study was 2D and 3D seismic data in the time domain. The areal coverage of 3D data was around 2800 sq km. near the international boundary of India and Pakistan (Figure 20.1). The 2D data was available in the entire Jaisalmer Basin and the data volume was around 20,000 lkm (line kilometer). This work was completed using seismic data volume, which was loaded, and quality checked using a seismic interpretation software. In this case, Decision Space Geosciences (Halliburton, Landmark) software was used. Manual mapping of the major regional faults was done in 2D lines. The 3D seismic volume was steered to improve its quality by increasing the signal to noise ratio. Apart from the Coherency map, structure attributes such as structure cube, dip, azimuth, and curvature were extracted from the steered cube. The extracted attributes were analyzed and it was observed that similarity showed a better response to faults than any other attribute used for the study.

In the Open Work (DSG) window environment, the calculation of similarity is favored by the determination of the direction of the best match at every position, which is a result by itself: the dip. The similarity is defined using fundamental mathematics, where trace samples are analyzed as component vectors. The point about using similarity is that it is mathematically simple; it is very clear what is going on. Then, by combining different kinds of similarities and other attributes, we can always get much better results with lots less computing time. The trace is considered as a vector in hyperspace, where similarity is defined. Coherency map on top of the Jaisalmer Formation led to the identification of prominent faults and their trends (Figure 20.2). Additionally, paleostructural analysis using 2D and 3D seismic lines and well log correlations were carried out.

20.3 Results and Discussions

The initial stratigraphy of the Jaisalmer Basin was established by Blanford (1876) and later on by Oldham (1886, 1893). Subsequently, the ONGC workers also worked and revised the stratigraphy of the basin (Table 20.1).

The faults present in Jurassic successions (Lathi, Jaisalmer, Baisakhi, and Bhadasar Formations) have penetrated the basement as well. In Cretaceous time (Pariwar, Goru, and Parh Formation), some additional normal faults have been observed in seismic data. During Paleocene and Eocene (Khuiala and Bandah Formations) strike-slip regime (Transpression and Transtension) has been identified from seismic. The fault attributes are best seen on top of the Jaisalmer Formation of the middle Jurassic, due to limestone

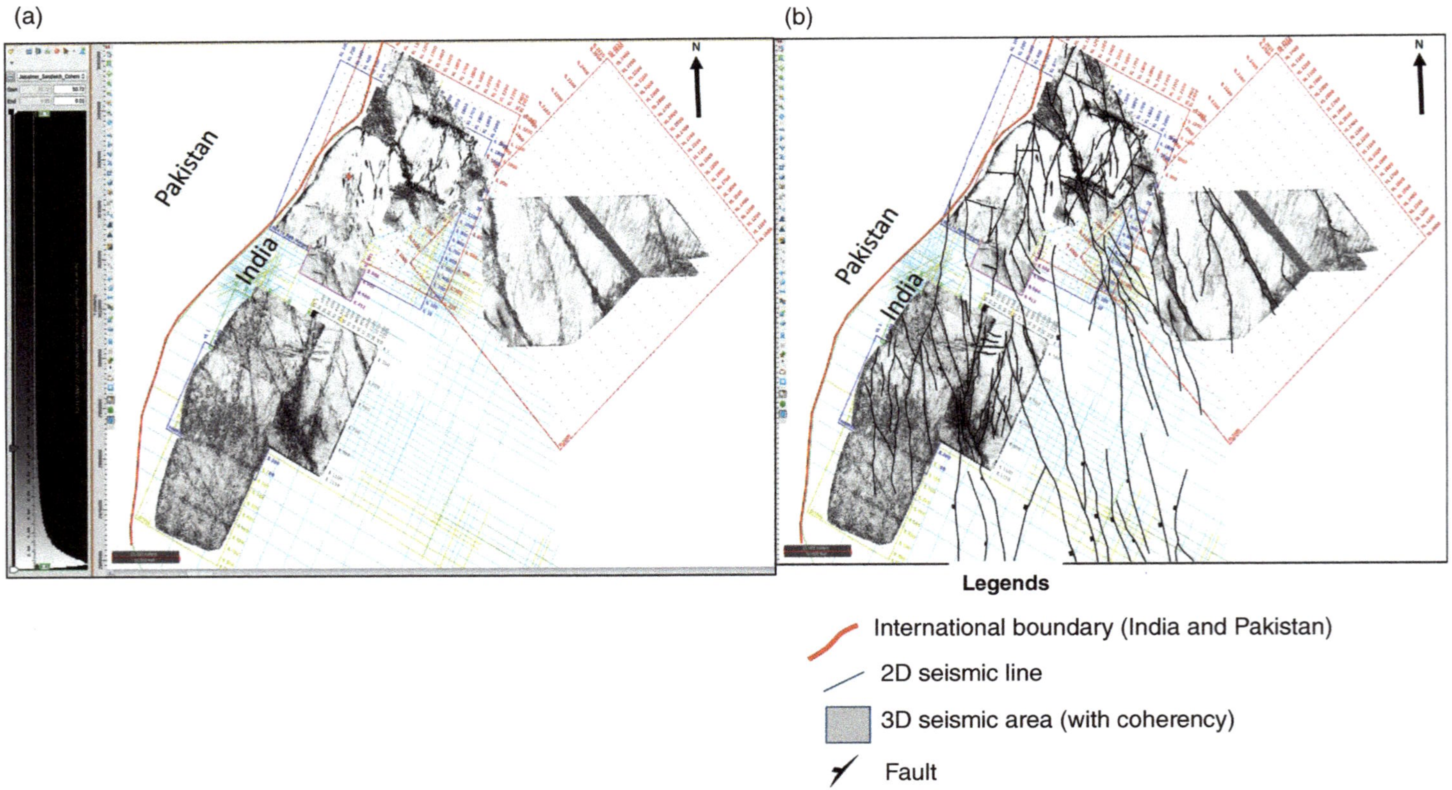

Figure 20.2 Coherency Map of the available 3D seismic volume at the Jaisalmer Top (a) without Fault and (b) with Fault.

Table 20.1 Generalized litho-stratigraphy of the Jaisalmer Basin.

Eon/Era	Period/Epoch	Formation	Lithology
Cenozoic	Holocene	Wind-blown Sand/ Alluvium	Loose sand and alluvial materials
	Holocene to Pleistocene	Shumar	Dune sands, gravels with ferruginous nodules
	Middle Eocene	Bandha	Foraminiferal limestone clayey at the base
	Early Eocene	Khuiala	Shales with limestone beds and calcareous silts.
	Paleocene	Sanu	Sandstone with minor clays
Mesozoic	Late Cretaceous	Parh	Marls and arenaceous limestone
	Middle-Early Cretaceous	Goru	Arenaceous limestones and calcareous sandstones
	Early Cretaceous	Pariwar	Sandstone, shale, lignite
	Late Jurassic	Baisakhi/Bhadasar	Sandstone and shale
	Middle Jurassic	Jaisalmer	Limestone and sandstone
	Early Jurassic	Lathi	Sandstone, shale and lignite
	Triassic	Sumarwali	Sandstone and claystone
Paleozoic	Permian	Karampur	Shale and sandstone
	Cambrian	Birmania	Sandstone and shale
Neo proterozoic	Tonian/ Cryogenian	Randha	Sandstone and shale
		Basement	Malani igneous suite/ Metamorphic basement

Sources: Misra and Singh (1993); Khatri et al. (2015); Khan and Khan (2015).

lithology. The coherency maps on top of the Jaisalmer Formation have brought out major fault trends oriented in NE–SW, E–W, and NNW–SSE directions. These faults can be traced up to more than 10 km. It has been observed from this study that the study area was subjected to extension up to Cretaceous age trending NW–SE direction along with minor compressional tectonic forces in post-Cretaceous time. The paleo-tectonics of western Rajasthan have been discussed next:

20.3.1 Late Proterozoic Rifting (Main Development Phase of the Bikaner-Nagaur Basin)

The first rifting event involving the north-western part of the Indian Plate occurred on the northeastern edge of the incipient Gondwana supercontinent adjacent to the Panthallassic ocean margin in the Late Proterozoic time (Gorin et al. 1982; Husseini and Husseini 1990; McKerrow et al. 1992). This event gave birth to the Bikaner Basin trending northeast, which mainly hosts a thick sequence of Infra-Cambrian sediments over the Pre-Cambrian basement. The offset of basement-involved deep-seated normal faults in the Bikaner-Nagaur

Basin is the outcome of the rifting event mentioned here (Mazumdar and Bhattacharya 2004) trending NNE–SSW. Torsvik et al. (2001) while working on Rodinia Supercontinent formation based on a match of paleomagnetic poles from ca. 750 Ma magmatic rocks in Seychelles and NW India (Malani) placed these regions only 600 km apart to strengthen the Proterozoic rifting phase.

20.3.2 Permo-Triassic Rifting (Inception Phase of Jaisalmer Sub-Basin)

During Permo-Triassic time, the northwestern part of the Indian Plate was subjected to second rifting, which is related to rifting of East Africa from Madagascar-India (Mazumder et al. 2017). During this period, development of Neo Tethys continued. The Tethys existed along the northern margin of India and extended westward between Arabia and Iran. The NW arm of this rift extended southward between Africa and Madagascar (Misra and Mukherjee 2015; Mazumder et al. 2017). It is pertinent to mention that there is no evidence for the development of ocean-floor of this age between Africa and Madagascar (Siawal and Kaul 2006). In the Majunga embayment of Madagascar, outcrops of cross-bedded continental sandstone and intercalated clays of the Isalo Formation (Kamen-Kaye 1982) are the representative sedimentary deposits. The subcrop occurrence of Wulgai Formation (shale and limestone of the outer shelf) encountered in well Nabisar-A in Pakistan part (South Indus) and the time equivalent Bhuana Formation (fluvial to shallow marine) as encountered in a well in the Jaisalmer Basin represent the Triassic age in the Indian part (Kadri 1995; Pandey et al. 2019). This rift is related to the initial rifting phase of Gondwana and the same event gave birth to the Indus basin (including the Jaisalmer Basin) rift trending the NE–SW extension (Asim et al. 2014). In the Jaisalmer Basin, this phase is not observed on seismic data, due to higher sediment thickness (more than 10 km) in the deeper part of the basin. Only in Miajlar and Ramgarh area (shallowest part of the Basin), few 2D seismic lines display the existence of Permo-Triassic rifting (Figures 20.3 and 20.4). The surface exposures of the rift deposit are also located in oval-shaped isolated exposure in further south of the Miajlar sub-basin of Jaisalmer (Maheshwari et al. 2002). The main formations are Randha and Birmania-trending NE–SW exposure. The Triassic were not encountered in the drilled wells of the Bikaner-Nagaur Basin where Palana Formation of Early Paleocene unconformably overlies the Bhadaura Formation of the Permian age (Prasad et al. 2010). Triassic sediments are shown to be restricted to the north of (proto) Allah Bund Fault (Raza et al. 2008). Bhuana Formation of Permo-Triassic age toward north in Rajasthan vis-à-vis with Karoos of Madagascar represents a continuous depositional realm. The Kutch-Saurashtra region was not yet under sedimentation during this period (Mazumder et al. 2017). But a continuous sedimentary realm prevailed from north to south covering northwestern part of the Seychelles and the western part of Madagascar, which were under continental sedimentation (Siawal et al. 2015). Nagar Parker hill is a basement high and it acted as a provenance during Jurassic period (Wynne 1867; Biswas 1987). Figure 20.5a, b shows the relative paleo-geographic positions of India, Laxmi Ridge, Seychelles, Madagascar, and Africa and the regional facies distribution reconstructed in the present work.

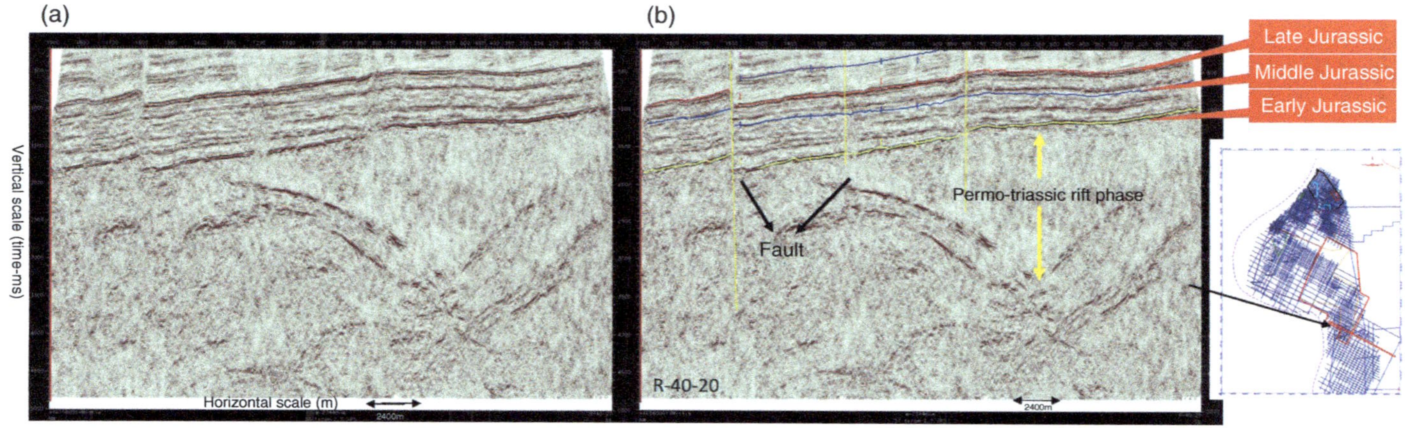

Figure 20.3 Seismic line (2D) R-40-20 passing through the Miajlar sub-basin representing Permo-Triassic rifting phase.

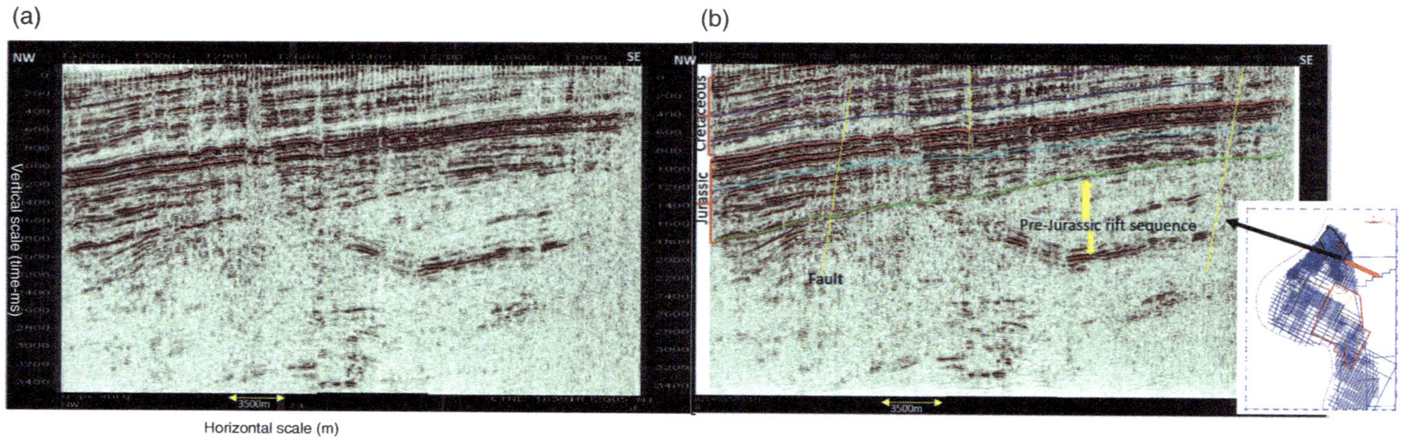

Figure 20.4 Seismic line passing through Ramgarh Structure represents Pre-Jurassic rift sequence.

20.3.3 Early Jurassic Rifting (Main Development Phase of the Jaisalmer Basin)

This phase is related to the initial rifting phase of Gondwana trending N–S to NNW–SSE (Salman and Abdula 1995). This rifting phase continued till 157 Ma. The Indus/Jaisalmer Basin was further developed during this rift phase with a number of N–S and NNW–SSE trending faults (Khan et al. 2018). Early Jurassic rift is well documented in the deeper part of the basin (Shahgarh and Jaisalmer Mari High) and has been seen within 3D seismic data acquired in the area (Figure 20.5). During this rifting, the Miajlar sub-basin/Ramgarh area

(a)

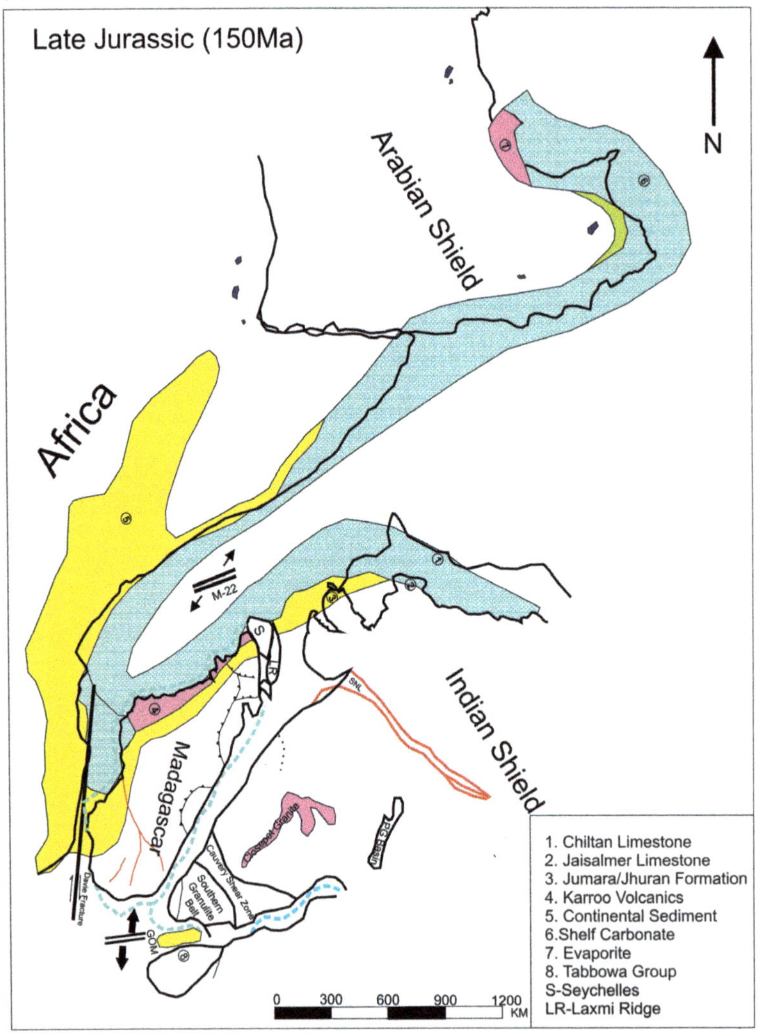

Figure 20.5 (a) Paleogeographic distribution of continents during Late Jurassic. *Source:* Modified after Smith et al. (1981); Raza et al. (2008); Dash et al. (2015); Siawal et al. (2015). (b) Paleogeographic distribution of continents during Middle Jurassic. *Source:* Modified after Smith et al. (1981); Raza et al. (2008; Dash et al. (2015); Siawal et al. (2015).

(b)

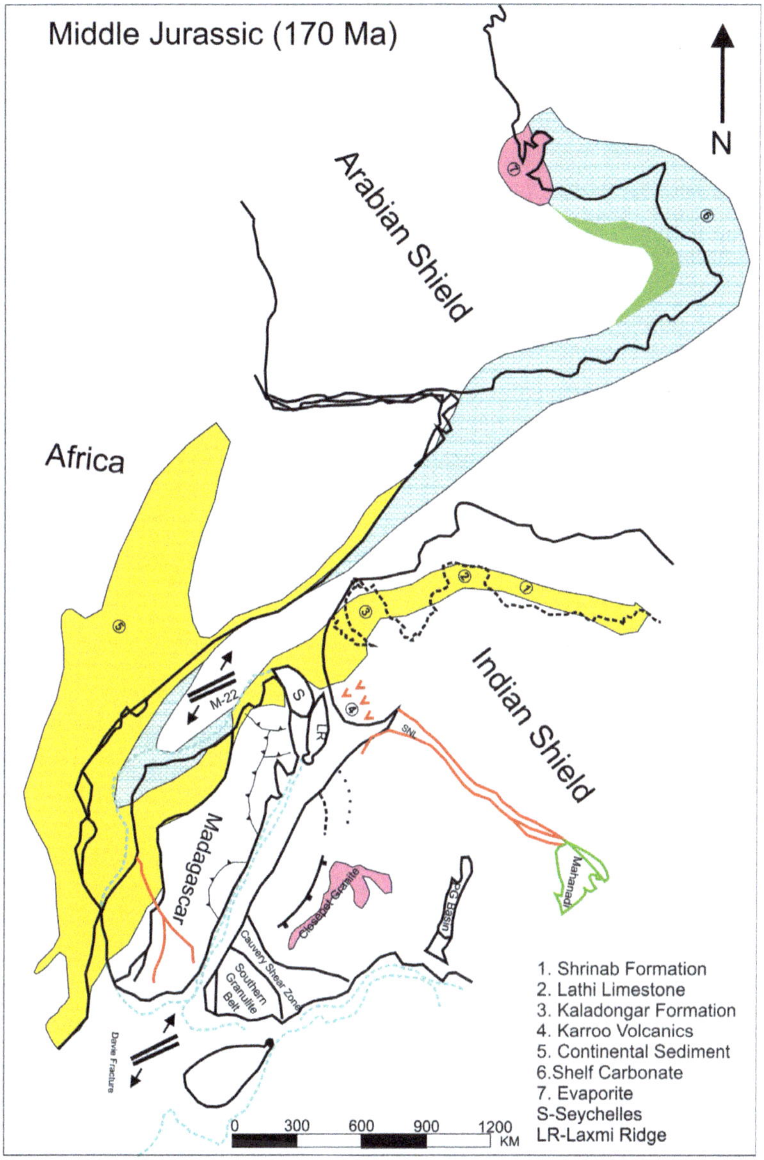

Figure 20.5 (Continued)

acted as a passive margin setting, which means the initial rifting was in the deeper part, while, at the same time, the shallower part received continuous sediment without rifting evidence (Subimal et al. 2010). During Middle Jurassic, the Jaisalmer Basin and Miajlar sub-basin of Rajasthan and Shirinab Formation (clastics/marginal marine) in the Pakistan part (well Nabisar-A), indicates the development of the Rajasthan shelf (Passive Margin) (Siawal et al. 2015). The Kutch region received the first continental sedimentation in the form of Nirona and Lodai Formations (Guha et al. 2005). However, the Saurashtra region did not experience any sedimentation during this period (Siawal et al. 2006, 2015).

20.3.4 Passive Margin Phase

The NW margin of India came under a passive margin phase from Late Jurassic to the Cretaceous (Pandey et al. 2019). The parallel to subparallel reflection geometries are the characteristic of uniform deposition on a subsiding surface within a passive margin setting (Figure 20.6; Boggs 2001). Mid-Cretaceous-rifting event between Seychelles/India from Madagascar was associated with a short period of counter-clockwise (Sinistral) rotation of India relative to Africa, leading to the N–S convergence between NW India and SE Arabia between ~95 and 84 Ma (Plummer and Belle 1995). From 84 to 60 Ma, a sinistral translation motion occurred between the Indian plate and the Arabian plate parallel to their passive margin along the left-lateral Chaman fault zone (Gaina et al. 2007, 2013; Eagles and Hoang 2014). The separation of Madagascar from the Indian subcontinent led to the modification and reactivation of the existing fault (Plummer 1996). The effect is well documented in seismic data. The Cretaceous sequences have more number of faults as compared to underlying Jurassic and overlying Tertiary (Figure 20.6). During the late Cretaceous, the Jaisalmer basin was subdivided into four tectonic units, viz. Kishan Garh sub-basin, Jaisalmer Mari High, Shahgarh sub-basin, and Miajlar sub-basin (Misra and Singh 1993).

Kishangarh shelf is a relatively stable, unfaulted area between Ramgarh fault to the west and the Bikaner-Nagaur Basin to the east. The area has an appreciable thickness of Mesozoic and Tertiary sediments and has a NE–SW depositional strike. It is separated from the Shahgarh sub-basin by the Jaisalmer Mari high area, which is a present-day gravity-high feature located along the shoulder zone of the Kanoi fault and is affected by upthrusting and wrench-faulting during late Cenozoic time (Misra and Sharma 1986).

Shahgarh depression is relatively better explored and represents the area west of Ramgarh fault that runs almost North–South and is structurally disturbed with numerous N–S-trending faults forming north to south-trending horsts and grabens. The Miajalar sub-basin is structurally simpler, being a homoclinally, westerly, gently dipping shelf.

The paleo-structural study was carried out using well correlation of the drilled wells in the basin passing from Shahgarh depression to Jaisalmer Mari High (SC-A, ISH-A, CWD-A, LT-A, CT-B, KT-E, and SD-D). For this study, the Formation tops within the wells were flattened from the bottom-most Formation top to the topmost, to see the thickness variation. It was observed that up to K-T (Cretaceous-Tertiary) boundary (Parh Top), the thickness is nearly the same in Shahgarh depression and Jaisalmer Mari High, while in Paleocene and Eocene sequences, less thickness has been observed at Jaisalmer Mari High, which demonstrates the time of development of Jaisalmer Mari High (Figure 20.7). Paleo-structural analyses were also performed in 3D seismic volume to understand the fault kinematics of the basin. It suggests that well-preserved rift sequences are clearly visible below Cambrian. Half grabens are trending North–South. Jurassic top flattening suggests that sequences from Middle Jurassic to Late Jurassic were deposited without major tectonic activity. Tectonic activity occurred during the Early Jurassic. Early Cretaceous flattening suggests that deposition continues from Late Jurassic to Early Cretaceous with readjustment of existing faults and some

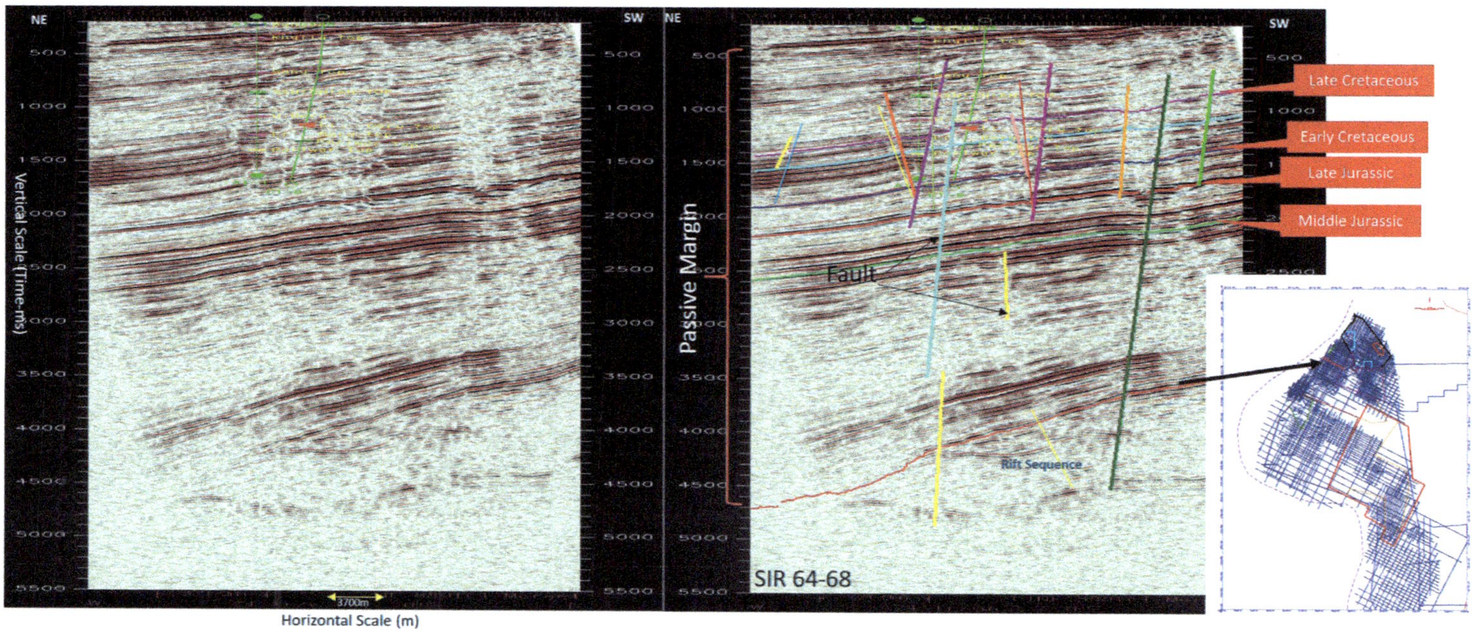

Figure 20.6 Seismic line passing through the Chinnewala Tibba Structure represents passive margin deposits above rift sequences.

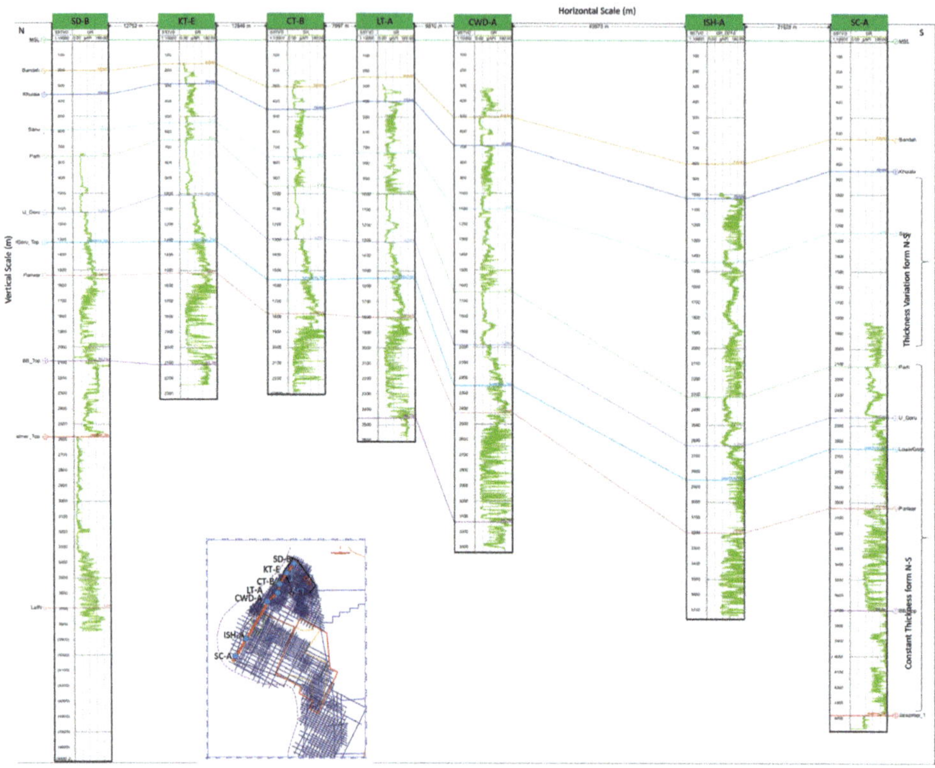

Figure 20.7 Log correlation from the Shahgarh Low to the Jaisalmer Mari High.

additional faults were developed in Early Cretaceous, due to post-deposition extensional tectonic in deeper (westward) part of the basin.

20.3.5 Collision Phase (Tertiary Period)

This is the time when the passive margin was superimposed by the foreland phase and involved with complex sutures between the Indian plate and the Afghan block due to the collision of the Indian plate with Kohistan-Ladakh arc and then with the Eurasian plate (Khan and Clyde 2013). During this time, the regional drainage pattern within the basin was SE–NW directed, while after collisional tectonics, the drainage pattern changed from the SE–NW to the NE–SW direction (Sharma et al. 2016). Between ~65 and 45 Ma, a sinistral transpression occurred followed by a minor E–W shortening after 45 Ma, and a minor right-lateral transform motion in the last ~8 Ma (Gaina et al. 2015). This is evidenced by the fold thrust structure overprinted by the sinistral flower structure in the Indus basin, it is a part of the peripheral Jaisalmer basin (Figure 20.8). Simultaneously, during this time (65–45 Ma), the northward movement of the Indian plate caused the northwestern edge of the Indian plate to pass over the reunion mantle plume (Pandey et al. 2019). This caused the extension of Tertiary and the birth of many rift basins in the

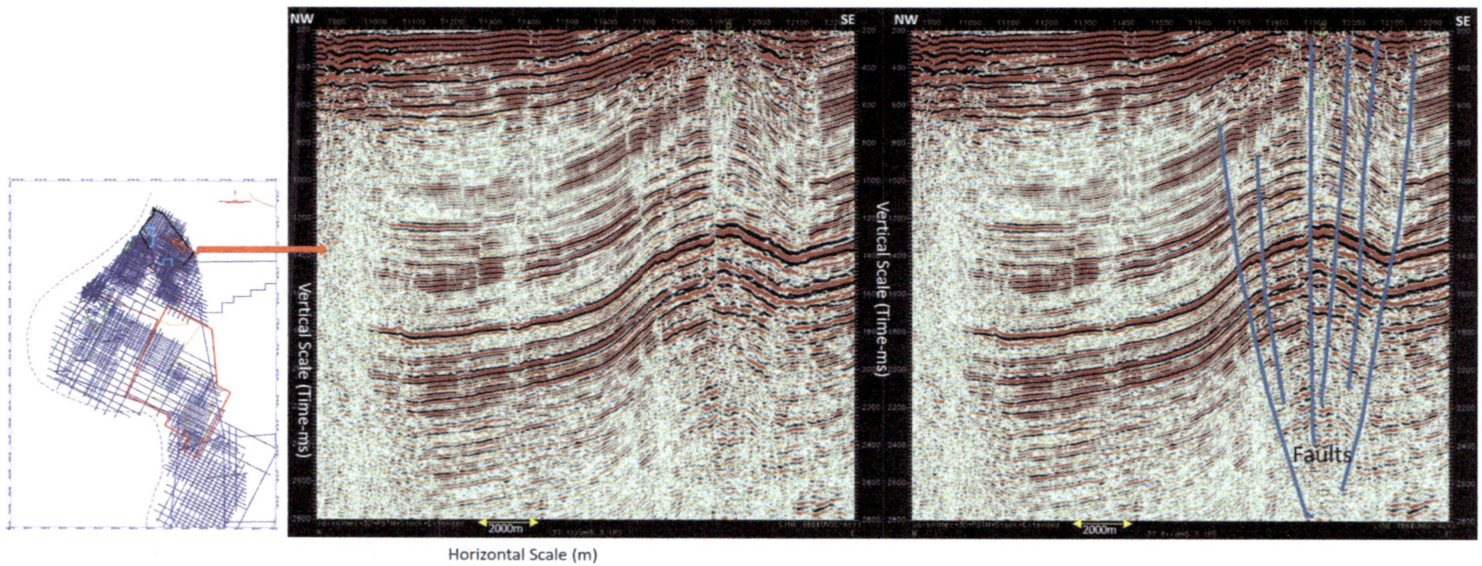

Figure 20.8 Seismic section displaying the positive flower structures in the Manhera Tibba field, the Jaisalmer Basin.

western continental margin. During the Late Paleocene and Early Eocene, the Indian Plate had started colliding with the Asian Plate resulting in the final phase of regression of the sea during the Middle Eocene from the western flank of the Indus Shelf (Alberti et al. 2017). Finally, during Lower Miocene, the axial belt was completely uplifted and the sea retreated totally from the foredeep and the Indus Shelf (Ahmad et al. 2020). During the fourth phase of tectonic evolution, the region witnessed further uplift of the Sindh-Balochistan Fold Belt resulting in the filling up of the Indus Shelf and the foredeep zone with molasse deposits (Khan and Clyde 2013).

The major NNW–SSE trending fault-set comprising the Ramgarh-Kanoi fault zone shows a distinct strike-slip movement during Paleocene-Eocene. The fault profiles on the sections are near vertical. The main sense of strike-slip movement along the two major faults bounding the Ramgarh-Kanoi fault zone appears to be sinistral (left-lateral), as is evident from NNE–SSW trending structure contours veering toward NNW in the fault zone and then again showing NNE–SSW trend (Pandey et al. 2019). The interplay of stresses, however, has created a right-lateral (clockwise) sense of movement in the area, confined between the bounding faults of the zone (Pandey et al. 2018). The fault throw in many cases shows wrenching effects, similar to Ramgarh Fault, mapped as the eastern bounding fault of the zone that shows an increase away from a point, where there appears to be little or no throw ("Null Point"), which is an indication of wrench faulting. The long-distance faults, as mapped in the area, often cover different segments. The fault toes tend to link together, where there is an overlap between two segments, giving rise to overstep or restraining bends. The over-steps or restraining bends are sites where transpressional stresses are in play. An array of "positive" flower structures was mapped in the areas of stepovers or restraining bends, where the transpressional stresses have led to the formation of upfaulted blocks. The highs thus formed in such sectors, where the fault toes link together (oversteps/stepovers), host some of the main gas fields in and around the Jaisalmer-Mari High area (Pandey et al. 2018). The interaction between the NNW–SSE and NNE–SSW-trending fault trends, as mapped in the Shahgarh Depression, has led to the creation of an array of segments where entrapment of gaseous hydrocarbons is noticed as in the Shahgarh field and the area in its vicinity.

Acknowledgments

Soumyajit Mukherjee (IIT Bombay) invited, handled, and reviewed this article. In addition, an anonymous reviewer provided review comments. The authors are grateful to the management of ONGC for giving permission to publish this work. The authors also acknowledge ED-Basin Manager, Frontier Basin, ONGC, Dehradun for providing an opportunity and all necessary facilities to prepare this paper. The authors wish to acknowledge the support given by all colleagues in the group. Merryl Le Roux, Frank Otmar Weinreich, Joss Everett, Andrew Harrison, and the proofreading team (Wiley Blackwell) are thanked for their assistance. This chapter has been encapsulated in Mukherjee et al. (2023).

References

Ahmad, F., Quasim, M.A., and Ahmad, A.H.M. (2020). Lithofacies characteristics and depositional environment interpretations of the middle Jurassic fort member rocks, Jaisalmer formation, Western Rajasthan, India, *J. Sediment. Environ.*, doi:https://doi.org/10.1007/s43217-020-00023-6.

Ahmad, F., Quasim, M.A., Ghaznavi, A.A., Khan, Z., Ahmad, A.H.M. (2017). Depositional environment of the fort member of the Jurassic Jaisalmer formation (western Rajasthan, India), as revealed from lithofacies and grain-size analysis, *Geol. Acta*, 15 (3) 153–167, DOI: https://doi.org/10.1344/GeologicaActa2017.15.3.1.

Ahmad, A.H.M., Upadhyay, G., and Noufal, K.N. (2013). Petrography and diagenetic history of the Jajiya member sediments, Jaisalmer formation (middle to late Jurassic), Western India. *Proc. Nat. Acad. Sci. India Sect. A* 79 (4): 895–901.

Alberti, M., Pandey, D.K., Sharma, J.K. et al. (2017). Slumping in the upper Jurassic Baisakhi formation of the Jaisalmer Basin, western India: sign of synsedimentary tectonics? *J. Palaeogeogr.* 6 (4): 321–332.

Asim, S., Qureshi, S.N., and Khan, N. (2014). Study of an uplift of Sargodha high by stratigraphical and structural interpretation of an east–west seismic profile in central Indus Basin, Pakistan. *Int. J. Geosci.* 5: 1027–1036.

Aswal, H.S., Murthy, M.S., Shanmukhappa, M. (2008). High resolution bio-chrono-stratigraphic framework for application in basin analysis and sequence stratigrphy of Jaisalmer Basin, Rajasthan, India. GEO India Expo XXI, Noida, New Delhi, India 2008©AAPG Search and Discovery.

Blanford, W.T. (1876). On the physical geography of the great Indian Desert with special reference to the former existence of the sea in the Indus Valley and on the origin and mode of formation of the sand hills. *J. Asiat. Soc. Bengal* 45 (2): 86–103.

Bhattacharya, B., Bandyopadhyay, S., Mahapatra, S., and Banerjee, S. (2012). Record of tide-wave influence on the coalbearing Permian Barakar formation, Raniganj Basin, India. *Sediment. Geol.* 267–268: 25–35.

Bhattacharya, H.N. and Bhattacharya, B. (2015). Lithofacies architecture and paleogeography of late Paleozoic glaciomarine Talchir formation, Raniganj Basin, India. *J. Palaeogeogr.* 4: 269–283.

Biswas, S.K. (1987). Regional tectonic framework, structure and evolution of the western marginal basin of India. *Tectonophysics* 135: 307–327.

Biswas M, Puniya M.K., Gogoi M.P., Dasgupta S., Mukherjee S., Kar N.R. (2022a). Morphotectonic analysis of petroliferous Barmer rift basin (Rajasthan, India). *Journal of Earth System Science.* 131: 140. https://doi.org/10.1007/s12040-022-01871-8

Biswas M., Gogoi M.P., Mondal B., Sivasankar T., Mukherjee S., Dasgupta S. (2022b). *Geomorphic assessment of active tectonics in Jaisalmer basin (western Rajasthan, India)*, Geocarto International. https://doi.org/10.1080/10106049.2022.2066726.

Boggs, S. Jr. (2001). *Principles of Sedimentology and Stratigraphy*, 3rde, 726. New Jersey: Prentice-Hall, Inc.

Boruah, K. and Deb, S.S. (2012). Heavy oil discoveries in Bikaner–Nagaur Basin: opportunities and challenges. *APG J.* 3: 40–43.

Casshyap, S.M. (1979). Pattern of sedimentation in Gondwana Basins, Proceedings of Volume Vth International Gondwana Symposium, pp. 525–551.

Casshyap, S.M. and Tewari, R.C. (1988). Depositional models and tectonic evolution of Gondwana basins of peninsular India. *Palaeobotanist* 36: 59–66.

Dash, P.P., Prakash, R., Gyanandra, S.P. K. (et al). 2015. Structure Map near Top of Jurassic, Kutch Saurashtra area, Western Offshore Group, KDMIPE, ONGC internal report, Unpublished.

Dasgupta, S.K. (1975). Revision of the Mesozoic–Tertiary stratigraphy of the Jaisalmer Basin Rajasthan. *Indian J. Earth Sci.* 2: 77–94.

Dasgupta, S. and Mukherjee, S. (2019). Remote sensing in lineament identification: examples from western India. In: *Problems and Solutions in Structural Geology and Tectonics, Developments in Structural Geology and Tectonics Book Series*, vol. 5 (ed. A. Billi and A. Fagereng), 205–221.

Dasgupta, S. and Mukherjee, S. (2017). Brittle shear tectonics in a narrow continental rift: asymmetric nonvolcanic Barmer Basin (Rajasthan, India). *J. Geol.* 125 (5): 561–591.

Eagles, G. and Hoang, H.H. (2014). Cretaceous to present kinematics of the Indian, African and Seychelles plates. *Geophys. J. Int.* 196: 1–14.

Farshori, M.Z. (1972). *The Geology of Sindh*. Jamshoro: University of Sindh.

Gaina, C., Hinsbergen, D.J.J., and Spakman, W. (2015). Tectonic interactions between India and Arabia since the Jurassic reconstructed from marine geophysics, ophiolite geology, and seismic tomography. *Tectonics* 34 (5): 875–906.

Gaina, C., Torsvik, T.H., Hinsbergen, D.J.J. et al. (2013). The African plate: a history of oceanic crust accretion and subduction since the Jurassic. *Tectonophysics* 604: 4–25.

Gaina, C., Muller, R.D., Brown, B. et al. (2007). Breakup and early seafloor spreading between India and Antarctica. *Geophys. J. Int.* 170: 151–169.

Gombos, A.M., Powell, W.C., and Norton, I.O. (1995). The tectonic evolution of western India and its impact on hydrocarbon occurrences: an overview. *Sediment. Geol.* 96: 119–129.

Gorin, G.E., Racz, L.G., and Walter, M.R. (1982). Late Precambrian-Cambrian sediments of Huqf group, Sultanate of Oman. *AAPG Bull.* 66: 2628–2648.

Guha, D., Das, S., Srikarni, C., and Chakarborty, S.K. (2005). Alkali basalt of Kachchh: its implication in tectonic framework of Mesozoic of western India. *J. Geol. Soc. India* 66: 599–608.

Husseini, M.I. and Husseini, S.I. (1990). Origin of the Infracambrian salt basins of the Middle East, classic petroleum provinces. In: *Geological Society, Special Publications*, vol. 50 (ed. J. Brookes), 279–292. London, U.K: Geological Society of London.

Kadri, I.B. (1995). *Petroleum Geology of Pakistan*, 275. Pakistan: Pakistan Petroleum Limited, Karachi.

Kamen-Kaye, M. (1982). Mozambique-Madagascar geosyncline: deposition and architecture. *J. Pet. Geol.* 5 (1): 3–30.

Kar, N.K., Mani, D., Mukherjee, S. et al. (2022). Source rock properties and kerogen decomposition kinetics of Eocene shales from petroliferous Barmer basin, western Rajasthan, India. *Journal of Natural Gas Science and Engineering* 100: 104497.

Khan, I.H. and Clyde, W.C. (2013). Lower Paleogene Tectonostratigraphy of Balochistan: evidence for time-transgressive late Paleocene-early Eocene uplift. *Geosciences* 3 (3): 466–501.

Khan, Z. and Khan, A.A. (2015). A review on lithostratigraphy and biostratigraphy of Jaisalmer basin, western Rajasthan, India. *Int. res. J. earth sci.* 3 (8): 37–45.

Khan, M., Liu, Y., Farid, A., and Owais, M. (2018). Characterizing seismo-stratigraphic and structural framework of late cretaceous-recent succession of offshore Indus Pakistan. *Open Geosci.* 10 (1): 174–191.

Khan, N., Konaté, A.A., Zhu, P., 2013, Integrated geophysical study of the lower Indus Platform Basin area of Pakistan, *Int. J. Geosci.*, 4, 1242–1247, doi:https://doi.org/10.4236/ijg.2013.49118.

Khatri, A., Chauhan, P., Rao, P.L.S., and Rao, P.H. (2015). Hydrocarbon prospectivity of indian sedimentary basins – a review. *Int. J. Sci. Res.* 4: 2277–8179.

Kothari, V., Naidu B., Sunder, V.R. et al. (2015). Discovery and Petroleum System of Barmer Basin, India, AAPG International Conference and Exhibition, Istanbul, Turkey.

Leggett, M. (2012). *In Search of a New Play in the Infracambrian Petroleum System of the Bikaner-Nagaur Basin in Rajasthan.* AAPG Search and Discovery Article #10393.

Maheshwari, A., Sial, A.N., and Mathur, S.C. (2002). Carbon isotope fluctuations through the Neoproterozoic-lower Cambrian Birmania Basin, Rajasthan, India. *Carbonates Evaporites* 17 (1): 53–59.

Mazumdar, A. and Bhattacharya, S.K. (2004). Stable isotopic study of late Neoproterozoic-early Cambrian (?) sediments from Nagaur-Ganganagar Basin, western India: possible signatures of global and regional C-isotopic events. *Geochem. J.* 38: 163–175.

Mazumder, S., Tep, B., Pangtey, K.K.S. et al. (2017). Probable existence of a Gondwana transcontinental rift system in western India: implications in hydrocarbon exploration in Kutch and Saurashtra offshore: a GIS-based approach. *J. Earth Syst. Sci.* 126 (81): 1–21.

McKerrow, W.S., Scotese, C.R., and Brasier, M.D. (1992). Early Cambrian continental reconstruction. *J. Geol. Soc. Lond.* 149: 599–606.

Misra, A.A. and Mukherjee, S. (2015). *Tectonic Inheritance In Continental Rifts and Passive Margins.* Heidelberg, New York: springer briefs in earth sciences 978-3-319-20576-2.

Misra, P.C. and Sharma, P.C. (1986). Geological analysis of Lower Goru Formation with Special Reference to Hydrocarbon Occurrence, Jaisalmer basin, Rajasthan. ONGC report (unpublished).

Misra, P.C., Singh, N.P. 1993. Lithostratigraphy of Indian Petroliferous Basin, Document-II, West Rajasthan Basin, ONGC report (unpublished).

Mukherjee, S., Dasgupta, S., Majumdar, C. et al. (2023). Introduction to: Handbook of Petroleum Geoscience: Exploration, Characterization, and Exploitation of Hydrocarbon Reservoirs. In: *Handbook of Petroleum Geoscience: Exploration, Characterization, and Exploitation of Hydrocarbon Reservoirs*, Petroleum Industry Updates: Geoscientific and Economic Issues (ed. S. Mukherjee, S. Dasgupta, C. Majumdar, et al.). John Wiley & Sons, Hoboken, ISBN 978-1-119-68003-1.

Oldham, R.D. (1886). Preliminary note on the geology of northern Jaisalmer. *Rec. Geol. Surv. Ind.* 19 (3).

Oldham R D, 1893 *A Manual of Geology of India.*

Pangtey, K.K.S., Mazumder, S., Tep, B., Mitra, D.S. (2013). A GIS based plate tectonic reconstruction of western India during neo-Proterozoic-Cambrian and its implication in hydrocarbon exploration in Ganga Basin, Proceedingsof National Conference.in Earth Science India: challenges and emerging trends (ESICET-2013), pp. 65–66.

Pandey, R., Maurya, A S, 2020, Hydrocarbon uncertainty based on Facies analysis: middle Jurassic sequence (Jaisalmer formation), Jaisalmer Basin, Rajasthan. *J. Geol. Soc. India*, 95, 301–307. doi:https://doi.org/10.1007/s12594-020-1429-9

Pandey, R., Kumar, D., Maurya, A.S., and Pandey, P. (2019). Facies analysis, depositional environment and reservoir characterization of tertiary sequences, Jaisalmer Basin, Rajasthan. *ONGC Bull.* 54 (1): 102–116.

Pandey, R., Kumar, D., Maurya, A.S., and Pandey, P. (2019). Evolution of gas bearing structures in Jaisalmer Basin (Rajasthan), India. *J. indian geophys. union* 23 (5): 398–407.

Pandey, R., Nonia, B.P., Mahanti, S. et al. (2018). Geocellular model for tertiary reservoirs in manhera tibba gas field, Jaisalmer Basin, Rajasthan, India; AAPG international conference and exhibition, Cape Town, South Africa, 1–20.

Pandey, D.K., Swami, N.K., Sharma, J.K., and Alberti, M. (2018). Callovian ammonites from a condensed horizon in the Kuldhar member of the northeastern Jaisalmer Basin (Rajasthan, India) and their biostratigraphic implications. *J. Palaeontol. Soc. India* 63: 37–52.

Plummer, P.S., Belle, E.R., 1995, Mesozoic tectono-stratigraphic evolution of the Seychelles microcontinent, *Sediment. Geol.*, 96 (1), 73–91. doi:https://doi.org/10.1016/0037-0738(94)00127-G

Plummer, P.S. (1996). The Amirante ridge/trough complex: response to rotational transform rift drift between Seychelles and Madagascar. *Terra Nova* 8: 34–47.

Prasad, B., Asher, R., and Borgohain, B. (2010). Late Neoproterozoic (Ediacaran)–early Paleozoic Cambrian acritarch from Marwar Suprgroup, Bikaner Nagaur basin, Rajasthan, India. *J. Geol. Soc. India* 75: 415–431.

Raju, S.V., Mathur, N., and Sarmah, M.K. (2014). Geochemical characterization of Neoproterozoic heavy oil from Rajasthan, India: implications for future exploration of hydrocarbons. *Curr. Sci.* 107 (8): 1298–1305.

Raza, H.A., Ahmad, W., Ali, S.M. et al. (2008). Hydrocarbon prospects of Punjab Platform Pakistan, with special reference to Bikaner–Nagaur Basin of India. *Pak. J. Sci. Res* 18: 1–33.

Raval, U. and Veeraswamy, K. (2000). The radial and linear modes of interaction between mantle plume and continental lithosphere: a case study from western India. *J. Geol. Soc. India* 56 (5): 525–536.

Roy, A.B. (2003). Geological and geophysical manifestations of the Reunion Plume-Indian lithosphere interactions-evidence from northwest India. *Gondwana Res.* 6 (3): 487–500.

Roy, A.B. and Jakhar, S.R. (2002). *Geology of Rajasthan (Northwest India, Precambrian to Recent).* Jodhpur: Scientific Publishers (India).

Sharma, P., Vaishnav, K., and Bhu, H. (2016). Tectono-geomorphic features around Jaisalmer (Rajasthan). *Int. J. Sci. Res.* 5 (4): 153–157.

Singh, A.K. and Tewari, P.K. (2011). Infracambrian hydrocarbon systems and emerging hydrocarbon potential in Bikaner Nagaur and Jaisalmer basins (Miajlar sub basin), of Rajasthan. Extended abstract, GeoIndia, 2nd south Asian Geoscience Conference and Exhibition, Greater Noida, New Delhi, India.

Salman, G. and Abdula, I. (1995). Development of the Mozambique and Ruvuma sedimentary basins, offshore Mozambique. *Sediment. Geol.* 96 (1–2): 7–41.

Siawal, A. and Kaul, A.K. (2006). Tectonic Evolution of Western Margin of India; Unpublished ONGC Report.

Smith, A.G., Hurley, A.M., and Briden, J.C. (1981). *Phanerozoic Paleocontinental World Maps.* Cambridge: Cambridge University Press.

Siawal, A., Rawat, C.P.S., Kishan, R., Saha, D., Gabba, P., 2015, *Tectonic Map of India*; ONGC Document.

Singh, N.P. (1984). Addition to the tertiary biostratigraphy of Jaisalmer Basin. *Pet. Asia. J. II* 1: 106–128.

Singh, N.P. (2000). Relevance of laboratory studies in geological modelling and field geology, ONGC report (unpublished).

Singh, N.P. (2006). Mesozoic Lithostratigraphy of the Jaisalmer Basin, Rajasthan. *J. Palaeontol. Soc. India* 51 (2): 1–25.

Subimal, K., Phukan, R.K., Karvannan, K. et al. (2010). Tectono-sedimentation of Kutch-Jaisalmer basins during Proterozoic and Mesozoic period and their relationship with Madagascar within the framework of plate tectonics. Studies in particular to the Jaisalmer Basin (Mari high, Shahgarh and Miajlar sub basins) to understand hydrocarbon prospectivity, ONGC report (unpublished).

Torsvik, T.H., Ashwal, L.D., Tucker, R.D., and Eide, E.A. (2001). Neoproterozoic geochronology and paleogeography of the Seychelles microcontinent: the India link. *Precambrian Res.* 110: 47–59.

Verma, D., Jadhav, G.N., Biswal, T.K. et al. (2012). Characterization of hydrocarbon-bearing fluid inclusion in sandstones of Jaisalmer basin, Rajasthan. *J. Geol. Soc. India* 80: 505–514.

Wynne, A.B. (1867). Memoir on the Geology of Kutch, to accompany a map complied by A.B. Wynne and F. Fedden during field season of 68 1867–1868 and 1868–1869. *Memoirs of the Geological Survey of India* 9 (1): 1868–1869.

21

Improving Insights Into Petrophysics using Geophysical Data for the Habiganj Structure, Surma Basin, Bangladesh

Kamruzzaman, Delwar Hossain, Mizanur Rahman Sarker, Mahmuda Khatun, and Md. Upal Shahriar*

Department of Geological Sciences, Jahangirnagar University, Dhaka, Bangladesh

Author for correspondence: bullbull.htc113a@gmail.com

Abstract

Habiganj gas field is one of the largest fields in Bangladesh. However, a very few amounts of analysis has been done on it. A study based on gravity and magnetic data of the Habiganj and adjoining areas has been made to delineate the structure and infer the general geology of the region. This study primarily encompasses formation evaluation and determination of petrophysical parameters of the gas-bearing zones in the Habiganj well no. 11 (HB # 11). As a result of detailed interpretation of various wireline logs of this well and using the cross-plots of logs, it is evident that there are two hydrocarbon (HC)-bearing zones in this field. The computed averages of shale volume, porosity, permeability, and gas and water saturation have been used to determine the reservoir interval pay zone, net, gross, net to gross, bulk volume of water, pore thickness, and HC pore thickness.

Keywords *Hydrocarbon; reservoir; fluid contacts; geophysical data; HB # 11; Habiganj Gas Field; Bangladesh*

21.1 Introduction

Reservoir characterization is essential for sound petroleum management aimed at improving the estimation of reserves and making decisions regarding the development of an oil and gas field. Economic and efficient oil and gas production is highly dependent on understanding key properties of reservoir rock, such as porosity, permeability, and wettability (Zeba and Murrellb 2015). Reservoir characterization plays an important role in different parts of an industrial project, which gives an insight into the rock and fluid properties to optimize the choice of drilling locations and reduce risk and uncertainty (Torres and Sen 2004). Petrophysical parameters are the most useful characteristics of a reservoir for the development and production of the drill well and estimation of reserves

Handbook of Petroleum Geoscience: Exploration, Characterization, and Exploitation of Hydrocarbon Reservoirs, First Edition. Edited by Soumyajit Mukherjee, Swagato Dasgupta, Chandan Majumdar, Subhandip Mandal and Troyee Dasgupta.
© 2023 John Wiley & Sons Ltd. Published 2023 by John Wiley & Sons Ltd.

in any gas field. Reservoir characterization is the process of mapping a reservoir's thickness (T), net to gross ratio (N/G), pore fluid resistivity (ρ_f) and density, porosity (ϕ), permeability (k), water saturation (S_w), structure, extent, and volume (Cooke and Muryanto 1999; Ajisafe and Ako 2013). In HC exploration, the interest lies mainly in the identification of porous and permeable formation and their thickness. Well logging is exactly such a device which aims at acquiring this information by measuring various physical, chemical, and lithological properties of the formations (Alger 1980). They also give the depth of formation, its nature, fluid types, the extent of fluid contact, porosity, permeability, saturation, moveable hydrocarbon index (MHI), and a score of other details with a high degree of precision (Serra 1983; Rider 1986; Darling 2005).

All these properties are analyzed under the heading of formation evaluation, which, therefore, involves the use of well logs. The well logs measurements are subject to irregularities due to dissimilarities that vary in spatial distribution, scale, and their relationship to reservoir properties (Roy 1997; Xu and Chacko 2008; Shekaili et al. 2012). The advancement in electronics and instrumentation and the advent of computers have found their wide applicability in well logging. The well logs fill a critical gap between core and seismic data and are routinely used for reservoir characterization (Rider 2000).

The objective of the present study has been mainly the general seismic interpretation of the Habiganj structure and determination of its petrophysical properties (well no. 11) implying HC potentiality from the available well log data.

Bengal Basin of Bangladesh is a young prolific HC basin in the world (Alam et al. 2003). It is bordered by Precambrian Indian shield in the west and north, Indo-Burman ranges (Orogen) to the east, and is open for a considerable distance into the Bay of Bengal to the south (Imam 2013). The Bengal Basin is a peripheral foreland basin formed due to continent–continent collision after the subduction of the oceanic part of the Indian Plate (Mukherjee et al. 2009). The Sylhet Trough might have been formed at the late orogenic stage due to subsidence along the Dauki Fault system and Plio-Pleistocene uplift of the Shillong Massif (BAPEX 2007a). The Sylhet trough is mainly composed of Jainta, Barail, Surma, Tipam, Dupi Tila, and Dihing groups with ages ranging from Paleocene to Pleistocene (Reimann 1993). This structure appeared as a reversibly faulted asymmetrical anticline with an NNE-SSW trending axis (Hiller and Elahi 1984; Islam et al. 2014).

The Miocene Surma Group is a diachronous unit consisting of a succession of alternating mudrock, sandstone, siltstone, and sandy shales with occasional thin conglomerates (Imam and Shaw 1985). The variety of sediment facies indicates a range of depositional environment during Neogene time (Evans 1964). The Surma Group accumulated in interdeltaic to open marine depositional environments (Banerji 1984; Reimann 1993). Overlying the Surma Group, the Upper Marine Shale (UMS) represents a regional marine transgression in the region (Holtrop and Keizer 1970). Evans (1932) has established most of the stratigraphic terms presently used to describe the tertiary sequence of the Bengal Basin, in the Lower and the Upper Assam.

21.2 Location of the Study Area

The Habiganj gas field is the second-largest gas-producing field in Bangladesh. The approximate area of Habiganj district is 2636.58 square kilometers (km^2) (BAPEX 2007b)

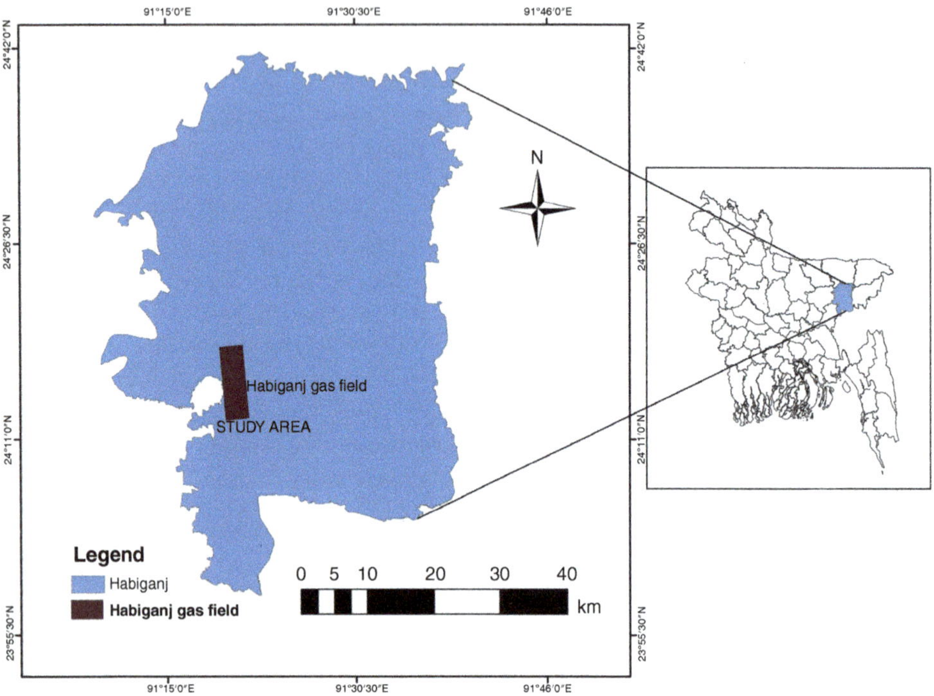

Figure 21.1 Location map of the Habiganj Gas Field.

(Figure 21.1). The structure measures 12 × 5 square kilometers (km^2) with a vertical closure of 300 m (BAPEX 2007b). Habiganj gas field lies in Madhabpurupazila under Habiganj district, which is 121 km north of Dhaka and is located at 24.3750 °N to 91.4167 °E (Rahman et al. 2017). It is just adjacent to the Rashidpur gas field (12 km) (Shikder 2005). The gas field was discovered by Pakistan Shell Oil Company (PSOC) Limited in 1963 (BAPEX 2007b). The gas field is operated by Bangladesh Gas Field Company Ltd. (BGFCL), a subsidiary of Bangladesh Petroleum Exploration and Production Company Limited (BAPEX) (BAPEX 2007b). Zonation as well as reservoir delineation based on geophysical logs are described in section Well Data.

21.3 Data and Methods

To study the HC potentiality of the structure, seismic sections, maps, wireline logs, and other information were collected from BAPEX.

Techlog (2015.0) software has been used to perform formation evaluation using well data to determine the zonation, lithology, and petrophysical properties and predict the HC presence in the reservoirs of HB # 11. The net pay assessments were made by using cutoff parameters for shale volume (V_{sh}), porosity (ϕ), and water saturation (S_w). The final results were summarized for each zone in the well. Some seismic sections were analyzed in order to delineate the reflectors and HC indicators and determine seismic stratigraphy.

21.4 Results and Discussion

21.4.1 Gravity and Magnetic Data

The physical rock property that relates gravity anomalies to rocks is density, and gravity anomalies represent lateral variations in the density of rocks (Nettleton 1971). The physical rock property that relates magnetic anomalies to rocks is total magnetization (Nettleton 1971; Reynolds et al. 1990), and magnetic anomalies represent lateral variations in the total magnetization of rocks.

The gravity anomaly at any point, in fact, represents the superimposed effect of mass distributions at various depths involving variations in thickness and densities of formations (Telford et al. 1990; Amiri et al. 2011; Khattach et al. 2013). The Bouguer anomaly (gB) pattern of the study area extracted from a Geological Survey of Bangladesh map (Rahman et al. 1990b) and shown in Figure 21.2 depicts that this area is the southern limb of the Surma Basin, characterized by a strong negative anomaly trending east-west. Producing the residual anomaly map (Figure 21.3) was based on using the Bouguer anomaly. A comparison of

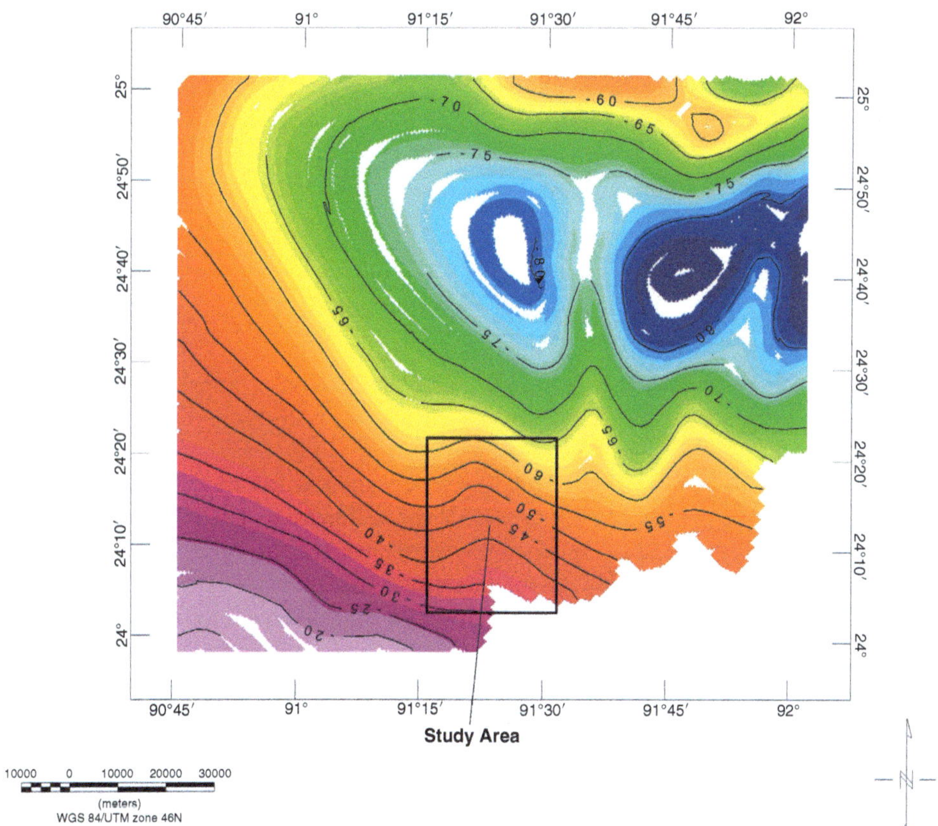

Figure 21.2 Bouguer anomaly map of the Habiganj and adjoining areas.

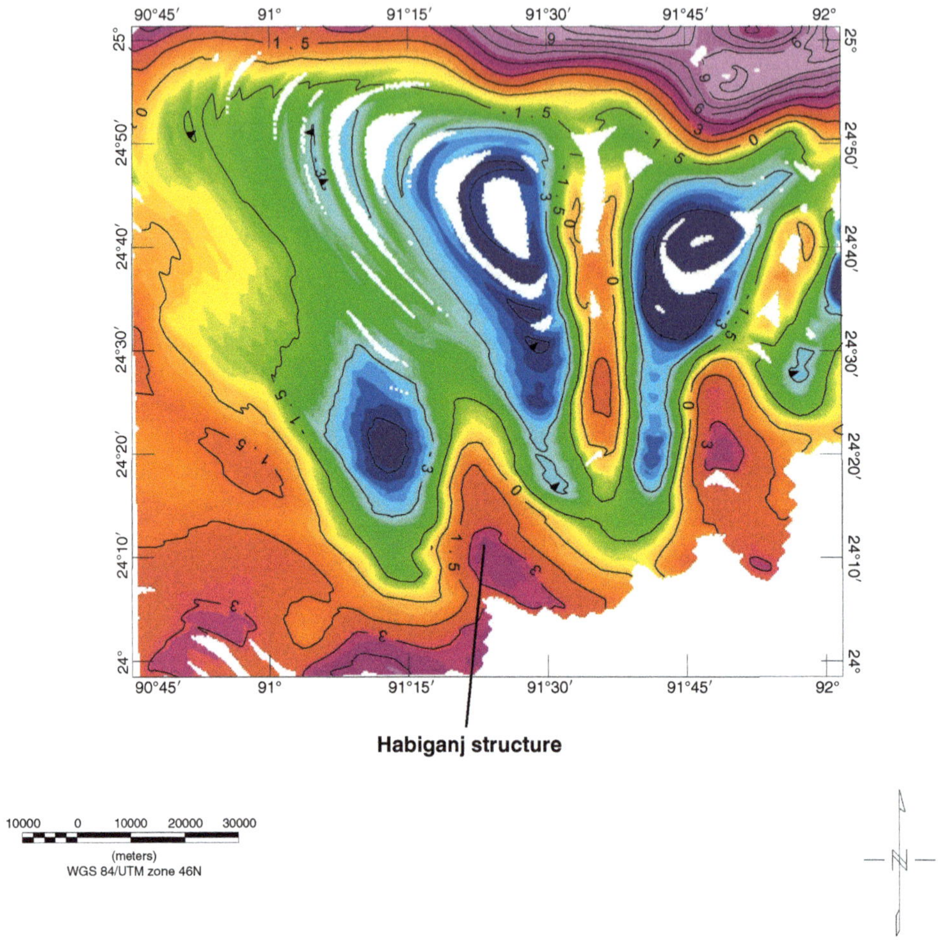

Figure 21.3 High-pass filtered Bouguer anomaly map of the Habiganj and adjoining areas.

gB with aeromagnetics (anomalies of long wavelength and low amplitude) (Figure 21.4) as well as surface geology revealed the existence of the sedimentary basin. The steep gravity gradients on the north and south separate this prominent gravity low associated with the Basin from the neighboring highs. The total Bouguer anomaly relief in this region is about 73 mGal, with values ranging from about −13 to −86 mGal. Except for north–south and north–northeast trending flexures in contours, there are no obvious local anomalies in the southern flank of the basin, i.e. they are significantly distorted by the broader regional trend. Similar feature is also apparent in the residual magnetic map (Figure 21.5). The area under investigation (Habiganj) is characterized by field intensities of −30 to −60 mGal.

Local details are better resolved on the residual anomaly map (Figure 21.3) prepared following the removal of the dominant regional trend by high-pass filtering using Oasis Montaj (2007). The regional gravity varies from about −82 to −16 mGal and brings out clearly the general basinal shape. The local anomaly pattern with structural trend NNW–SSE to N–S

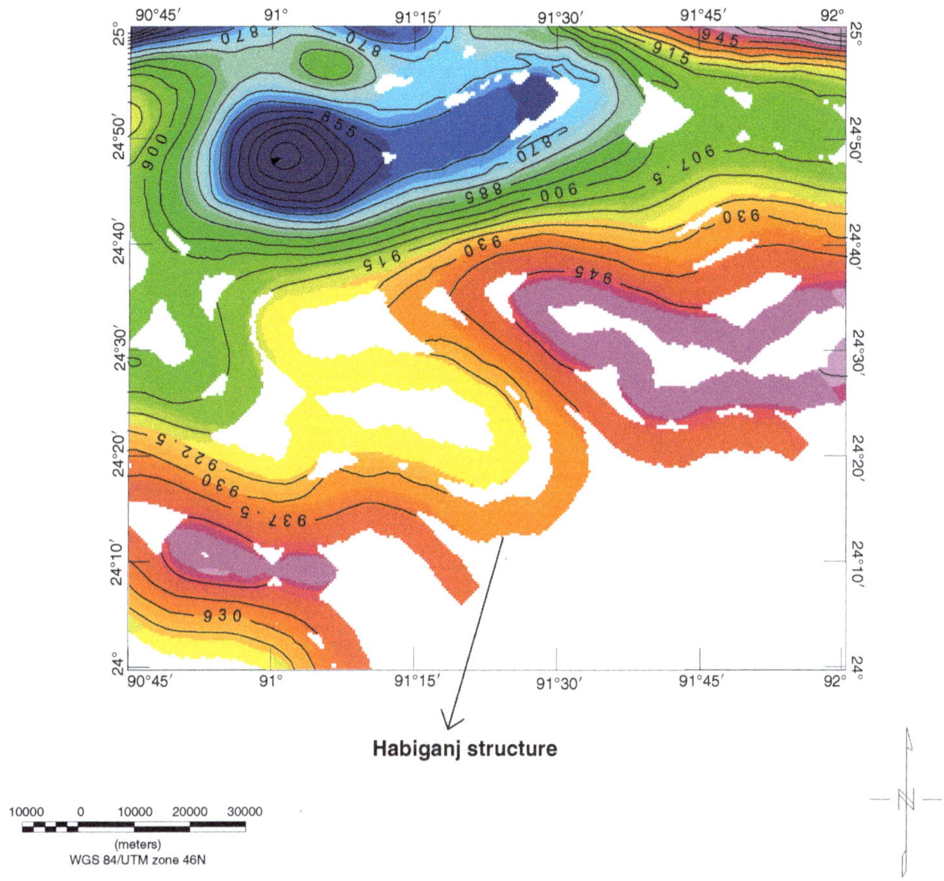

Figure 21.4 Aeromagnetic anomaly map of the Habiganj and adjoining areas.

and nonuniform gradient is essentially different from that of the Bouguer gravity. The peak anomaly value of the Habiganj structure appears to be about 3.5 mGal. The local patterns in this region are caused by density contrasts associated mainly with folding.

The aeromagnetic data (Figure 21.4) extracted from a Geological Survey of Bangladesh map (Rahman et al. 1990a) vary from 820 to 970 nT. At the Surma Basin, we have the inclination $I = 34°$, declination $D = -0.7°$ and regional total field of the Earth's field $F_e = 46\,175$ nT. As mentioned earlier, the magnetic anomaly pattern depicts the presence of a thick sedimentary cover. The depth to the magnetic basement in this region derived from radially averaged power spectrum using wavenumber 0.04 km^{-1} appears not to exceed 14 km.

Zones of contrasting magnetic characteristics are apparent on the magnetic map with the general magnetic trend of E–W. As usual, the complicated magnetic pattern is due to the overlapping effects of several sources.

The strong negative anomaly with a peak-to-peak of 67 nT (ranging from 821 to 888 nT) seen in the northwest corner on the total magnetic intensity (TMI) map is not brought out by the gravity anomaly data implying a lack of density variation for the source with respect

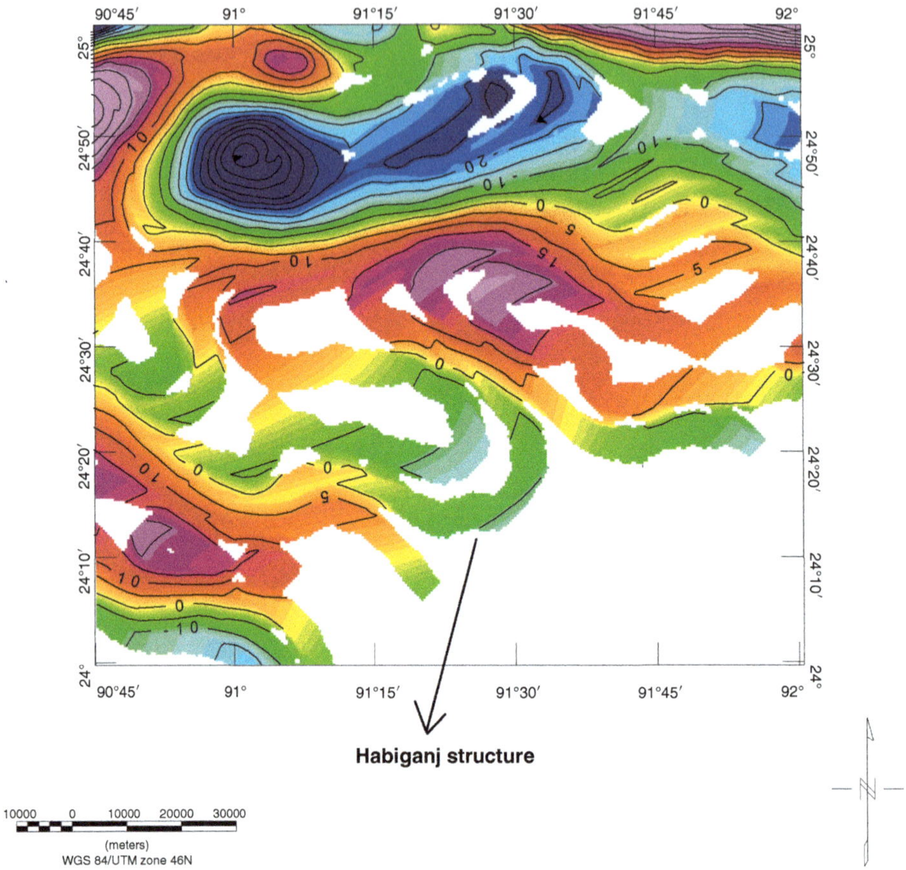

Figure 21.5 High-pass filtered magnetic anomaly map of Habiganj and adjoining areas.

to the surrounding deeper formation. However, the general trend for both cases is E–W. The same feature is also well expressed on the residual magnetics (Figure 21.5). The residual magnetic trend is almost perpendicular to that of the residual gravity.

21.4.2 Seismic Data

The seismic dataset used in this study is 2-D post-stack time migration data. Visualization and identification of many structures are sometimes easily achieved using seismic attributes (Misra and Mukherjee 2018). The seismic line HB-6, which is about 9.5 km long and trends SW–NE, passes through the vertical well HB # 11 (Figure 21.6). Vertical seismic profiling (VSP) data has been used to relate seismic data to well logs. Relating well logs to seismic data helps find events that correspond to geological formations and can be considered to be one of the first steps in interpreting seismic data. The seismically derived stratigraphy is given in Figure 21.6 and Table 21.1. Correlation between well logs and seismic data indicates that R2

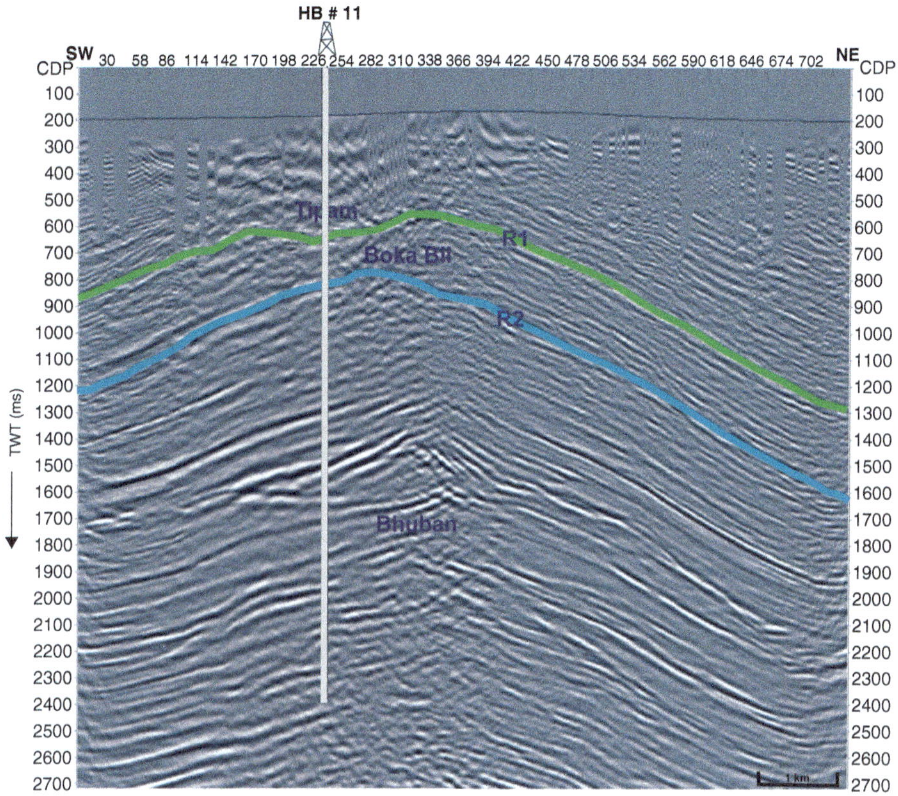

Figure 21.6 The reflectors along the seismic line HB-6.

Table 21.1 Seismically derived stratigraphy of the Habiganj area along the line HB-6.

Age	Group	Formation	Reflectors	Lithology	Thickness approx. (m)
Pliocene	Tipam	Tipam-sandstone	R1	Medium to coarse-grained sand or loosely consolidated sandstone	170+
Late-Miocene	Surma	BokaBil	R2	Shale or Claystone with medium to fine-grained sandstone	480
Early–Middle Miocene		Bhuban		Alternating sand, shale, sandy shale and silty shale	1850+

corresponds to the boundary between the Bhuban and BokaBil Formations, and R1 corresponds to the boundary between the BokaBil and the Tipam Formations.

According to Reimann (1993), the maximum thicknesses of the Surma Group and Tipam Group are 3350 m (Atgram 1X) and 3500 m (Goyain trough), respectively. As found in this study, the thicknesses of the Surma Group and Tipam Group are above 2330 m and above 170 m, respectively. The lithological characteristics of these groups are almost the same in

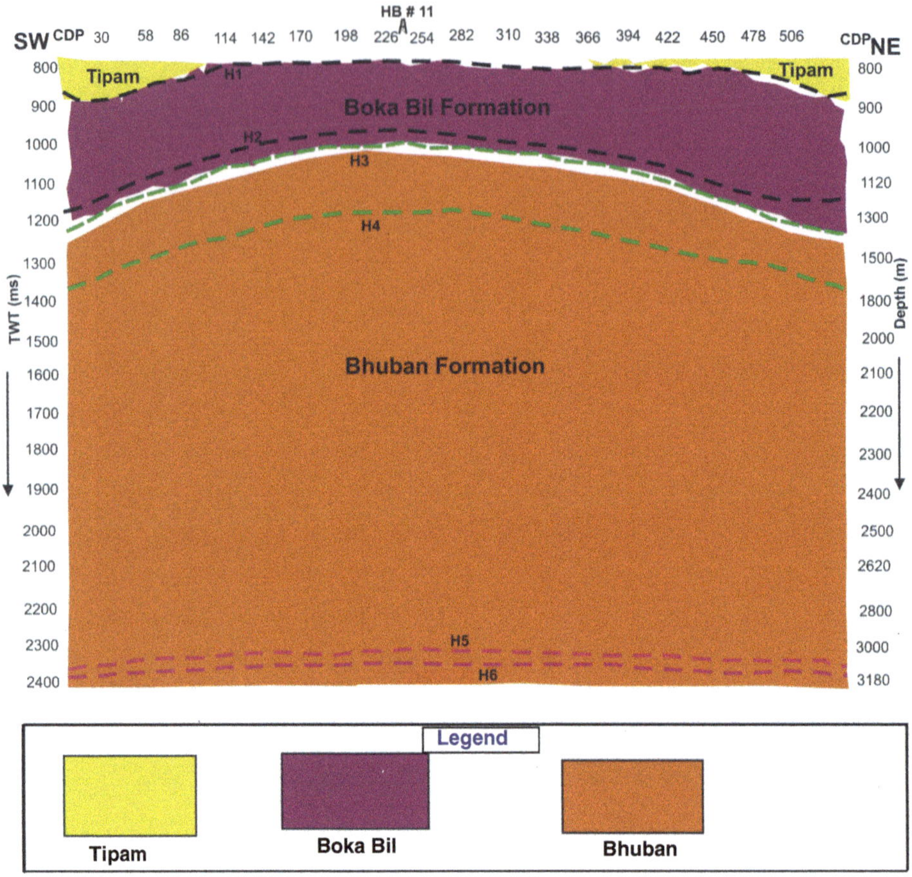

Figure 21.7 Geological interpretation of seismic sequences in terms of TWT and depth along the seismic section HB-6.

these two studies. The thicknesses of Bhuban Formation, BokaBil Formation, and Tipam Formation are above 1850 m (at depths 1300–3150 m), 480 m (at depths 820–1300 m), and above 170 m (at depths 650–820 m), respectively (Kamruzzaman, 2017).

The seismo-geological cross-section in terms of TWT and depth of the Habiganj structure area has been constructed on the basis of seismic sections and the time-depth information with a maximum depth of 3200 m. To interpret a structure, a seismic line in the dip direction is of good use. In this case, the depths have been taken from the seismic section by using the time-depth information of HB # 11. The events H1, H2, H3, H4, H5, and H6 show rather high impedance contrast in the seismic section (Kamruzzaman 2017). The structural and seismo-stratigraphic analysis of line HB-6 is shown in Figure 21.7.

21.4.3 Well Data

The entire section of HB # 11 was divided into ten (10) zones based on the signature of geophysical logs, especially GR, SP, and Caliper logs; this allowed to determine changes in lithology (Kamruzzaman 2017). The zones for HB # 11 are given in Table 21.2.

Table 21.2 Well zones based on geophysical logs.

Well zones	Depth range (m)	Lithology
1	165–1182	Sand with some shale layers
2	1182–1360	Mainly shale
3	1360–1720	Shaly sand
4	1720–1895	Sandy shale
5	1895–2525	Sandy shale with shaly sand lenses
6	2525–2930	Sandy shale
7	2930–3120	Mainly shale
8	3120–3140	Mainly sand
9	3140–3170	Shaly sand
10	3170–3206	Mainly shale

Geophysical logs were then used to identify the formation fluids (gas/oil/water) and possible fluids contact; pressure data to be used for this purpose was not available. HC-bearing zone is indicated by high resistivity readings (R_t), whereas water-bearing zones are shown by a resistivity (R_t) decrease. There are two gas pools found in these zones. The thicknesses of the sealing units of these pools are about 178 m (mainly shale) and 190 m (mainly shale) for the upper one and lower one, respectively. Table 21.3 shows the fluid contacts for well HB # 11 (Kamruzzaman 2017).

After determining lithology and fluid contacts for HB # 11, three HC-bearing zones have been delineated based on resistivity, NPHI, RHOB, and sonic logs (Table 21.4). The large crossover between neutron-density curves and high resistivity (R_t) values are the reflection

Table 21.3 The fluid contacts delineated for HB # 11.

Contact type	Contact depth (m)
Gas–water	1473.9
Gas–water	3138

Table 21.4 The reservoir identification for HB # 11.

HC-bearing zones	Depth range (m)	Thickness (m)
Upper gas zone (UGZ)	1365–1473.9 (Most prominent HC-bearing reservoir ranging at depths from 1418–1473.9 m)	108.9
Lower gas zone (LGZ)	3122–3138	16

Source: Adapted from Kamruzzaman (2017).

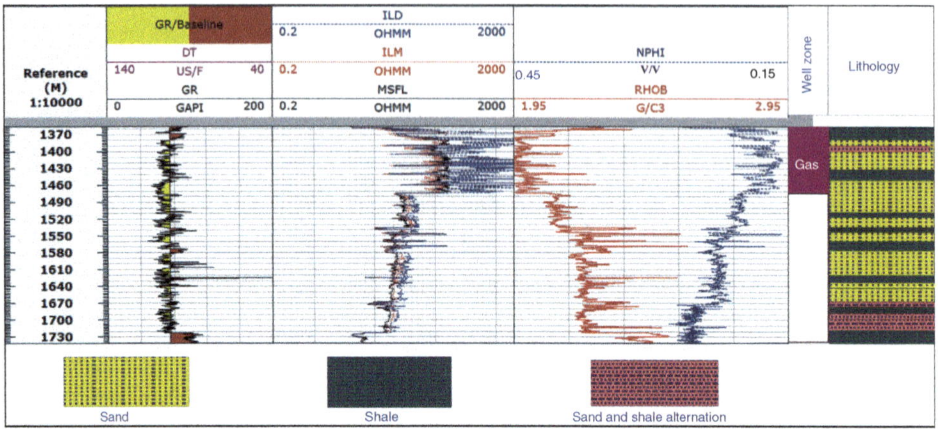

Figure 21.8 UGZ detection from well logs. The sonic log data was not available for this zone.

of the gas-bearing HC zone (Rider 2002). The depth of 1418–1473.9 m represents the most prominent HC-bearing zone within the upper gas zone.

The well log signatures are shown in Figures 21.8 and 21.9.

The relationship between effective porosity and permeability for the reservoirs under study has been obtained from the cross-plot of these parameters that helps understand the nature of correlation (positive or negative) between them.

It has been found that effective porosity and permeability cross-plots for the reservoirs show a linear relationship and are positive in nature. The example cross-plot for a zone with low GR values and other well log signatures for a reservoir at depths from 865 to 1175.9 m (Figure 21.10) shows a very strong correlation with a regression coefficient value of 0.961.

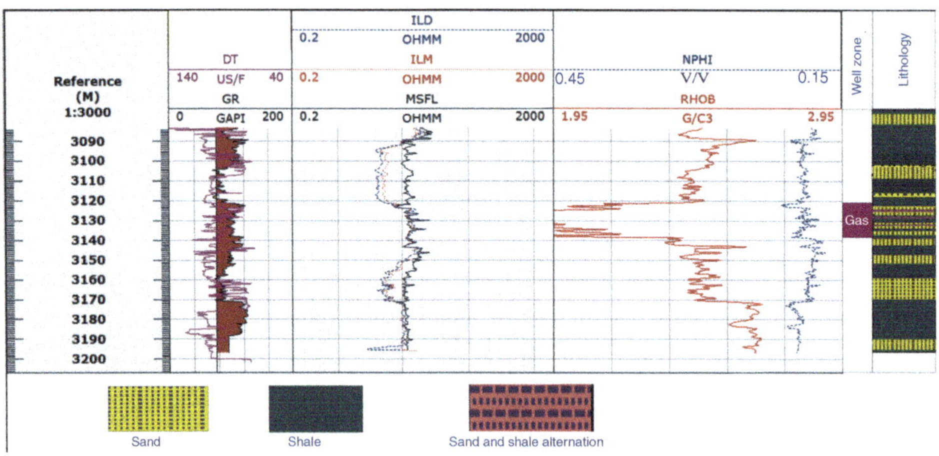

Figure 21.9 LGZ detection from well logs.Using reservoir statistics, reservoir parameters for the HC-bearing zones studied were computed (Tables 21.5 and 21.6). The cutoffs used were based on the log measurements as well as experience.

Table 21.5 Summary results for UGZ.

Dataset	Flag	Top (m)	Bottom (m)	Gross (m)	Net (m)	Net to gross (fr)	BVW (%)	POR-TH (m)	HC POR-TH (m)	Av_effective porosit	Av_water saturation	Av_shale volume	Av_Porosity	Av_Perme-ability (mD)
UGZ	RES	1365	1473.9	108.9	72.1	0.662	4.838	47.18	42.348	0.251	0.103	0.23	0.303	166.612

Table 21.6 Summary results for LGZ.

Dataset	Flag	Top (m)	Bottom (m)	Gross (m)	Net (m)	Net to gross (fr)	BVW (%)	POR-TH (m)	HC POR-TH (m)	Av_ Effective porosity	Av_ Water saturation	Av_ Shale volume	Av_ porosity	Av_ Perme-ability (mD)
LGZ	RES	3122	3138	16	11.30	0.706	1.16	3.91	2.746	0.278	0.299	0.301	0.346	12.504

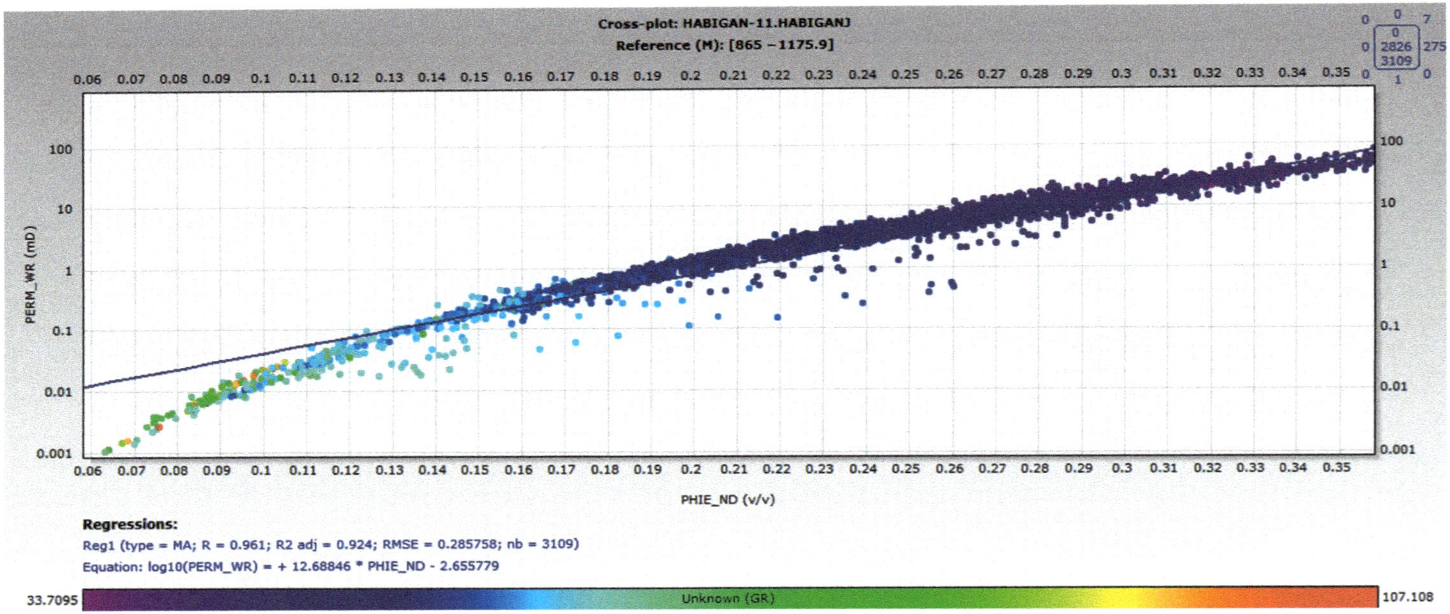

Figure 21.10 Effective porosity vs. permeability cross-plot for a clean zone at depths from 865 to 1175.9 m.

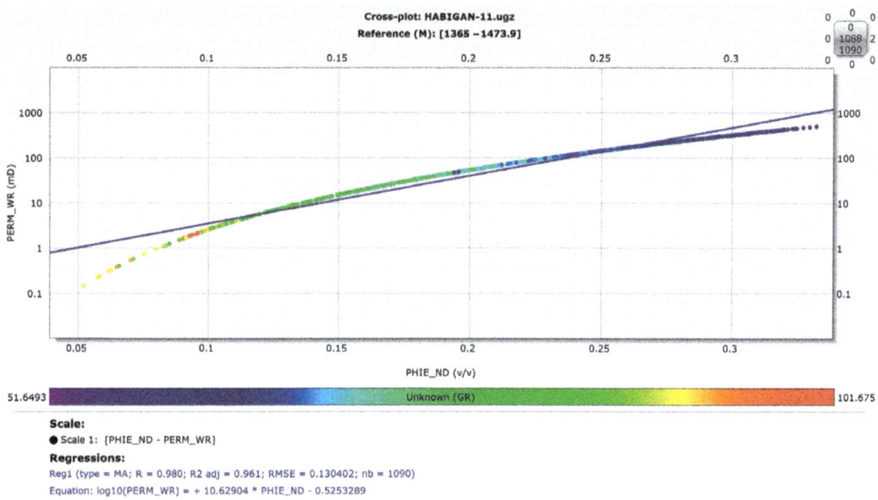

Figure 21.11 Effective porosity vs. permeability cross-plot for UGZ.

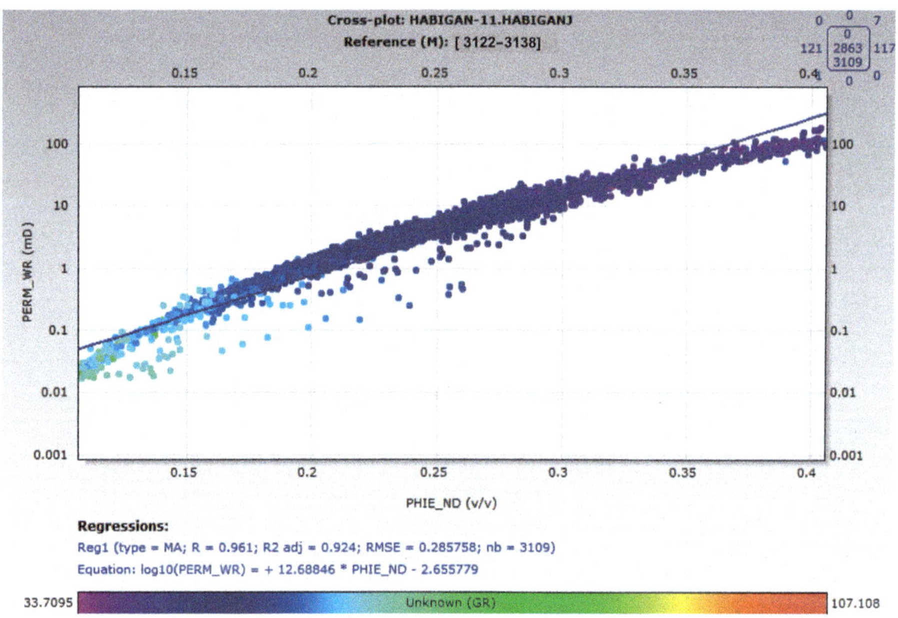

Figure 21.12 Effective porosity vs. permeability cross-plot for LGZ.

This model has a very good match with the dataset. This zone comprises clean sand. Figure 21.11 illustrating the cross-plot for reservoir UGZ (with a gross thickness of 108.9 m) indicates that there is a very strong correlation between the parameters with a regression coefficient value of 0.980. This zone does not consist of clean sand, but rather shaly/silty sand. The cross-plot for reservoir LGZ (with a gross thickness of 16 m) shown in Figure 21.12 shows a very strong correlation with a regression coefficient value of

0.961 between the parameters. This zone likely comprises sand with some shale lenses. Cross-plotting effective porosity and permeability appears to act as a measure for inferring lithology.

21.5 Correlations

21.5.1 Relating Well Logs to Seismic Data

Better correspondence between seismic and well data can be achieved depending on the location of well position with respect to 2D seismic line and the distance between them. If the offset distance from well to the seismic line is minimum, then a better correspondence can be expected. It was possible to convert seismic data from the time domain into the depth domain by using VSP and check the correspondence between seismic data and well log data including DT and RHOB. The correlation between seismic section HB-12 and well HB # 11 is shown in Figure 21.13. As mentioned earlier, well log signatures aided in delineating

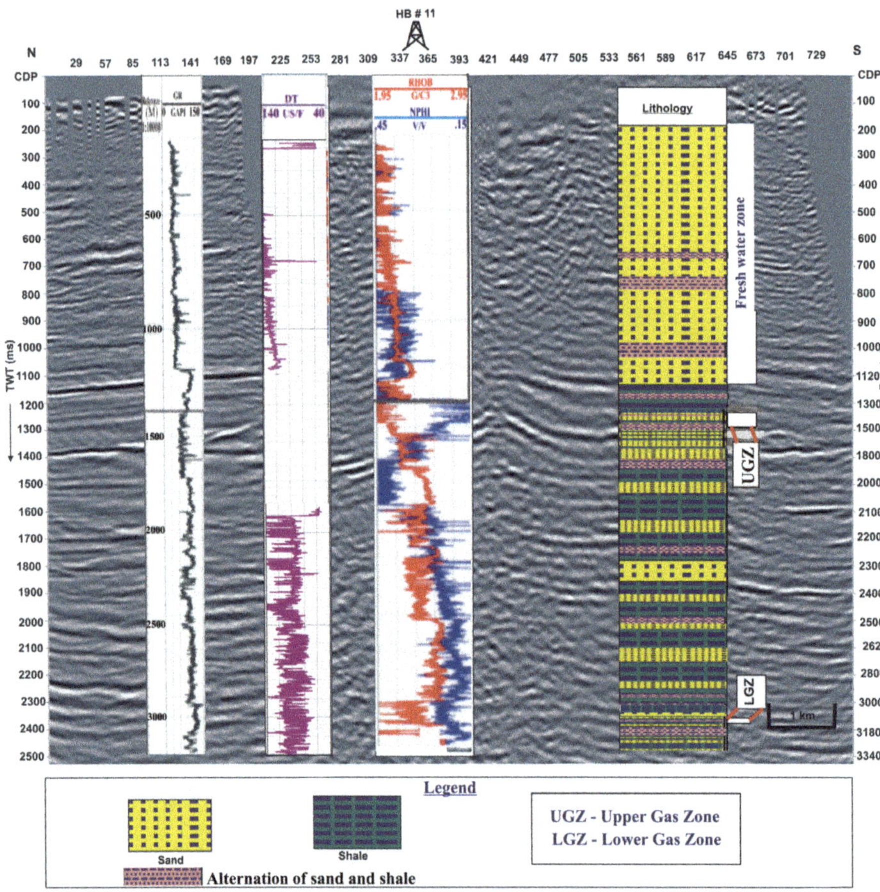

Figure 21.13 Correlation between seismic section HB-12 and HB # 11.

zones with contrasting lithology and fluid content. The events with increased reflection strength relate to contacts between different zones and changes in fluid types. There are two prominent interfaces found at two-way time (TWT) 1160 and 1390 ms in this seismic section (Kamruzzaman 2017).

21.5.2 Correlation of Well HB # 11 to Sylhet Trough

The Bhuban, BokaBil, and Tipam Sandstone Formations are found (based on log data) in HB # 11, but Dupi Tila Formation is absent. The lower boundary of the Bhuban Formation cannot be delineated because of the lack of log data below the depth of 3195 m. It is too difficult to say about the presence of Jaintia and Barail Groups without geophysical investigations. But as the Habiganj Gas Field is within Surma Basin, so we can expect that Jaintia and Barail Groups will be found by further deeper exploration. UGZ and LGZ have been found in the Bhuban Formation (Kamruzzaman 2017). The stratigraphic correlation of HB # 11 to Sylhet Trough is shown in Figure 21.14.

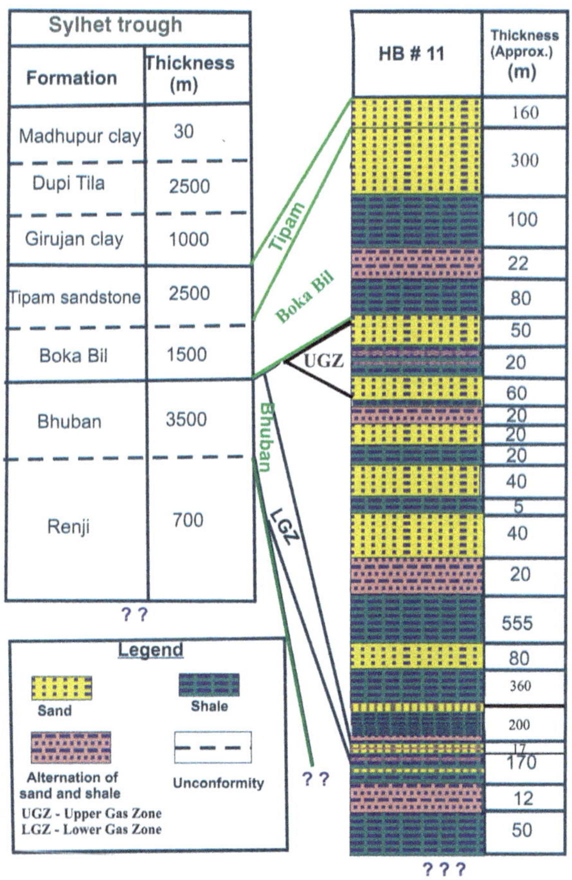

Figure 21.14 Stratigraphic correlation of HB # 11 to Sylhet Trough. *Source:* Adapted from Khan and Muminullah (1980); Reimann (1993); Uddin and Lundberg (1998).

21.6 Conclusions

Both log interpretations and neutron-density cross-plots confirm that the reservoir consists of sand mixed with shale lithology. The average shaliness values measured for UGZ and LGZ are 23 and 30%, respectively. The average effective porosity values measured for UGZ and LGZ are 25 and 28%, respectively. The average permeability values measured for UGZ and LGZ are 166.61 and 12.50 mD, respectively. The UGZ and LGZ are of very high reservoir permeability and high reservoir permeability, respectively, according to classification (Glover 2011). The average gas saturations measured for UGZ and LGZ are 90 and 70%, respectively.

Acknowledgments

One of the authors (Kamruzzaman) sincerely thanks BAPEX for logistic support and encouragement. Soumyajit Mukherjee (IIT Bombay) invited, handled, and reviewed this article in several rounds. Sincere thanks are due to the anonymous reviewer for comments and suggestions that have improved the paper. Merryl Le Roux, Frank Otmar Weinreich, Joss Everett, Andrew Harrison, and the proofreading team (Wiley Blackwell) are thanked for their assistance. This chapter has been encapsulated in Mukherjee et al. (2023).

References

Ajisafe, Y.C. and Ako, B.D. (2013). 3-D seismic attributes for reservoir characterization of 'Y' field Niger Delta, Nigeria. *IOSR Journal of Applied Geology and Geophysics* 1 (2): 23–31.

Alam, M.M., Curray, J.R., Chowdhury, M.L.R., and Gani, M.R. (2003). An overview of the sedimentary geology of the Bengal Basin in relation to the regional tectonic framework and basinfill history. *Sedimentary Geology* 155: 179–208.

Alger, R.P. (1980). Geological use of wireline logs. In: *Developments in Petroleum Geology-2* (ed. G.D. Hobson), 207–272. London: Applied Science Publication.

Amiri, A., Chaqui, A., Nasr, I.H. et al. (2011). Role of preexisting faults in the geodynamic evolution of Northern Tunisia, insights from gravity data from the Medjerda valley. *Tectonophysics* 506 (1-4): 1–10.

Banerji, R.K. (1984). Post-Eocene biofacies, palaeoenvironments and palaeogeography of the Bengal Basin, India. *Palaeogeography, Palaeoclimatology and Palaeoecology* 45: 49–73.

BAPEX (2007a). *Interpretation Report on the Greater Bakhrabad Structure*. Petrobangla, Dhaka: Data Centre.

BAPEX (2007b). Well completion report on Habiganj well no. 11. Dhaka: Development Geology Department (Unpublished).

Cooke, D.A., and Muryanto, T. (1999). Reservoir Quantification of B Field. In Java Sea via Statistical and Theoretical Methods, Submitted for Presentation at the SEG International Exposition and Meeting, Houston, TX USA. pp. 14–26.

Darling, T. (2005). *Well Logging and Formation Evaluation*. Elsevier.

Evans, P. (1932). Tertiary succession in Assam. *Trans. Min. Geol. Inst. India* 27 (3): 155–260.

Evans, P. (1964). The tectonic framework of Assam. *Journal Geological Society of India* 5: 80–96.

Glover, P. (2011). *Introduction to Petrophysics and Formation Evaluation*. MSc Course Note. Retrieved from http://homepages.see.leeds.ac.uk/~earpwjg/PG_EN/CD%20Contents/GGL66565%20Petrophysics%20English/Chapter%201.PDF.

Hiller, K., and Elahi, M. (1984). Structural development and hydrocarbon entrapment in the Surma basin/Bangladesh (northwest Indo Burman fold belt). In *Southeast Asia Show*. OnePetro.

Holtrop, J.F. and Keizer, J. (1970). Some aspects of the stratigraphy and correlation of the Surma Basin wells, East Pakistan. *ECAFE Mineral Resources Development Series* 36: 143–154.

Imam, B. (2013). *Energy Resources of Bangladesh*, seconde. University Grants Commission of Bangladesh. ISBN: 984-809-021-1.

Imam, M.B. and Shaw, H.F. (1985). The diagenesis of Neogene clastic sediments from the Bengal Basin, Bangladesh. *Journal of Sedimentary Research* 55 (5): 665–671.

Islam, S.M.A., Islam, M.S., and Hossain, M.M. (2014). Investigation of fluid properties and their effect on seismic response, A case study of Fenchuganj Gas Field, Surma Basin, Bangladesh. *International Journal of Oil, Gas and Coal Engineering* 2 (3): 36–54.

Kamruzzaman, (2017). Formation evaluation and reservoir characterization using geophysical data of the Habiganj structure, surma basin, Bangladesh. Dissertation. Jahangirnagar University, Savar, Dhaka. 93.

Khan, M.R. and Muminullah, M. (1980). Stratigraphy of Bangladesh. In: *Petroleum and Mineral Resources of Bangladesh. Seminar and Exhibition*, 35–40. Bangladesh, Dhaka: Ministry of Petroleum and Mineral Resources.

Khattach, D., Houari, M.R., Corchete, V. et al. (2013). Main crustal discontinuities of Morocco derived from gravity data. *Journal of Geodynamics* 68: 37–48.

Misra, A.A. and Mukherjee, S. (ed.) (2018). *Atlas of Structural Geological Interpretation from Seismic Images*. John Wiley & Sons.

Mukherjee, S., Dasgupta, S., Majumdar, C. et al. (2023). Introduction to: Handbook of Petroleum Geoscience: Exploration, Characterization, and Exploitation of Hydrocarbon Reservoirs. In: *Handbook of Petroleum Geoscience: Exploration, Characterization, and Exploitation of Hydrocarbon Reservoirs*, Petroleum Industry Updates: Geoscientific and Economic Issues (ed. S. Mukherjee, S. Dasgupta, C. Majumdar, et al.). John Wiley & Sons, Hoboken, ISBN 978-1-119-68003-1.

Mukherjee, A., Fryar, A.E., and Thomas, W.A. (2009). Geologic, geomorphic and hydrologic framework and evolution of the Bengal basin, India and Bangladesh. *Journal of Asian Earth Sciences* 34 (3): 227–244.

Nettleton, L.L. (1971). *Elementary Gravity and Magnetics for Geologists and Seismologists*, 121. Tulsa, OK, USA: Society of Exploration Geophysicists.

Montaj, O. (2007). *Computer software*. Toronto: Geosoft Incorporated.

Rahman, A., Seddique, A.A., and Khan, M.H.R. (2017). Present gas reserve estimation using wireline logging data of Habiganj Gas Field, Bangladesh. *International Journal of Oil, Gas and Coal Engineering* 5 (7): 44–50.

Rahman, M.A., Blank, H.R., Kleinkope, M.D., and Kucks, R.P. (1990a). *Aeromagnetic Anomaly Map of Bangladesh*. Bangladesh, Dhaka: Geol. Surv.

Rahman, M.A., Mannan, M.A., Blank, H.R. et al. (1990b). *Bouguer Gravity Anomaly Map of Bangladesh*. Bangladesh, Dhaka: Geol. Surv.

Reimann, K.U. (1993). *Geology of Bangladesh*, 160. GebruderBorntraeger, Berlin, Stuttgart, Germany: Beitraegezurregionalen geologic der eipde.

Reynolds, R.L., Rosenbaum, J.G., Hudson, M.R., and Fishman, N.S. (1990). Rock magnetism, the distribution of magnetic minerals in the Earth's crust, and aeromagnetic anomalies. In: *Geological Applications of Modern Aeromagnetic Surveys*, U.S. Geological Survey, Bulletin 1924 (ed. W.F. Hanna), 24–45. United states government printing office.

Rider, M.H. (1986). *The Geological Interpretation of Well Logs*. Glasgow and London: Blackie and Son Limited.

Rider, M.H. (2000). *The Geological Interpretation of Well Logs*, 280. Whitens Publishing.

Rider, M.H. (2002). *The Geological Interpretation of Well Logs*. Second edition. Malta: interpret Ltd.

Roy, W. (1997). The accuracy of well ties: Practical procedures and examples. *SEG Expanded Abstracts* 16: 816.

Serra, O. E. (1983). Fundamentals of well-log interpretation.

Shekaili, F.A., Chacko, S., Din, S.S.E., et al. (2012). Well log conditioning for quantitative seismic interpretation. Abu Dhabi International Petroleum Conference and Exhibition, Abu Dhabi, UAE (11–14 November 2012).

Shikder, M.H. (2005). Study of natural gas processing in Bangladesh. M.Sc. Project. Petroleum and Mineral Resources Engineering Department, BUET, Dhaka.

Techlog (2015). *Computer software*. Montpellier (South of France): Schlumberger.

Telford, W.M., Geldart, L.P., Sheriff, R.E., and Keys, D.A. (1990). *Applied Geophysics*, 2nde. Cambridge.

Torres, C. and Sen, M. (2004). *Integrated Approach for the Petrophysical Interpretation of Post- and Pre-stack 3D Seismic Data, Well Log and Core Data, Geological Information and Production Data via Bayesian Stochastic Inversion; Third Annual Report*. Texas: Institute of Geophysics.

Uddin, A. and Lundberg, N. (1998). Unroofing history of the eastern Himalaya and the Indo-Burman Ranges: Heavy-mineral study of Cenozoic sediments from the Bengal Basin. *Bangladesh. Journal of Sedimentary Research* 68 (3).

Xu, S. and Chacko, S. (2008). well log data conditioning using a rock physics modeling approach: examples from the Banyu Urip Field, East Java Basin. International Petroleum Technology Conference, Kuala Lumpur, Malaysia (3–5 December 2008).

Zeba, J., &Murrellb, R. (2015). Evaluation of Rock Properties from Logs Affected by Deep Invasion–A Case Study.

22

Assessment of Efficacy of "*b*" Value as a Seismic Precursor for Select Major Seismic Events

Deepali Gadkari[1] *and Soumyajit Mukherjee*[2,*]

[1] *Department of Geography, University of Mumbai, Kalina, Santacruz (E), Mumbai, Maharashtra, India*
[2] *Department of Earth Sciences, Indian Institute of Technology Bombay, Mumbai, Maharashtra, India*

**Author for Correspondence: soumyajitm@gmail.com*

Abstract

The "*b*" value is a seismic long-term precursor that indicates the state of crustal stress condition. The magnitude of "*b*" is the slope of the line of best fit when log (to the base 10) of earthquake frequency over a certain time period is plotted against the earthquake magnitude, following the Richter-Gutenberg equation. We demonstrate an inverse relationship between earthquake frequency and magnitude at specific locations. This means that the frequency of earthquakes with low magnitude can be more than that of higher-magnitude earthquakes. The decrease in the frequency of low-magnitude earthquakes, particularly in the seismically active areas, indicates stress accumulation. This is reflected in terms of lower "*b*" values before the major earthquakes. This article assesses the efficacy of temporal variations of *b*-values by applying it to a number of recent major earthquakes in Nepal, Sumatra, Japan, and Chile. The "*b*" values before the earthquakes drop significantly. It is also found that the "*b*" per se varies from region to region, but the temporal pattern of the value invariably shows lower trends before an impending earthquake. If a seismically active region is constantly monitored for the "*b*" value variations, it can reveal the stress regimes.

Keywords *Tectonics; Geodynamics; Deformation mechanism; Tectonic instability*

22.1 Introduction

> "... the problem of earthquake prediction remained one of the most important unsolved problems in geophysics ..."—Zalohar (2018)

Terrains that accumulate strain slowly can be vulnerable to seismicity (Landgraf et al. 2017). Earthquake prediction has always been a fascinating branch of seismology having rather limited success. The US National Academy of Sciences (1976) defines such prediction as

Handbook of Petroleum Geoscience: Exploration, Characterization, and Exploitation of Hydrocarbon Reservoirs, First Edition. Edited by Soumyajit Mukherjee, Swagato Dasgupta, Chandan Majumdar, Subhandip Mandal and Troyee Dasgupta.
© 2023 John Wiley & Sons Ltd. Published 2023 by John Wiley & Sons Ltd.

the process that specifies the expected magnitude range, the geographical area within which it will occur, and the time interval within which it will happen, with sufficient accuracy so that the degree of success of the prediction can be gauged (Iyer 1976). The definition reveals that there are three important aspects of earthquake prediction, viz. location, magnitude, and time. Earthquake prediction, as a scientific discipline, started after the theory of plate tectonics, got well established around the 1960s. Japan, China, the USA, and the erstwhile USSR, undertook several earthquake prediction-related programs.

Seismicity falling within the omega-sequence is stated to be predictive in terms of timing and magnitude of seismicity. Also, the Cosserat model for Båth's law speculates periodic and geometric seismic sequences (Zalohar 2018). Earthquake prediction stands on three important milestones. (i) The concept of "seismic gap" (McCann et al. 1978). (ii) The observation that the ratio between P-velocity to S-velocity (V_P/V_S) drops ~15% and then recovers before medium and large earthquakes. (iii) Successful prediction of the Haicheng earthquake (4 February, 1975, $M = 7.3$).

Earthquake prediction depends on other factors such as availability and authenticity of seismic data/seismic records, velocities of seismic waves, electrical resistivity, and chemical composition of water in the source region, changes in atmospheric parameters, crustal deformations, etc. All these parameters can be used for earthquake prediction and therefore can be termed seismic precursors.

The present work uses the data published online by the National Earthquake Information Centre (NEIC) and United States Geological Survey (USGS). This data is not declustered (i.e. the foreshocks and aftershocks are not removed from the records). The "*b*" value depends on the frequency of earthquakes in a particular region over a unit time period. It is observed (Gutenberg and Richter 1944) that the frequency of earthquakes decreases significantly before a major earthquake occurs. The time period in which this decrease will take place varies from place to place. While analyzing the records for the "*b*" value, it is found that this period ranges widely from six months to five years. The frequency of earthquakes again starts increasing just before the main earthquake in the form of foreshocks. This period may vary from one year to a few hours. It is also noted that the magnitude of completeness (Woessner and Wiemer 2005) of the NEIC catalog is for $M > 4.0$. The catalog is not complete for $M < 4.0$. For some years, sufficient data has not been available and so "*b*" values could not be computed for those time periods (Japan Region: 1991–2000; Chilean Region: 1986–1992).

USGS catalog considers all earthquake magnitudes as *mb* (body wave magnitude), but the source magnitude is also mentioned for many earthquakes in the catalog (Mueller et al. 1997; Petersen et al. 2008).

The "*b*" values are computed by the least squares method. The slope of the line relating earthquake magnitude and the logarithm of the number of earthquakes (log frequency in the unit time period) of that magnitude is the "*b*" value for that particular time period.

$$\text{Log}\, N = a - b\, M \tag{22.1}$$

where *N*, number of earthquakes; *M*, magnitude (on Richter scale); constant *a*, intercept; *b*, slope of the line.

Rather than the actual value of *b*, the general trend of the *b*-value over the specified time period is important. This is because the actual values of "*b*" for a region depend on the

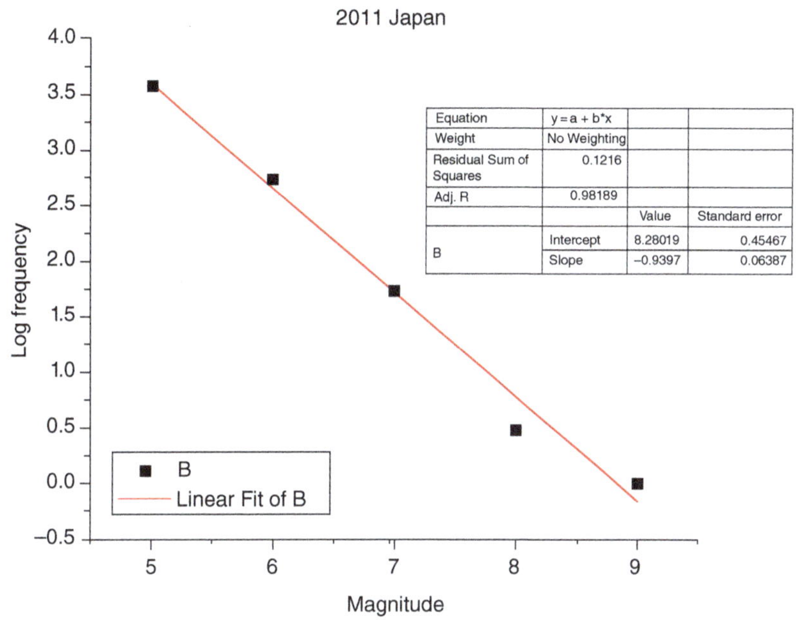

Figure 22.1 Computation of "*b*" values. Compare with Equation (2.1).

seismic characteristics of the areas. The frequency of earthquakes in each magnitude class (e.g. *M* 4.0–5.0) in each unit time period (one or five years) is the main input required to compute the "*b*" values (Figure 22.1). As this frequency differs from terrain to terrain, the "*b*" values of two different areas with different seismotectonics are not comparable. Computing a single "*b*" value for a region may not quantify how stress accumulates. The stress conditions vary with respect to the major impending earthquakes (Liu and Zhu 2010). Stress accumulates before major earthquakes over any region. This being a long-term process needs to be assessed over a considerable time span (>25 years). Therefore, the "*b*" values should be computed for consecutive unit time periods and then the trend be observed.

The "*b*" value remains to be one of the most important inferential statistical seismic precursors that indicate stress conditions in the area under consideration. Gutenberg and Richter (1944) demonstrated the relation between the earthquake frequency and its magnitude. Earthquakes though presumably momentary, are actually a process of slow accumulation and quick release of (crustal) stresses. An earthquake of a certain magnitude occurs when a region reaches the threshold beyond which crustal stress cannot be accumulated. A major earthquake demands a considerable accumulation of stress, which can imply an overall decrease in the frequency of earthquakes in the region. This can be reflected by lowering "*b*" values when observed over a particular time period.

The sensitivity of the *b*-valuc as a seismic precursor has been used by a number of researchers (e.g. Gibowicz 1973; Guha 1983; Patil et al. 1986; Utsu 1965). The 1967 Caracas earthquake occurred when the *b*-value crossed a maximum and when the standard deviation crossed a minimum (Fiedler 1974). Smith (1981) noted that before specific seismicity, the *b*-value elevates and then falls. Likewise, Huang and Turcotte (1988) referred seismicity to take place when the *b*-value crossed a threshold magnitude from several terrains. Post-earthquake studies confirmed a good correlation between a lowering trend in *b*-value and a major seismic event in the area. The examples include seven New Zealand earthquake sequences studied by Gibowicz (1973) in which *b*-values reduce initially before the main earthquake, increase after the main shock, and decrease again until the largest aftershock. He also noted during the earthquake swarm that the coefficient *b* decreases logarithmically with time. Profound lowering/variation (from 0.55 to 2.40) of *b*-value before seismicity, such as that reported from the Andaman-Sumatra region (Nuannin et al. 2005a, 2012), also remains valid for artificially induced seismicity (Nuannin et al. 2005b). Such lowering is conspicuous over a time scale of months or years, and a further time resolution is not possible (El Nader et al. 2016). Note that the *b*-value anomaly can also be caused by natural reasons other than seismicity, viz. significant heterogeneity of rocks, weak stress field, elevated thermal gradient, and reduced effective stress. The *b*-value can vary depth-wise as well in a single terrain (review in El Nader et al. 2016). Besides its temporal variation, a spatial disparity of *b*-values has been used to predict seismicity (Sobiesiak et al. 2007).

Ayyub and McCuen (2011) referred to a statistical cumulative transformation function to describe earthquake magnitudes (also see Zuniga and Wiemer 1999). A similar approach was found to be fruitful by Zuniga and Wyss (1995).

The present work assesses the efficacy of temporal variation of the *b*-value as a seismic precursor for four major seismic events from: (i) Nepal Himalaya; (ii) Sumatra region; (iii) Japan trench; and (iv) Chilean Coast (Figure 22.2).

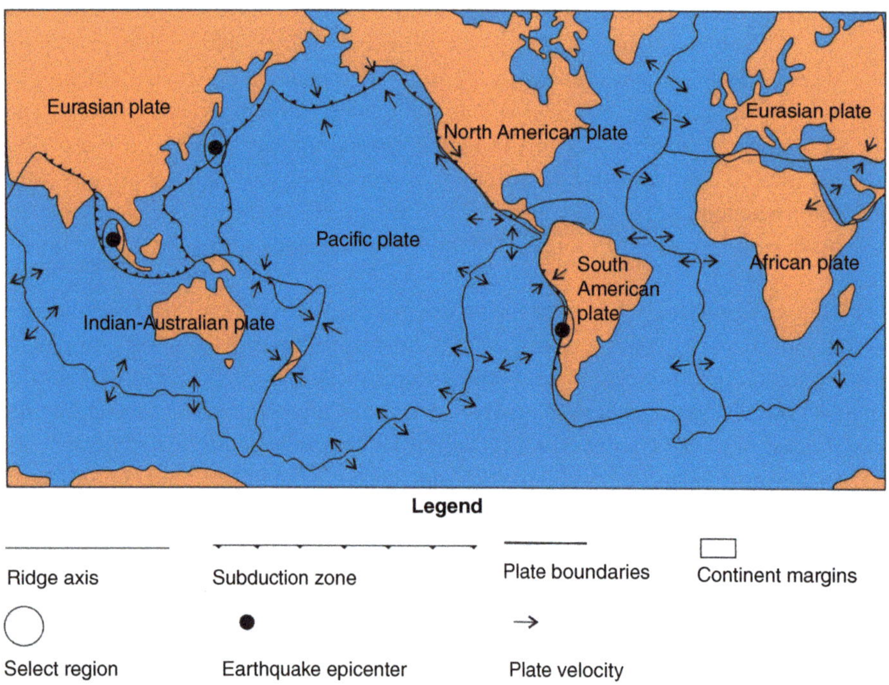

Figure 22.2 Locations of select earthquakes: (1) Nepal Himalayas, (2) Sumatra region, (3) Japan trench, and (4) Chilean Coast.

22.2 Tectonics of the Select Regions

22.2.1 Nepal Himalayas

Three major earthquakes in the Himalayas took place in the geological record: the 1905 Kangra (Mw 7.8), the 1934 Bihar-Nepal (Mw 8.2), and the 1950 Assam (Mw 8.6) earthquakes. Seismicity in the last ~100 years in the Himalayas has been well studied including in the Nepal region (review in Rajendran et al. 2017). The arc-parallel major faults, such as progressively northward the Main Frontal Thrust, the Main Boundary Thrust, and the Main Central Thrust are the active tectonic regions in the Himalayas. Whether an earthquake of magnitude >9 can strike the Himalayas has been recently debated geophysically (review in Wyss 2014). Out-of-sequence/active deformation zones pass through the Nepal Himalayas (review in Mukherjee 2015). Therefore, the Nepal Himalayas is considered seismogenic in general.

India-Eurasia collision-induced crustal shortening of the Greater-, Lesser-, and Siwalik regions of the far eastern Nepal Himalayas so far has been ~185–245 km (Schelling and Arita 1991). This estimate of shortening is, however, at least ~3 times less than what was proposed by DeCelles et al. (2001). The Siwalik equivalent Churia Zone is the most tectonically active part of the Nepalese Himalayas (Upreti 1999). However, seismicity in Nepal is also well-spread in the Lesser and the Greater Himalayas. Noting from a different viewpoint, seismicity in the Nepal Himalayas has been compiled in terms of its informally defined segments, viz. western, central, and eastern Himalayas. Singh et al. (2010) analyzed seismicity in the western Nepal Himalayas and noted that quiescence followed by seismicity is not ubiquitous in this region. They also predicted medium-sized earthquakes in this Himalayan segment from 2011 onward. Further, microseismicity has been documented from the Greater Himalayas, with ramp and flat geometry of major thrusts at the subsurface, since 1985 (Pandey et al. 1995). Pandey et al. (1999) pointed out that both the far eastern and the far western parts of the Nepal Himalayas are more seismic. Bollinger et al. (2007) reported that the small- and moderate-sized earthquakes in Nepal, in general, during 1995–2000 took place more in winter than during the summer. They postulated that the water contributed by the rainy seasons before the winter seasons augmented the Coulomb failure.

22.2.2 Sumatra Region

A rather up-to-date review of tectonics of the Sundaland is available in Metcalfe (2011). A sinistral slip separated the Australian and the Bird's Head along with the convergence in the New Guinea fold-and-thrust belt (McCaffrey 1989). The Bird's Head lithosphere is subducted below the Seram trough at ~300 km depth. Overall the Bird's head has been moving westward (Das 2004). Tectonic inversion in Sumatra that started in Late Miocene and has been continuing today is linked with a strike-slip deformation (Daly et al. 1991). The Sumatran Fault is a megascopic dextral strike-slip fault. Prawirodirdjo et al. (2000) deduced a 20-km depth of this fault at a specific location. Lasitha et al. (2006) reported that from 1900 to 2000, the slip rate for the Sumatra-Java arc varied unusually from 1 to 29 mm year^{-1}, and that the slip rate of the Sumatran Fault Zone has been 30–50 mm year^{-1}.

Regions nearby to this fault have recently been affected by the aftershocks of the 2004 Sumatra-Andaman quake (Mw 9; Engdahl et al. 2007). Sukmono et al. (1997) established a link between earthquake recurrence related to the Sumatran Fault and a fault segment geometric parameter. The relationship between the Sumatran Fault and the subduction zone has been debated (Lange et al. 2010) and still requires more study to comment on the relation between this fault and the nearby seismicity (McCaffrey 2009).

22.2.3 Japan Trench

The Japanese island represents an active subduction-related orogenic front (Isozaki et al. 2010). These islands were formed by the accretion of Permian, Jurassic, and Cretaceous rocks. As the Japan Sea opened 20–15 Ma ago, the islands attained the present position (review in Shikazono 2012). The Japan trench margin has been described as a ...*homogeneous wedge of deformable, noncohesive Coulomb material during seamount subduction* (Lallemand and Le Pichon 1987). The Pacific Plate has been subducting below the Japanese islands along the Japan Trench with variable intraplate coupling and producing mainly thrust-type earthquakes (Miura et al. 2003 and references therein; also see Mogi 1990). Seismicity (>M 7.5) commonly displays an uneven epicenter distribution along this structurally varying trench with an intraplate sedimentary unit governing the release of shear stress (Tsuru et al. 2002). Specifically, the Nankai trough has been a place of vigorous seismicity with a recurrence interval of 100–200 years (Kodaira et al. 2002). The Philippine Sea Slab subducts with 10–40° dip along the Nankai trough (Xu and Kono 2002). Subduction followed by erosion of the seamounts along the Nankai trough might hinder segmentation of the seismogenic rupturing (Bangs et al. 2006), or even acted as a barrier to earthquakes (Kodaira et al. 2002). The mode of earthquakes varies along the Japan trench due to the differing roughness of the subducting plate (Tanioka 1997).

22.2.4 Chilean Region

The Chilean coast usually shows along-strike variation in tectonics (Gerbault et al. 2009, and its references). The maximum principal compressive stress axis (σ_1) in central Chile trends NW and E (Barrientos et al. 2004). Seismicity studies so far, however, indicated that the Chilean coast is basically a two-layered Benioff zone (Malgrange et al. 1981). Tassara and Echaurren (2012) referred to ~ 66 mm year^{-1} of convergence rate at the Andean Subduction Zone, which is the longest subduction zone in the world.

Stauder (1973) categorized and analyzed seismicity in the Chilean region but that did not include the b-values. The southern Chilean region has been divided into several seismic areas, but here again, the b-values have not yet been analyzed (Perucca and Bastias 2008). Along with other subsurface geophysical details, note the intermediate-depth seismicity does not belong to the Andean Subduction Zone (The ANCORP Working Group 1999). Principal stress axes orientation has been deduced from 20 to 50 km-deep locked zone in the northern Chile portion of the Andean Subduction Zone. Interestingly, this depth range is the zone where the hypocenters occur (Delouis et al. 1996). Lange et al. (2007) linked the spatial distribution of seismicity in the southern portion of the Chilean region fundamentally with the age and temperature of the oceanic plate.

The Chilean coast parallels the Andes Mountains chain. This chain varies in dip, age, and obliquity of the subducting slab. The Andean Subduction zone and the Chilean margin are characterized by a ramp and flat structure at depth (Farias et al. 2010). NW-striking steeply dipping reverse faulting has been deciphered based on fault plane solutions from the Andes Mountains (Suarez et al. 1983). Gerbault et al.'s (2009) numerical modeling demonstrated that the strength of the continental crust and that of the subduction channel governs the crustal shortening in this mountain chain (Gerbault et al. 2009).

Deformation in the central and southern Andes (22 °S–42 °S) is characterized by a seismicity cycle and an interseismic phase (Klotz et al. 2001). A nearly aseismic region exists in the Andes between 36 °S and 40 °S latitudes in between the ~NW trending Bio-Bio and the Gastre Fault zones – a zone of very high uplift by antiformal stacking (Bohm et al. 2002 and its references).

22.2.5 Data and Methods

Computation of "*b*" values requires considering two parameters of seismicity: (i) time interval, and (ii) magnitude interval/classes. The trend of "*b*" values over a time span is assessed with respect to the major seismic events in the region. Therefore, the optimum time interval for *b*-values computation is to be selected. This interval would depend on the seismotectonics of the areas. If frequently recurred and well documented, the time interval should be taken as <1 year. But in case of unavailability of seismic records, particularly those of the lower magnitudes, the interval can range 2–5 years.

A large number of earthquake records are required to link *b*-values with seismicity. Figure 22.3 represents the methodology followed for the "*b*" value computation and how

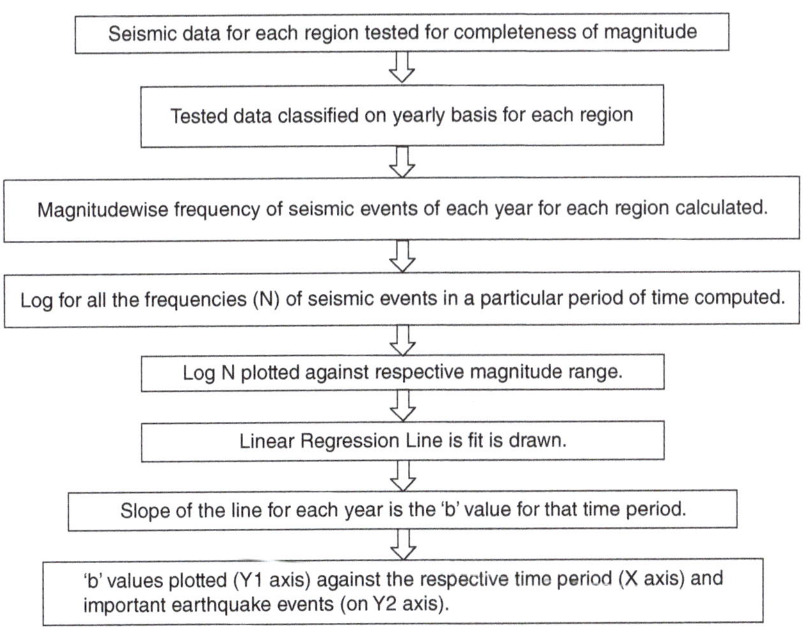

Figure 22.3 Methodology of computation of temporal "*b*" values in this work.

to correlate the same with major seismic events in the specific region. Accurate knowledge of magnitude of completeness (M_c) is essential for many seismicity-based studies, and particularly for mapping seismicity parameters (Woessner and Wiemer 2005). The M_c is defined as the lowest magnitude at which all the events in a space–time volume are detected (Rydelek and Sacks 1989; Taylor et al. 1990; Wiemer and Wyss 2000). After checking M_c, respective threshold magnitudes are decided and seismic data above a particular magnitude are analyzed.

At the same time, there should be a clear understanding of whether to include or exclude foreshocks and aftershocks into the *b*-value analyses. The number of aftershocks and the area affecting the aftershock are related to the main shock magnitude (review and original work by Christophersen and Smith 2000). The foreshocks and aftershocks of seismicity are clubbed together as the accessory shocks. No universally accepted physical basis exists at present to accurately separate the accessory shocks from the main shock (Musson 2000).

The "*b*" values are computed by the least squares method. The USGS earthquake catalog is used for this purpose. The earthquake records provide data of earthquake epicenters for location, date, magnitude, depth, region name, and others. The first three parameters are used to compute the "*b*" values. The catalog is found to be complete for seismicity with $M >$ 3.0. For some years, especially in the Chilean and Japanese regions (1986, 1989, 1991, 2009 for the Chilean region and 1991, 1998 Japanese region), the catalog is not complete and "*b*" values could not be computed. Such problems of incomplete catalog, in general, have also been pointed out by previous researchers (e.g. Zuniga and Wyss 1995).

22.2.6 Analyses and Results

The chosen four seismo-tectonically active regions recently experienced large earthquakes ($M > 8.0$). Seismic data for each region is analyzed independently for computation of "*b*" values and then correlated with major seismic events in the respective regions. The select extent of each region represents 10° × 10° longitudinal and latitudinal areas around the recent major earthquakes.

1) Nepal Region: (Extent: 23 °N to 32 °N and 79 °E to 89 °E; location of the recent major earthquake: 28 °9′N × 84 °42′E)

   Nepal experienced 20 earthquakes of magnitude >6.00 from 1965 to 2015 (Figure 22.4). Earthquake records from 1976 to 2015 are considered for "*b*" value analysis considering the completeness of the earthquake catalog (USGS) for the region. The pattern of "*b*" values (Figure 22.5) shows that there are two conspicuous drops during 1986–1990 and 2011–2015. Both these drops are followed by the earthquakes of $M = 6.9$ in 1988 and $M = 7.8$ in 2015. Note that after the year 2000, the downward trend in "*b*" values is more prominent and there is no earthquake with $M \geq 7.0$. This indicates that crustal stresses accumulated, which were released in 2015 in the form of the Gorkha earthquake ($M = 7.8$).
2) Indonesian Region: (2 °S to 8 °N and 91 °E to 100 °E; location of the recent major earthquake: 3 °N × 96 °E).

   The Indonesian region has not experienced earthquakes of $M \geq 8.0$ from 1966 to 2004 (Figure 22.6). This clearly indicates the phase of accumulation of crustal stresses.

   An inverse relation between "*b*" value and seismic events is revealed for the Indonesian region as well (Figure 22.7). The region experienced an earthquake of $M = 7.7$ in

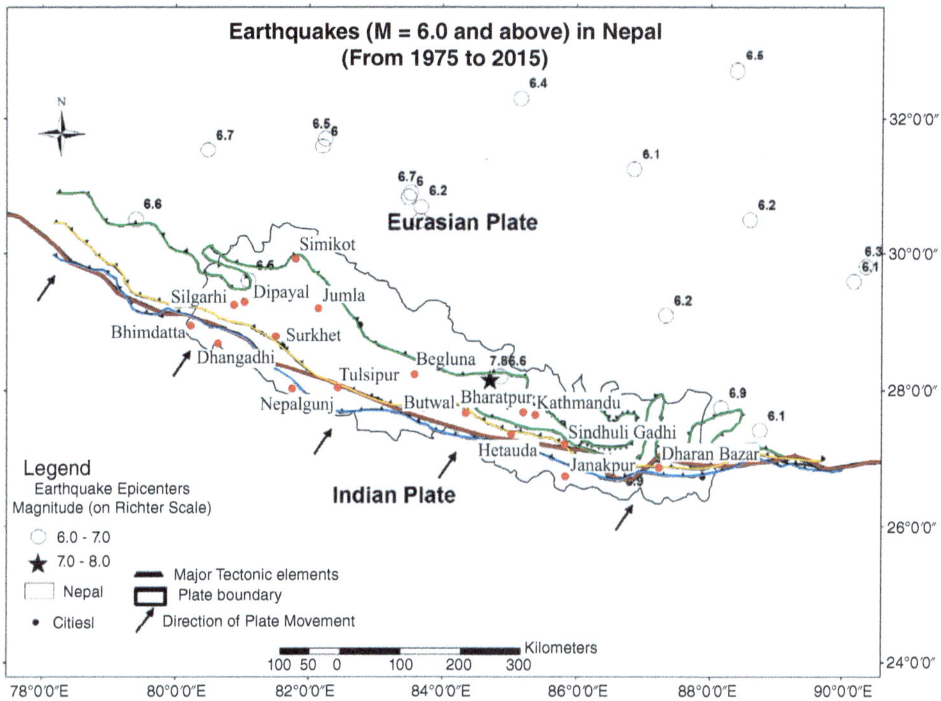

Figure 22.4 Major seismic events in Nepal from 1975 to 2005.

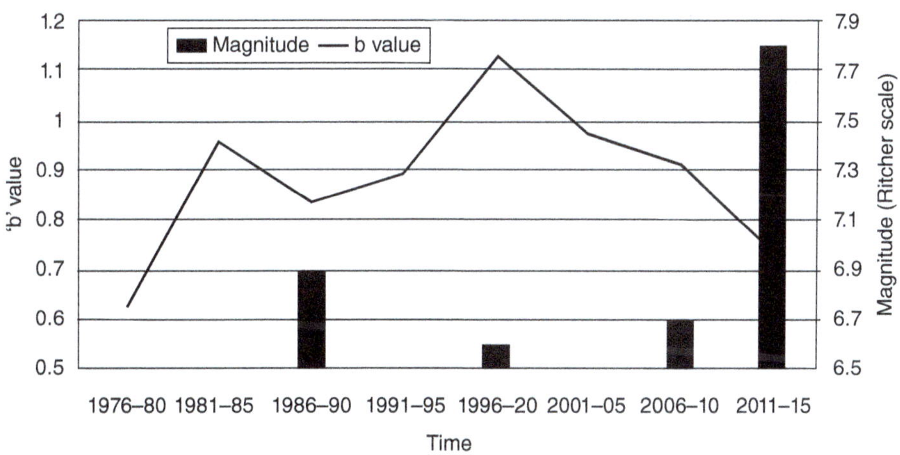

Figure 22.5 Temporal variations (1976 to 2015) in "*b*" values: Nepal Region. *Source:* Data source: USGS Earthquake catalog.

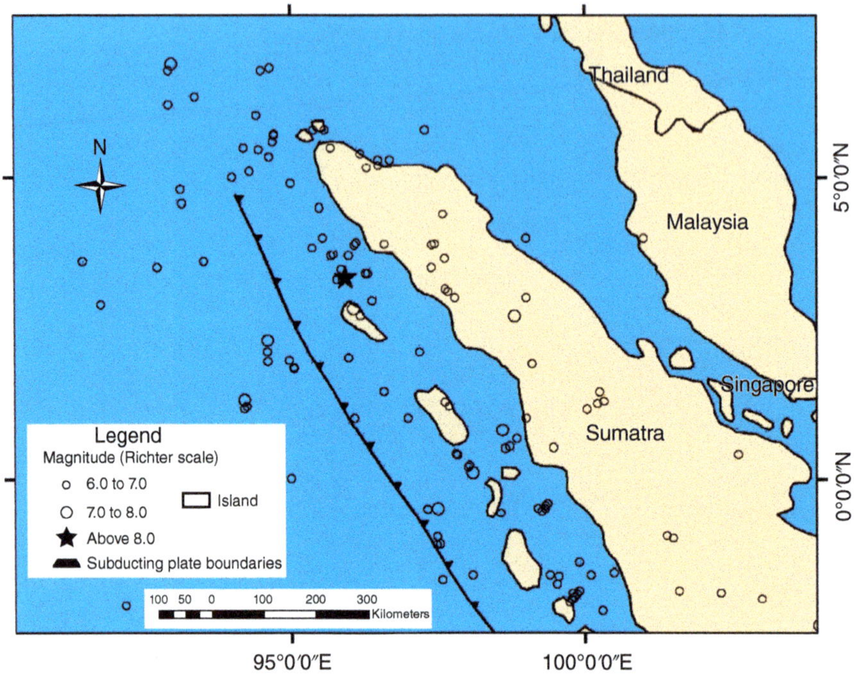

Figure 22.6 Major seismic events in the Indonesian region from 1966 to 2014.

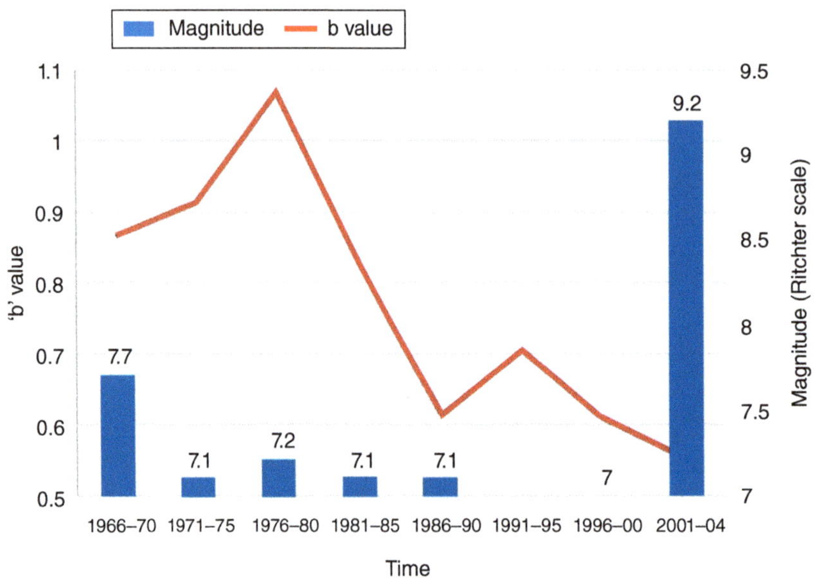

Figure 22.7 Temporal variations (1966–2004) in "*b*" values in the Indonesian region. *Source:* Data source: USGS Earthquake catalog.

1969. Thereafter, there has not been an earthquake of $M > 7.5$ till 2002, except for one event of $M = 7.6$. Since 1981, a falling trend of "b" value is observed, indicating that the region has been accumulating stress as the frequency of earthquakes is decreasing in the region. The lowest "b" value is observed during 2001–2005, the period in which the major earthquake ($M = 9.2$, dated 26 December 2004) struck.

3) Japanese Region: (extent: 131 °E to 144 °E longitude and 29 °N to 45 °N latitude, location of the recent major earthquake: 38.4 °N × 142.4 °E).

Figure 22.8 presents the earthquakes of $M > 7.0$ (from 1973 to 2013) in the Japanese region. The earthquakes are concentrated more in the region off the eastern coast of

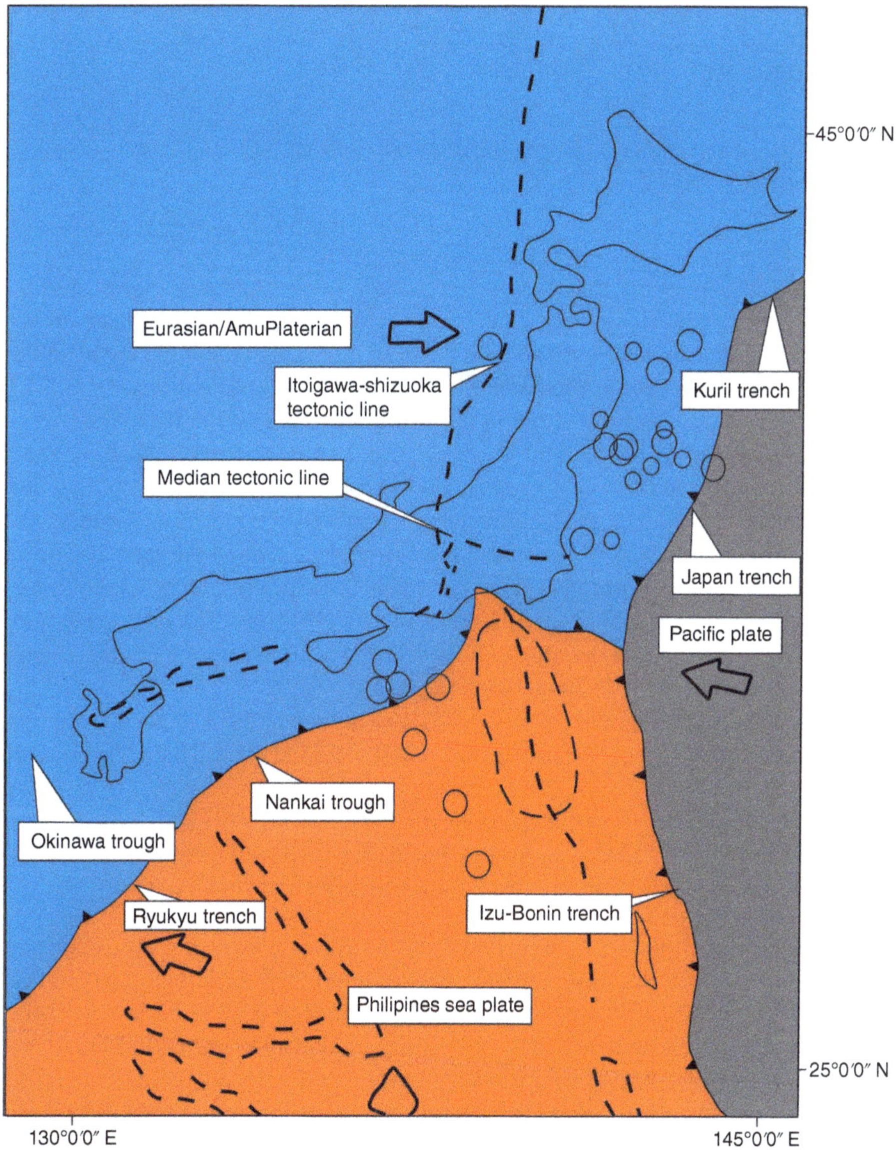

Figure 22.8 Major seismic events in the Japanese region from 1973 to 2013.

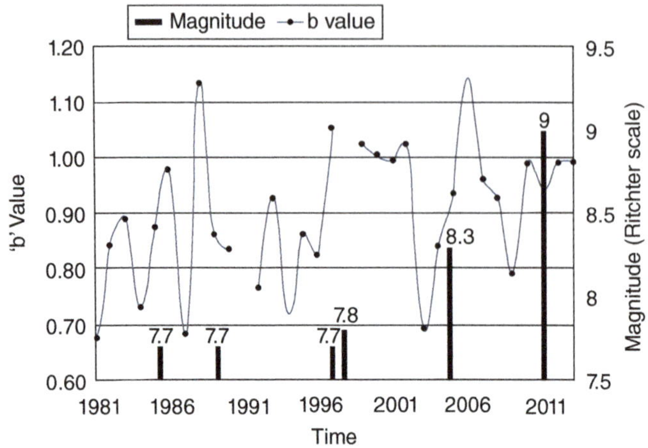

Figure 22.9 Temporal variations (1973–2013) in "*b*" values in the Japanese region. *Source:* Data source: USGS Earthquake catalog.

Japan. Figure 22.9 presents the trend of "*b*" values in the Japanese region during 1973–2013. The Y1-axis shows variations in "*b*" values and Y2-axis the major earthquakes ($M > 7.5$) The "*b*" values range from 0.6 to 1.1. It is noted that drop in "*b*" value from ~1 to <0.8 is followed by a major earthquake in the region (Figure 22.6).

Lower "*b*" values are observed either in the year prior to the year of the major earthquake or in the same year, wherein the major earthquake has occurred. There have been two large earthquakes in the region in this period: the first with $M = 8.3$ in 2003 and the second with $M = 9.0$ in 2011. Both earthquakes are preceded by lower "*b*" values. An increase in "*b*" value is also marked after the earthquakes of $M > 7.0$ have occurred in a group of three/four consecutive years identified. Some groups of years with major earthquake events can be prominently identified – (i) 1981–1984, (ii) 2003–2005, and (iii) 2008–2012, except for the year 2010 – no event of $M > 7.0$, seems to be compensated by a large earthquake of $M = 9.0$ on 11 March 2011.

4) Part of Chilean Coast (Extent: 65 °W to 76 °W longitudes and 25 °S to 36 °S latitudes; location of the recent major earthquake: 31°.57 S × 71°.65 W, M = 8.0, date: 16 September 2015.

   The Chilean region shows earthquakes of $M < 8.0$ in the select region from 1981 to 2016 (Figure 22.10). As in other regions, the very absence of a major earthquake in such a seismically active region connotes stress accumulation.

Figure 22.11 shows the trend of "*b*" values and major earthquakes in the part of Chilean coast. Since 1981, this area has experienced three earthquakes of magnitude >8.0, in the years 1985, 2010, and 2015, which are always preceded by a drop in "*b*" values. Note there have been gaps in acquiring sufficient data for the computation of "*b*" values in 1986, 1989, 1991, and 2009. Notwithstanding the downward trend of "*b*" values evident since 1996, there were also occasional mild rises in 2005, 2012, and 2015. From 1981 to 1985, there have been two earthquakes of magnitude 7.2 in 1981 and 1983. This period must be of

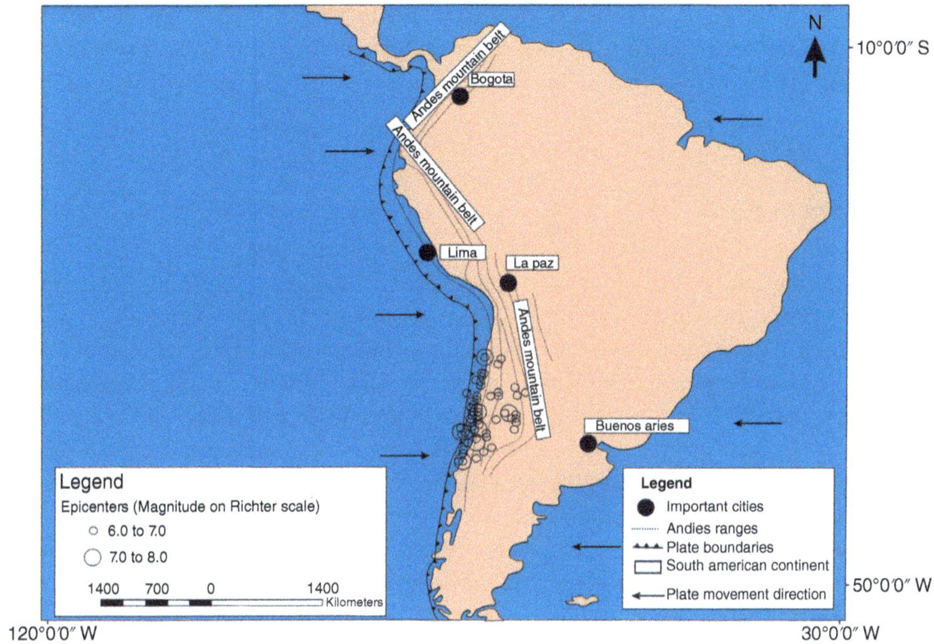

Figure 22.10 Major seismic events in the Chilean Region from 1981 to 2016.

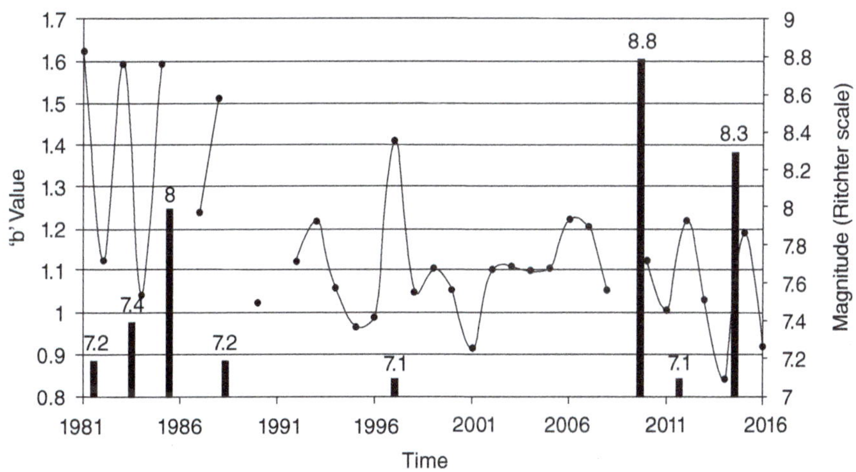

Figure 22.11 Temporal variations (1981 to May 2016) in "*b*" values: Chilean 772 region. *Source:* Data source: USGS Earthquake catalog.

accumulating stresses as there are only two of these events and that too $M < 7.5$. The major earthquake of 1985 – $M = 8.0$ (on 3 March 1985) can be seen as the result of accumulated stresses. It would have been far more interesting if the seismic records before the year 1981 were available. Since the 1985 event, there have been only two earthquakes of $M = 7.2$

(5 February 1988) and $M = 7.1$ (15 December 1997) till the next major earthquake ($M = 8.8$; 27 February 2010) occurred. Note that all the major events ($M = 8.0$, 1985; $M = 8.8$, 2010; and $M = 8.3$, 2015) are preceded by lower "*b*" values.

22.3 Discussions

The data gaps and the dearth of lower magnitude ($M < 3.0$) data are the main limitations. So, the rationale of this work can be applied effectively solely to higher magnitude ($M > 7.5$) events. If lower-magnitude data were available, the technique would apply for medium- and low-magnitude earthquake events as well. Due to the same limitation, it was not possible to compute the annual "*b*" values from the Indonesian and the Nepalese regions. The annual variations in "*b*" values might be able to predict seismicity. The time period for which the complete data is available also varies from region to region. The uniformity of the time period for data analysis could not be maintained perfectly.

In this work, only one section from each of the four seismically active zones has been considered to analyze the "*b*" values. One may also go for computing "*b*" values for the entire seismically active zones, section by section. This can give a better idea of the stress conditions in the regions and temporally rhythmic patterns in stress conditions if any. Along with the "*b*" value, other precursors such as seismic gap, V_p/V_s ratio, the electrical conductivity of groundwater, ground tilt, etc., demand attention as the future work.

Lowering of *b*-value connotes high-stress regime and vice versa. The fall of *b*-value indicates greater stress and subsequent seismicity (Andaman-Sumatra region: Nuannin 2006).

What we conclude from the chosen terrain may not hold true in other places. For example, Jafari (2008) compiled *b*-values from Iran but did not find any clear-cut relation between seismicity and *b*-value variations. Seo et al. (2010)-compiled *b*-values in the Korean peninsula do not show any conclusive pattern between *b*-values, which has an average magnitude of 0.96, and seismicity. Amorese et al. (2010) pointed out that depthwise *b*-value variation may not yield unequivocal conclusions. More specifically, the *b*-value can reduce linearly with depth in the continental crust, and can be dependent on the age of rocks in subduction zones (review in Scholz 2015; also see Enescu et al. 2011). Also, the *b*-value falls linearly as the differential stress increases, and vice versa (Scholz 2015).

Okal and Romanowicz (1994) reported an increase in *b*-value with seismicity size. A laboratory study revealed that the subduction zones are characterized by a linear relationship between the *b*-values and the slab pull force (Scholz 2015). A low *b*-value at ~30 km depth was linked with a specific ongoing tectonic deformation (Khan et al. 2011). Cao and Gao (2002) correlated the fall in *b*-value in the Japan island arc region from 0.86 during 1984–1990 to 0.73 during 1991–1995 to not just the seismicity but also the subduction rate.

Enescu et al. (2011) pointed out that *b*-value reduced before the 1995 Kobe seismicity (Japan). A similar report of *b*-value fall during a major quake has been made also from the Pollino range (southern Italy, Passarelli et al. 2015). Patwardhan (2012) reached the same conclusion in general from several terrains. However, recently, Nanjo and Yoshida (2017) pointed out an increase in *b*-value by an amount of 0.3–0.5 preceding the *M*7.3 earthquake in Kumamoto (16 April 2016; Japan). The *b*-value may decrease (Imoto 1991) or even increase before seismicity, and therefore cannot be used reliably in predicting seismicity (review in Burroughs and Tebbens 2002). Shi and Bolt (1982) through statistical analyses inferred that the *b*-value cannot be used to predict seismicity of magnitude >5 from central California. Henderson et al. (1994) deduced a negative correlation between *b*-value and fractal dimension of epicenter locations in Brazil. Yu (2016) recently referred that the "*b*-value ratio" would be a better predictor of seismicity than merely the *b*-value. Srivardhan and Srinu (2014) discussed the usefulness of predicting the aftershocks using the *b*-values. The *b*-values can fluctuate in a stressed region during crack propagation (Main et al. 1989).

The global average *b*-value is known to be ~1 (1.02 ± 0.03 to be more precise: El-Esa and Eaton 2014). The *b*-values < than the global average indicate the total energy released by earthquakes is < than the energy gathered. El-Esa and Eaton (2014) reviewed a global variation of *b*-value to be $0.3 \leq b \leq 2.5$. They concluded through their global review that *b*-value varies due to changes in effective stress. Less sampling, errors in magnitudes, and inhomogeneous capability for detection are the loopholes in predicting seismicity using *b*-values (review in Kamer 2014).

Even though the *b*-value has been researched for earthquake prediction (Manna et al. 2015), its physical significance has remained indeterminate (Cheng and Sun 2018).

22.4 Conclusions

In this article, the trend in "*b*" values and major seismic events are correlated with each other for sections of four seismically active zones in the world. After computing "*b*" values for >25 years of data for each of these regions, it is found that the range of "*b*" values varies from region to region, but a similar pattern in the case of a trend of "*b*" values and the major events is observed in all the four regions studied. The result can be terrain-specific since a different relationship has also been found in other places of the world.

In all the four chosen regions, the major seismic events are categorically preceded by lows in "*b*" values (Table 22.1). This in fact implies that the trend in "*b*" values in the respective region is more important than the absolute values of "*b*". Though the value of "*b*" per se cannot be correlated to a specific range of earthquake magnitude, the trend in "*b*" values

Table 22.1 The "*b*" value ranges, specific "*b*" values, and major events for the select regions.

Region	Highest "*b*" value	Lowest "*b*" value	Specific "*b*" value before the major event (year)	Magnitude of major earthquake on Richter scale (year)
Nepal	1.12	0.65	0.75 (2011–2015)	7.8 (2015)
Indonesia	1.05	0.55	0.55 (2001–2004)	9.2 (2004)
Japan	1.15	0.68	0.95 (2010)	9.0 (2011)
Chile coast	1.65	0.85	0.84 (2014)	8.3 (2015)

shows a correlation with the earthquake magnitude. The value of "*b*" itself is not indicative of the stress conditions and varies spatially and temporally, but the trend does indicate the stress conditions in the respective region and, thus, an impending earthquake.

It is also found that immediately before a major event, there is a slight increase in "*b*" values (Japan region – before the earthquake of 2005 [$M = 8.3$], the "*b*" value increased from 0.7 to 0.85 and in Chilean region from 0.85 to 9 before the earthquake of 2015 [$M = 8.3$]) in "*b*" value. This indicates increased frequency of small earthquakes in the region, just before the major event. Considering the results of the analysis, it is concluded that "*b*" value can be applied as an indicator that will give an idea about the relative stress conditions in a region. If "*b*" decreases in a region than the previous time periods, then it should be taken as an indication of an impending earthquake.

Reviewing the past trend of research, Sykes et al. (1999) expected that long-term (10–30 years) and intermediate-term (1 month to 10 years) seismicity might be possible to predict in tectonically very active areas. Interestingly, the study areas we have chosen in this article are tectonically active as per Sykes et al.'s (1999) requirement.

Acknowledgments

Late Dr. S. K. Guha (Central Water and Power Research Station, Khadakwasla, Pune) inspired Deepali Gadkari to undertake this work. His guidance and experience in the field of earthquake science made this work possible. Support by Prof. S. R. Jog (Department of Geography, S. P. University of Pune) is also thanked. The administrative help and support extended by the Head, Department of Geography, University of Mumbai, Mumbai is acknowledged. Merryl Le Roux, Frank Otmar Weinreich, Joss Everett, Andrew Harrison, and the proofreading team (Wiley Blackwell) are thanked for their assistance. CPDA grant (IIT Bombay) supported SM. This chapter has been encapsulated in Mukherjee et al. (2023).

References

Amorese, D., Grasso, J.-R., and Rydelek, P.A. (2010). On varying b-values with depth: results from computer-intensive tests for southern California. *Geophys J Int* 180: 347–360.

Ayyub, B.M. and RH, M.C. (2011). *Probablility, Statistics, and Reliability for Engineers and Scientists*, 3e, 105. Boca Raton, FL: CRC Press. ISBN: ISBN −13: 978–1–4398-9533-7.

Bangs, N.L.B., Gulick, S.P.S., and Shipley, T.H. (2006). Seamount subduction erosion in the Nankai trough and its potential impact on the seismogenic zone. *Geology* 34: 701–704.

Barrientos, S., Vera, E., Alvarado, P., and Monfret, T. (2004). Crustal seismicity in Central Chile. *J South Am Earth Sciences* 16: 759–768.

Bohm, M., Luth, S., Echtler, H. et al. (2002). The southern Andes between 36° and 40°S latitude: seismicity and average seismic velocities. *Tectonophysics* 356: 275–289.

Bollinger, L., Perrier, F., Avouac, J.-P. et al. (2007). Seasonal modulation of seismicity in the Himalaya of Nepal. *Geophys Res Lett* 34: L08304.

Burroughs, S.M. and Tebbens, S.F. (2002). The upper-truncated power law applied to earthquake cumulative frequency–magnitude distributions: evidence for a time-independent scaling parameter. *Bull. Seismol. Soc. Am.* 92: 2983–2993.

Cao, A. and Gao, S.G. (2002). Temporal variation of seismic *b*-values beneath northeastern Japan island arc. *Geophys Res Lett* 29: https://doi.org/10.1029/2001GL013775.

Cheng, Q. and Sun, H. (2018). Variation of singularity of earthquake-size distribution with respect to tectonic regime. *Geosci. Front.* 9: 453–458.

Christophersen A, Smith GC. 2000. A global model for aftershock behavior. *12WCEE conference* 0379.

Daly, M.C., Cooper, M.A., Wilson, I. et al. (1991). Cenozoic plate tectonics and basin evolution in Indonesia. *Mar. Pet. Geol* 8: 2–21.

Das, S. (2004). Seismicity gaps and the shape of the seismic zone in the Banda Sea region from relocated hypocenters. *J Geophys Res* 109: B 1203.

DeCelles, P., Robinson, D.M., Quade, J. et al. (2001). Stratigraphy, structure, and tectonic evolution of the Himalayan fold-thrust belt in western Nepal. *Tectonics* 20: 487–509.

Delouis, B., Cisternas, A., Dorbath, L. et al. (1996). The Andean subduction zone between 22 and 25°S (northern Chile): precise geometry and state of stress. *Tectonophysics* 259: 81–100.

El-Esa, Z.H. and Eaton, D.W. (2014). Spatiotemporal variations in the b-value of earthquake magnitude–frequency distributions: classification and causes. *Tectonophysics* 615-616: 1–11.

El-Nader, I.F.A., Shater, A., and Hussein, H.M. (2016). Mapping b-values beneath Abu Dabbab from June to august 2004 earthquake swarm. *NRIAG J.Astron.Geophys* 5: 403–412.

Enescu, B., Enescu, D., and Eto, K. (2011). Values of b and p: their variations and relation to physical processes for earthquakes in Japan and Romania. *Rom.J.Phys* 56: 590–608.

Engdahl, E.R., Villasenor, A., DeShon, H.R., and Thurber, C.H. (2007). Teleseismic relocation and assessment of seismicity (1918–2005) in the region of the 2004 *M*w 9.0 Sumatra–Andaman and 2005 *M*w 8.6 Nias Island great earthquakes. *Bull. Seismol. Soc. Am* 97: S43–S61.

Farias, M., Comte, D., Charrier, R. et al. (2010). Crustal-scale structural architecture in central Chile based on seismicity and surface geology: implications for Andean mountain building. *Tectonics* 29: TC3006.

Fiedler, G.B. (1974). Local b-values related to seismicity. *Tectonophysics* 23 (3): 277–282.

Gerbault, M., Cembrano, J., Mpodozis, C. et al. (2009). Continental margin deformation along the Andean subduction zone: thermo-mechanical models. *Phys. Earth Planet. Inter* 177: 180–205.

Gibowicz, S.J. (1973). Variation of the frequency-magnitude relation during earthquake sequences in New Zeland. *Bull. Seism. Soc. Am.* 63: 517–528.

Guha, S.K. (1983). Seismo-statistical modelling of tectonic stress-building process. In: *Proc, Workshop on Geomodelling* (ed. B.K. Sahu), 138–143. Bombay: IIT.

Gutenberg, B. and Richter, C.F. (1944). Frequency of earthquakes in California. *Bull. Seism. Soc. Am* 34: 185–188.

Henderson, J., Main, I.G., Pearce, R.G., and Takeya, M. (1994). Seismicity in North-Eastern Brazil: fractal clustering and the evolution of the *b* value. *Geophys. J. Int* 116: 217–226.

Huang, J. and Turcotte, D.L. (1988). Fractal distributions of stress and strength and variations of b-value. *Earth Planet. Sci. Lett* 91: 223–230.

Imoto, M. (1991). Changes in the magnitude-frequency b-value prior to large (M > 6.0) earthquakes in Japan. *Tectonophysics* 193: 311–325.

Isozaki, Y., Aoki, K., Nakama, T., and Yanai, S. (2010). New insight into a subduction-related orogen: a reappraisal of the geotectonic framework and evolution of the Japanese Islands. *Gondwana Research* 18: 82–105.

Iyer, H.M. (1976). Earthquake prediction – present status. In: *Earthquake Prediction: How Close Is it to Reality* (ed. S.K. Guha and A.M. Patwardhan), 1–11. Pune, India: University of Poona.

Jafari, M.A. 2008. The distribution of b-values in different seismic provinces of Iran. *The 14th World Conference on Earthquake Engineering*. (12–17 October). Beijing. China.

Kamer, Y. (2014). Minimum sample size for detection of Gutenberg-Richter's b-value. *arXiv* 1410.1815v1.

Khan, P.K., Ghosh, M., Chakraborty, P.P., and Mukherjee, D. (2011). Seismic *b*-value and the assessment of ambient stress in Northeast India. *Pure Appl. Geophys* 168: 1693–1706.

Klotz, J., Khazaradze, G., Angermann, D. et al. (2001). Earthquake cycle dominates contemporary crustal deformation in central and southern Andes. *Earth Planet. Sci. Lett* 193: 437–446.

Kodaira, S., Kurashimo, E., Park, J.-O. et al. (2002). Structural factors controlling the rupture process of a megathrust earthquake at the Nankai trough seismogenic zone. *Geophys J Int* 149: 815–835.

Lallemand, S. and Le Pichon, X. (1987). Coulomb wedge model applied to the subduction of seamounts in the Japan trench. *Geology* 15: 1065–1069.

Landgraf, A., Kübler, S., Hintersberger, E. & Stein, S. 2017. Active tectonics, earthquakes and palaeoseismicity in slowly deforming continents. In: Landgraf, A., Kübler, S., Hintersberger, E. & Stein, S. (Eds) *2017. Seismicity, Fault Rupture and Earthquake Hazards in Slowly Deforming Regions*. Geological Society, London, Special Publications, 432, 1–12.

Lange, D., Rietbrock, A., Haberland, C. et al. (2007). Seismicity and geometry of the south Chilean subduction zone (41.5°S-43.5°S): implication for controlling parameters. *Geophys Res Lett* 34: L06311.

Lange, D., Tilmann, F., Rietbrock, A. et al. (2010). The fine structure of the subducted investigator fracture zone in Western Sumatra as seen by local seismicity. *Earth Planet. Sci. Lett* 298: 47–56.

Lasitha, S., Radhakrishna, M., and Sanu, T.D. (2006). Seismically active deformation in the Sumatra-Java trench-arc region: geodynamic implications. *Curr. Sci* 90: 690–696.

Liu, Y. and Zhu, S.B. (2010). Comparison of the 1999Chi-Chi Mw7.6 earthquake and the 2008 Wenchuan MS 8.0 earthquake. In: *Rock Stress and Earthquakes* (ed. F. Xie), 659–664. London: Taylor & Francis Group.

Main, I.G., Meredith, P.G., and Jones, C. (1989). A reinterpretation of the precursory seismic b-value anomaly from fracture mechanics. *Geophys. J* 96: 131–138.

Malgrange, M., Deschamps, A., and Madariaga, R. (1981). Thrust and extensional faulting under the Chilean coast: 1965, 1971 Aconcagua earthquakes. *Geophys J Int* 66: 313–331.

Manna, S., Paul, H., and Mitra, S. (2015). B-value characterization of earthquakes in the eastern Himalayan and indo-Burman plate boundary system. *AGU Abstract.*.

McCaffrey, R. (1989). Seismological constraints and speculations on Bandra arc tectonics. *Neth.J. Sea Res* 24: 141–152.

McCaffrey, R. (2009). The tectonic framework of the Sumatran subduction zone. *Annu. Rev. Earth Planet. Sci* 37: 345–366.

McCann, W. R., Nishenko, S. P., Sykes, L. R. Krause, J. (1978) Seismic gaps and plate tectonics: seismic potential for major plate boundaries in proceedings of conference VI – methodology for identifying seismic gaps and soon – to – break gaps. U.S. Geological Survey Open-File Report 78-943: 441–584.

Metcalfe, I. (2011). Tectonic framework and Phanerozoic evolution of Sundaland. *Gondwana Research* 19: 3–21.

Miura, S., Kodaira, S., Nakanishi, A. et al. (2003). Structural characteristics controlling the seismicity of southern Japan trench fore-arc region, revealed by ocean bottom seismographic data. *Tectonophysics* 363: 79–102.

Mogi, K. (1990). Seismicity before and after large shallow earthquakes around the Japanese islands. *Tectonophysics* 175: 1–33.

Mueller C.S, Frankel A.D, Petersen M.D, Leyendecker, E.V. 1997. Documentation for 2003 USGS seismic Hazard maps for Puerto Rico and the U.S. Virgin Islands, open-file report 03-379, U.S. Geological Survey.

Mukherjee, S. (2015). A review on out-of-sequence deformation in the Himalaya. In: *Tectonics of the Himalaya*, vol. 412 (ed. S. Mukherjee, R. Carosi, P. van der Beek, et al.), 67–109. Geological Society, London. Special Publications.

Mukherjee, S., Dasgupta, S., Majumdar, C. et al. (2023). Introduction to: Handbook of Petroleum Geoscience: Exploration, Characterization, and Exploitation of Hydrocarbon Reservoirs. In: *Handbook of Petroleum Geoscience: Exploration, Characterization, and Exploitation of Hydrocarbon Reservoirs*, Petroleum Industry Updates: Geoscientific and Economic Issues (ed. S. Mukherjee, S. Dasgupta, C. Majumdar, et al.). John Wiley & Sons, Hoboken, ISBN 978-1-119-68003-1.

Musson, R.M.W. (2000). Generalized seismic hazard maps for the Pannonian basin using probabilistic methods. *Pure Appl. Geophys* 157: 147–169.

Nanjo, K.Z. and Yoshida, A. (2017). Anomalous decrease in relatively large shocks and increase in the *p* and *b* values preceding the April 16, 2016, *M*7.3 earthquake in Kumamoto, Japan. *Earth, Planets and Space* 69: 13.

Nuannin P. 2006. The Potential of *b*-value variations as earthquake precursors for small and large events. Ph.D. thesis. Uppsala University.

Nuannin, P., Kulhanek, O., and Persson, L. (2005a). Spatial and temporal b value anomalies preceding the devastating off coast of NW Sumatra earthquake of December 26, 2004. *Geophys. Res. Lett* 32: L11307.

Nuannin, P., Kulhanek, O., Persson, L., and Askemur, T. (2005b). Inverse correlation between induced seismicity and b-value, observed in the Zingruvan mine. *Sweden. Acta Geodyn. Geomater* 2: 5–13.

Nuannin, P., Kulhanek, O., and Persson, L. (2012). Variations of b-values preceding large earthquakes in the Andaman–Sumatra subduction zone. *J. Asian Earth Sci* 61: 237–242.

Okal, E.A. and Romanowicz, B.A. (1994). On the variation of b-values with earthquake size. *Phys. Earth Planet. Inter* 87: 55–76.

Pandey, M.R., Tandukar, R.P., Avouac, J.P. et al. (1995). Interseismic strain accumulation on the Himalayan crustal ramp (Nepal). *Geophys Res Lett* 22: 751–754.

Pandey, M.R., Tandukar, R.P., Avouac, J.P. et al. (1999). Seismotectonics of the Nepal Himalaya from a local seismic network. *J Asian Earth Sci* 17: 703–712.

Passarelli, L., Hainzl, S., Cesca, S. et al. (2015). Aseismic transient driving the swarm-like seismic sequence in the Pollino range, southern Italy. *Geophys. J. Int* 201: 1553–1567.

Patil, N., Guha, S.K., and Powar, K.B. (1986). Seismic risk evaluation of Northeast India. In: *Earthquake Prediction: Present Status* (ed. S.K. Guha and A.M. Patwardhan), 235–246. Pune, India: University of Poona.

Patwardhan, A.M. (2012). *The Dynamic Earth System*, 3e, 484. New Delhi: PHI Learning Private Ltd.

Perucca, L. and Bastias, H. (2008). Neotectonics, seismology and paleoseismology. In: *The Late Cenozoic Pantagonia and Tierra Del Fuego. Developments in Quaternary Sciences 11* (ed. J. Rabassa), 73–94. Amsterdam: Elsevier. ISBN: 978-0-444-52954-1.

Petersen M.D, Frankel A.D, Harmsen S.C, Mueller C.S, Haller K.M, Wheeler R. L, Wesson R.L, Zeng Y, Boyd O S, Perkins D. M, Luco N, Field E.H, Wills C. J, and Rukstales K.S. 2008. Documentation for the 2008 update of the United States National Seismic Hazard Maps, open file report - 2008-1128; U.S. Department of the Interior U.S. Geological Survey.

Prawirodirdjo, L., Bock, Y., Genrich, J.F. et al. (2000). One century of tectonic deformation along the Sumatran fault from triangulation and Global Positioning System surveys. *J Geophys Res* 105: 28343–28361.

Rajendran, K., Parameshwaran, R.M., and Rajendran, C.P. (2017). Seismotectonic perspectives on the Himalayan arc and contiguous areas: inferences from past and recent earthquakes. *Earth Sci Rev* 173: 1–30.

Rydelek, P.A. and Sacks, I.S. (1989). Testing the completeness of earthquake catalogues and the hypothesis of self-similarity. *Nature* 337: 251–253.

Schelling, D. and Arita, K. (1991). Thrust tectonics, crustal shortening, and the structure of the far-eastern Nepal Himalaya. *Tectonics* 10: 851–862.

Scholz, C.H. (2015). On the stress dependence on the earthquake b value. *Geophys.Res.Lett* 42: 1399–1402.

Seo, J.M., Rhee, H.M., and Choi, I.K. (2010). A study on the best estimate of seismicity parameter b-value and its application for the seismic Hazard analysis of the Korean peninsula. *J.Earthq. Eng.Soc.Korea* 35: 42. https://doi.org/10.5000/EESK.2010.14.1.035.

Shi, Y. and Bolt, B.A. (1982). The standard error of the magnitude-frequency b-value. *Bull. Seismol. Soc. Am* 72: 1677–1687.

Shikazono, N. (2012). *Introduction to Earth and Planetary System Science: New View of Earth Planets and Humans. Springer*, 71. ISBN: 978-4-431-54058-8.

Singh HN, Paudyal H, Shanker D, Panthi A, Kumar A, Sinh VP. 2010. Anomalous seismicity and earthquake forecast in western Nepal Himalaya and its adjoining Indian region. *Pure Appl. Geophys.* DOI: https://doi.org/10.1007/s00024-010-0072-6

Smith, W.D. (1981). The b-value as an earthquake precursor. *Nature* 289: 136–139.

Sobiesiak, M., Meyer, U., Schmidt, S. et al. (2007). Asperity generating upper crustal sources revealed by b value and isostatic residual anomaly grids in the area of Antofagasta, Chile. *J. Geophys. Res.* 112: B12.

Srivardhan V, Srinu U. 2014. Potential of fractal analysis of earthquakes through wavelet analysis and determination of b value as an aftershock precursor: a case study using earthquake data between 2003 and 2011 in Turkey. *J. Earthq.*2014, article id 123092.

Stauder, W. (1973). Mechanism and spatial distribution of Chilean earthquakes with relation to subduction of the oceanic plate. *J Geophys Res* 78: 5033–5062.

Suarez, G., Molnar, P., and Burchfiel, B.C. (1983). Seismicity, fault plane solutions, depth of faulting, and active tectonics of the Andes of Peru, Ecuador, and southern Columbia. *J Geophys Res* 88: 10403–10428.

Sukmono, S., Zen, M.T., Hendrajaya, L. et al. (1997). Fractal pattern of the Sumatra fault seismicity and its possible application to earthquake prediction. *Bull. Seismol. Soc. Am.* 87: 1685–1690.

Sykes, L.R., Shaw, B.E., and Scholz, C.H. (1999). Rethinking earthquake prediction. *Pure and Applied Geophysics* 155: 207–232.

Tanioka, Y., Ruff, L., and Satake, K. (1997). What controls the lateral variation of large earthquake occurrence along the Japan trench? *Island Arcs* 6: 261–266.

Tassara, A. and Echaurren, A. (2012). Anatomy of the Andean subduction zone: three-dimensional density model upgraded and compared against global-scale models. *Geophys. J. Int.* 189: 161–168.

Taylor, D.W.A., Snoke, J.A., Sacks, I.S., and Takanami, T. (1990). Non-linear frequency-magnitude relationship for the Hokkaido corner, Japan. *Bull. Seism. Soc. Am.* 80: 340–353.

The ANNCORP Working Group (1999). Seismic refection image revealing offset of Andean subduction-zone earthquake locations into oceanic mantle. *Nature* 397: 341–344.

Tsuru, T., Park, J.-O., Miura, S. et al. (2002). Along-arc structural variation of the plate boundary at the Japan trench margin: implication of interplate coupling. *J Geophys Res* 107: 2357.

U.S. National Academy of Sciences (1976). *Predicting Earthquakes – a Scientific and Technical Evaluation with Implications for Society*, 62. Washington, D.C.: Natnl. Acdm. of Scis.

Upreti, B.N. (1999). An overview of the stratigraphy and tectonics of the Nepal Himalaya. *J Asian Earth Sci* 17: 577–606.

Utsu, T. (1965). A method for determining the value of b in the formula log n = a-bM showing the magnitude frequency relation for earthquakes. Geophys. Bull. *Hokaido Univ.* 13: 99–103.

Wiemer, S. and Wyss, M. (2000). Minimum magnitude of complete reporting in earthquake catalogs: examples from Alaska, the Western United States, and Japan. *Bull. Seism. Soc. Am* 90: 859–869.

Woessner, J. and Wiemer, S. (2005). Assessing the quality of earthquake catalogues: estimating the magnitude of completeness and its uncertainty. *Bull. Seism. Soc. Am* 95: 684–698.

Wyss, M. (2014). *Earthquake Hazards, Risks and Disasters. Hazards and Disaster Series*, 123–124. Amsterdam: Elsevier. ISBN: 978-0-12-394848-9.

Xu, J. and Kono, Y. (2002). Geometry of slab, intraslab stress field and its tectonic implication in the Nankai trough, Japan. *Earth Planets Space* 54: 733–742.

Yu, D. (2016). Investigations of the b-value and its variations on possible earthquake prediction in the North-South China Seismic Belt. Doctoral Dissertation, Massachusetts Institute of Technology.

Zalohar, J. (2018). *The Omega-Theory: A New Physics of Earthquakes*, Developments in Structural Geology & Tectonics. Series Editor: S. Mukherjee, 131. Amsterdam: Elsevier. ISBN: 978-0-12-814580-7.

Zuniga, F.R. and Wiemer, S. (1999). Seismicity patterns: are they always related to natural causes? Pure Appl. *Geophys.* 155: 713–726.

Zuniga, F.R. and Wyss, M. (1995). Inadvertent changes in magnitude reported in earthquake catalogs: their evaluation through b-value estimates. *Bull SeismolSoc Am* 85: 1858–1866.

Web References

1. http://plcmets.pbworks.com/f/Tectonics+rate+map.jpg

2. http://montessorimuddle.org/2011/03/11/plate-tectonics-and-the-earthquake-in-japan